POWER CONVERSION AND CONTROL OF WIND ENERGY SYSTEMS

POWER CONVERSION AND CONTROL OF WIND ENERGY SYSTEMS

Bin Wu
Yongqiang Lang
Navid Zargari
Samir Kouro

IEEE Press

A JOHN WILEY & SONS, INC., PUBLICATION

Library of Congress Cataloging-in-Publication Data:

Power conversion and control of wind energy systems / Bin Wu ... [et al.].
   p. cm. — (IEEE Press series on power engineering ; 74)
 Summary: "The book covers a wide range of topics on wind energy conversion and control from the electrical engineering aspect. It includes wind generators and modeling, power converters and modulation schemes, operating principle of fixed and variable speed wind turbines, advanced generator control schemes, active and reactive power controls of individual wind and is a valuable reference book for academic researchers, practicing engineers, and other professionals. The book can also be used as a textbook for graduate level and final year undergraduate-level courses"— Provided by publisher.
 Includes bibliographical references and index.
 ISBN 978-0-470-59365-3 (hardback)
 1. Wind energy conversion systems. I. Wu, B. (Bin), 1957–
 TK1541.P685 2011
 621.31'2136—dc22   2010045226

Printed in the United States of America.

oBook ISBN: 978-1-118-02900-8
ePDF ISBN: 978-1-118-02899-5
ePub ISBN: 978-1-118-02898-8

10 9 8 7 6 5 4

CONTENTS

IEEE Press Series on Power Engineering

PREFACE

Wind energy is clean and sustainable. It is one of the fastest growing renewable energy resources. The conversion of wind kinetic energy into electric energy is of a multidisciplinary nature, involving aerodynamics, mechanical systems, electric machines, power electronics, control theory, and power systems. In the past, a number of books have addressed some of these subjects. This book explores the power conversion and control of wind energy conversion systems (WECS) from the electrical engineering perspective. It provides a comprehensive and in-depth analysis of wind generators, system configurations, power converters, control schemes, and dynamic/steady-state performance of various practical wind energy systems.

The book contains nine chapters. Chapter 1 provides a market survey and an overview of wind turbine technology, wind energy system classifications, costs, and grid codes for wind power integration. Chapter 2 introduces the fundamentals and control principles of wind energy systems, including wind turbine components, aerodynamics, stall and pitch controls, and maximum power point tracking schemes. Chapter 3 presents commonly used wind generators, including squirrel cage induction generators, doubly fed induction generators, and synchronous generators. The dynamic and steady-state models of these generators are also derived to facilitate the analysis of wind energy systems in the subsequent chapters.

Chapter 4 discusses various power converters and PWM schemes used in wind energy systems. Both voltage and current source converters are presented with an emphasis on high-power wind energy system. Chapter 5 presents a general overview of configurations and characteristics of major practical WECS. Chapter 6 focuses on fixed-speed, inductor generator based wind energy systems; important issues such as grid connection, two-speed operation, and reactive power compensation are discussed.

Chapter 7 deals with wind energy systems with variable-speed, squirrel cage induction generators (SCIG), in which typical system configurations and advanced control schemes such as field oriented controls (FOC) and direct torque control (DTC) are elaborated. Chapter 8 discusses doubly fed induction generator (DFIG) systems, where the subsynchronous and supersynchronous modes of operation are investigated. Chapter 9 is dedicated to variable-speed, synchronous generator wind systems, in which various control schemes, including zero d-axis current (ZDC) control, maximum torque per ampere (MTPA) control, and unity power factor (UPF) control, are analyzed in detail.

To help the reader understand the principle and operation of various wind energy conversion systems, we have developed more than 30 case studies in the body of the book and more than 100 solved problems in Appendix C—Problems and Solutions. Therefore, this book is not only for academic researchers and practicing engineers as a reference book, but also suitable for graduate students and final-year undergraduate students as a textbook.

We would like to express our sincere appreciation to Mr. Venkata Yaramasu, our Ph.D. student in the Laboratory for Electric Drive Applications and Research (LEDAR) at Ryerson University, for his great assistance in preparing the appendices and the first three sections of Chapter 4. We would like to thank our postdoctoral fellows and graduate students in LEDAR, in particular, Dr. Victor F. Liu, Dr. Moya J. Dai, Mr. Ning Zhu, and Mr. Ehsan Al-Nabi, for their kind assistance in preparing the manuscript. Our special thanks go to Wiley/IEEE Press editors for their great help and support. We also wish to acknowledge the support and inspiration of our families during the preparation of this book.

BIN WU
YONGQIANG LANG
NAVID ZARGARI
SAMIR KOURO

Toronto, Ontario, Canada
January 2011

LIST OF SYMBOLS

A	sweep area of turbine rotor blades
C_i	filter capacitor of current source inverter
C_p	power coefficient of blade
$C_{p,\max}$	maximum power coefficient of blade
C_r	filter capacitor of current source rectifier
D	duty cycle of boost converter
D_1	coefficient $(D_1 = L_s L_r - L_m^2)$ in the model of induction generators
E	DC capacitor voltage of three-level NPC converter
	battery voltage in DC link circuit
f	fundamental frequency of inverter output voltage
f_{cr}	frequency of carrier wave in sinusoidal pulse width modulation
f_g	frequency of the grid
f_m	frequency of modulating wave in sinusoidal pulse width modulation
f_o	cut-off frequency of a first-order low-pass filter
f_s	stator frequency of induction or synchronous generators
f_{sp}	sampling frequency in space vector modulation
f_{sw}	switching frequency of solid-state switching device
$f_{sw,inv}$	equivalent switching frequency of inverter
i_a	phase-a current of AC voltage controller
i_{ag}	phase-a current of the grid
i_{ar}	phase-a rotor current of induction generator
i_{as}	phase a stator current of induction or synchronous generator
i_{awi}	phase-a PWM current of current source inverter
i_{awr}	phase-a PWM current of current source rectifier
i_b	phase-b current of AC voltage controller
i_{bg}	phase-b current of the grid
i_{br}	phase-b rotor current of induction generator
i_{bs}	phase-c stator current of induction or synchronous generator
i_{bwi}	phase-b current of current source inverter
i_c	phase-c current of AC voltage controller
i_{cid}	d-axis filter capacitor current in current source inverter
i_{ciq}	q-axis filter capacitor current of current source inverter

i_{cg}	phase-c current of the grid
i_{cr}	phase-c rotor current of induction generator
$\vec{i}_{cr}$	filter capacitor current vector of current source rectifier
i_{crd}	d-axis filter capacitor current in current source rectifier
i_{crq}	q-axis filter capacitor current in current source rectifier
i_{cs}	phase-c stator current of induction or synchronous generator
i_{cw}	phase-c PWM current of current source inverter
i_{dc}	DC link current
i_{dc}^{*}	DC link reference current
i_{dci}^{*}	DC link reference current of current source inverter
i_{dcr}^{*}	DC link reference current of current source rectifier
i_{dg}	d-axis current of the grid
i_{dg}^{*}	d-axis reference current of the grid
i_{dr}	d-axis rotor current of induction generator
i_{dr}^{*}	d-axis rotor reference current of DFIG
i_{ds}	d-axis stator current of induction or synchronous generator
i_{ds}^{*}	d-axis stator reference current of induction or synchronous generator
i_{dwi}^{*}	d-axis PWM reference current of current source inverter
i_{dwr}	d-axis PWM current of current source rectifier
i_{dwr}^{*}	d-axis PWM reference current of current source rectifier
i_g	current of the grid
i_{g1}	gate signal for thyristor T_1 in AC voltage controller
i_{g2}	gate signal for thyristor T_2 in AC voltage controller
i_{g3}	gate signal for thyristor T_3 in AC voltage controller
i_{g4}	gate signal for thyristor T_4 in AC voltage controller
i_{g5}	gate signal for thyristor T_5 in AC voltage controller
i_{g6}	gate signal for thyristor T_6 in AC voltage controller
i_i	input current of boost converter
i_{L1}	current through inductor L_1 in boost converter
i_{L2}	current through inductor L_2 in boost converter
i_o	output current of boost converter
i_{qg}	q-axis current of the grid
i_{qg}^{*}	q-axis reference current of the grid
i_{qr}	q-axis rotor current of induction generator
i_{qr}^{*}	q-axis rotor reference current of DFIG
i_{qs}	q-axis stator current of induction or synchronous generator
i_{qs}^{*}	q-axis stator reference current of induction or synchronous generator
i_{qwi}^{*}	q-axis PWM reference current of CSI
i_{qwr}	q-axis PWM current of current source rectifier
i_{qwr}^{*}	q-axis PWM reference current of current source rectifier
i_r	rotor current of induction generator or doubly fed induction generator
$\vec{i}_r$	rotor current vector of induction generator
i_{ref}	magnitude (length) of reference current vector $\vec{i}_{\mathrm{ref}}$
$\vec{i}_{\mathrm{ref}}$	reference current vector of current source converter
$i_{\mathrm{ref,max}}$	maximum magnitude of reference current vector $\vec{i}_{\mathrm{ref}}$

i_s	stator current of induction or synchronous generator
$\vec{i}_s$	stator current vector of induction or synchronous generator
i^*_{wi}	PWM reference current of current source inverter
i_{wr}	PWM current of current source rectifier
$\vec{i}_{wr}$	PWM current vector in current source rectifier
i^*_{wr}	PWM reference current in current source rectifier
i_α	α-axis current
i_β	β-axis current
I_{ag}	rms value of grid phase-a current
I_{awn}	rms value of the nth-order harmonic current in current source inverter
$I_{aw1,\max}$	maximum rms fundamental-frequency current in current source inverter
$\bar{I}_{cr}$	filter capacitor current in current source rectifier (complex)
I_{dc}	average value of DC link current
I_{ds}	rms value of d-axis stator current of induction or synchronous generator
I_f	field current of synchronous generator
I_g	rms value of grid phase current
I_i	average value of boost converter input current
$\vec{I}_k$	current space vector in current source converter ($k = 0, 1, \ldots, 6$)
I_{L1}	average value of inductor L_1 current in boost converter
I_{L2}	average value of inductor L_2 current in boost converter
I_{LB}	inductor boundary current between the CCM and DCM
$I_{LB,\max}$	maximum inductor boundary current
$\bar{I}_m$	magnetizing current of induction generators (complex)
I_o	average value of output current of boost converter
I_{oB}	boundary output current of boost converter
$I_{oB,\max}$	maximum boundary output current
I_{qs}	rms value of q-axis stator current of induction or synchronous generator
I_r	rms value of rotor current of induction generator
$\bar{I}_r$	rotor current of induction generator (complex)
I_s	rms value of stator current of induction or synchronous generator
$\bar{I}_s$	stator current of induction generator (complex)
$\bar{I}_{wr}$	PWM current of current source rectifier (complex)
J	moment of inertia
k	number of harmonics to be eliminated by SHE scheme
K_T	armature constant $\left(K_T = \dfrac{2PL_m}{3L_r}\right)$
K_1	coefficient of boost converter $\left(K_1 = \dfrac{2L_1}{K_1 T_s D} I_0\right)$
K_m	coefficient $K_m = \dfrac{T_{m,R}}{\omega_{m,R}}$
L	inductance of each interleaved boost converter
L_d	d-axis self inductance of synchronous generator
L_{dc}	DC link inductance
L_{eq}	equivalent inductance of rotor-side converter in DFIG

L_g	grid-side line inductance
L_{lr}	rotor leakage inductance of induction or synchronous generator
L_{ls}	stator leakage inductance of induction or synchronous generator
L_L	inductance of three-phase RL load
L_m	magnetizing inductance of induction generator
L_q	q-axis self inductance of synchronous generator
L_r	rotor self-inductance of induction generator
L_s	stator self-inductance of induction or synchronous generator
m_a	amplitude modulation index of SPWM or SVM
$m_{a,max}$	maximum modulation index of SPWM or SVM
m_f	frequency modulation index in sinusoidal pulse-width modulation
m_i	modulation index of current source inverter
$m_{i,max}$	maximum modulation index of current source inverter
m_{max}	maximum modulation index
m_r	modulation index of current source rectifier
$m_{r,max}$	maximum modulation index of current source rectifier
n_m	generator speed in rpm
n_M	turbine speed in rpm
n_r	rotor speed of induction or synchronous generator in rpm
N	number of channels of interleaved boost converter
N_P	number of pulses per half-cycle of current source converter
p	derivative operator $\left(p = \dfrac{d}{dt} \right)$
P	number of pole pairs
	system active power
P_{ac}	ac-side active power of converter
P_{ag}	air-gap power of induction generator
$P_{cu,s}$	stator copper loss of induction generator
$P_{cu,r}$	rotor copper loss of induction generator
PF	power factor
PF_s	stator power factor
PF_L	load power factor
P_{dc}	DC power of converter
P_g	active power delivered to the grid
P_m	mechanical power of generator
P_M	mechanical power captured by turbine
P_L	load active power
P_r	rotor power of DFIG
P_{rot}	rotational power losses of generator
P_s	stator active power of generator
P_w	wind power
Q	reactive power
Q_g	grid-side reactive power
Q_g^*	grid-side reactive power reference

$Q_{g,\max}$	maximum grid-side reactive power
Q_{GSC}^*	reactive power reference of grid-side converter in DFIG
Q_L	load reactive power
Q_s	stator reactive power of DFIG
Q_s^*	stator reactive power reference
r_{gb}	gearbox conversion ratio (gear ratio)
r_T	radius of turbine rotor (blade length)
R	load resistance of boost converter
	resistance of simplified generator-side converter
R_{eq}	equivalent resistance of rotor-side converter in DFIG
R_r	rotor winding resistance of induction generator
R_s	stator winding resistance of induction or synchronous generator
R_L	resistance of three-phase RL load
s	slip of induction generator
$s_{T,\max}$	slip at the maximum torque
S	Laplace operator
	apparent power of system
S_g	grid-side apparent power
t_{on}	turn-on time of boost converter
t_{off}	turn-off time of boost converter
T_0	dwell time for voltage vector $\vec{V}_0$ and current vector $\vec{I}_0$ in space vector modulation
T_a	dwell time for voltage vector $\vec{V}_1$ in space vector modulation
T_b	dwell time for voltage vector $\vec{V}_2$ in space vector modulation
T_1	dwell time for current vector $\vec{I}_1$ in space vector modulation
T_2	dwell time for current vector $\vec{I}_2$ in space vector modulation
T_e	electromagnetic torque of induction and synchronous generators
T_e^*	electromagnetic torque reference
T_L	load torque
T_m	mechanical torque from generator shaft
T_M	mechanical torque generated by turbine
$T_{\max}$	maximum torque
T_s	switching period of solid switching device
	sampling period of space vector modulation
v_a	phase-a voltage
v_a^*	phase-a reference voltage
v_{ab}	inverter line-to-line voltage between phase a and phase b
v_{AB}	line-to-line supply voltage of three-phase AC voltage controller
v_{ab1}	fundamental frequency component of inverter line-to-line voltage v_{ab}
v_{abs}	line-to-line stator voltage or rectifier PWM voltage
v_{abs1}	fundamental component of line-to-line voltage v_{abs}
v_{ag}	phase-a grid voltage
v_{ai}^*	phase-a reference voltage of PWM inverter
v_{an}	phase-a load voltage of three-phase AC voltage controller
v_{aN}	phase-a load voltage of two-level voltage source converter

v_{AN}	phase-A supply voltage of three-phase AC voltage controller
v_{ar}	phase-a rotor voltage of DFIG
v_{ar}^{*}	phase-a reference rotor voltage
v_{ar1}	fundamental frequency component of phase-a rotor voltage of DFIG
v_{as}	phase-a stator voltage of induction and synchronous generator
v_{as}^{*}	phase-a reference voltage of PWM rectifier
v_{as1}	fundamental frequency component of phase-a stator voltage
v_{aZ}	phase-a terminal voltage of three-level NPC inverter
v_{b}	phase-b voltage
v_{b}^{*}	phase-b reference voltage
v_{bc}	line-to-line voltage between phase b and phase c
v_{BC}	line-to-line supply voltage between phase B and phase C of AC voltage controller
v_{bg}	phase-b grid voltage
v_{bi}^{*}	phase-b reference voltage of PWM inverter
v_{bn}	phase-b load voltage of three-phase AC voltage controller
v_{bN}	phase-b load voltage of two-level voltage source converter
v_{BN}	phase-B supply voltage of three-phase AC voltage controller
v_{br}	phase-b rotor voltage of DFIG
v_{br}^{*}	phase-b rotor reference voltage
v_{bs}	phase-b stator voltage of induction and synchronous generator
v_{bs}^{*}	phase-b reference voltage for PWM rectifier
v_{bZ}	phase-b terminal voltage of three-level NPC inverter
v_{c}	phase-c voltage
v_{c}^{*}	phase-c reference voltage
v_{ca}	line-to-line voltage between phase c and phase a
v_{CA}	line-to-line supply voltage between phase C and phase A
v_{ci}	capacitor voltage in current source inverter
v_{ci}^{*}	phase-c reference voltage of PWM inverter
v_{cg}	phase-c voltage of the grid
v_{cn}	phase-c load voltage of three-phase AC voltage controller
v_{cN}	phase-c load voltage of two-level voltage source converter
v_{cr}^{*}	phase-c rotor reference voltage
v_{cs}	phase-c stator voltage of induction and synchronous generator
v_{cs}^{*}	phase-c reference voltage for PWM rectifier
v_{cr}	triangular carrier wave in sinusoidal pulse-width modulation phase-c rotor voltage of DFIG
v_{cZ}	phase-c terminal voltage of three-level NPC inverter
v_{dc}	DC link voltage
v_{dc}^{*}	reference DC link voltage
v_{dc1}	DC voltage of boost converter
v_{dc2}	DC voltage of boost converter
v_{dci}^{*}	inverter-side DC voltage reference
v_{dcr}	DC output voltage of current source rectifier
v_{dg}	d-axis voltage of the grid

v_{di}	d-axis voltage of grid-tied inverter
v_{di}^*	d-axis reference voltage of PWM inverter
v_{dr}	d-axis rotor voltage
v_{dr}^*	d-axis rotor reference voltage
v_{ds}	d-axis stator voltage
v_{ds}^*	d-axis reference voltage of PWM rectifier
v_g	grid phase voltage
$\vec{v}_g$	grid voltage vector
v_{g1}	gating signal of switch S_1
v_{g2}	gating signal of switch S_2
v_i	input voltage of boost converter
	output voltage of grid-tied inverter
v_{L1}	inductor voltage
v_{ma}	phase-a modulating wave in SPWM scheme
v_{mb}	phase-b modulating wave in SPWM scheme
v_{mc}	phase-c modulating wave in SPWM scheme
v_{qg}	q-axis voltage of the grid
v_{qi}	q-axis voltage of grid-tied inverter
v_{qi}^*	q-axis reference voltage of PWM inverter
v_{qr}	q-axis rotor voltage
v_{qr}^*	q-axis rotor reference voltage
v_{qs}	q-axis stator voltage
v_{qs}^*	q-axis reference voltage of PWM rectifier
v_r	rotor voltage of DFIG
$\vec{v}_r$	rotor current vector
v_{ref}	magnitude (length) of reference voltage vector $\vec{v}_{\mathrm{ref}}$
$\vec{v}_{\mathrm{ref}}$	reference voltage vector
$v_{\mathrm{ref,max}}$	maximum magnitude of reference voltage vector
v_s	supply voltage of single-phase AC voltage controller
	stator voltage of induction or synchronous generator
$\vec{v}_s$	stator voltage vector
v_{s1}	voltage across switch S_1 in boost covnerter
v_w	wind velocity/speed
v_α	α-axis voltage
v_β	β-axis voltage
$\hat{V}_a$	peak value of inverter phase-a voltage
V_{abn}	the nth-order rms harmonic voltage of inverter line-to-line voltage v_{ab}
$V_{ab1,\mathrm{max}}$	maximum fundamental rms voltage of inverter line-to-line voltage v_{ab}
$V_{ab,\mathrm{max}}$	maximum rms voltage of inverter line-to-line voltage v_{ab}
V_{ag}	rms value of grid phase-a voltage
V_{an}	rms value of load phase-a voltage
V_{aN}	rms inverter terminal voltage (phase-a)
V_{bN}	rms inverter terminal voltage (phase b)
V_{cN}	rms inverter terminal voltage (phase-c)
$\hat{V}_{cr}$	peak value of carrier wave in sinusoidal pulse-width modulation

V_{dc}	average value of DC link voltage
V_{dc}^*	DC link reference voltage
V_{ds}	d-axis rms stator voltage of induction or synchronous generator
V_g	rms grid phase voltage
V_i	average value of input DC voltage in boost converter
V_{i1}	rms fundamental-frequency component of converter phase voltage v_i
$\hat{V}_{i1}$	peak value of fundamental-frequency component of converter phase voltage
$\vec{V}_k$	voltage space vector (for two-lever $k = 0, 1, \ldots, 6$, three-lever $k = 0, 1, \ldots, 18$)
$\hat{V}_m$	peak value of modulating wave in sinusoidal pulse width modulation
$\overline{V}_m$	magnetizing voltage of induction generator (complex)
V_o	average value of output DC voltage in boost converter
	rms value of single-phase AC voltage controller output voltage
V_{qs}	q-axis rms stator voltage of induction or synchronous generator
$\overline{V}_r$	rotor voltage of induction generators (complex)
V_s	rms value of supply-phase voltage
	rms stator voltage
$\overline{V}_s$	stator voltage of induction and synchronous generator (complex)
x_a	phase-a variable in the abc stationary reference frame
x_b	phase-b variable in the abc stationary reference frame
x_c	phase-c variable in the abc stationary reference frame
x_d	d-axis variable in an arbitrary dq reference frame
x_q	q-axis variable in an arbitrary dq reference frame
x_α	α-axis variable in a two-phase stationary frame
x_β	β-axis variable in a two-phase stationary frame
X_c	reactance of capacitor in parallel with induction generator
X_{lr}	rotor leakage reactance of induction generator
X_{ls}	stator leakage reactance of induction or synchronous generators
X_{eq}	equivalent reactance of rotor-side converter in DFIG
X_m	magnetizing impedance of induction generators
Z_{eq}	equivalent impedance of the rotor-side converter in DFIG
$\overline{Z}_{eq}$	equivalent impedance of the rotor-side converter in DFIG (complex)
Z_m	magnetizing impedance of induction generator
Z_r	rotor impedance of induction generator
Z_{rm}	equivalent impedance of rotor and magnetizing branch of induction generator
$\overline{Z}_s$	total input complex impedance of generator (complex)
Z_{sr}	equivalent stator-rotor impedance of induction generators
α	angle of attack of blade
	delay angle of thyristor
	delay angle of current source converter
α_i	delay angle of current source inverter
α_r	delay angle of current source rectifier
α_R	rated (optimal) angle of attack of blade

δ	angle of stator current vector with respect to q-axis
φ_g	grid power factor angle
φ_L	load power factor
φ_r	rotor power factor angle
φ_s	stator power factor angle
η	efficiency of generator
λ_{ar}	rotor phase-a flux linkage
λ_{dr}	d-axis rotor flux linkage
λ_{ds}	d-axis stator flux linkage
λ_{qr}	q-axis rotor flux linkage
λ_{qs}	q-axis stator flux linkage
λ_r	rotor flux linkage
$\vec{\lambda}_r$	rotor flux linkage vector
λ_r^*	reference rotor flux linkage
$\vec{\lambda}_s$	stator flux linkage vector
$\vec{\lambda}_s^*$	reference stator flux linkage vector
λ_T	tip speed ratio
$\lambda_{T,\mathrm{opt}}$	optimal tip speed ratio
$\theta(t)$	angular displacement between $\vec{v}_{\mathrm{ref}}$ and the α-axis of the stationary reference frame
	angular displacement between $\vec{i}_{\mathrm{ref}}$ and the α-axis of the stationary reference frame
θ_0	initial angle of space vector
θ_{ci}	angle of capacitor voltage vector in current source inverter
θ_f	rotor flux angle of squirrel cage induction generator
θ_g	angle of grid voltage vector
θ_i	angle of stator current vector
θ_k	switching angle of SHE modulation scheme
θ_r	rotor position angle of induction generator
θ_s	synchronous angle of the dq frame
	angle of stator voltage vector of DFIG
θ_{sl}	slip angle of induction generator
θ_T	torque angle between stator and rotor flux vectors in DTC control
θ_v	angle of stator voltage vector
θ_{wi}	firing angle of current source inverter
θ_{wr}	firing angle of current source rectifier
ρ	air density
σ	total leakage factor of induction generator $\left(\sigma = 1\dfrac{L_m^2}{L_s L_r}\right)$
τ_r	rotor time constant of induction generator
ω	rotating speed of an arbitrary reference frame
ω_g	angular frequency of the grid
	speed of grid synchronous frame
ω_m	rotor mechanical angular speed

ω_m^*	rotor mechanical reference angular speed
ω_M	rotating speed of blade or turbine
ω_r	rotor electrical angular speed
ω_s	synchronous speed of the dq frame
	stator angular frequency
ω_{sl}	angular slip frequency
ΔI_i	input ripple current of boost converter (peak to peak)
$\Delta I_{i,\max}$	maximum input ripple current of boost converter (peak to peak)
ΔI_L	inductor ripple current (peak to peak)
Δi_{L1}	inductor ripple current of boost converter
$\Delta I_{L1,\max}$	maximum inductor ripple current of boost converter (peak to peak)
ΔT_e	torque difference between reference and feedback
ΔV_o	ripple voltage (peak to peak)
$\Delta \lambda_s$	stator flux error between reference and calculation
Λ_m	rms value of rotor flux linkage
$\overline{\Lambda}$	magnetizing flux linkage vector (complex)
Λ_r	rms value of rotor flux linkage
$\overline{\Lambda}_r$	rotor flux linkage vector (complex)
Λ_s	rms value of stator flux linkage
$\overline{\Lambda}_s$	stator flux linkage vector (complex)

ACRONYMS AND ABBREVIATIONS

APC	Aerodynamic power control
CCM	Continuous current mode
CSC	Current source converter
CSI	Current source inverter
CSR	Current source rectifier
DCM	Discontinuous current mode
DFIG	Doubly fed induction generator
DTC	Direct torque control
FC	Flux controller
FOC	Field-oriented control
GSC	Grid-side converter
HAWT	Horizontal-axis wind turbine
HVAC	High-voltage alternating current
HVDC	High-voltage direct current
IG	Induction generator
LVRT	Low-voltage ride through
MPP	Maximum power point
MPPT	Maximum power-point tracking
MTPA	Maximum torque per ampere control
MW	Megawatts
MV	Medium voltage
NPC	Neutral point clamped converter
OPC	Optimal power control
OTC	Optimal torque control
PCC	Point common coupling
PF	Power factor
PLL	Phase-locked loop
PMSG	Permanent magnet synchronous generator
RSC	Rotor-side converter
SCIG	Squirrel cage induction generator

SG	Synchronous generator
SHE	Selective harmonic elimination
SMPM	Surface-mounted permanent magnet
SVM	Space vector modulation
SVOC	Stator voltage oriented control
TC	Torque controller
THD	Total harmonic distortion
THM	Top head mass
TSR	Tip speed ratio
UPF	Unity power factor control
VAWT	Vertical-axis wind turbine
VOC	Voltage oriented control
VSC	Voltage source converter
VSI	Voltage source inverter
WECS	Wind energy conversion system(s)
WRIG	Wound rotor induction generator
WRSG	Wound-rotor synchronous generator
ZDC	Zero d-axis current control

1

INTRODUCTION

1.1 INTRODUCTION

Over the last twenty years, renewable energy sources have been attracting great attention due to the cost increase, limited reserves, and adverse environmental impact of fossil fuels. In the meantime, technological advancements, cost reduction, and governmental incentives have made some renewable energy sources more competitive in the market. Among them, wind energy is one of the fastest growing renewable energy sources [1].

Wind energy has been used for hundreds of years for milling grains, pumping water, and sailing the seas. The use of windmills to generate electricity can be traced back to the late nineteenth century with the development of a 12 kW DC windmill generator [2]. It is, however, only since the 1980s that the technology has become sufficiently mature to produce electricity efficiently and reliably. Over the past two decades, a variety of wind power technologies have been developed, which have improved the conversion efficiency of and reduced the costs for wind energy production. The size of wind turbines has increased from a few kilowatts to several megawatts each. In addition to on-land installations, larger wind turbines have been pushed to offshore locations to harvest more energy and reduce their impact on land use and landscape.

This chapter provides an overview of wind energy conversion systems (WECS) and their related technologies. The aim of the chapter is to provide a background on sever-

Power Conversion and Control of Wind Energy Systems. By B. Wu, Y. Lang, N. Zargari, and S. Kouro
© 2011 the Institute of Electrical and Electronics Engineers, Inc. Published 2011 by John Wiley & Sons, Inc.

1

al aspects related to this exciting technology and market trends such as installed capacity, growth rate, and costs. The details of turbine components, system configurations, and control schemes are analyzed in depth in the subsequent chapters.

1.2 OVERVIEW OF WIND ENERGY CONVERSION SYSTEMS

1.2.1 Installed Capacity and Growth Rate

Installed wind power capacity has been progressively growing over the last two decades. Figure 1-1 shows the evolution of cumulative installed capacity worldwide as of 2009 [3]. The installed capacity of global wind power has increased exponentially from approximately 6 GW in 1996 to 158 GW by 2009. The wind industry has achieved an average growth rate of over 25% since 2000, and is expected to continue this trend in the coming years. This impressive growth has been spurred by the continuous cost increase of classic energy sources, cost reduction of wind turbines, governmental incentive programs, and public demand for cleaner energy sources.

Figure 1-2 shows the cumulative installed wind power capacity of the top ten countries in the world as of 2009 [3]. Although Europe has maintained its role as the largest wind power producer as a region, the Unites States has surpassed the long-time world leader Germany by increasing its installed capacity of almost 50% in just two years. It has now an installed capacity of 35 GW, equivalent to 22.3% of the global installed capacity. Asian countries are catching up, mainly driven by the markets in China and India. In fact, China doubled its installed capacity in one year, and is expected to continue to grow at a fast pace in the next few years.

Although this sustained growth is impressive, the real challenge is to increase the share of wind power in relation to total power generation. For example, Germany generated 6.4% of its total power demand from the wind in 2009, whereas in the United States this ratio is only 1.8%. In contrast, Demark, Portugal, and Spain lead in this aspect, with over 20%, 15%, and 13% of wind power penetration, respectively.

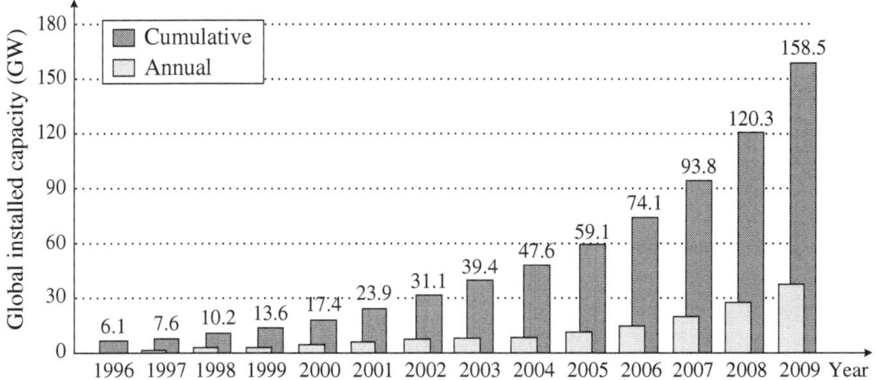

Figure 1-1. Global annual and cumulative installed wind power capacity.

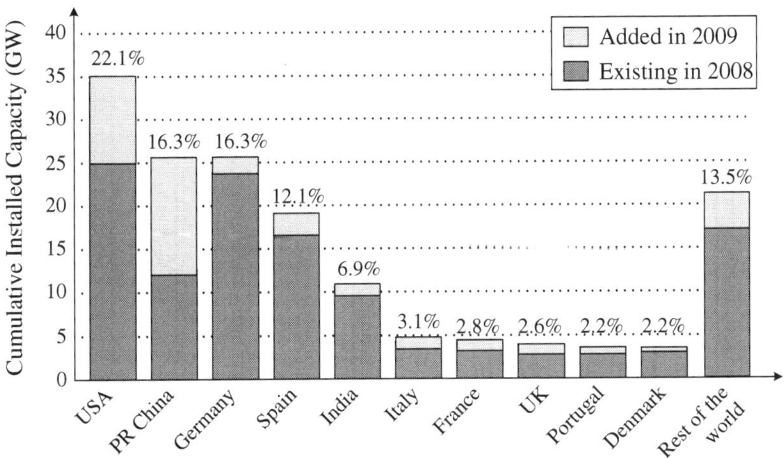

<u>Figure 1-2.</u> Top ten countries in cumulative installed wind power capacity as of 2009.

1.2.2 Small and Large Wind Turbines

Wind turbines range from a few kilowatts for residential or commercial use to several megawatts in large wind farms. Small-to-medium-size wind turbines are normally below 300 kW, and can be installed at homes, farms, and businesses to offset the consumption of utility power. Small wind power units can be used in combination with other energy sources such as photovoltaic power and diesel generators to form a stand-alone/off-grid generation system for remote areas, where access to the power grid is difficult or costly.

On the other hand, the size of large wind turbines has steadily increased over the years. The evolution in turbine size can be clearly appreciated in Figure 1-3. Starting with a 50 kW power rating and a 15 m rotor radius in the early 1980s, wind turbines can be found today up to 7.5 MW with a rotor diameter of 126 m. It is expected that a 10 MW wind turbine will be developed in the future with a turbine rotor diameter of 145 m, which is approximately twice the length of a Boeing 747 airplane.

The increase in wind turbine size implies more power output since the energy captured is a function of the square of the rotor radius. This can be observed by correlating the rotor diameter and tower height to the power rating of the wind turbines shown in Figure 1-3. Larger wind turbines often result in reduced cost since their production, installation, and maintenance costs are lower than the sum of smaller wind turbines achieving the same power output.

1.2.3 Stand-Alone and Grid-Connected Applications

The wind turbines can operate as stand-alone units of small power capacity to power villages, farms, and islands where access to the utility grid is remote or costly. Since the power generated from the wind is not constant, other energy sources are normally

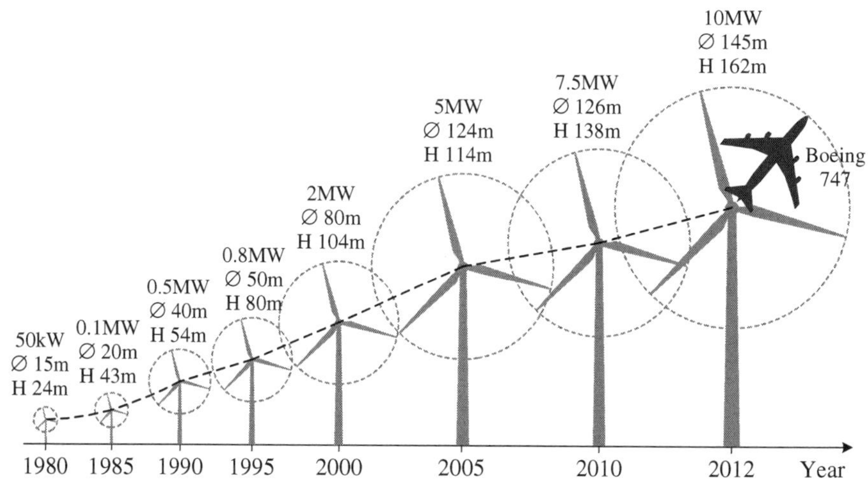

<u>Figure 1-3.</u> Evolution of wind turbine size (Ø: rotor diameter; H: tower height).

required in stand-alone systems. It is common that a stand-alone wind energy system operates with diesel generators, photovoltaic energy systems, or energy storage systems to form a more reliable distributed generation (DG) system [4]. Due to its limited applications, stand-alone wind power constitutes only a small fraction of the total installed wind capacity in the world.

The majority of wind turbines operating in the field are grid-connected, and the power generated is directly uploaded to the grid. As most generators operate at a few hundred volts (typically 690 V), transformers are used to increase the generator voltage to tens of kilovolts, for example, 35 kV, for wind farm substations. This voltage is stepped up further by the substation transformer, which connects the wind farm to the grid as shown in Figure 1-4.

1.2.4 On-Land and Offshore Applications

Large capacity wind farms have traditionally been placed on land for several reasons: easy construction, low maintenance cost, and proximity to transmission lines. On the other hand, offshore wind farms are also commercially viable. One of the main reasons for the offshore wind farm development is the lack of suitable wind resources on land. This is particularly the case in densely populated areas such as in some European countries. Another important reason is that the offshore wind speed is often significantly higher and steadier than that on land. Considering that the energy obtained by wind turbines is proportional to the cube of the wind speed, the turbines can capture more energy when operating offshore. Moreover, the environmental impact, such as audible noise and visual impact, is minimal in offshore applications. These factors are the primary drivers for the development of offshore wind turbine technology.

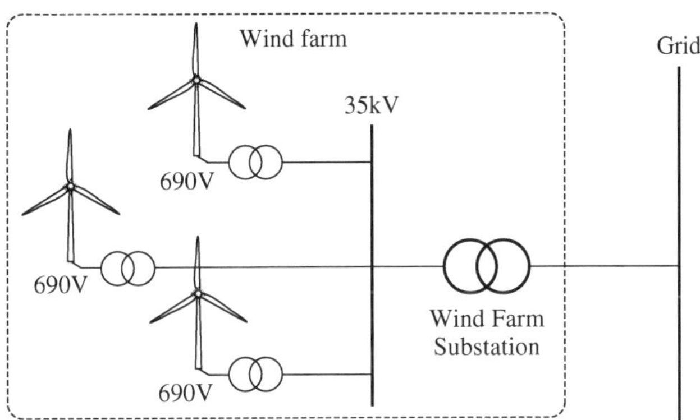

Figure 1-4. Example of a grid-connected wind farm.

Although offshore turbine prototypes can be found around the world, only a limited number of countries (mostly in Europe) have operational offshore wind farms. The biggest players in this area are the United Kingdom, Denmark, and the Netherlands, as shown in Figure 1-5 [5]. Although the total European offshore capacity of 2056 MW as of 2009 accounts for only 1.3% of total installed wind power capacity, offshore wind energy is a fast-growing market, as shown in Figure 1-6 [5]. The installed capacity increased from less than 0.1 GW in 2000 to more than 2 GW in 2009.

In addition, offshore wind resources are enormous. For instance, the offshore wind resources in Europe are several times the total European electricity consumption [6]. Some offshore wind farms scheduled to be commissioned in the near future will reach a capacity of more than 1 GW each. Examples of on-land and offshore wind farms are given in Tables 1-1 and 1-2, respectively, where the location, number of turbines, turbine size, suppliers, and production dates are provided. The photographs of some of these wind farms are shown in Figure 1-7.

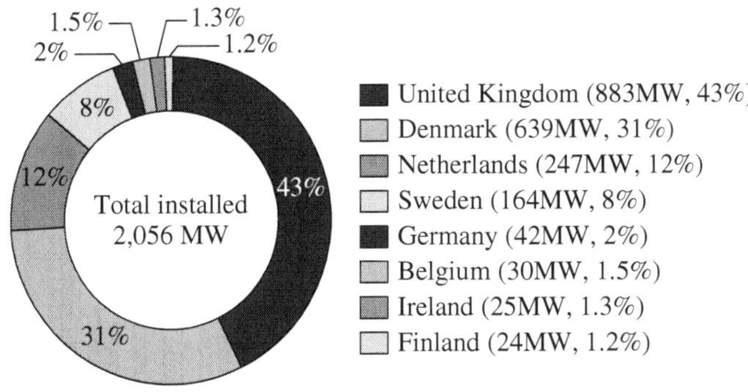

Figure 1-5. Installed European offshore wind power capacity as of 2009.

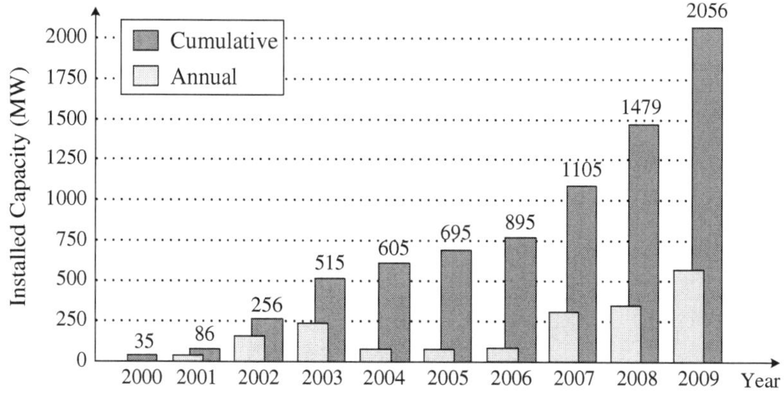

<u>Figure 1-6.</u> Annual and cumulative installed European offshore wind power capacity.

Table 1-1. Examples of on-land wind farms

Parameter	Roscoe	Whitelee	Bowbeat
Location	Texas, USA	Glasgow, Scotland	Moorfoot Hills, Scotland
Number of turbines	627	140	24
Power rating	781.5 MW	322 MW	31.2MW
Area	404.7 km^2	72.52 km^2	
Turbine model/supplier	Mitsubishi 1000A, GE, and Siemens	Siemens 2.3 MW	Nordex N60
Production date	2009	2009	2002

Table 1-2. Examples of offshore wind farms

Parameter	London Array	Horns Rev	Nysted/Rodsand I
Location	London, UK	Jutland, Denmark	Lolland, Denmark
Distance from shore	20 km	14–24 km	10.8 km
Number of turbines	341	80	72
Power rating	1 GW	160 MW	166 MW
Area	245 km^2	20 km^2	26 km^2
Internal bus voltage	33 kV	34 kV	33 kV
Turbine model/supplier	Siemens SWT-3.6	Vestas V80 2MW	Siemens SWT-2.3
Transmission line	150 kV subsea cable	150 kV subsea cable	132 kV subsea cable
Offshore substations	Two	One	One
Production date	2012 (expected)	2002	2003

(a) Bowbeat wind farm
(photo courtesy E.ON)

(b) Rodsand I wind farm
(photo courtesy E.ON)

<u>Figure 1-7.</u> Examples of on-land and offshore wind farms.

The increase of turbine power capacity and reduction of maintenance costs are crucial for offshore wind farms. The average power rating of installed offshore wind turbines was around 2.9 MW as of 2009 [5], and the power rating of the generators for offshore applications is expected to increase in the next decade. To reduce the maintenance cost, direct-driven wind turbines using low-speed permanent magnet synchronous generators (PMSGs) is a viable technology. The maintenance costs for these turbines are reduced due to elimination of the gearbox and brushes. For offshore wind farms, the foundation and transmission cable add significantly to the total project costs. A comparison summary of on-land and offshore farms is given in Table 1-3. The wind resources, capital/maintenance costs, and energy production are the critical factors to be considered in the development of offshore wind farms.

Table 1-3. Comparison between offshore and on-land wind turbines

	Description	On-land	Offshore
Resources	Wind speed	Adequate	Higher and steadier
	Limits in available land/area	Yes	No
Power transmission	HVDC/HVAC	Location dependent	Required
Environmental impact	Visual impact	Yes for nearby residents	No
	Acoustic noise	Yes for nearby residents	No
Operation	Access	Convenient	Inconvenient
	Erosion	Low	High
	Capital cost	Low	High
	Maintenance cost	Low	High
	Energy production	Good	Better

1.2.5 Costs of Wind Energy Conversion Systems

The cost of electricity from wind power has declined steadily over the past two decades. When the first utility-scale turbines were installed in the early 1980s, wind-generated electricity cost $0.3 per kWh. Today, wind power plants can generate electricity for $0.07 to $0.12 per kWh [7]. Compared with other clean energy resources, such as photovoltaic (PV) energy and solar thermal energy, wind energy is one of the most economically viable renewable energy resources, as illustrated in Figure 1-8 [8]. Note that for a given energy source, the cost for energy production is not constant, but varies with the power rating, operating condition, location, and technology used.

Table 1-4 gives the cost breakdown of a typical 2 MW wind turbine [9]. Around 75% of the total cost is directly related to the turbine, which includes rotor blades, gearbox, generator, power converters, nacelle, and tower. The other costs include grid connection, foundation construction, land rent, electric installation, and road construction.

One of the most effective methods for reduction in cost of per installed kilowatt is to increase the turbine size. As the swept area covered by the rotor blades grows proportionally to the square of the blade length, there is a favorable nonlinearity between the blade length and the captured wind power. A clear example has been given in Figure 1-3, in which a 50 kW wind turbine in 1980 had a rotor diameter of 15 m, and the turbine size was increased to 7.5 MW in 2010 with a rotor diameter of 126 m. The turbine rotor diameter and the power rating have been increased by 8.4 and 150 times, respectively. The increase in the power rating is also facilitated by the increase in tower height for larger turbines.

Table 1-5 gives typical costs for 10 kW, 50 kW, and 1.7 MW wind turbines installed in the mid-2000s [10]. The cost per installed kilowatt for these turbines was

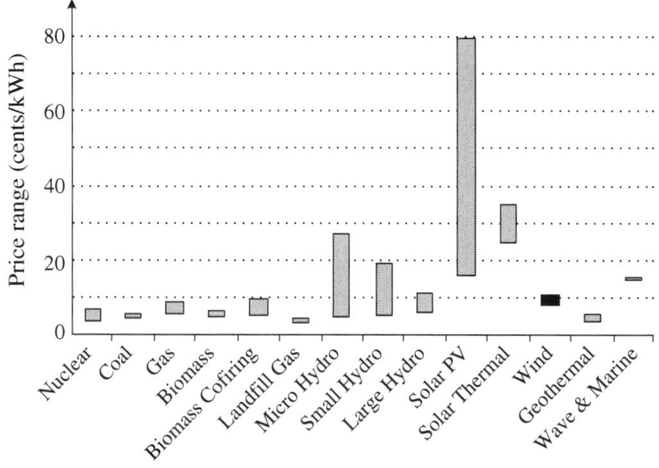

Figure 1-8. Cost range of different energy sources.

Table 1-4. Cost breakdown of a typical 2 MW wind turbine

Component	Share of total cost, %
Turbine	75.6
Grid connection	8.9
Foundation	6.5
Land rent	3.9
Electric installation	1.5
Consultancy	1.2
Financial costs	1.2
Road construction	0.9
Control systems	0.3

$5760, $3300, and $1680, respectively. Obviously, the larger the wind turbines, the lower the cost per kilowatt installed. This also explains the rapid development of large megawatt turbine technology in the past years.

The total cost of wind energy systems is also affected by the location of wind turbines. Offshore wind turbines are normally more costly than on-land turbines mainly due to higher turbine installation and power transmission costs. Figure 1-9 gives an estimated cost of offshore and on-land turbines from year 2000 to 2020 [11]. With the technological advancements, total cost of both offshore and on-land turbines will decrease in the coming years. Although the offshore wind turbines cost more than the on-land turbines, the greater energy output of offshore turbines can compensate for the higher initial costs. This is one of the reasons making offshore wind power generation attractive.

1.3 WIND TURBINE TECHNOLOGY

The wind turbine is one of the most important elements in wind energy conversion systems. Over the years, different types of wind turbines have been developed [12]. This section provides an overview of wind turbine technologies, including horizontal/vertical-axis turbines and fixed/variable-speed turbines.

Table 1-5. Costs of small and large wind turbines installed in the mid-2000s

Item	Small wind turbine		Large wind turbine
Rated output power	10 kW	50 kW	1.7 MW
Turbine cost	$32,500	$110,000	$2,074,000
Installation	$25,100	$55,000	$782,000
Total installed cost	$57,600	$165,000	$2,856,000
Total cost per kW installed	$5,760	$3,300	$1,680

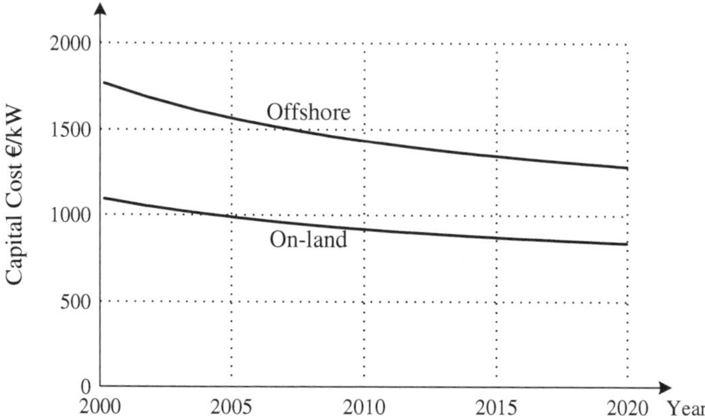

Figure 1-9. Comparison of total installed costs for on-land and offshore wind energy systems.

1.3.1 Horizontal- and Vertical-Axis Wind Turbines

Wind turbines can be categorized based on the orientation of their spin axis into horizontal-axis wind turbines (HAWT) and vertical-axis wind turbines (VAWT) [12], as shown in Figure 1-10.

In horizontal-axis wind turbines, the orientation of the spin axis is parallel to the ground as shown in Figure 1-10a. The tower elevates the nacelle to provide sufficient

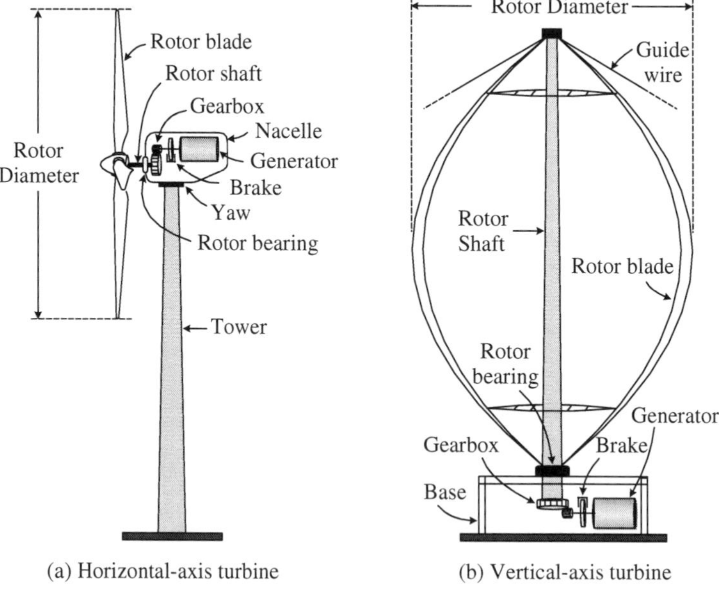

Figure 1-10. Horizontal- and vertical-axis wind turbines.

space for the rotor blade rotation and to reach better wind conditions. The nacelle supports the rotor hub that holds the rotor blades and also houses the gearbox, generator, and, in some designs, power converters. The industry standard HAWT uses a three-blade rotor positioned in front of the nacelle, which is known as upwind configuration. However, downwind configurations with the blades at the back can also be found in practical applications. Turbines with one, two, or more than three blades can also be seen in wind farms.

In vertical-axis wind turbines, the orientation of the spin axis is perpendicular to the ground. The turbine rotor uses curved vertically mounted airfoils. The generator and gearbox are normally placed in the base of the turbine on the ground, as shown in Figure 1-10b. The rotor blades of the VAWT have a variety of designs with different shapes and number of blades. The design given in the figure is one of the popular designs. The VAWT normally needs guide wires to keep the rotor shaft in a fixed position and minimize possible mechanical vibrations.

A comparison between the horizontal- and vertical-axis turbine technologies is summarized in Table 1-6. The HAWT features higher wind energy conversion efficiency due to the blade design and access to stronger wind, but it needs a stronger tower to support the heavy weight of the nacelle and its installation cost is higher. On the contrary, the VAWT has the advantage of lower installation costs and easier maintenance due to the ground-level gearbox and generator installation, but its wind energy conversion efficiency is lower due to the weaker wind on the lower portion of the blades and limited aerodynamic performance of the blades. In addition, the rotor shaft is long, making it prone to mechanical vibrations. It is these disadvantages that hinder the practical application of vertical-axis turbines for large-scale wind energy conversion. Horizontal-axis turbines dominate today's wind market, especially in large commercial wind farms.

Table 1-6. Comparison between horizontal- and vertical-axis wind turbines

Turbine type	Advantages	Disadvantages
HAWT	• Higher wind energy conversion efficiency • Access to stronger wind due to high tower • Power regulation by stall and pitch angle control at high wind speeds	• Higher installation cost, stronger tower to support heavy weight of nacelle • Longer cable from the top of tower to ground • Orientation required (yaw control)
VAWT	• Lower installation cost and easier maintenance due to the ground-level gearbox and generator • Operation independent of wind direction • Suitable for rooftops (stronger wind without need of tower)	• Lower wind energy conversion efficiency • Higher torque fluctuations and prone to mechanical vibrations • Limited options for power regulation at high wind speeds

1.3.2 Fixed- and Variable-Speed Turbines

Wind turbines can also be classified into fixed-speed and variable-speed turbines [13]. As the name suggests, fixed-speed wind turbines rotate at almost a constant speed, which is determined by the gear ratio, the grid frequency, and the number of poles of the generator. The maximum conversion efficiency can be achieved only at a given wind speed, and the system efficiency degrades at other wind speeds. The turbine is protected by aerodynamic control of the blades from possible damage caused by high wind gusts. The fixed-speed turbine generates highly fluctuating output power to the grid, causing disturbances to the power system. This type of turbine also requires a sturdy mechanical design to absorb high mechanical stresses [14].

On the other hand, variable-speed wind turbines can achieve maximum energy conversion efficiency over a wide range of wind speeds. The turbine can continuously adjust its rotational speed according to the wind speed. In doing so, the tip speed ratio, which is the ratio of the blade tip speed to the wind speed, can be kept at an optimal value to achieve the maximum power conversion efficiency at different wind speeds [12].

To make the turbine speed adjustable, the wind turbine generator is normally connected to the utility grid through a power converter system [13]. The converter system enables the control of the speed of the generator that is mechanically coupled to the rotor (blades) of the wind turbine. As shown in Table 1-7, the main advantages of the variable-speed turbine include increased wind energy output, improved power quality, and reduced mechanical stress [14]. The main drawbacks are the increased manufacturing cost and power losses due to the use of power converters. Nevertheless, the additional cost and power losses are compensated for by the higher energy production.

Furthermore, the smoother operation provided by the controlled generator reduces mechanical stress on the turbine, the drive train and the supporting structure. This has enabled manufacturers to develop larger wind turbines that are more cost-effective. Due to the above reasons, variable-speed turbines dominate the present market.

1.3.3 Stall and Pitch Aerodynamic Power Controls

Turbine blades are aerodynamically optimized to capture the maximum power from the wind in normal operation with a wind speed in the range of about 3 to 15 m/s. In

Table 1-7. Advantages and drawbacks of fixed- and variable-speed wind turbines

Speed mode	Advantages	Disadvantages
Fixed speed	• Simple, robust, reliable • Low cost and maintenance	• Relatively low energy-conversion efficiency • High mechanical stress • High power fluctuations to the grid
Variable speed	• High-energy conversion efficiency • Improved power quality • Reduced mechanical stress	• Additional cost and losses due to use of converters • More complex control system

order to avoid damage to the turbine at a high wind speed of approximately 15 to 25 m/s, aerodynamic power control of the turbine is required. There are a number of different ways to control aerodynamic forces on the turbine blades, the most commonly used methods being pitch and stall controls [2].

The simplest control method is the passive stall control, in which the blades of the turbine are designed such that when the wind speed exceeds the rated wind speed of about 15 m/s, air turbulence is generated on the blade surface that is not facing the wind. The turbulence reduces the lift force on the blade, resulting in a reduction in captured power, which prevents turbine damage. Since there are no mechanical actuators, sensors, or controllers, the power control by passive stall is robust and cost-effective. The main disadvantage of this method is the reduction in power-conversion efficiency at low wind speeds. Passive stall is normally used in small-to-medium-size WECS.

Pitch control is normally used for large wind turbines. During normal operating conditions with the wind speed in the range from 3 to 15m/s, the pitch angle is set at its optimal value to capture the maximum power from the wind. When the wind speed becomes higher than the rated value, the blade is turned out of the wind direction to reduce the captured power [12]. The blades are turned in their longitudinal axis, changing the pitch angle through a hydraulic or electromechanical device located in the rotor hub attached to a gear system at the base of each blade. As a result, the power captured by the turbine is kept close to the rated value of the turbine.

In cases in which the wind speed is higher than the limit of about 25 m/s, the blades are pitched completely out of the wind (fully pitched or feathered), and thus no power is captured. This method is effective in protecting the turbine and the supporting structure from damage caused by strong wind gusts. When the blades are fully pitched, the rotor is locked by a mechanical brake, and the turbine is in the parking mode. The major disadvantages of pitch control include the extra complexity and cost due to the pitch mechanism, and the power fluctuations during strong wind gusts due to slow pitch-control dynamics.

Another aerodynamic power control method is active stall control, which is essentially a pitch-control mechanism with the difference that the angle of attack of the blade is turned into the wind, causing stall (turbulence on the back of the blade), instead of being turned out of the wind. Active stall mechanism is an improvement over the passive stall, and can improve the power conversion efficiency at low wind speeds and limit the maximum captured power in high wind gusts. However, as with pitch-controlled WECS, it is a complex system. Active stall methods are normally used in medium-to-large-size WECS.

1.4 WIND ENERGY CONVERSION SYSTEM CONFIGURATIONS

The generator and power converter in a wind energy conversion system are the two main electrical components. Different designs and combinations of these two components lead to a wide variety of WECS configurations [13], which can be classified into three groups: (1) fixed-speed WECS without power converter interface, (2) WECS using reduced-capacity converters, and (3) full-capacity converter operated WECS.

1.4.1 Fixed-Speed WECS without Power Converter Interface

A typical configuration of WECS without a power converter interface is illustrated in Figure 1-11, where the generator is connected to the grid through a transformer. A squirrel cage induction generator (SCIG) is exclusively used in this type of WECS, and its rotational speed is determined by the grid frequency and the number of poles of the stator winding. For a four-pole megawatt generator connected to a grid of 60 Hz, the generator operates at a speed slightly higher than 1800 rpm. At different wind speeds, the generator speed varies within 1% of its rated speed. The speed range of the generator is so small that this system is often known as a fixed-speed WECS, as mentioned earlier.

A gearbox is normally required to match the speed difference between the turbine and generator such that the generator can deliver its rated power at the rated wind speed. This configuration requires a soft starter to limit high inrush currents during system start-up, but the soft starter is bypassed by a switch after the system is started. During normal operation, the system does not need any power converter. A three-phase capacitor bank is usually required to compensate for the reactive power drawn by the induction generator.

This wind energy system features simplicity, low manufacturing/maintenance costs, and reliable operation. The main drawbacks include: (1) the system delivers the rated power to the grid only at a given wind speed, leading to low energy conversion efficiency at other wind speeds; and (2) the power delivered to the grid fluctuates with the wind speed, causing disturbances to the grid. Despite its disadvantages, this wind energy system is still widely accepted in industry with a power rating up to a couple of megawatts. Examples of practical fixed-speed WECS are given in Table 1-8.

1.4.2 Variable-Speed Systems with Reduced-Capacity Converters

Variable-speed operation has a series of advantages over fixed-speed wind systems. It increases the energy conversion efficiency and reduces mechanical stress caused by wind gusts. The latter has a positive impact on the design of the structure and mechanical parts of the turbine and enables the construction of larger wind turbines. It also re-

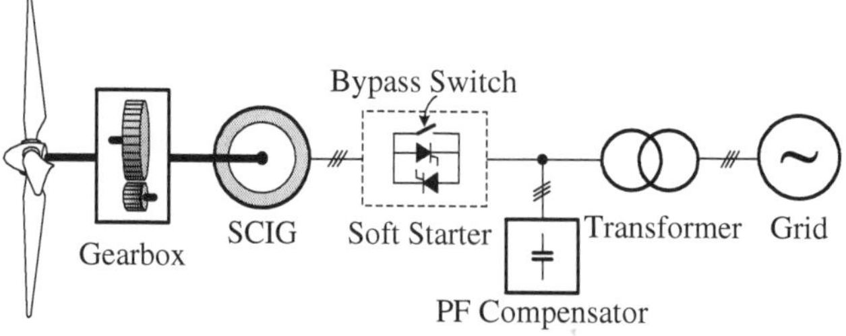

<u>Figure 1-11.</u> Wind energy conversion system without power converter interface.

Table 1-8. Example of commercial fixed-speed WECS

Parameter	Vestas V82-1.65
Power rating	1.65 MW
Turbine diameter	82 m
Numbers of blades	3
Turbine speed	14.4 rpm
Wind speed (cut-in/rated/cut-out)	3.5/13/20 m/s
Generator	SCIG
Gearbox	Planetary/helical stages
Pitch/stall mechanism	Active stall

Photo courtesy Vestas Wind Systems A/S.

duces the wear and tear on the gearbox and bearings, expanding the life cycle and reducing the maintenance requirements. The main drawback of variable-speed WECS is the need for a power converter interface to control the generator speed, which adds cost and complexity to the system. However, the power converter decouples the generator from the grid, which enables the control of the grid-side active and reactive power [13].

Variable-speed WECS can be further divided into two types based on the power rating of the converter with respect to the total power of the system: reduced-capacity power converter and full-capacity power converter. The variable-speed WECS with reduced-capacity converters are only feasible with wound-rotor induction generators (WRIG) since variable-speed operation can be achieved by controlling the rotor currents without the need to process the total power of the system. There are two designs for the WRIG configurations: one with a converter-controlled variable resistance, and the other with a four-quadrant power converter system.

Wound Rotor Induction Generator with Variable Rotor Resistance.
Figure 1-12 shows a typical block diagram of the WRIG wind energy system with a variable resistance in the rotor circuit. The change in the rotor resistance affects the torque/speed characteristic of the generator, enabling variable-speed operation of the turbine. The rotor resistance is normally made adjustable by a power converter. The speed adjustment range is typically limited to about 10% above the synchronous speed

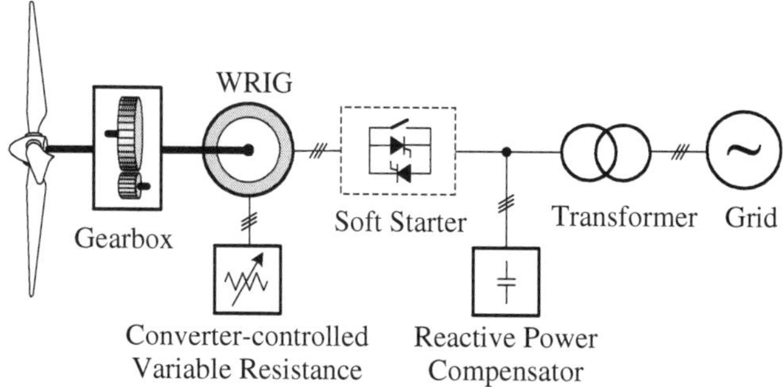

<u>Figure 1-12.</u> Variable-speed configuration with variable rotor resistance.

of the generator [15]. With variable-speed operation, the system can capture more power from the wind, but also has energy losses in the rotor resistance. This configuration also requires a soft starter and reactive power compensation. The WRIG with variable rotor resistance has been in the market since the mid 1990s with a power rating up to a couple of megawatts. A practical example of this configuration and its respective parameters are given in Table 1-9.

Doubly Fed Induction Generator with Rotor Converter. A typical block diagram of the doubly fed induction generator (DFIG) wind energy system is shown in

Table 1-9. Example of commercial WECS with variable rotor resistance

Parameter	Vestas V80-1.8US
Power rating	1.8 MW
Turbine diameter	80 m
Turbine speed	15.5 or 16.8 rpm
Wind speed (cut-in/rated/cut-out)	4/15/25 m/s
Generator	WRIG
Gearbox	Planetary/parallel axle
Pitch/stall mechanism	Pitch

Photo courtesy Vestas Wind Systems A/S.

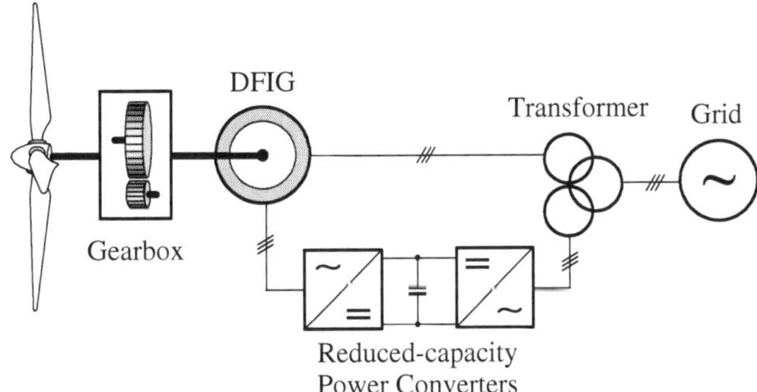

Figure 1-13. Variable-speed configuration with reduced-capacity converters.

Figure 1-13. The configuration of this system is the same as that of the WRIG system except that (1) the variable resistance in the rotor circuit is replaced by a grid-connected power converter system, and (2) there is no need for the soft starter or reactive power compensation. The power factor of the system can be adjusted by the power converters. The converters only have to process the slip power in the rotor circuits, which is approximately 30% of the rated power of the generator, resulting in reduced converter cost in comparison to the wind energy systems using full-capacity converters [13]. Examples of this configuration are given in Table 1-10.

The use of the converters also allows bidirectional power flow in the rotor circuit and increases the speed range of the generator. This system features improved overall power conversion efficiency, extended generator speed range ($\pm30\%$), and enhanced dynamic performance as compared to the fixed-speed WECS and the variable resistance configuration. These features have made the DFIG wind energy system widely accepted in today's market.

1.4.3 Variable-Speed Systems with Full-Capacity Power Converters

The performance of the wind energy system can be greatly enhanced with the use of a full-capacity power converter. Figure 1-14 shows such a system in which the generator is connected to the grid via a full-capacity converter system [13]. Squirrel cage induction generators, wound rotor synchronous generators, and permanent magnet synchronous generators (PMSG) have all found applications in this type of configuration with a power rating up to several megawatts. The power rating of the converter is normally the same as that of the generator. With the use of the power converter, the generator is fully decoupled from the grid, and can operate in full speed range. This also enables the system to perform reactive power compensation and smooth the grid connection. The main drawback is a more complex system with increased costs.

It is noted that the wind energy system can operate without the need for a gearbox if

Table 1-10. Examples of commercial DFIG WECS

Parameter	Model		
	Nordex N100	Vestas V90	Repower 5M
Power rating	2.5 MW	3 MW	5 MW
Turbine diameter	100 m	9 0m	126 m
Turbine speed	9.6 ~ 14.9 rpm	8.6 ~ 18.4 rpm	7.7 ~ 12.1 rpm
Wind speed (cut-in/rated/cut-out)	3/13/20 m/s	3.5/15/25 m/s	3.5/14/25 m/s
Generator	6-pole WRIG	4-pole WRIG	6-pole WRIG
Gearbox	Planetary/spur stages	Planetary/helical stages	Planetary/spur stages
Pitch/stall mechanism	Pitch	Pitch	Pitch

Photos courtesy (from left to right) Nordex, Vestas Wind Systems A/S, and REpower Systems AG.

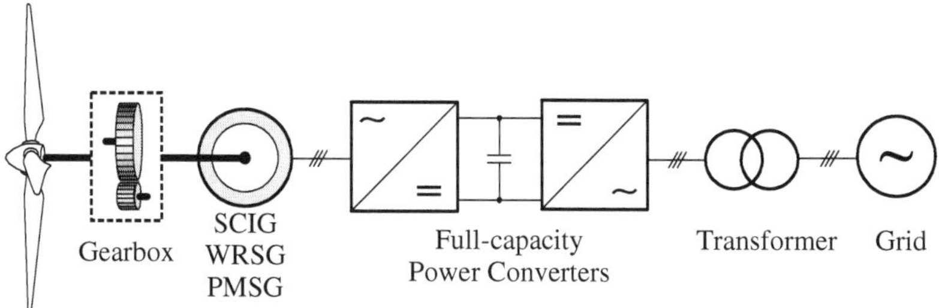

Figure 1-14. Variable-speed configurations with full-capacity converters.

a low-speed synchronous generator with a large number of poles is used. The elimination of the gearbox improves the efficiency of the system and reduces initial costs and maintenance. However, a low-speed generator has a substantially larger diameter to accommodate the large number of poles on the perimeter, which may lead to an increase in generator and installation costs.

Examples of commercial full-capacity power converter operated WECS with geared and gearless designs are listed in Table 1-11, where the system power rating, turbine speed, generator type, and power converter topology are provided. Some of the

Table 1-11. Examples of variable-speed configurations with full-capacity converters

Parameter	Enercon E-82E3	Vestas V-112	Avantis AV928
	Model		
Power rating	3 MW	3 MW	2.5 MW
Turbine diameter	82 m	112 m	93.2 m
Turbine speed	6 ~ 18.5 rpm	4 ~ 17.7 rpm	6 ~ 18 rpm
Wind speed (cut-in/ rated/cut-out)	na/na/28–34 m/s	3/12/25 m/s	3/11.3/25–30 m/s
Gearbox	Gearless (direct drive)	Planetary/spur stages	Gearless (direct drive)
Generator	Multipole WRSG	PMSG	120-pole PMSG
Converter system	Diode + boost + 2L-VSC	na	4-quadrant 2L-VSC
Pitch/stall mechanism	Pitch	Pitch	Pitch

Parameter	Clipper Liberty 2.5	Siemens SWT 3.0-101	Goldwind 2.5 PMDD
Power rating	2.5 MW	3 MW	2.5 MW
Turbine diameter	100 m (C-100 model)	101 m	99.8 m
Turbine speed	9.6 ~ 15.5 rpm	6 ~ 16 rpm	6.5 ~ 14.5 rpm
Wind speed (cut-in/ rated/cut-out)	4/na/25 m/s	3/13/25 m/s	3/13.5/25 m/s
Gearbox	Distributed gearbox	Gearless (direct drive)	Gearless (direct drive)
Generator	4 x PMSG	Multipole PMSG	Multipole PMSG
Converter system	4 × diode + 2L-VSC	4-quadrant	diode + boost + 2L VSC
Pitch/stall mechanism	Pitch	Pitch	Pitch

Photos courtesy (left to right, top to bottom) ENERCON GmbH, Vestas Wind Systems A/S, Avantis, Clipper Windpower Inc., Siemens Wind Power A/S, and Goldwind.

most common converter topologies used for this type of WECS include two-level voltage source converter (2L-VSC) in back-to-back configuration, diode-bridge rectifier plus DC-DC boost stage and 2L-VSC, and three-level neutral point clamped converter (3L-NPC) in back-to-back configuration.

1.5 GRID CODE

Grid codes have been developed and enforced in many countries for many years. They ensure applications of uniform standards for power systems and provide a framework for manufacturers to develop their equipment. Grid codes are usually based on the experience acquired through the operation of power systems and may vary from one utility to another. Differences in various grid codes also stem from regional and geographical conditions. However, the key elements in the different grid codes remain similar across the globe since their ultimate goal is to ensure safe, reliable, and economic operation of the power system.

Due to the rapid development of renewable energies and their integration into the grid, the grid codes in many countries have been updated to address issues related to renewable energy power generation [16]. According to the updated grid codes, wind farms tend to be considered as power generation plants, which should perform in a similar manner to conventional power-generation plants. The main elements in the grid codes include fault ride-through requirements, active/reactive power control, frequency/voltage regulation, power quality, and system protection. The following subsection only provides a brief overview of fault ride-through requirements and reactive power control.

1.5.1 Fault Ride-Through Requirements

Grid disturbances such as severe voltage dips caused by short-circuit faults can lead to power-generating units disconnected from the grid, which may cause instability in the grid. To avoid this, the grid code requires power-generating units to remain connected and continuously operated even if the voltage dips reaches very low values. The depth and duration of the voltage dips are usually defined by a voltage–time diagram. Figure 1-15 shows an example of low-voltage ride-through (LVRT) requirements during grid faults, where V_N is the nominal voltage of the grid [17]. Above the limit line, a power-generating system must remain connected during the fault even when the grid voltage falls to zero with duration of less than 150 ms. The system is allowed to be disconnected from the grid only when the voltage dips are in the area below the limit line. The grid codes also require the system to supply a certain amount of reactive current to support the grid voltage during the fault [18]. It is noted that the limits and ranges for the LVRT requirements vary with the grid operators in different countries, but they all share a common background and purpose. Most WECS equipped with full-capacity converters are capable of meeting the requirements.

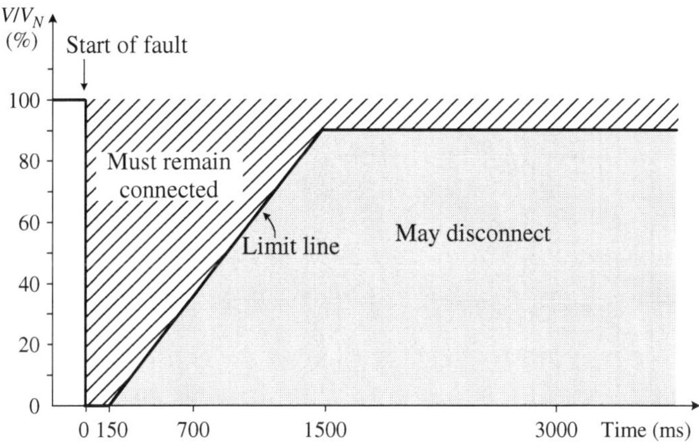

Figure 1-15. Example of grid requirements for low-voltage ride-through.

1.5.2 Reactive Power Control

Like the conventional power plants, wind turbines or wind farms are required to provide reactive power to the grid. Figure 1-16 shows an example of the range of the reactive power versus the active power for a power-generating unit [16]. Take a large megawatt wind turbine as an example. When the wind turbine deliveries the rated active power (1 pu) to the grid, it should be able to produce a maximum reactive power of ±0.33 pu to support the grid voltage. This corresponds to a 0.95 lagging and leading

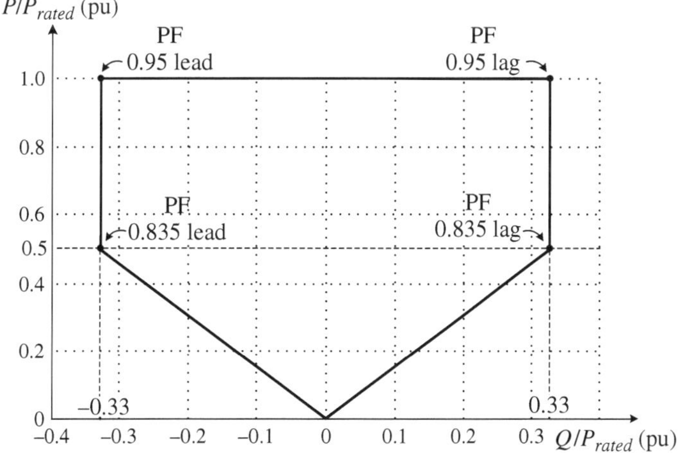

Figure 1-16. Example of reactive power requirements during system normal operation.

power factor, respectively. Similarly, when the wind turbine produces an active power of 0.5 pu, it should be capable of providing a reactive power up to ±0.33 pu, which corresponds to a 0.835 lagging and leading power factor, respectively. This requirement can be fulfilled by a properly designed variable-speed WECS. Note that the reactive power requirements given in Figure 1-16 are only examples, and different requirements may be specified in various grid codes across the globe.

1.6 SUMMARY

This chapter provided an overview of wind energy conversion systems. The basic concepts, facts, current state, and market trends of wind power technology were presented. The fundamentals of wind energy systems were discussed, including stand-alone and grid-connected operations, on-land and offshore applications, horizontal- and vertical-axis wind turbines, fixed- and variable-speed operations, stall and pitch angle controls, and grid codes. The technical background provided in this chapter complements the in-depth analysis of wind energy systems covered in the other chapters of the book.

REFERENCES

1. Renewable Energy Policy Network for the 21st Century (REN21), Renewables Global Status Report 2009 Update, available at www.ren21.net.

2. T. Burton, D. Sharpe, N. Jenkins, and E. Bossanyi, *Wind Energy Handbook,* Wiley, 2001.

3. Global Wind Energy Council (GWEC), Global Wind 2009 Report, available at www.gwec.net.

4. R. Cardenas, R. Pena, G. Asher, J. Clare, and R. Blasco-Gimenez, Control Strategies for Power Smoothing Using a Flywheel Driven by a Sensorless Vector-Controlled Induction Machine Operating in a Wide Speed Range, *IEEE Transactions on Industrial Electronics,* Vol. 51, No. 3, pp. 603–614, 2004.

5. European Wind Energy Association (EWEA), The European Offshore Wind Industry—Key Trends and Statistics 2009, available at www.ewea.org, 2010.

6. European Wind Energy Association (EWEA), Oceans of Opportunity: Harnessing Europe's Largest Domestic Energy Resource, available at www.ewea.org, 2009.

7. International Energy Association, IEA 2006 Wind Energy Annual Report, July 2006.

8. Canadian Nuclear Association, Canada's Nuclear Energy: Reliable, Affordable and Clean Electricity, available at http://www.cna.ca/, 2007.

9. European Wind Energy Association (EWEA), The Economics of Wind Energy, available at www.ewea.org, 2009.

10. Natural Resources Canada (NRCan), Survey of the Small Wind (300 W to 300 kW) Turbine Market in Canada—Executive Summary, 2005.

11. European Commission, Renewable Energy Roadmap, available at http://ec.europa.eu, 2007.

12. E. Hau, *Wind Turbines: Fundamentals, Technologies, Application, Economics,* 2nd edition, Springer, 2005.

13. F. Blaabjerg and Z. Chen, *Power Electronics for Modern Wind Turbines,* Morgan & Claypool Publishers, 2006.

14. T. Ackermann, *Wind Power in Power System,* Wiley, Ltd, 2005.

15. D. Burnham, S. Santoso, and E. Muljadi, Variable Rotor-Resistance Control of Wind Turbine Generators, in *IEEE Power and Energy Society General Meeting (PES),* 2009.

16. M. Tsili and S. Papathanassiou, A Review of Grid Code Technical Requirements for Wind Farms, *IET Renewable Power Generation,* Vol. 3, No. 3, pp. 308–332, 2009.

17. E.ON Netz, Grid Code: High and Extra High Voltage, www.eon-netz.com, 2006.

18. M. Altin, O. Goksu, R. Teodorescu, P. Rodríguez, B.-B. Jensen, and L. Helle, Overview of Recent Grid Codes for Wind Power Integration, in *12th International Conference on Optimization of Electrical and Electronic Equipment (OPTIM),* pp. 1152–1160, 2010.

2

FUNDAMENTALS OF WIND EVERGY CONVERSION SYSTEM CONTROL

2.1 INTRODUCTION

A wind energy conversion system (WECS) transforms wind kinetic energy to mechanical energy by using rotor blades. This energy is then transformed into electric energy by a generator. The system is made up of several components, participating directly in the energy conversion process. There are also other components that assist the system to achieve this task in a controlled, reliable, and efficient way. In order to better understand the process of wind energy conversion, descriptions of the major parts of a wind turbine are given in this chapter.

Since the energy source for a WECS is wind kinetic energy, wind speed plays a key role in several aspects of the conversion process, especially in relation to the maximum power output. Therefore, this chapter introduces basic concepts of and relations between wind speed and power captured by the blades. This provides the necessary insight to explain how the power output of a wind turbine can be regulated by adjusting the blade pitch angle or by controlling the generator's torque or speed. These power control methods are essential to ensure a maximum power output over a wide range of wind speeds. They also enable reliable and safe operation, protecting the mechanical and structural parts of the wind turbine from damage during strong wind gusts.

Power Conversion and Control of Wind Energy Systems. By B. Wu, Y. Lang, N. Zargari, and S. Kouro **25**
© 2011 the Institute of Electrical and Electronics Engineers, Inc. Published 2011 by John Wiley & Sons, Inc.

This chapter provides a brief overview of wind turbines, blade aerodynamic power control, and maximum power tracking schemes. The basic concepts introduced in this chapter lay the background for better understanding of materials in the subsequent chapters.

2.2 WIND TURBINE COMPONENTS

A wind turbine is composed of several parts to achieve kinetic-to-electric energy conversion. The side view of a typical wind turbine is shown in Figure 2-1. There are several variants to this layout of components, particularly for direct-drive (gearless) wind turbines. Nonetheless, the figure serves as a general reference to locate and describe the different parts in modern wind turbines.

The wind kinetic energy is converted to mechanical energy by the blades mounted on the rotor hub. The rotor hub is installed on the main shaft, also known as the low-speed shaft. The mechanical energy is transmitted through the drive train (shafts, bearings, and gearbox) to the generator, which converts mechanical energy into electric energy. This conversion is usually assisted by a power converter system which delivers the power from the generator to the grid. Most of the wind turbine components are enclosed in a nacelle on top of the tower.

There are other parts that are not directly involved in the power conversion but are important to ensure the proper, efficient, and reliable operation of the system. Examples include the pitch system, yaw system, mechanical brake, wind speed and direction

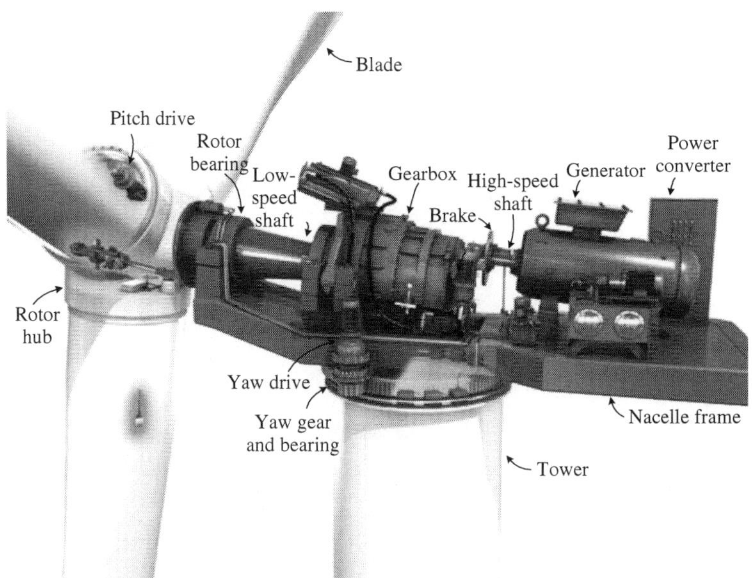

Figure 2-1. Main components of a wind turbine (photo courtesy Bosch Rexroth AG).

sensors, power distribution cables, heat dissipation/exchange system, lightning protection system, and structural components such as the tower, foundation, and nacelle enclosure. Large wind turbines are also equipped with an uninterruptable power supply or backup energy system that ensures uninterrupted operation of essential parts such as the control system, pitch drive, and brakes.

In direct-drive (gearless) turbines, the absence of the gearbox and high-speed shaft leads to a more compact drive train and, hence, a shorter nacelle. However, the wider diameter of low-speed generators requires a taller nacelle structure. This phenomenon is more evident in wound rotor synchronous generators than permanent-magnet synchronous generators.

2.2.1 Turbine Blade

The blade is the most distinctive and visible component of a wind turbine. It is also responsible for carrying out one of the most essential tasks of the energy conversion process: transforming the wind kinetic energy into rotational mechanical energy. Blades have greatly evolved in aerodynamic design and materials from the early windmill blades made of wood and cloth. Modern blades are commonly made of aluminum, fiberglass, or carbon-fiber composites that provide the necessary strength-to-weight ratio, fatigue life, and stiffness while minimizing the weight [1].

Although single- and two-bladed wind turbines have found practical applications, the three-blade rotor is considered the industry standard for large wind turbines. Turbines with fewer blades operate at higher rotational speeds. This is an advantage from the drive train point of view since they require a gearbox with a lower gear ratio, which translates into lower cost. In addition, fewer blades imply lower costs. However, acoustic noise increases proportionally to the blade tip speed. Therefore, acoustic noise is considerably higher for single- and two-bladed turbines, which is considered an important problem, particularly in populated areas.

Singe-blade turbines have an asymmetrical mechanical load distribution. The turbine rotors are aerodynamically unbalanced, which can cause mechanical vibrations. Moreover, higher rotational speed imposes more mechanical stress on the blade, turbine structure, and other components, such as bearings and gearbox, leading to more design challenges and lower life span.

Rotors with more than three blades are not common since they are more expensive (more blades). Operating at lower rotational speeds requires a higher gear ratio. The lagging wind turbulence of one blade can affect the other blades since they are closer to each other. Hence, the three-blade rotor presents the best trade-off between mechanical stress, acoustic noise, cost, and rotational speed for large wind turbines.

The aerodynamic operating principle of the turbine blade is similar to the wings of an airplane. It can be explained by Bernoulli's principle, which states that as the speed of a moving fluid (liquid or gas) increases, the pressure within the fluid decreases [2]. The curved shape of the blade creates a difference between the wind speed above (v_{w1}) and below (v_{w2}) the blade, as illustrated in Figure 2-2. The airflow above the blade is faster than the one below ($v_{w1} > v_{w2}$), which, according to the Bernoulli's principle, has the inverse effect on the pressure ($p_{w2} > p_{w1}$). The pressure difference between the top

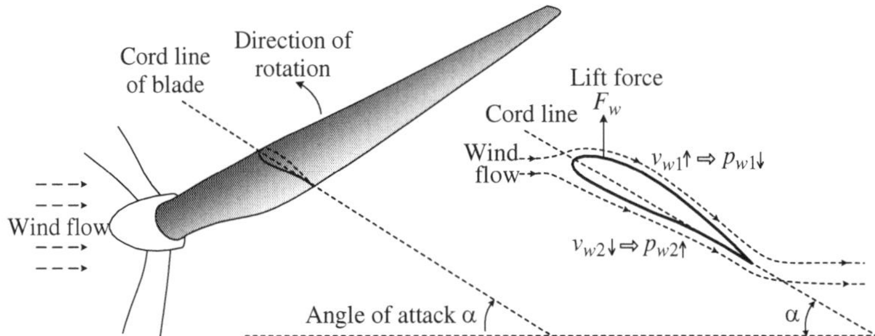

Figure 2-2. Wind turbine blade aerodynamics and angle of attack.

and bottom of the blade results in a net lift force F_w on the blade. The force applied at a certain distance of a pivot (the turbine shaft) produces torque, which creates the rotational movement of the wind turbine.

One of the important parameters for controlling the lift force of the blade is the angle of attack α, which is defined as the angle between the direction of the wind speed v_w and the cord line of the blade as shown in Figure 2-2. For a given blade, its lift force F_w can be adjusted by α. When this angle is equal to zero, no lift force or torque will be produced, which often occurs when the wind turbine is stopped (parked) for maintenance or repair.

The power of an air mass flowing at speed v_w through an area A can be caldulated by

$$P_w = \frac{1}{2}\, \rho A v_w^3 \tag{2.1}$$

where ρ is the air density in kg/m^3, A is the sweep area in m^2, and v_w is the wind speed in m/s. The air density ρ is a function of air pressure and air temperature. At sea level and temperature of 15°C, air has a density of approximately 1.2 kg/m^3.

The wind power captured by the blade and converted into mechanical power can be calculated by

$$P_M = \frac{1}{2}\, \rho A v_w^3 C_p \tag{2.2}$$

where C_p is the power coefficient of the blade. This coefficient has a theoretical maximum value of 0.59 according to the Betz limit. With today's technology, the power coefficient of a modern turbine usually ranges from 0.2 to 0.5, which is a function of rotational speed and number of blades. For a three-blade turbine with a rotor diameter of 82 m and power coefficient of $C_p = 0.36$, the captured power is 2 MW at a wind speed of 12 m/s and air density of $\rho = 1.225$ kg/m^3.

As can be observed from Equation (2.2), there are three possibilities for increasing the power captured by a wind turbine: the wind speed v_w, the power coefficient C_p, and the sweep area A. Since wind speed cannot be controlled, the only way to increase wind speed is to locate the turbines in regions with higher average wind speeds. An example is the offshore wind farm, where the wind speed is usually higher and steadier than that on land. The captured power is a cubic function of the wind speed. Doubling the average wind speed would increase the wind power by eight times.

Second, the wind turbine can be designed with larger sweep area (i.e., longer blades) to capture more power. The sweep area is given by $A = \pi l^2$, where l is the blade length. An increase in the blade length has a quadratic effect on the sweep area and the captured power. This explains the trend of increasing the rotor diameter experienced during the last decade. Finally, the third way of increasing the captured power is by improving the power coefficient of the blade through a better aerodynamic design.

Additional blade requirements, such as lightning protection, audible noise reduction, transportation, optimum shape and weight, as well as manufacturability, make the blade design a challenging task. Blade manufacturing is a tedious process and requires careful planning and large factory space. It is a combination of automated and manual processes. Considering that the blade length of a 5 MW wind turbine is as long as a Boeing 747, as shown in Figure 1-3, one can appreciate the complexity and potential manufacturing issues. The blade transportation and installation are other great challenges. The transportation and related costs are one of the reasons why the size of the blade are expected to increase at a slower pace in the future compared to the rapid developments experienced in the last decade.

2.2.2 Pitch Mechanism

The pitch mechanism in large wind turbines enables the rotation of the blades on their longitudinal axis. It can change the angle of attack of the blades with respect to the wind, by which the aerodynamic characteristics of the blade can be adjusted. This provides a degree of control over the captured power to improve conversion efficiency or to protect the turbine. When the wind speed is at or below its rated value, the angle of attack of the blades is kept at an optimal value, at which the turbine can capture the maximum power available from the wind. When the wind speed exceeds the rated value, the pitch mechanism is activated to regulate and limit the output power, thus keeping the power output within the designed capability. For this purpose, a pitch range of around 20 to 25 degrees is usually sufficient. When the wind speed increases further and reaches the limit of the turbine, the blades are completely pitched out of the wind (fully pitched or feathering), and no power will be captured by the blades. The wind turbine is then shut down and protected.

The pitch mechanism can be either hydraulic or electric. Electric pitch actuators are more common nowadays since they are simpler and require less maintenance. Traditionally, all blades on the rotor hub are pitched simultaneously by one pitch mechanism. Modern wind turbines are often designed to pitch each blade individually, allowing an independent control of the blades and offering more flexibility. The pitch system is usually placed in the rotor hub together with a backup energy storage system for safety purposes (an accumulator for the hydraulic type or a battery for the electric type). An electric pitch mechanism is shown in Figure 2-3, where the three motor drives and pitch gears can be seen inside the rotor hub.

<u>Figure 2-3.</u> Blade pitch system with three pitch drives and gears (photo courtesy Bosch Rexroth AG).

2.2.3 Gearbox

The rotor of a large three-blade wind turbine usually operates in a speed range from 6–20 rpm. This is much slower than a standard 4- or 6-pole wind generator with a rated speed of 1500 or 1000 rpm for a 50 Hz stator frequency and 1800 or 1200 rpm for a 60 Hz stator frequency. Therefore, a gearbox is necessary to adapt the low speed of the turbine rotor to the high speed of the generator.

The gearbox conversion ratio (r_{gb}), also known as the gear ratio, is designed to match the high-speed generator with the low-speed turbine blades. For a given rated speed of the generator and turbine, the gearbox ratio can be determined by

$$r_{gb} = \frac{n_m}{n_M} = \frac{(1-s) \cdot 60 \cdot f_s}{P \cdot n_M} \tag{2.3}$$

where n_m and n_M are the generator and turbine rated speeds in rpm, s is the rated slip, f_s is the rated stator frequency in Hz, and P is the number of pole pairs of the generator. The rated slip is usually less than 1% for large induction generators, and zero for synchronous generators. Considering the rated slip of 1% for an induction generator, the gear ratio as a function of the rated turbine speed is given in Figure 2-4 for different pole numbers and rated stator frequencies.

The wind turbine gearboxes normally have multiple stages to achieve the high conversion ratio needed to couple the turbine rotor and generator. For example, with a rated turbine rotor speed of 15 rpm and a 4-pole, 50 Hz induction generator, a gear ratio of 100 is needed according to Figure 2-4, which is difficult to achieve by one gear stage. Among the type of gear stages are the planetary, helical, parallel shaft, spur, and worm types. Two or more gear types may be combined in multiple stages. The gear-

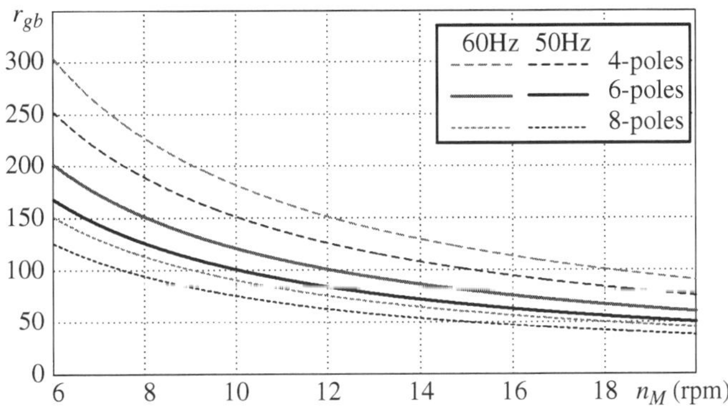

<u>Figure 2-4.</u> Gear ratio versus the rated turbine speed.

boxes are generally made of superior quality aluminum alloys, stainless steel, and cast iron. An example of a multistage gearbox is shown in Figure 2-5, which has two planetary stages and one parallel shaft stage with a gear ratio of 78:1 or 136:1 designed for a megawatt wind turbine.

The gearbox usually generates a high level of audible noise. The noise mainly arises from the meshing of individual teeth. The efficiency of the gearbox normally varies between 95% and 98%. The gearbox is a major contributor to the cost of the wind turbine in terms of initial investment and maintenance.

Random changes in wind speed and strong wind gusts result in sudden load variations on the gearbox. These sudden changes produce wear and tear on the gearbox, reducing its life span. As a result, the gearbox needs regular maintenance. The elimination of the gearbox contributes to reliability improvements and cost reduction. In order to eliminate the gearbox, a generator with the same rated rotational speed of the tur-

<u>Figure 2-5.</u> Gearbox of a large wind turbine (photo courtesy GE Drivetrain Technologies).

bine rotor is required. This can be achieved by a low-speed generator. Gearless config-urations have been adopted by several manufacturers due to the reduced cost, mainte-nance, audible noise, and power losses.

Single-stage gearboxes have also been used in practical wind turbines. This is achieved by using a special medium-speed generator that has a proper number of poles and operates at certain stator frequencies. Compared with the multistage gearbox, the single-stage gearbox is more reliable and costs less.

2.2.4   Rotor Mechanical Brake

A mechanical brake is normally placed on the high-speed shaft between the gearbox and the generator, as shown in Figure 2-1, but there are some turbines in which the brake is mounted on the low-speed shaft between the turbine and gearbox. The main advantage of placing the brake on the high-speed shaft is that it handles much lower braking torque. The brake is normally used to aid the aerodynamic power control (stall or pitch) to stop the turbine during high speed winds or to lock the turbine into a parking mode during maintenance. Hydraulic and electromechanical disc brakes are often used [2]. Figure 2-6 illustrates an electromechanical disc brake used on a high-speed shaft.

To minimize the wear and tear on the brake and reduce the stress on drive train dur-ing the braking process, most large wind turbines use the aerodynamic power control to reduce the turbine speed to a certain level or zero, and then the mechanical brake to stop or lock the wind turbine. However, the mechanical brake should be able to bring

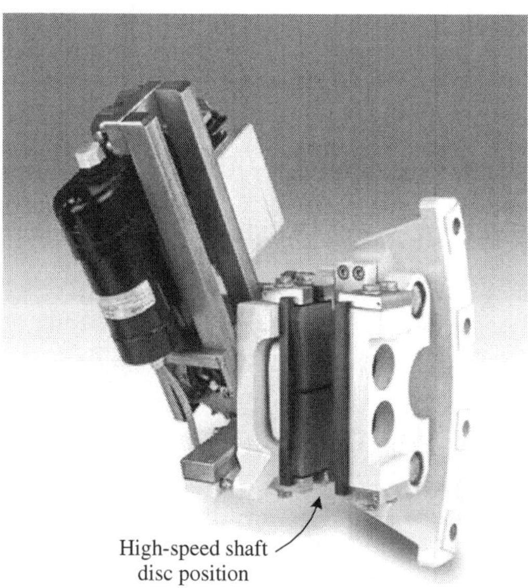

Figure 2-6. Electromechanical disc brake for high-speed shaft (photo courtesy Hanning & Kahl).

the turbine rotor to a complete stop at any wind speeds, as required by some standards such as IEC61400-1.

2.2.5 Generator

The conversion of rotational mechanical energy to electric energy is performed by the generator. Different generator types have been used in wind energy systems over the years. These include the squirrel cage induction generator (SCIG), doubly fed induction generator (DFIG), and synchronous generator (SG) (wound rotor and permanent magnet) with power ratings from a few kilowatts to several megawatts [4]. The operating principles, characteristics, and mathematical models of these generators will be presented in Chapter 3.

The SCIG is simple and rugged in construction. It is relatively inexpensive and requires minimum maintenance. Traditional direct grid-connected wind energy systems are still available in today's market. All these turbines use SCIGs and operate at a fixed speed. Two-speed SCIGs are also commercially available, in which a tapped stator winding can be adapted to change the pole pairs to allow two-speed operation. An example of a 2 MW two-speed SCIG with 6/4 poles rated at 1000/1500 rpm is shown in Figure 2-7a. The SCIGs are also employed in variable-speed wind energy systems. To date, the largest SCIG wind energy systems are around 3.5 MW in offshore wind farms.

The DFIG is the current workhorse of the wind energy industry. The stator of the generator is connected to the grid directly, while the rotor is interfaced with the grid through a power converter system with reduced power capacity. The DFIG typically operates about 30% above and below synchronous speed, sufficient for most wind speed conditions. It also enables generator-side active power control and grid-side reactive power control. The reduced-capacity converter is less expensive and requires less space, which makes the DFIG WECS popular in today's market. An example of a large DFIG is shown in Figure 2-7b.

The synchronous generator is very well suited for direct-drive wind turbines. Wound rotor synchronous generators (WRSGs) and permanent magnet synchronous generators (PMSGs) are used in wind energy systems with a maximum power rating up to 7.5 MW. Permanent magnet generators have higher efficiency and power density as compared to wound rotor generators. Recent trends indicate a move toward direct-drive turbines with PMSG. Although most SG-based turbines are direct driven, some manufacturers have developed SG turbines with gearbox drive trains. Examples are given in Figure 2-7 c and d for 4-pole and 120-pole PMSGs, respectively. A multipole WRSG for a direct-drive WECS is shown in Figure 2-7e.

2.2.6 Yaw Drive

The main function of the yaw drive is to maximize the captured wind energy by keeping the turbine facing into the wind. It usually consists of more than one electric motor drive, yaw gear, gear rim, and bearing, as can be observed in Figure 2-8, where a four-

(a) SCIG (photo courtesy ABB)

(b) DFIG (photo courtesy ABB

(c) Low-pole number PMSG
(photo courtesy ABB)

(d) High-pole number PMSG
(photo courtesy Avantis)

(e) High-pole number WRSG (photo courtesy Enercon GmbH)

Figure 2-7. Induction and synchronous generators used in modern wind turbines.

drive yaw system is illustrated. A set of yaw brakes are disposed around the yaw rim to lock the position of the turbine when facing the wind or during maintenance. The yaw drive uses a planetary gear to lower the rotating speed of the yaw gear. All the motors are commanded by the same signals and lock after turning the wind turbine into the desired position. The yaw system typically needs to generate torque from 10,000 to 70,000 Nm to turn the nacelle.

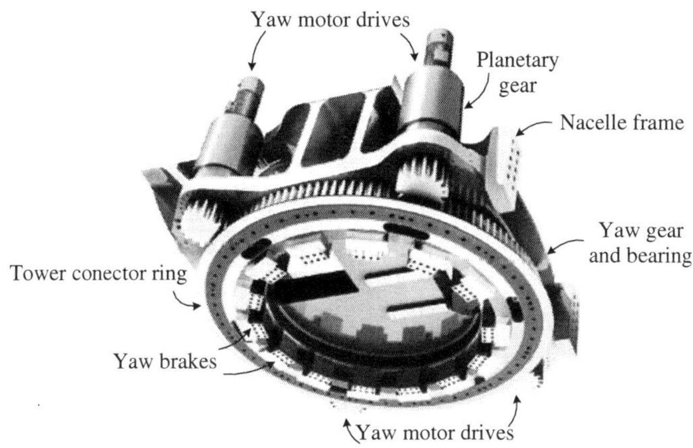

<u>Figure 2-8.</u> Yaw drive system of a large wind turbine (source Nordex).

In older wind turbines, the yaw control is also used for power regulation. For example, to limit the power captured by the turbine during high wind gusts, the turbine can be horizontally turned out of the wind. However, this technology is no longer in use since the power regulation by means of yaw control is very limited for three reasons. First, the large moment of inertia of the nacelle and turbine rotor along the yaw axis reduces the speed of response of the yaw system. Second, the cosine relationship between the component of the wind speed perpendicular to the rotor disc and the yaw angle makes the power capture insensitive to the yaw angle. For example, 15 degrees of yaw change only brings power reduction of a few percent. Third, yaw control imposes mechanical stress on different parts of the turbine, causing vibrations that could reduce the life span of the turbine.

2.2.7 Tower and Foundation

The main function of the tower is to support the nacelle and the turbine rotor, and provide the rotor with the necessary elevation to reach better wind conditions. Most towers for wind turbines are made of steel. Concrete towers or towers with a concrete base and steel upper sections are sometimes used as well. The height of the tower increases with the turbine power rating and rotor diameter. In addition, the tower must be at least 25 to 30 m high to avoid turbulence caused by trees and buildings. Small wind turbines have towers as high as a few blade rotor diameters. However, the towers of medium and large turbines are approximately equal to the turbine rotor diameter as shown in Figure 1-3 in Chapter 1. The highest tower to date is a 160 m steel lattice tower for a 2.5 MW wind turbine.

The tower also houses the power cables connecting the generator or power converters to the transformer located at the base of the tower. In some cases, the transformer is also included in the nacelle and the cables connect the transformer to the wind farm substation. In large multimegawatt turbines, the power converters may be located at

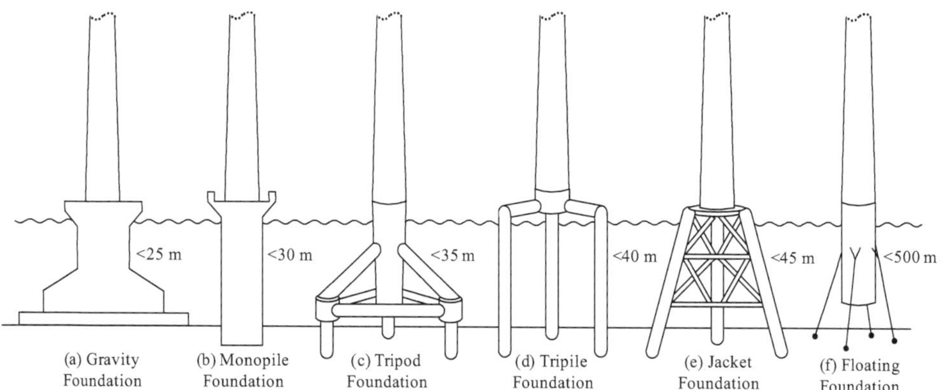

Figure 2-9. Foundations for offshore wind turbines.

the base of tower to reduce the weight and size of the nacelle. The stairs to the nacelle for maintenance are often attached along the inner wall of the tower in large wind turbines.

Special attention should be given to the structural dynamics in order to avoid vibration caused by the mechanical resonance modes of the wind turbine. The top head mass (THM) of the nacelle and the turbine rotor has a significant bearing on the dynamics of the tower and foundation. In practice, low THM is generally a measure of design for reduction of manufacturing and installation costs.

The wind-turbine foundation is also a major component in a wind energy system. The types of foundations commonly used for on-land wind turbines include slab, multipile, and monopile types. Foundations for offshore wind turbines are particularly challenging since they are located at variable water depths and in different soil types. They have to withstand harsh conditions as well. This explains the wide variety of foundations developed over the years for offshore turbines, some more proven than others. Figure 2-9 shows the most commonly adopted foundations for different water depths [2]. The gravity foundation and monopile foundation are more common in shallow waters. The tripod, tripile, and jacket foundations can reach greater depths, and are usually located farther away from the coastline. New technologies such as the floating (anchored) foundation are currently under development for more challenging water depths.

2.2.8 Wind Sensors (Anemometers)

The pitch/stall and yaw control systems require wind speed and direction measurements, respectively. The pitch/stall control needs the wind speed to determine the angle of attack of the blade for optimal operation. The yaw control requires the wind direction to face the turbine into the wind for maximum wind power capture. In addition, in variable speed turbines, the wind speed is needed to determine the generator speed for maximum power extraction.

Most large wind turbines are equipped with sensors, also referred to as anemome-

Figure 2-10. Ultrasonic anemometer wind speed sensor (photo courtesy Clipperwind).

ters, for wind data collection and processing. The wind speed sensor is usually made of a three-cup vertical-axis microturbine driving an optoelectronic rotational speed transducer. The wind direction is measured by a wind vane connected to an optoelectronic angle transducer. These are the main components of a wind measurement system, and are usually located on the top back part of the nacelle. More than one sensor system may be used in a wind turbine for more reliable and accurate measurements.

Ultrasonic anemometers are also used in practical wind turbines. They measure the wind speed by emitting and receiving acoustic signals through the air and monitoring the transmission time. Several emitters and receptors are disposed in such a way that a three-dimensional measurement can be made. The transmission time is affected by both wind speed and direction. With a given physical distribution of the sensors, the wind speed and direction can be computed from the propagation times. The ultrasonic anemometers are more accurate and reliable than the mechanical ones with moving parts. However, they are more expensive. Figure 2-10 shows an ultrasonic anemometer on the top of the nacelle of a modern wind turbine.

2.3 WIND TURBINE AERODYNAMICS

The aerodynamic design of the turbine blade has a significant influence on the amount of energy captured from the wind. The design should consider the means to limit the power and rotating speed of the turbine rotor for wind speeds above the rated value in order to keep the forces on the mechanical components (blade, gearbox, shaft, etc.) and the output power of the generator within the safety margins. This becomes critical for larger turbines as they would have narrower safety margins due to cost and size constraints.

2.3.1 Power Characteristic of a Wind Turbine

The power characteristics of a wind turbine are defined by the power curve, which relates the mechanical power of the turbine to the wind speed. The power curve is a wind turbine's certificate of performance that is guaranteed by the manufacturer. The International Energy Association (IEA) has developed recommendations for the definition of the power curve. The recommendations have been continuously improved and adopted by the International Electrotechnical Commission (IEC). The standard, IEC61400-12, is generally accepted as a basis for defining and measuring the power curve.

A typical power curve is characterized by three wind speeds: cut-in wind speed, rated wind speed, and cut-out wind speed, as described in Figure 2-11, where P_M is the mechanical power generated by the turbine and v_w is the wind speed. The cut-in wind speed, as the name suggests, is the wind speed at which the turbine starts to operate and deliver power. The blade should be able to capture enough power to compensate for the turbine power losses. The rated wind speed is the speed at which the system produces nominal power, which is also the rated output power of the generator. The cut-out wind speed is the highest wind speed at which the turbine is allowed to operate before it is shut down. For wind speeds above the cut-out speed, the turbine must be stopped, preventing damage from excessive wind.

As can be seen from Figure 2-11, the wind turbine starts to capture power at the cut-in wind speed. The power captured by the blades is a cubic function of wind speed [see Equation (2.2)] until the wind speed reaches its rated value. To deliver captured power to the grid at different wind speeds, the wind generator should be properly controlled with variable speed operation. As the wind speed increases beyond the rated speed,

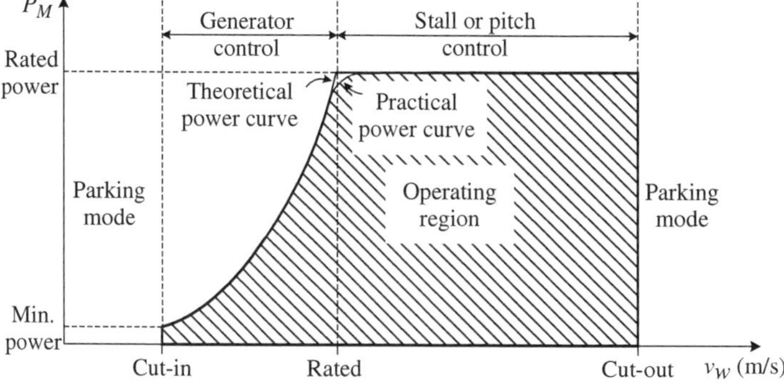

Figure 2-11. Qualitative turbine mechanical power versus wind speed curve.

aerodynamic power control of blades is required to keep the power at the rated value. This task is performed by three main techniques: passive stall, active stall, and pitch control [3]. The wind turbine should stop generating power and be shut down when the speed is higher than the cut-out wind speed. Note that the theoretical curve of Figure 2-11 has an abrupt transition from the cubic characteristic to the constant power operation at higher speeds. However, practical turbines do not exhibit this behavior, and the transition is smoother [2].

2.3.2 Aerodynamic Power Control: Passive Stall, Active Stall, and Pitch Control

As discussed at the beginning of the chapter, the aerodynamics of wind turbines are very similar to that of airplanes. The blade rotates in the wind because the air flowing along the surface that is not facing the wind moves faster than that on the surface against the wind. This creates a lift force to pull the blade to rotate. The angle of attack of the blade plays a critical role in determining the amount of force and torque generated by the turbine. Therefore, it is an effective means to control the amount of captured power. There are three aerodynamic methods to control the capture of power for large wind turbines: passive stall, active stall, and pitch control.

Passive-Stall Control. In passive-stall-controlled wind turbines, the blade is fixed onto the rotor hub at an optimal (rated) angle of attack. When the wind speed is below or at the rated value, the turbine blades with the rated angle of attack can capture the maximum possible power from the wind. With the wind speed exceeding the rated value, the strong wind can cause turbulence on the surface of the blade not facing the wind. As a result, the lifting force will be reduced and eventually disappear with the increase of the wind speed, slowing down the turbine rotational speed. This phenomenon is called stall. The stall phenomenon is undesirable for airplanes, but it provides an effective means to limit the power capture to prevent turbine damage. The operating principle of the passive-stall control is illustrated in Figure 2-12, where the lift force produced by higher than rated wind, which is the stall lifting force $F_{w,\text{stall}}$, is lower than the rated force $F_{w,\text{rated}}$ [2].

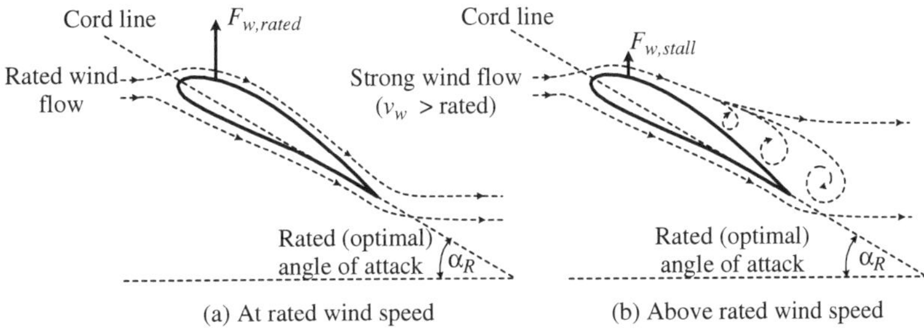

Figure 2-12. Passive stall with rated and above rated wind speeds.

The blade profile is aerodynamically designed to ensure that stall occurs only when the wind speed exceeds the rated value. To ensure that the blade stall occurs gradually rather than abruptly, the blades for large wind turbines are usually twisted along the longitudinal axis by a couple of degrees. The passive-stall-controlled wind turbines do not need complex pitch mechanisms, but the blades require a complex aerodynamic design. The passive stall may not be able to keep the captured power P_M at a constant value, as shown in Figure 12-13a. It may exceed the rated power at some wind speeds, which is not a desirable feature.

Active-Stall Control. In active-stall turbines, the stall phenomenon can be induced not only by higher wind speeds, but also by increasing the angle of attack of the blade. Thus, active-stall wind turbines have adjustable blades with a pitch control mechanism. When the wind speed exceeds the rated value, the blades are controlled to turn more into the wind, leading to the reduction of captured power. The captured

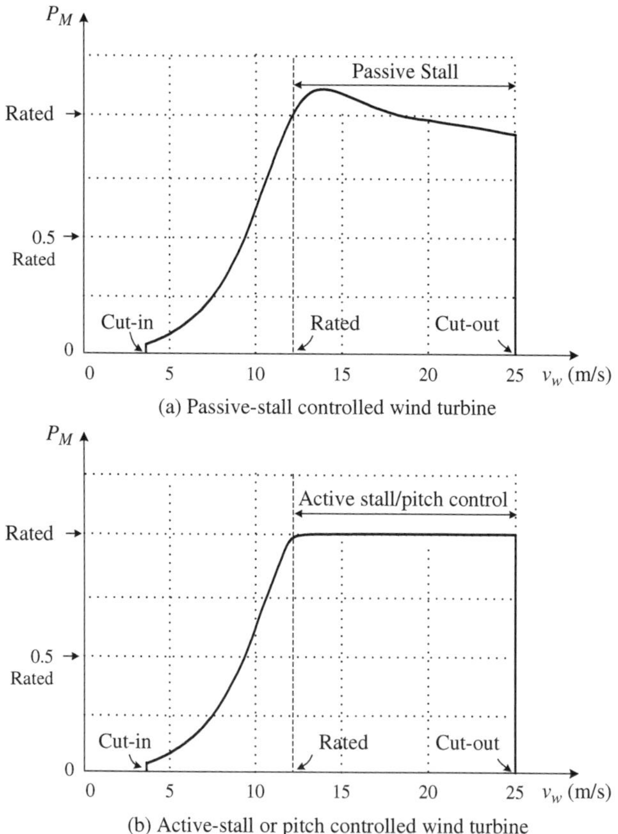

Figure 2-13. Typical turbine mechanical power versus wind speed curves with stall and pitch control.

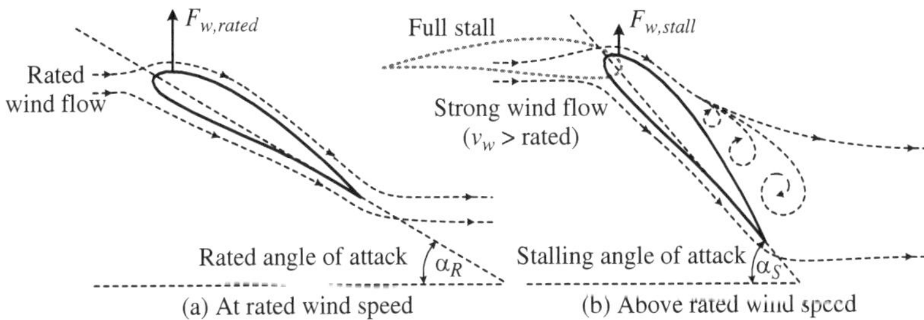

Figure 2-14. Active-stall control with rated and above-rated wind speeds.

power can, therefore, be maintained at the rated value by adjusting the blade angle of attack [2].

A qualitative example of the active-stall principle is illustrated in Figure 2-14. When the blade is turned completely into the wind, as shown in the dashed blade, the blade loses all interaction with the wind and causes the rotor to stop. This operating condition can be used above the cut-out wind speed to stop the turbine and protect it from damage.

With active-stall control, it is possible to maintain the rated power above the rated wind speed, as can be appreciated in Figure 2-13b. Active-stall controlled large megawatt wind turbines are commercially available.

Pitch Control. Similar to the active-stall control, pitch-controlled wind turbines have adjustable blades on the rotor hub. When the wind speed exceeds the rated value, the pitch controller will reduce the angle of attack, turning the blades (pitching) gradually out of the wind. The pressure difference in front and on the back of the blade is reduced, leading to a reduction in the lifting force on the blade.

The operating principle of the pitch control is illustrated in Figure 2-15. When the wind is below or at the rated speed, the blade angle of attack is kept at its rated (optimal) value α_R. With higher than the rated wind, the angle of attack of the blade is reduced, causing a reduction in lift force, $F_{w,\text{pitch}}$. When the blade is fully pitched, the blade angle of attack is aligned with the wind, as shown by the dashed blade in Figure 2-15b, and no lift force will be produced. The turbine will stop rotating and then be locked by the mechanical brake for protection. The performance of the pitch control is given in Figure 2-15b, where the mechanical power of the turbine operating at above the rated wind speed can be tightly controlled.

Both pitch and active-stall controls are based on rotating actions on the blade, but the pitch control turns the blade out of the wind, leading to a reduction in lift force, whereas the active-stall control turns the blades into the wind, causing turbulences that reduce the lift force. Pictures of wind turbines with fully-stalled and fully-pitched blades are shown in Figure 2-16.

The passive-stall technology was mainly used in the early fixed-speed wind turbines. This technology was further developed into the active-stall technology. The

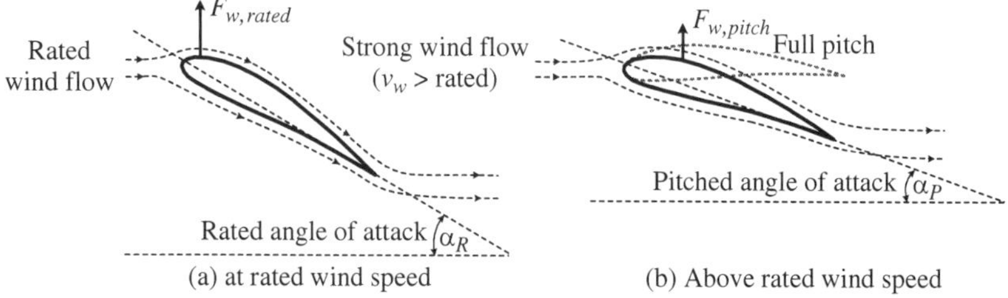

Figure 2-15. Aerodynamic pitch-control principle.

pitch control reacts faster than the active-stall control and provides better controllability. It is widely adopted in today's large wind energy systems.

2.3.3 Tip Speed Ratio

The tip speed ratio (TSR) is an important parameter in wind energy systems. It is defined as the ratio of the blade tip speed to the speed of the incoming wind, given by

$$\lambda_T = \frac{\omega_M\, r_T}{v_w} \tag{2.4}$$

where r_T is the radius of the turbine rotor (blade length), v_w is wind speed, and ω_M is the rotating speed of the blade. Another important parameter is the power conversion efficiency C_p of the blade since the mechanical power is directly proportional to C_p, as described by Equation (2.1). The power coefficient is a function of λ_T and the angle of

(a) Full stall
(photo courtesy accionsustentable.cl)
(b) Full pitch

Figure 2-16. Parked wind turbines with fully stalled and fully pitched blades.

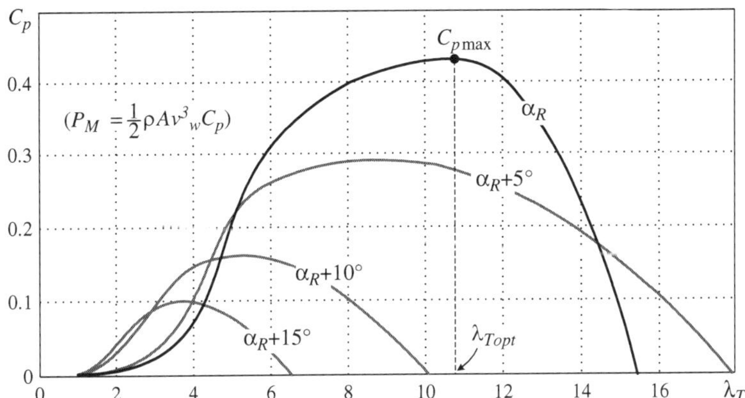

<u>Figure 2-17.</u> Power coefficient C_p versus TSR with pitch angle as a parameter.

attack α as shown in Figure 2-17. The maximum power coefficient occurs at the optimal tip speed ratio $\lambda_{T,\mathrm{opt}}$ with the rated (optimal) angle of attack α_R.

The optimal tip speed ratio $\lambda_{T,\mathrm{opt}}$ is a constant for a given blade. The speed of the turbine that produces that maximum power is related to $\lambda_{T,\mathrm{opt}}$ and wind speed v_w by

$$\omega_M = \lambda_{T,\mathrm{opt}}\,\frac{v_w}{r_T} \tag{2.5}$$

The above equation indicates that in order to obtain the maximum power and conversion efficiency, the turbine speed must be made adjustable according to the wind speed.

2.4 MAXIMUM POWER POINT TRACKING (MPPT) CONTROL

The control of a variable-speed wind turbine below the rated wind speed is achieved by controlling the generator. The main goal is to maximize the wind power capture at differenct wind speeds, which can achieved by adjusting the turbine speed in such a way that the optimal tip speed ratio $\lambda_{T,\mathrm{opt}}$ is maintained.

Figure 2-18 shows the typical characteristics of a wind turbine operating at different wind speeds, where P_M and ω_M are the mechanical power and mechanical speed of the turbine, respectively. The P_M versus ω_M curves are obtained with the blade angle of attack set to its optimal value. For the convenience of analysis and discussion, the mechanical power, turbine speed, and the wind speed are all expressed in per-unit terms. The definition of the per-unit system is given in Appendix A.

For a given wind speed, each power curve has a maximum power point (MPP) at which the optimal tip speed ratio $\lambda_{T,\mathrm{opt}}$ is achieved. To obtain the maximum available power from the wind at different wind speeds, the turbine speed must be adjusted to ensure its operation at all the MPPs. The trajectory of MPPs represents a power curve, which can be described by

$$P_M \propto \omega_M^3 \tag{2.6}$$

The mechanical power captured by the turbine can also be expressed in terms of the torque:

$$P_M = T_M \omega_M \tag{2.7}$$

where T_M is the turbine mechanical torque. Substituting (2.7) into (2.6) yields

$$T_M \propto \omega_M^2 \tag{2.8}$$

The relations between the mechanical power, speed, and torque of a wind turbine can be used to determine the optimal speed or torque reference to control the generator and achieve the MPP operation. Several control schemes have been developed to perform the maximum power point tracking (MPPT), and a brief description of three MPPT methods is given in the next subsections.

According to the power curve illustrated in Figure 2-18, the operation of the wind turbine can be divided into three modes: parking mode, generator-control mode, and pitch-control mode:

- Parking mode. When the wind speed is below cut-in speed, the turbine system generates less power than its internal consumption and, therefore, the turbine is

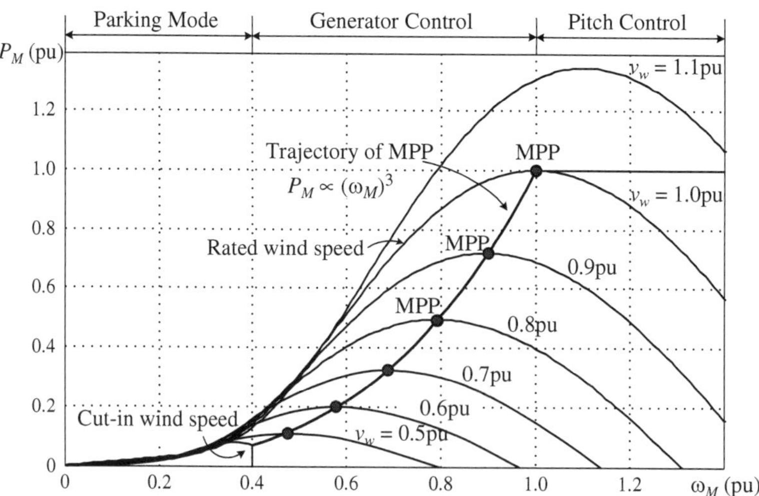

Figure 2-18. Wind turbine power–speed characteristics and maximum power point (MPP) operation.

kept in parking mode. The blades are completely pitched out of the wind, and the mechanical brake is on.

- Generator-control mode. When the wind speed is between the cut-in and rated speed, the blades are pitched into the wind with its optimal angle of attack. The turbine operates with variable rotational speeds in order to track the MPP at different wind speeds. This is achieved by the proper control of the generator.
- Pitch-control mode. For higher than rated wind speeds but below the cut-out limit, the captured power is kept constant by the pitch mechanism to protect the turbine from damage while the system generates and delivers the rated power to the grid. The blades are pitched out of the wind gradually with the wind speed, and the generator speed is controlled accordingly.

When the wind speed reaches or exceeds the cut-out speed, the blades are pitched completely out of the wind. No power is captured, and turbine speed is reduced to zero. The turbine will be locked into the parking mode to prevent damage from the strong wind.

2.4.1 MPPT with Turbine Power Profile

One of the maximum power point tracking methods is based on the power versus wind speed curve provided by the manufacturer for a given wind turbine. The power curve defines the maximum power that can be produced by the turbine at different wind speeds. A simplified control block diagram with this method is illustrated in Figure 2-19. The wind speed is measured in real time by a wind speed sensor. According to the MPPT profile provided by the manufacturer, the power reference P_m^* is generated and sent to the generator control system, which compares the power reference with the measured power P_m from the generator to produce the control signals for the power

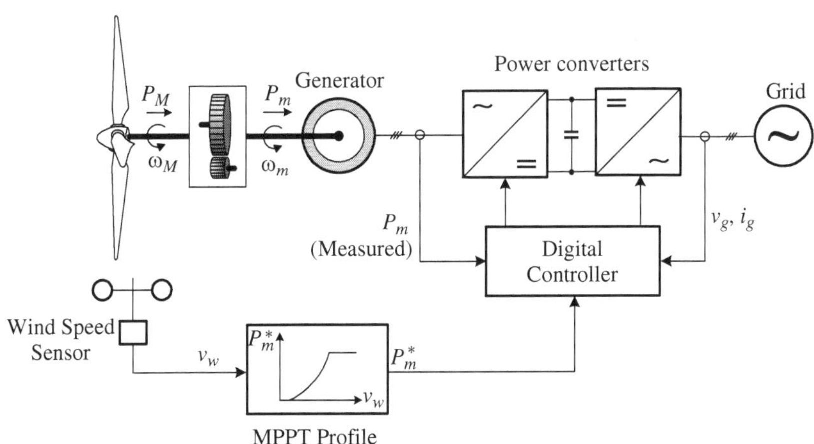

Figure 2-19. Maximum power control with wind turbine power profile.

converters. Through the control of power converters and generator, the mechanical power P_m of the generator will be equal to its reference in steady state, at which the maximum power operation is achieved. It is noted that the power losses of the gearbox and drive train in the above analysis are neglected and, therefore, the mechanical power of the generator P_m is equal to the mechanical power P_M produced by the turbine.

2.4.2 MPPT with Optimal Tip Speed Ratio

In this method, the maximum power operation of the wind turbine is achieved by keeping the tip speed ratio to its optimal value $\lambda_{T,\text{opt}}$. The principle of this control scheme is shown in Figure 2-20, where the measured wind speed v_w is used to produce the generator speed reference ω_m^* according to the optimal tip speed ratio $\lambda_{T,\text{opt}}$. The generator speed ω_m is controlled by the power converters and will be equal to its reference in steady state, at which the MPPT is achieved.

2.4.3 MPPT with Optimal Torque Control

The maximum power operation can also be achieved with optimal torque control according to Equation (2.8), where the turbine mechanical torque T_M is a quadratic function of the turbine speed ω_M. For a given gear ratio and with the mechanical power losses of the gearbox and drive train neglected, the turbine mechanical torque T_M and speed ω_M can be easily converted to the generator mechanical torque T_m and speed ω_m, respectively. Figure 2-21 shows the principle of the MPPT scheme with optimal torque control, where the generator speed ω_m is measured and used to compute the desired torque reference T_m^*. The coefficient for the optimal torque K_{opt} can be calculated according to the rated parameters of the generator. Through the feedback control, the generator torque T_m will be equal to its reference T_m^* in steady state, and the MPPT is realized. It is noted that there is no need to use the wind speed sensors in this scheme.

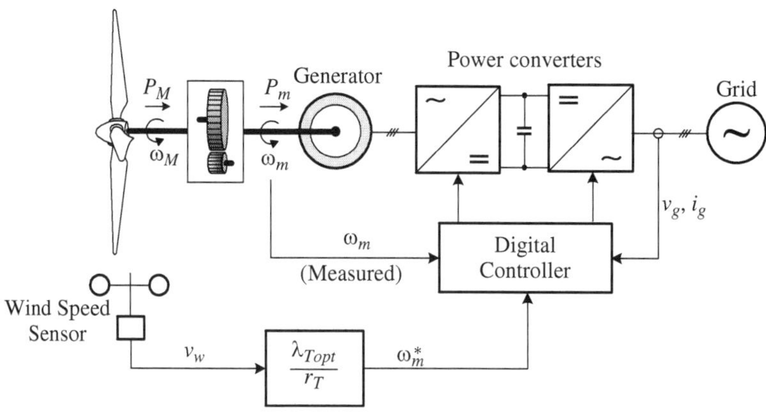

Figure 2-20. Optimal TSR control of wind turbine.

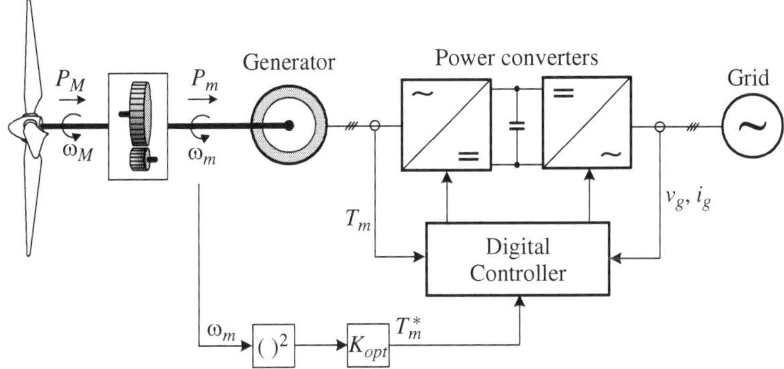

Figure 2-21. MPPT with optimal torque control of wind turbines.

2.5 SUMMARY

This chapter introduced the major components of a wind energy system, including the blade, pitch mechanism, gearbox, brakes, generators, and yaw drives. The wind turbine power curves and characteristics were discussed. The blade aerodynamics, stall and pitch controls, tip speed ratio, and maximum power point tracking schemes were also presented. The basic concepts and turbine fundamentals given in this chapter provide a technical background for the more advanced discussions in the other chapters of the book.

REFERENCES

1. T. Burton, D. Sharpe, N. Jenkins, and E. Bossanyi, *Wind Energy Handbook,* Wiley, 2001.
2. E. Hau, *Wind Turbines: Fundamentals, Technologies, Application, Economics,* 2nd edition, Springer, 2005.
3. T. Ackermann, *Wind Power in Power System,* Wiley, 2005.
4. I. Boldea, *The Electric Generators Handbook: Variable Speed Generators,* CRC Press, 2005.

3

WIND GENERATORS AND MODELING

3.1 INTRODUCTION

The evolution of wind power conversion technology has led to the development of different types of wind turbine configurations that make use of a variety of electric generators [1]. A classification of most common electric generators in large wind energy conversion systems (WECS) is presented in Figure 3-1. Depending on their construction and operating principle, the wind generators are divided in two main groups: induction generators (IGs) and synchronous generators (SGs). Both induction and synchronous generators have wound rotors, which are fed by slip rings through brushes or by a brushless electromagnetic exciter. The wound-rotor induction generator, also known as the doubly fed induction generator (DFIG), is one of the most commonly used generators in the wind energy industry [2]. The wound-rotor synchronous generator (WRSG) is also found in practical WECSs with high numbers of poles operating at low rotor speeds.

Squirrel-cage induction generators (SCIGs) are also widely employed in wind energy systems where the rotor circuits (rotor bars) are shorted internally and therefore not brought out for connection with external circuits. In permanent-magnet synchronous generators (PMSGs), the rotor magnetic flux is generated by permanent magnets. Two types of PMSG are used in the wind energy industry: surface mounted and inset magnets. Some examples of the generators used in practical wind turbines are listed in Table 3-1, where the voltage/power ratings, speed ranges, and manufacturers of these wind generators are provided.

Power Conversion and Control of Wind Energy Systems. By B. Wu, Y. Lang, N. Zargari, and S. Kouro   **49**
© 2011 the Institute of Electrical and Electronics Engineers, Inc. Published 2011 by John Wiley & Sons, Inc.

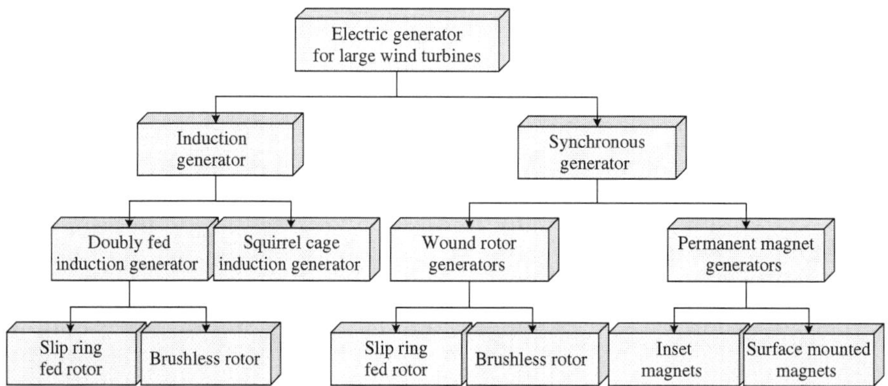

Figure 3-1. Classification of commonly used electric generators in large wind turbines.

In this chapter, the construction and operating principles of induction and synchronous generators are reviewed, and their dynamic and steady-state models are presented. Case studies are provided to illustrate important concepts and performance of the generators. The wind generator models and the equations derived in this chapter will be used in subsequent chapters.

3.2   REFERENCE FRAME TRANSFORMATION

The reference frame theory can be used to simplify the analysis of electric machines and also to facilitate the simulation and digital implementation of control schemes in wind energy conversion systems. A number of reference frames have been proposed

Table 3-1. Examples of large generators for wind energy conversion systems

Generator type	Doubly fed induction generator (DFIG)	Squirrel-cage induction generator (SCIG)	Wound-rotor synchronous generator (WRSG)	Permanent-magnet synchronous generator (PMSG)	
Rated voltage (line to line)	690 V	690 V	400 V	700 V	3000 V
Rated power	2MW	2.3 MW	2.3 MW	2.75 MW	5.32 MW
Rated stator frequency	50 Hz	50 Hz	na	na	19.6 Hz
Speed range	900–1900 rpm	600–1600 rpm	6–21.5 rpm	6–18 rpm	58.6–146.9 rpm
Number of poles	4	4	72	120	28
Brand/model	Gamesa G90	Siemens SWT-2.3-101	Enercon E-70	Avantis AV928	Multibrid M5000

over the years, of which the three-phase stationary frame (also known as *abc* frame), the two-phase stationary frame ($\alpha\beta$ frame), and the synchronous frame (*dq* rotating frame) are most commonly used [3]. The transformation of variables between these reference frames is presented below.

3.2.1 *abc/dq* Reference Frame Transformation

Consider generic three-phase electrical variables, x_a, x_b, and x_c, which can represent either voltage, current, or flux linkage. The three-phase variables can be represented by a space vector $\vec{x}$ in a three-phase (*abc*) stationary reference frame (coordinate system). The relationship between the space vector and its three-phase variables is illustrated in Figure 3-2, where the space vector $\vec{x}$ rotates at an arbitrary speed ω with respect to the *abc* stationary frame. Its phase values, x_a, x_b, and x_c, can be obtained by projecting $\vec{x}$ to the corresponding *a-*, *b-*, and *c-* axes that are $2\pi/3$ apart in space. Since the *abc* axes are stationary in space, each of the three-phase variables varies one cycle over time when $\vec{x}$ rotates one revolution in space. Assuming that the length (magnitude) and the rotating speed of space vector $\vec{x}$ are constant, the waveforms of x_a, x_b, and x_c are sinusoidal with a phase displacement of $2\pi/3$ between any two waveforms, as shown in Figure 3-2. The space vector diagram and its corresponding waveforms indicate that at the instant of ωt_1, x_b is greater than x_a, and x_c is negative.

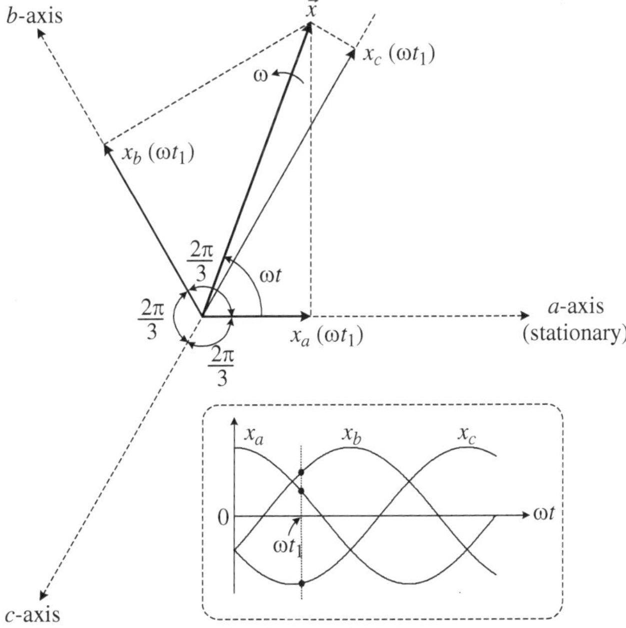

Figure 3-2. Space vector $\vec{x}$ and its three-phase variables x_a, x_b, and x_c.

The three-phase variables in the *abc* stationary frame can be transformed into two-phase variables in a reference frame defined by the *d* (direct) and *q* (quadrature) axes that are perpendicular to each other as shown in Figure 3-3. The *dq*-axis frame has an arbitrary position with respect to the *abc*-axis stationary frame given by the angle θ between the *a*-axis and *d*-axis. The *dq*-axis frame rotates in space at an arbitrary speed ω, which relates to θ by $\omega = d\theta/dt$.

To transform variables in the *abc* stationary frame to the *dq* rotating frame, simple trigonometric functions can be derived from the orthogonal projection of the x_a, x_b, and x_c variables to the *dq*-axis as shown in Figure 3-3, where only the projections to the *d*-axis are illustrated. The sum of all projections on the *d*-axis corresponds to the transformed x_d, given by $x_d = x_a \cos\theta + x_b \cos(2\pi/3 - \theta) + x_c \cos(4\pi/3 - \theta)$, which can be rewritten as $x_d = x_a \cos\theta + x_b \cos(\theta - 2\pi/3) + x_c \cos(\theta - 4\theta/3)$. Similarly, the transformation of the *abc* variables into the *q*-axis can be performed. The transformation of the *abc* variables to the *dq* frames, referred to as *abc/dq* transformation, can be expressed in a matrix form:

$$\begin{bmatrix} x_d \\ x_q \end{bmatrix} = \frac{2}{3}\begin{bmatrix} \cos\theta & \cos(\theta - 2\pi/3) & \cos(\theta - 4\pi/3) \\ -\sin\theta & -\sin(\theta - 2\pi/3) & -\sin(\theta - 4\pi/3) \end{bmatrix} \cdot \begin{bmatrix} x_a \\ x_b \\ x_c \end{bmatrix} \qquad (3.1)$$

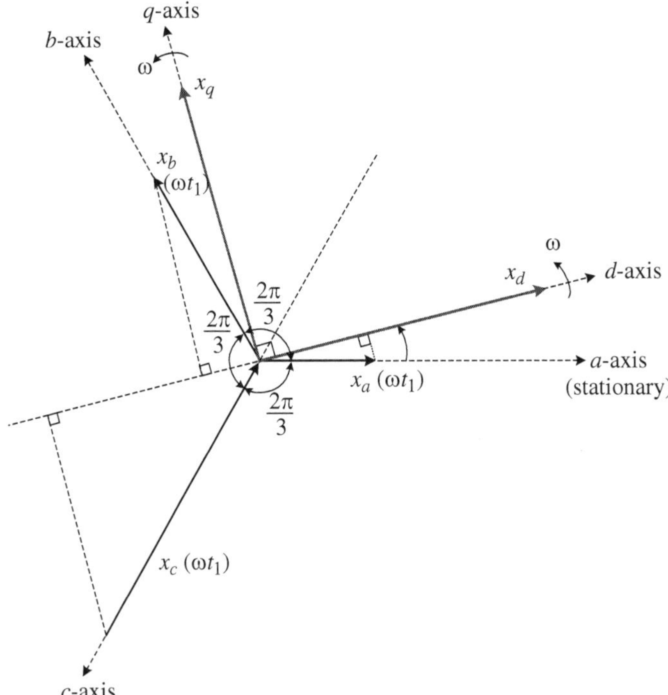

Figure 3-3. Transformation of variables in the three-phase (*abc*) stationary frame to the two-phase (*dq*) arbitrary frame.

It is noted that in the above *abc*/*dq* transformation:

- A coefficient of 2/3 is arbitrarily added to the equation. The commonly used value is 2/3 or $\sqrt{2/3}$. The main advantage of using 2/3 is that the magnitude of the two-phase voltages is equal to that of the three-phase voltages after the transformation. The reason why the coefficient can be arbitrarily selected will be explained in Section 3.3.5.
- The two-phase *dq* variables after the transformation contain all the information of the three-phase *abc* variables. This is under the condition that the system must be three-phase balanced. Of the three variables in a three-phase balanced system, only two are independent. Given two independent variables, the third one can be calculated by

$$x_a + x_b + x_c = 0 \tag{3.2}$$

The equations for an inverse transformation can be obtained through matrix operations, by which the *dq* variables in the rotating frame can be transformed back to the *abc* variables in the stationary frame. The transformation, referred to as *dq*/*abc* transformation, can be performed by

$$\begin{bmatrix} x_a \\ x_b \\ x_c \end{bmatrix} = \begin{bmatrix} \cos\theta & -\sin\theta \\ \cos(\theta - 2\pi/3) & -\sin(\theta - 2\pi/3) \\ \cos(\theta - 4\pi/3) & -\sin(\theta - 4\pi/3) \end{bmatrix} \cdot \begin{bmatrix} x_d \\ x_q \end{bmatrix} \tag{3.3}$$

Figure 3-4 illustrates the decomposition of the space vector $\vec{x}$ into the *dq* rotating reference frame. Assuming that $\vec{x}$ rotates at the same speed as that of the *dq* frame, the vector angle ϕ between $\vec{x}$ and the *d*-axis is constant. The resultant *dq*-axis components, x_d and x_q, are DC variables. This is one of the advantages of the *abc*/*dq* transformation, whereby three-phase AC variables can be effectively represented by two-phase DC variables.

For the control of wind energy systems, the synchronous reference frame is often used. In this case, the rotating speed of the arbitrary reference frame ω is set to the synchronous speed ω_s of an induction or synchronous generator, given by

$$\omega_s = 2\pi f_s \tag{3.4}$$

where f_s is the stator frequency in Hertz. The angle θ can be found from

$$\theta(t) = \int_0^t \omega_s(t)\, dt + \theta_0 \tag{3.5}$$

where θ_0 is the initial angular position.

3.2.2 *abc*/$\alpha\beta$ Reference Frame Transformation

The transformation of three-phase variables in the stationary reference frame into the two-phase variables also in the stationary frame is often referred to as *abc*/$\alpha\beta$ transfor-

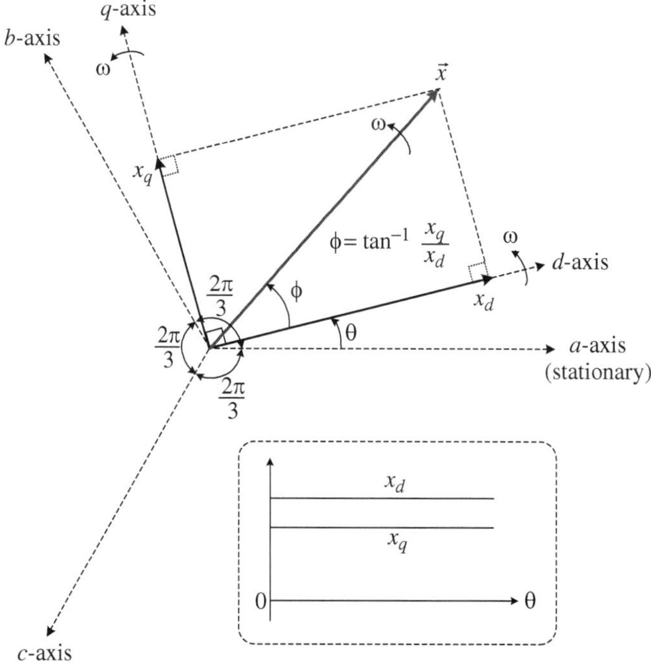

Figure 3-4. Decomposition of space vector $\vec{x}$ into the dq rotating reference frame.

mation. Since the $\alpha\beta$ reference frame does not rotate in space, the transformation can be obtained by setting θ in Equation (3.1) to zero, from which

$$\begin{bmatrix} x_\alpha \\ x_\beta \end{bmatrix} = \frac{2}{3}\begin{bmatrix} 1 & -1/2 & -1/2 \\ 0 & \sqrt{3}/2 & -\sqrt{3}/2 \end{bmatrix}\cdot\begin{bmatrix} x_a \\ x_b \\ x_c \end{bmatrix} \tag{3.6}$$

It is interesting to note that in a three-phase balanced system, where $x_a + x_b + x_c = 0$, the relationship between x_α in the $\alpha\beta$ reference frame and x_a in the abc frame is given by

$$x_\alpha = \frac{2}{3}\left(x_a - \frac{1}{2}x_b - \frac{1}{2}x_c \right) = x_a \tag{3.7}$$

Similarly, the two-phase to three-phase transformation in the stationary reference frame, known as $\alpha\beta/abc$ transformation, can be performed by

$$\begin{bmatrix} x_a \\ x_b \\ x_c \end{bmatrix} = \begin{bmatrix} 1 & 0 \\ -1/2 & \sqrt{3}/2 \\ -1/2 & -\sqrt{3}/2 \end{bmatrix}\cdot\begin{bmatrix} x_\alpha \\ x_\beta \end{bmatrix} \tag{3.8}$$

3.3 INDUCTION GENERATOR MODELS

3.3.1 Construction

As shown in the classification of Figure 3-1, there are two main types of induction generators in the wind energy industry: doubly fed induction generators (DFIGs) and squirrel-cage induction generators (SCIGs). These generators have the same stator structure and differ only in the rotor structure.

Figure 3-5a shows the construction of a squirrel cage induction generator. The stator is made of thin silicon steel laminations. The laminations are insulated to minimize iron losses caused by induced eddy currents. The laminations are basically flat rings with openings disposed along the inner perimeter of the ring. When the laminations are stacked together with the openings aligned, a canal is formed, in which a three-phase copper winding is placed.

The rotor of the SCIG is composed of the laminated core and rotor bars. The rotor bars are embedded in slots inside the rotor laminations and are shorted on both ends by end rings. When the stator winding is connected to a three-phase supply, a rotating magnetic field is generated in the air gap. The rotating field induces a three-phase voltage in the rotor bars. Since the rotor bars are shorted, the induced rotor voltage produces a rotor current, which interacts with the rotating field to produce the electromagnetic torque.

The rotor of the DFIG has a three-phase winding similar to the stator winding. The rotor winding is embedded in the rotor laminations but in the exterior perimeter. This winding is usually fed through slip-rings mounted on the rotor shaft. In DFIG wind energy systems, the rotor winding is normally connected to a power converter system that makes the rotor speed adjustable.

A simplified diagram of the induction generator is shown in Figure 3-5b, where the multiple coils in the stator and multiple bars in the rotor are grouped and represented

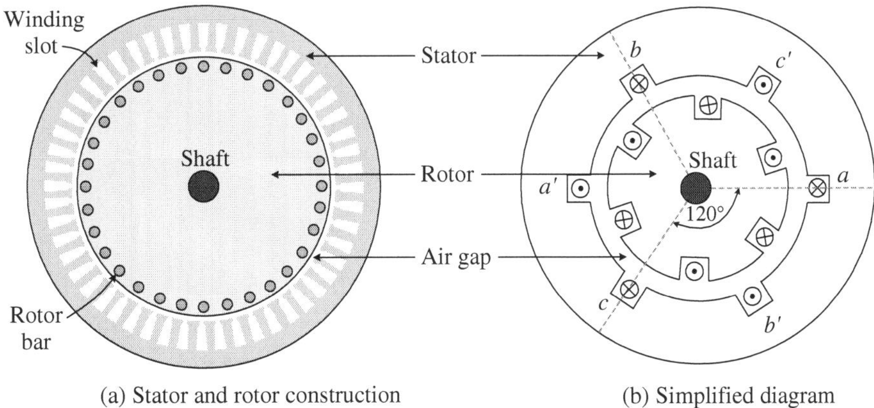

(a) Stator and rotor construction   (b) Simplified diagram

Figure 3-5. Cross-sectional view of an SCIG.

by a single coil for each phase. Practical examples of SCIGs and DFIGs are shown in Figure 2-7a and b in Chapter 2, respectively.

There are two commonly used dynamic models for the induction generator. One is based on space vector theory and the other is the dq-axis model derived from the space vector model. The space vector model features compact mathematical expressions and a single equivalent circuit but requires complex (real and imaginary part) variables, whereas the dq-frame model is composed of two equivalent circuits, one for each axis. These models are closely related to each other and are equally valid for the analysis of transient and steady-state performance of the induction generator. In the following sections, the two models are presented and their relationship is elaborated.

3.3.2 Space-Vector Model

In developing the IG space-vector model, it is assumed that (1) the induction generator is symmetrical in structure and three-phase balanced, and (2) the magnetic core of the stator and rotor is linear with negligible core losses. The IG space-vector model is generally composed of three sets of equations: voltage equations, flux linkage equations, and motion equation [3,4]. The voltage equations for the stator and rotor of the generator in the arbitrary reference frame are given by

$$\begin{cases} \vec{v}_s = R_s \vec{i}_s + p\vec{\lambda}_s + j\omega\vec{\lambda}_s \\ \vec{v}_r = R_r \vec{i}_r + p\vec{\lambda}_r + j(\omega - \omega_r)\vec{\lambda}_r \end{cases} \tag{3.9}$$

where

$\vec{v}_s$, $\vec{v}_r$—stator and rotor voltage vectors (V)
$\vec{i}_s$, $\vec{i}_r$—stator and rotor current vectors (A)
$\vec{\lambda}_s$, $\vec{\lambda}_r$—stator and rotor flux-linkage vectors (Wb)
R_s, R_r—stator and rotor winding resistances (Λ)
ω—rotating speed of the arbitrary reference frame (rad/s)
ω_r—rotor electrical angular speed (rad/s)
p—derivative operator ($p = d/dt$).

The terms $j\omega\vec{\lambda}_s$ and $j(\omega - \omega_r)\vec{\lambda}_r$ on the right-hand side of Equation (3.9) are referred to as speed voltages, which are induced by the rotation of the reference frame at the arbitrary speed of ω.

The second set of equations is for the stator and rotor flux linkages $\vec{\lambda}_s$ and $\vec{\lambda}_r$:

$$\begin{cases} \vec{\lambda}_s = (L_{ls} + L_m)\vec{i}_s + L_m\vec{i}_r = L_s\vec{i}_s + L_m\vec{i}_r \\ \vec{\lambda}_r = (L_{lr} + L_m)\vec{i}_r + L_m\vec{i}_s = L_r\vec{i}_r + L_m\vec{i}_s \end{cases} \tag{3.10}$$

where

$L_s = L_{ls} + L_m$—stator self-inductance (H)
$L_r = L_{lr} + L_m$—rotor self-inductance (H)
L_{ls}, L_{lr}—stator and rotor leakage inductances (H)
L_m—magnetizing inductance (H)

All the rotor-side parameters and variables, such as R_r, L_{lr}, $\vec{i}_r$, and $\vec{\lambda}_r$, in the above equations are referred to the stator side.

The third and final equation is the motion equation, which describes the dynamic behavior of the rotor mechanical speed in terms of mechanical and electromagnetic torque:

$$\begin{cases} J\dfrac{d\omega_m}{dt} = T_e - T_m \\[2mm] T_e = \dfrac{3P}{2}\operatorname{Re}\left(j\vec{\lambda}_s\,\vec{i}_s^{\,*} \right) = -\dfrac{3P}{2}\operatorname{Re}\left(j\vec{\lambda}_r\,\vec{i}_r^{\,*} \right) \end{cases} \tag{3.11}$$

where
J—moment of inertia of the rotor (kgm^2)
P—number of pole pairs
T_m—mechanical torque from the generator shaft (N $\cdot$ m)
T_e—electromagnetic torque (N $\cdot$ m)
ω_m—rotor mechanical speed, $\omega_m = \omega_r/P$ (rad/sec)

The above equations constitute the space-vector model of the induction generator, whose equivalent circuit representation is given in Figure 3-6. The generator model is in the arbitrary reference frame, rotating in space at the arbitrary speed ω.

It is important to note that the IG space-vector model of Figure 3-6 is based on the motor convention in terms of the direction of the stator current that flows into the stator. This convention is widely accepted since most induction machines are used as motors. Nevertheless, there is no loss of generality; the space-vector model and its associated equations can be used to model the induction machine either as a motor or a generator. More discussion will be provided in Case Study 3-2.

The induction generator model of Figure 3-6 in the arbitrary reference frame can be easily transformed into the other reference frames. For example, a synchronous frame model is very useful for simulation and digital implementation of IG WECS with advanced control systems. Such a model can be obtained by setting the arbitrary speed ω in Equation (3.9) and in Figure 3-6 to the synchronous speed ω_s. The derived model in

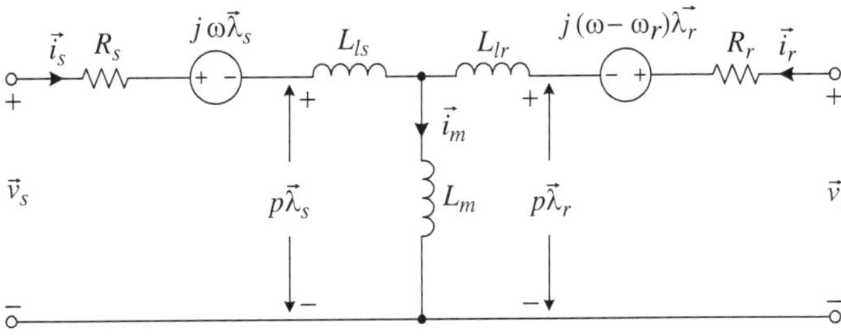

Figure 3-6. Space-vector equivalent circuit of an induction generator in the arbitrary reference frame.

the synchronous frame is given in Figure 3-7a, where ω_s is the synchronous speed and ω_{si} is the angular slip frequency of the generator, given by

$$\begin{cases} \omega_s = 2\pi f_s \\ \omega_{sl} = \omega_s - \omega_r \end{cases} \tag{3.12}$$

The synchronous speed of the reference frame ω_s corresponds to the stator angular frequency, which is proportional to the stator frequency f_s.

To obtain the IG model in the stationary reference frame, we can set the speed of the arbitrary frame ω in Figure 3-6 to zero since the stationary frame does not rotate in space. The resultant equivalent circuit is shown in Figure 3-7b. The IG space-vector models in Figures 3-6 and 3-7 are valid for both SCIG and DFIG. In the SCIG, the rotor circuit is shorted and, therefore, the rotor voltage is set to zero, whereas for the DFIG the rotor circuit is connected to a power converter system that controls the speed and torque of the generator.

3.3.3 *dq* Reference Frame Model

The *dq*-axis model of the induction generator can be obtained by decomposing the space-vectors into their corresponding *d*- and *q*-axis components, that is,

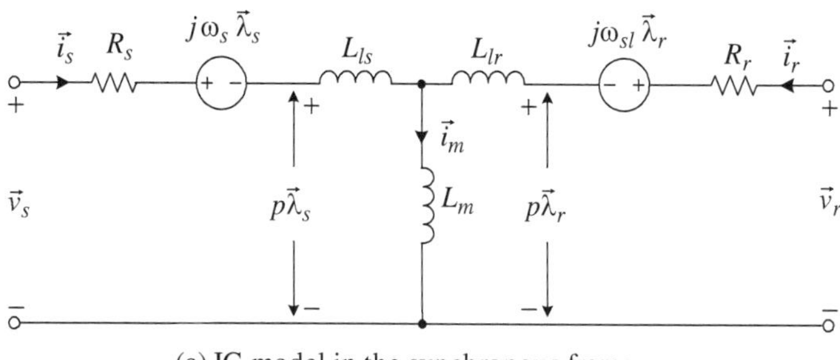

(a) IG model in the synchronous frame

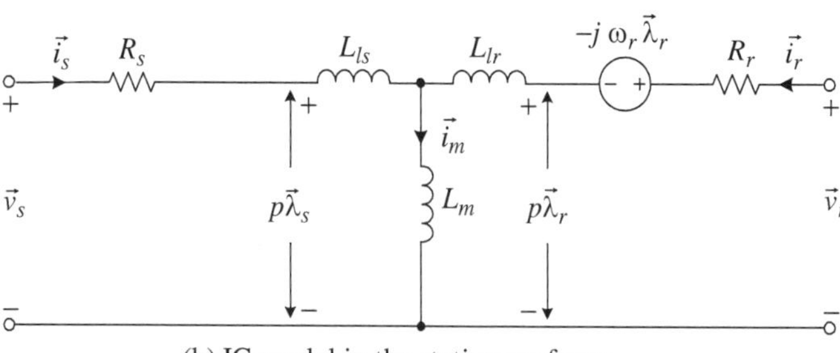

(b) IG model in the stationary frame

Figure 3-7. Space-vector models for induction generator in the synchronous and stationary reference frames.

$$
\begin{cases}
\vec{v}_s = v_{ds} + j v_{qs}; \ \vec{i}_s = i_{ds} + j i_{qs}; \ \vec{\lambda}_s = \lambda_{ds} + j \lambda_{qs} \\
\vec{v}_r = v_{dr} + j v_{qr}; \ \vec{i}_r = i_{dr} + j i_{qr}; \ \vec{\lambda}_r = \lambda_{dr} + j \lambda_{qr}
\end{cases}
\tag{3.13}
$$

Substituting Equation (3.13) into Equation (3.9) and grouping real and imaginary components on both sides of the equations, the dq-axis voltage equations for the induction generator are obtained:

$$
\begin{cases}
v_{ds} = R_s i_{ds} + p\lambda_{ds} - \omega\lambda_{qs} \\
v_{qs} = R_s i_{qs} + p\lambda_{qs} + \omega\lambda_{ds} \\
v_{dr} = R_r i_{dr} + p\lambda_{dr} - (\omega - \omega_r)\lambda_{qr} \\
v_{qr} = R_r i_{qr} + p\lambda_{qr} + (\omega - \omega_r)\lambda_{dr}
\end{cases}
\tag{3.14}
$$

Similarly, substituting Equation (3.13) into Equation (3.10), the dq-axis flux linkages are obtained:

$$
\begin{cases}
\lambda_{ds} = (L_{ls} + L_m)i_{ds} + L_m i_{dr} = L_s i_{ds} + L_m i_{dr} \\
\lambda_{qs} = (L_{ls} + L_m)i_{qs} + L_m i_{qr} = L_s i_{qs} + L_m i_{qr} \\
\lambda_{dr} = (L_{lr} + L_m)i_{dr} + L_m i_{ds} = L_r i_{dr} + L_m i_{ds} \\
\lambda_{qr} = (L_{lr} + L_m)i_{qr} + L_m i_{qs} = L_r i_{qr} + L_m i_{qs}
\end{cases}
\tag{3.15}
$$

The electromagnetic torque T_e in Equation (3.11) can be expressed by dq-axis flux linkages and currents as well. By mathematical manipulations, several expressions for the torque can be obtained. The most commonly used expressions are given by

$$
T_e = \begin{cases}
\dfrac{3P}{2}(i_{qs}\lambda_{ds} - i_{ds}\lambda_{qs}) & \text{(a)} \\[2ex]
\dfrac{3PL_m}{2}(i_{qs}i_{dr} - i_{ds}i_{qr}) & \text{(b)} \\[2ex]
\dfrac{3PL_m}{2L_r}(i_{qs}\lambda_{dr} - i_{ds}\lambda_{qr}) & \text{(c)}
\end{cases}
\tag{3.16}
$$

Equations (3.14) to (3.16) together with the motion equation (3.11) represent the dq-axis model of the induction generator in the arbitrary reference frame, and its corresponding dq-axis equivalent circuits are shown in Figure 3-8 [5]. To obtain the dq-axis model in the synchronous and stationary reference frames, the speed of the arbitrary reference frame ω can be set to the synchronous (stator) frequency ω_s of the generator and zero, respectively.

3.3.4 Simulation Model

To build the simulation model, the equations derived previously should be rearranged. Equation (3.14) can be rewritten as

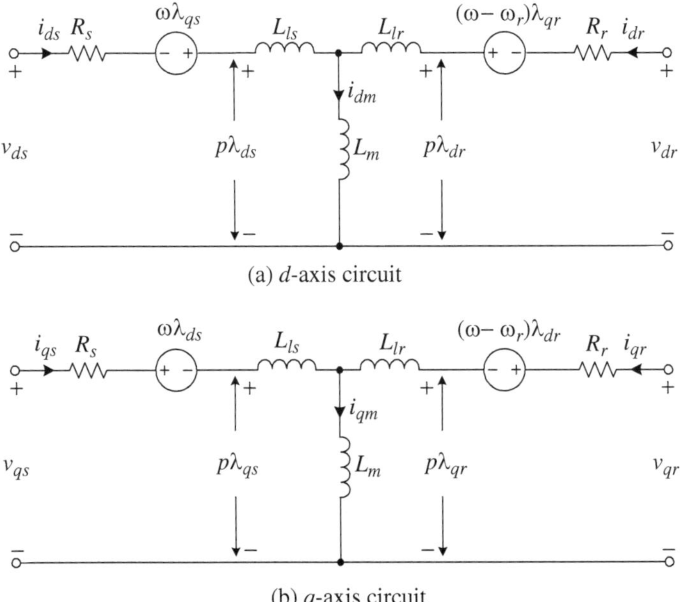

(a) *d*-axis circuit

(b) *q*-axis circuit

<u>Figure 3-8.</u> Induction generator *dq*-axis model in the arbitrary reference frame.

$$
\begin{cases}
\lambda_{ds} = \left(v_{ds} - R_s i_{ds} + \omega\lambda_{qs} \right)/S \\
\lambda_{qs} = \left(v_{qs} - R_s i_{qs} - \omega\lambda_{ds} \right)/S \\
\lambda_{dr} = \left(v_{dr} - R_r i_{dr} + (\omega-\omega_r)\lambda_{qr} \right)/S \\
\lambda_{qr} = \left(v_{qr} - R_r i_{qr} - (\omega-\omega_r)\lambda_{dr} \right)/S
\end{cases}
\tag{3.17}
$$

where the derivative operator p in Equation (3.14) is replaced by the Laplace operator S, and $1/S$ represents an integrator.

The flux linkage equations of (3.15) can be represented in a matrix form:

$$
\begin{bmatrix} \lambda_{ds} \\ \lambda_{qs} \\ \lambda_{dr} \\ \lambda_{qr} \end{bmatrix}
=
\begin{bmatrix}
L_s & 0 & L_m & 0 \\
0 & L_s & 0 & L_m \\
L_m & 0 & L_r & 0 \\
0 & L_m & 0 & L_r
\end{bmatrix}
\cdot
\begin{bmatrix} i_{ds} \\ i_{qs} \\ i_{dr} \\ i_{qr} \end{bmatrix}
\tag{3.18}
$$

The stator and rotor currents in the above equation can be expressed in terms of stator and rotor flux linkages. This can be obtained by applying the inverse inductance matrix on both sides of Equation (3.18), using the following matrix manipulation:

$$
[\lambda]=[L][i] \quad \rightarrow \quad [L]^{-1}[\lambda]=[L]^{-1}[L][i] \quad \rightarrow \quad [i]=[L]^{-1}[\lambda]
\tag{3.19}
$$

from which

$$
\begin{bmatrix} i_{ds} \\ i_{qs} \\ i_{dr} \\ i_{qr} \end{bmatrix} = \frac{1}{D_1} \begin{bmatrix} L_r & 0 & -L_m & 0 \\ 0 & L_r & 0 & -L_m \\ -L_m & 0 & L_s & 0 \\ 0 & -L_m & 0 & L_s \end{bmatrix} \cdot \begin{bmatrix} \lambda_{ds} \\ \lambda_{qs} \\ \lambda_{dr} \\ \lambda_{qr} \end{bmatrix} \qquad (3.20)
$$

where $D_1 = L_s L_r - L_m^2$ [3].

The motion and the torque equations for the simulation model are given by

$$
\begin{cases} \omega_r = \dfrac{P}{JS}\left(T_e - T_m\right) & \text{(a)} \\[2mm] T_e = \dfrac{3P}{2}\left(i_{qs}\lambda_{ds} - i_{ds}\lambda_{qs}\right) & \text{(b)} \end{cases} \qquad (3.21)
$$

Based on Equations (3.17), (3.20), and (3.21), the block diagram for simulation of an induction generator in the arbitrary reference frame can be developed and is shown in Figure 3-9.

The input variables of the model include the dq-axis stator voltages v_{ds} and v_{qs}, rotor voltages v_{dr} and v_{qr}, the mechanical torque T_m, and the speed of the arbitrary reference frame ω, whereas the output variables are dq-axis stator currents, i_{ds} and i_{qs}, the electromagnetic torque T_e, and the mechanical speed ω_m of the generator. To simulate the induction generator in the synchronous reference frame or the stationary reference frame, the speed of the arbitrary reference frame ω can be set to the synchronous speed ω_s or zero, respectively.

3.3.5 Induction Generator Transient Characteristics

The transient characteristics of an induction-generator-based WECS with direct grid connection can be investigated using the simulation block diagram of Figure 3-10. Assuming a three-phase balanced grid, the grid voltages v_{as}, v_{bs}, and v_{cs} in the stationary frame are transformed to the two-phase voltages $v_{\alpha s}$ and $v_{\beta s}$ in the $\alpha\beta$ stationary frame through the $abc/\alpha\beta$ transformation. In this case, the IG model in the stationary reference frame should be used, which can be realized by setting the speed of the arbitrary reference frame to zero ($\omega = 0$). The simulated dq-axis stator currents i_{ds} and i_{qs} are also in the stationary frame, which are transformed to the three-phase currents i_{as}, i_{bs}, and i_{cs} by the $\alpha\beta/abc$ transformation. The dq-axis rotor voltages are set to zero for simulation of squirrel-cage induction generators.

For the $abc/\alpha\beta$ transformation using Equation (3.6), the coefficient of 2/3 is arbitrarily selected, based on which Equation (3.8) for the $\alpha\beta/abc$ transformation is derived. The reason why the coefficient can be arbitrarily selected can be revealed by Figure 3-10, where the three-phase stator voltages are transformed to the two-phase voltages via the $abc/\alpha\beta$ transformation and the calculated two-phase stator currents are

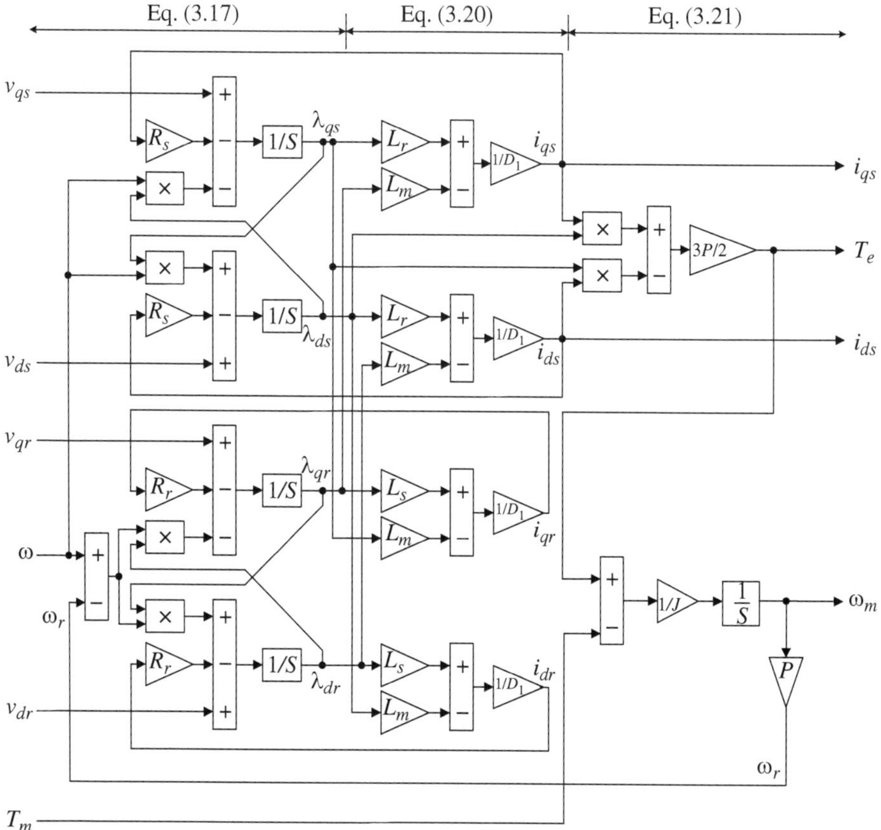

Figure 3-9. Block diagram for dynamic simulation of an induction generator in the arbitrary reference frame.

transformed back to the three-phase stator currents via the $\alpha\beta/abc$ transformation. If the $abc/\alpha\beta$ and $\alpha\beta/abc$ transformations in Figure 3-10 are replaced by those derived with a coefficient other than 2/3, the simulation results from the two systems, such as the three-phase stator currents, would be identical.

Case Study 3-1—Direct Grid Connection of SCIG during System Start-up.
This case study investigates the dynamic performance of a SCIG wind energy system during system start-up, and verifies that a large SCIG cannot be directly connected to the grid due to the excessive inrush current and torque oscillations.

Consider a 2.3 MW, 690 V, 50 Hz, 1512 rpm squirrel-cage induction generator. Its nameplate and parameters are listed in Table B-1 of Appendix B. The shaft of the generator is coupled to the wind turbine through a gearbox. During the system start-up, the turbine and generator are brought by the wind to a certain speed, at which the generator is connected to the grid of 690 V/50 Hz by the circuit breaker shown in Figure 3-10. The investigation is carried out in the following two cases.

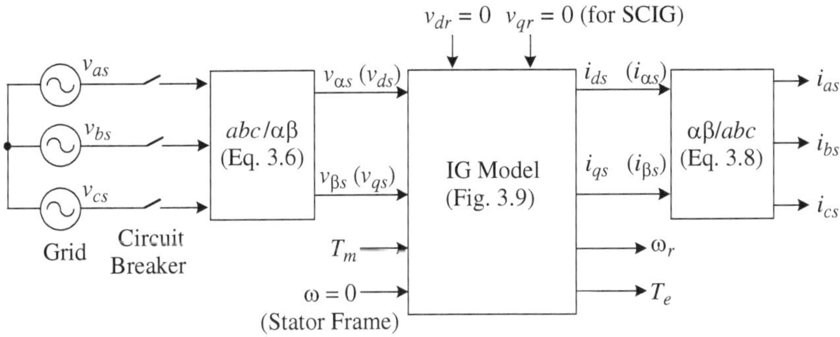

Figure 3-10. Block diagram for dynamic simulation of SCIG with direct grid connection.

DYNAMIC PERFORMANCE OF SCIG WITH DIRECT GRID CONNECTION. The wind turbine is initially in a parking mode with the blades pitched out of the wind. When the wind speed reaches an operative level, the blades are pitched into the wind slightly, and wind turbine and generator start to rotate slowly. When the generator is accelerated close to the rated speed—1450 rpm (0.959 pu)—the circuit breaker is closed and the generator is directly connected to the grid. The simulated waveforms for the generator are illustrated in Figure 3-11.

During the system transients, a high inrush current flows into the generator and a DC offset current appears in each of the stator currents i_{as}, i_{bs}, and i_{cs}, but the sum of these offset currents is zero due to a three-phase balanced system. As a rotating magnetic field is being built and generator core is being magnetized by the stator current, an electromagnetic torque T_e is produced. Since the generator operates below synchronous speed in motoring mode, it produces a positive torque that accelerates the turbine. The generator finally reaches the synchronous speed of 1500 rpm (0.992 pu) at $t = 0.84$ sec, at which it enters the steady-state operation with $T_e = T_m = 0$. With the start-up process completed, the turbine can now start capturing power by adjusting the pitch angle of the blades.

The direct connection of the generator to the grid during the system start-up causes excessive inrush currents with peak values of more than 10 per unit (pu), high electromagnetic torque (2 pu, peak), as well as high torque oscillations (refer to Appendix A for the definition of the per-unit system). The high inrush current may have adverse impact on the grid, especially for a weak grid, and the high torque oscillations may cause excessive mechanical stress on the drive train. It can be concluded that the direct grid connection of the SCIG during the system startup cannot be used in practice, especially for the large megawatt turbines. Soft-starter or power converter interfaces should be used in the SCIG wind energy systems, which will be discussed in Chapters 6 and 7.

DIRECT GRID CONNECTION OF SCIG WITH CONSTANT ROTOR SPEED. The second part of the case study investigates the transients during the direct grid connection of the generator when the speed of the generator is brought by the wind exactly to the synchronous speed of 1500 rpm (0.992 pu). It is assumed in the analysis that the combined

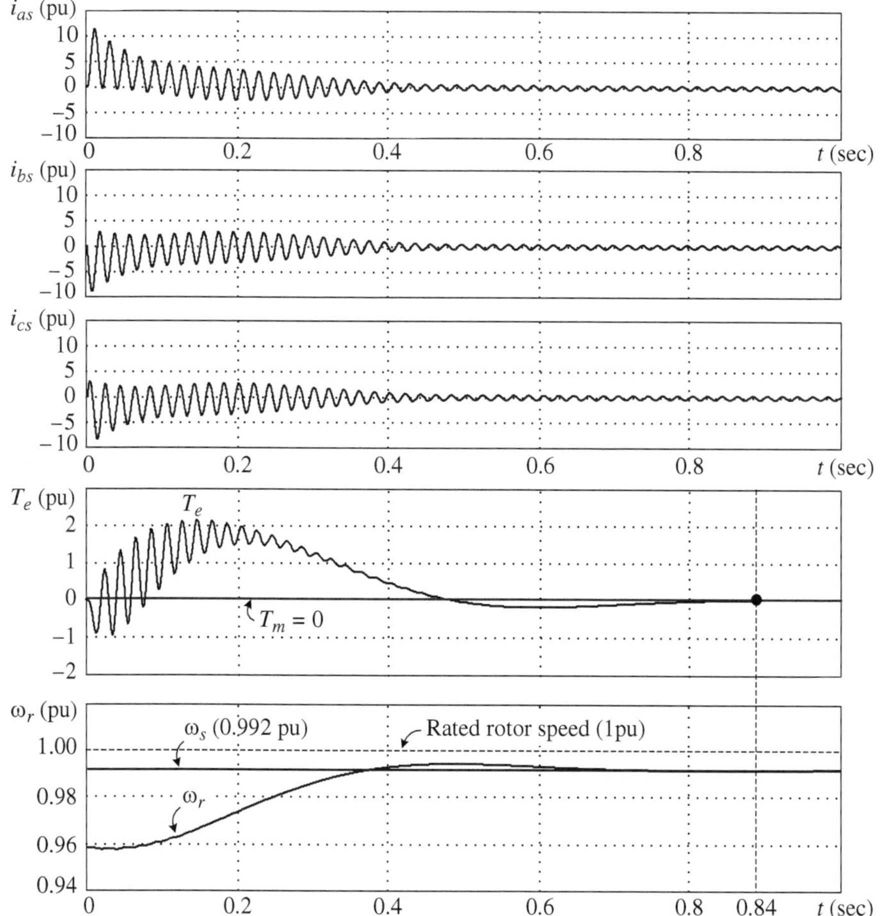

Figure 3-11. Dynamic response of SCIG with direct grid connection.

moment of inertia of the generator, gearbox, and blades are very large, such that the rotor speed is kept constant during the electric transients. In this case, the motion equation (3.21a) is simplified to $d\omega_r/dt = 0$ and $\omega_r = \omega_s$. The rotor speed ω_r is then used as one of the system input variables. As a result, the IG simulation block diagram of Figure 3-9 should be slightly modified to accommodate the changes.

The dynamic response at the moment the SCIG is connected to the grid is depicted in Figure 3-12. A high inrush current is drawn by the generator, and its peak value is more than 10 pu. The high amplitude of the stator currents causes oscillations in the generator torque T_e. Compared with the previous case, the transient process is faster due to the constant rotor speed that eliminates the motion equation in the simulation. It can be concluded that the direct connection of a SCIG to the grid is not allowed in practice due to the excessive stator current and torque oscillations.

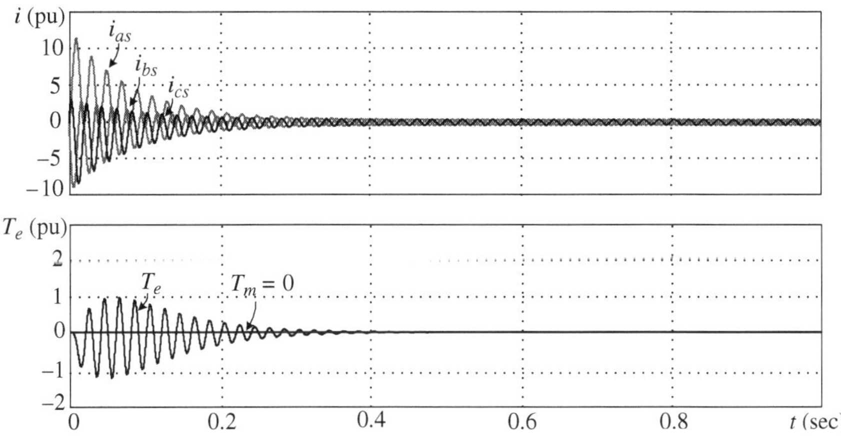

<u>Figure 3-12.</u> Dynamic response of SCIG with a fixed rotor speed during direct grid connection.

3.3.6 Steady-State Equivalent Circuit

To investigate the steady-state performance of induction generators, the steady-state equivalent circuit is a useful tool. The steady-state equivalent circuit can be derived from the IG space-vector model described by Equation (3.9). To obtain the steady-state equivalent circuit, the IG space-vector model in the synchronous frame is used, and the following steps are taken:

- Set the arbitrary ω in Equation (3.9) to the synchronous speed ω_s.
- Set the derivative terms, $p\vec{\omega}_s$ and $p\vec{\omega}_r$, in Equation (3.9) to zero (all variables of the IG in the synchronous frame are of DC quantity in steady state, and their derivatives are equal to zero).
- Replace all space vectors in Equation (3.9) with their corresponding phasors. For example, the stator voltage vector $\vec{v}_s$ is replaced with the stator voltage phasor $\overline{V}_s$, where $\vec{v}_s = v_{ds} + jv_{qs}$, $\overline{V}_s = \mathrm{Re}(V_s) + j\,\mathrm{Im}(V_s)$, and the relationship between V_s and v_s is given by $V_s = v_s/\sqrt{2}$.
- Reverse the rotor current direction, that is, the rotor current flows out of the rotor circuit instead of into the rotor circuit shown in Figure 3-6. This is not a must. However, the revised rotor current direction is consistent with the conventional steady-state equivalent circuit of induction machine. More importantly, it facilitates the analysis of a DFIG wind energy system, where the rotor circuit is connected to a power converter system with bidirectional power flow.

The equations for the steady-state analysis of the induction generator are then given by

$$\begin{cases} \overline{V}_s = R_s \overline{I}_s + j\omega\overline{\Lambda}_s \\ \overline{V}_r = -R_r \overline{I}_r + j(\omega_s - \omega_r)\overline{\Lambda}_r \end{cases} \tag{3.22}$$

where uppercase letters with a bar on top represent phasors. For instance, $\overline{\Lambda}_s$ and $\overline{\Lambda}_r$ are the phasors for the stator and rotor flux linkages λ_s and λ_r, respectively. The above equation can be rewritten as

$$\begin{cases} \overline{V}_s = R_s \overline{I}_s + j\omega_s(L_{ls}\overline{I}_s + L_m\overline{I}_m) \\ \overline{V}_r = -R_r \overline{I}_r + j\omega_{sl}(-L_{lr}\overline{I}_r + L_m\overline{I}_m) \end{cases} \tag{3.23}$$

where ω_{ls} is the angular slip frequency, given by $\omega_{ls} = \omega_s - \omega_r$. Dividing the rotor voltage equation by the slip

$$s = \frac{\omega_{sl}}{\omega_s} \tag{3.24}$$

and rearranging Equation (3.23), we have

$$\begin{cases} \overline{V}_s = R_s \overline{I}_s + j\omega_s(L_{ls}\overline{I}_s + L_m\overline{I}_m) = R_s \overline{I}_s + jX_{ls}\overline{I}_s + jX_m\overline{I}_m \\ \dfrac{\overline{V}_r}{s} = -\dfrac{R_r}{s}\overline{I}_r + j\omega_s(-L_{lr}\overline{I}_r + L_m\overline{I}_m) = -\dfrac{R_r}{s}\overline{I}_r - jX_{lr}\overline{I}_r + jX_m\overline{I}_m \end{cases} \tag{3.25}$$

where X_{ls} and X_{lr} are the stator and rotor leakage reactances and X_m is the magnetizing reactance, given by

$$\begin{cases} X_{ls} = \omega_s L_{ls} \\ X_{lr} = \omega_s L_{lr} \\ X_m = \omega_s L_m \end{cases} \tag{3.26}$$

Based on Equation (3.25), the steady-state equivalent for induction generator can be derived and is illustrated in Figure 3-13. In DFIG wind energy systems, the rotor circuit is normally connected to a rotor-side converter (RSC) that can be represented by an equivalent impedance as shown in Figure 3-13a, whereas for the SCIG the rotor circuit is shorted and the rotor voltage V_r is zero, as illustrated in Figure 3-13b.

Power Flow. To facilitate the power flow analysis for the induction generator, the rotor resistance R_r/s in Figure 3-13 is split into two components:

$$\frac{R_r}{s} = R_r + \frac{1-s}{s}R_r \tag{3.27}$$

The resultant steady-state equivalent circuit for the SCIG is given in Figure 3-14, where P_{in} is the total input power produced by the wind turbine, and P_{rot} is the total rotational losses of the mechanical system (the power losses of gearbox are neglected for simplicity). The mechanical power from the generator shaft can be calculated by [6]

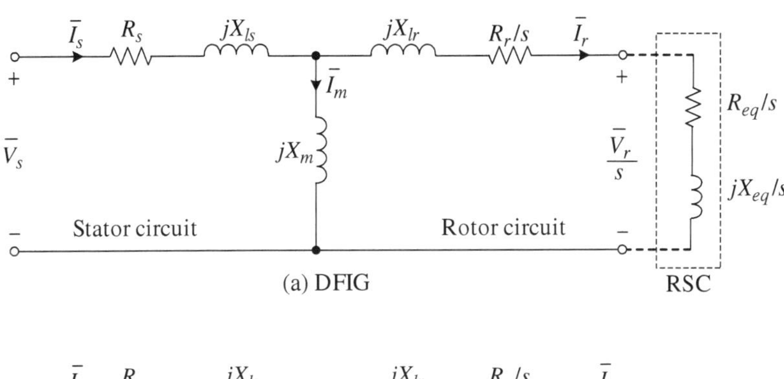

(a) DFIG

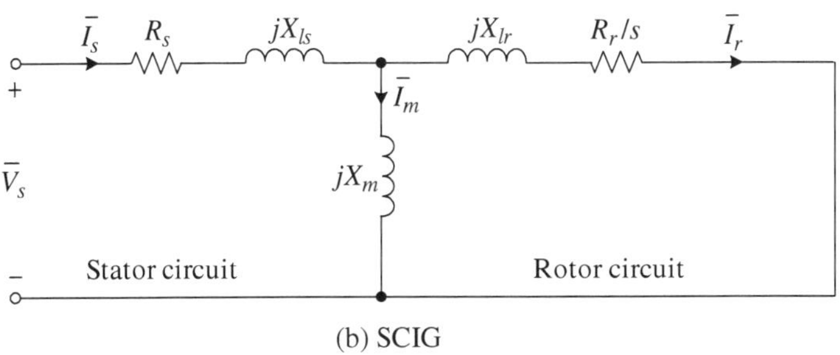

(b) SCIG

Figure 3-13. Steady-state equivalent circuits of an induction generator.

$$P_m = 3I_r^2 \frac{(1-s)}{s} R_r \tag{3.28}$$

The copper losses of the rotor and stator windings are

$$\begin{cases} P_{cu,r} = 3I_s^2 R_r \\ P_{cu,s} = 3I_s^2 R_s \end{cases} \tag{3.29}$$

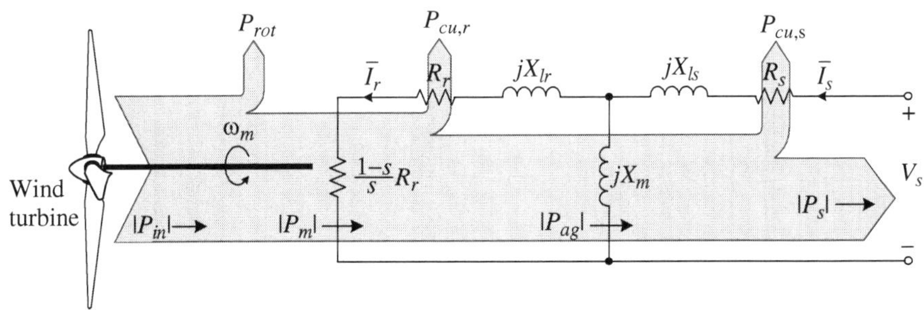

Figure 3-14. Power flow and losses in an induction generator.

The stator output power can be found from

$$|P_s| = |P_m| - P_{cu,r} - P_{cu,s} \qquad (3.30)$$

where absolute values are used to avoid confusion since the stator power and mechanical power are negative when the IG operates in the generating mode. The stator power can also be calculated by

$$P_s = 3V_s I_s \cos\varphi_s \qquad (3.31)$$

where φ_s is the stator power factor angle, which is the phase displacement between the stator voltage and current, defined by

$$\varphi_s = \angle \overline{V}_s - \angle \overline{I}_s \qquad (3.32)$$

Generator Torque-Speed Characteristics. The torque-speed curve provides insight on how the torque developed by the generator varies with the speed for a given stator voltage and frequency. In order to obtain the torque-speed curve, it is necessary to find an equation relating these variables. The mechanical power of the generator is given by

$$P_m = T_m \omega_m \qquad (3.33)$$

Substituting (3.28) into (3.33), the mechanical torque can be calculated by

$$T_m = \frac{1}{\omega_m}\left(3I_r^2 \frac{1-s}{s} R_r\right) = \frac{1}{\omega_r/P}\left(3I_r^2 \frac{1-s}{s} R_r\right) \qquad (3.34)$$

Substituting $(1-s) = \omega_r/\omega_s$ into (3.34) yields

$$T_m = \frac{1}{\omega_s/P}\left(3I_r^2 \frac{R_r}{s}\right) = \frac{P_{ag}}{\omega_s/P} \qquad (3.35)$$

where the air gap power is given by

$$P_{ag.} = 3I_r^2 \frac{R_r}{s} \qquad (3.36)$$

Neglecting the magnetizing branch in Figure 3-13b for simplicity, the rotor current can be calculated by

$$I_r = \frac{V_s}{\sqrt{\left(R_s + \dfrac{R_r}{s}\right)^2 + \left(X_{ls} + X_{lr}\right)^2}} \qquad (3.37)$$

substituting (3.37) into (3.35) yields

$$T_m = \frac{3P}{\omega_s} \cdot \frac{R_r}{s} \cdot \frac{V_s^2}{\left(R_s + \dfrac{R_r}{s}\right)^2 + \left(X_{ls} + X_{lr}\right)^2}$$

(3.38)

Equation (3.38) relates the mechanical torque T_m with slip s for a given stator voltage V_s and stator frequency ω_s.

A typical torque-slip curve of induction generator is given in Figure 3-15. There are two operating modes: motoring and generating modes. When the generator operates in the motoring mode, the rotor speed ω_r is below the synchronous speed ω_s, and both mechanical torque and slip are positive ($T_m > 0, s > 0$). On the contrary, when the generator operates in the generating mode, the rotor speed is higher than the synchronous speed and both torque and slip are negative ($T_m < 0, s < 0$). This is due to the use of the motor convention discussed earlier, whereby the direction of the stator current is into the stator, as shown in Figure 3-13. When the generator operates at the rated operating point, its rated mechanical torque is −1.0 pu, and its rated slip is normally in the range of −0.005 to −0.01 for large megawatt induction generators in wind energy systems.

Case Study 3-2—Power and Efficiency Analysis. This case study investigates the power losses and efficiency of the SCIG based on its steady-state equivalent circuit and also examines the differences between the generating and motoring modes of operation.

Consider the 2.3 MW, 690 V, 50 Hz, 1512 rpm SCIG in Case Study 3-1. The shaft of the generator is coupled to the wind turbine through a gearbox and its stator is directly connected to the grid of 690 V/50 Hz. For a given wind speed, the generator operates at a rotor speed of 1506 rpm, at which the rotational losses of the generator are 23 kW.

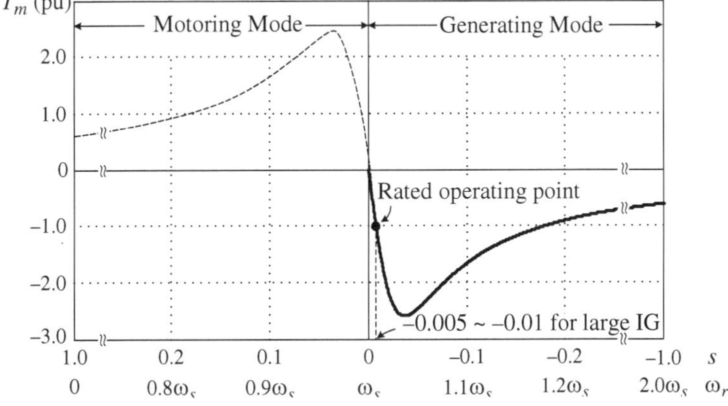

Figure 3-15. Typical torque-slip curve of squirrel-cage induction generator.

Based on the SCIG steady-state equivalent circuit of Figure 3-13b, the total input impedance of the generator is given by

$$\overline{Z}_s = R_s + jX_{ls} + jX_m // \left(\frac{R_r}{s} + jX_{lr} \right) = 0.339 \angle 145.3 - \Omega \qquad (3.39)$$

where the slip is

$$s = \frac{1500 - 1506}{1500} = -0.004 \qquad (3.40)$$

The negative slip indicates that the SCIG operates in the generating mode.
   The stator and rotor currents can be obtained from

$$\begin{cases} \overline{I}_s = \dfrac{\overline{V}_s}{\overline{Z}_s} = \dfrac{690/\sqrt{3}\angle 0°}{0.330\angle 145.3°} = 1206.9\angle -145.3° \text{A} \\[4mm] \overline{I}_r = \dfrac{jX_m \overline{I}_s}{jX_m + (R_r/s + jX_{lr})} = 1030.0\angle -173.8° \text{A} \end{cases} \qquad (3.41)$$

The stator current lags the stator voltage, which implies that the generator draws the reactive power from the grid.
   The stator power factor angle φ_s and the power factor PF_s can be calculated by

$$\begin{cases} \varphi_s = \angle \overline{V}_s - \angle \overline{I}_s = \angle 0° - \angle -145.3° = 145.3° \\[2mm] PF_s = \cos \varphi_s = -0.822 \end{cases} \qquad (3.42)$$

The stator power factor angle is greater than $90°$ and power factor is negative, indicating the machine operates in the generating mode.
   The stator power to the grid can be calculated by

$$P_s = 3V_s I_s \cos \phi_s = 3 \times 690/\sqrt{3} \times 1206.9 \times (-0.822) = -1186.2 \ kW \qquad (3.43)$$

The negative power implies that the generator delivers the active power to the grid.
   From the equivalent circuit of Figure 3-13b, the mechanical power of the generator can be calculated by

$$P_m = 3I_r^2 R_r (1-s)/s = -1195.78 \ \text{kW} \qquad (3.44)$$

based on which its mechanical torque is

$$T_m = \frac{P_m}{\omega_m} = \frac{-1195.8 \times 10^3}{(1506 \times 2 \times \pi / 60)} = -7.58 \ \text{kN} \cdot \text{m} \qquad (3.45)$$

The stator and rotor winding losses are given by

$$\begin{cases} P_{cu,s} = 3I_s^2 R_s = 4.82\,\text{kW} \\ P_{cu,r} = 3I_r^2 R_r = 4.76\,\text{kW} \end{cases} \tag{3.46}$$

The stator power to the grid is calculated by

$$\left|P_s\right| = \left|P_m\right| - P_{cu,s} - P_{cu,r} = 1186.2\,\text{kW} \tag{3.47}$$

The total input power from the shaft of the generator is

$$\left|P_{in}\right| = \left|P_m\right| + P_{rot} = 1218.8\,\text{kW} \tag{3.48}$$

from which the generator efficiency is calculated by

$$\eta = \left|P_s\right| / \left|P_{in}\right| = 97.33\% \tag{3.49}$$

It can be observed from the above analysis that with the IG operating in a generating mode, its stator power P_s, mechanical power P_m, mechanical torque T_m, stator power factor PF_s, and slip s are all negative. This is due to the IG equivalent circuit with the motor convention, whereby the stator current i_s is assumed to flow into the stator from the grid.

Table 3-2 provides a summary of the induction machine operating in the motoring and generating modes; a simplified connection diagram and its corresponding phasor diagram are also given. The stator power, power factor, electromagnetic and mechanical torque, and slip of the induction machine are all positive in the motoring mode and negative in the generating mode. The stator power factor angle is in the range of $0 \leq \varphi_s < 90°$ and $90° \leq \varphi_s < 180°$ for motoring and generating operations, respectively.

3.4 SYNCHRONOUS GENERATORS

Synchronous generators (SGs) are widely used in wind energy conversion systems of a few kilowatts to a few megawatts. As mentioned in the introduction of this chapter, the synchronous generators can be classified into two categories: wound-rotor synchronous generators (WRSGs) and permanent-magnet synchronous generators (PMSGs). In the WRSG the rotor flux is generated by the rotor field winding, whereas the PMSG uses permanent magnets to produce the rotor flux. Depending on the shape of the rotor and the distribution of the air gap along the perimeter of the rotor, synchronous generators can be categorized into salient-pole and nonsalient-pole types.

In this section, the construction of WRSGs and PMSGs in wind energy systems is presented, and the dynamic and steady-state models for both types of synchronous generators are derived. The block diagrams for the simulation of synchronous generators are developed, and case studies are provided for the dynamic and steady-state analysis of the generators.

Table 3-2. Summary for induction machine operating in the motoring and generating modes

Connection	I_s — V_s — **IM** — ω_m — T_m → Mechanical load or wind turbine; Electric Grid; Induction Machine		
Phasor diagram	$\bar{V}_s$; $P_s < 0$, Power Flow: Machine → Grid (Generating mode); φ_s; $\bar{I}_s$; $P_s > 0$, Power Flow: Grid → Machine (Motoring mode)		
Operating mode	Generating mode		Motoring mode	
Stator power	$P_s = 3V_s I_s \cos \varphi_s < 0$		$P_s = 3V_s I_s \cos \varphi_s > 0$	
Slip and torque (normal operation)	$s < 0$	$T_m < 0,\ T_e < 0$	$s > 0$	$T_m > 0,\ T_e > 0$
Power factor angle and power factor	$90° \leq \varphi_s < 180°$	$-1 \leq PF_s \leq 0$	$0 \leq \varphi_s < 90°$	$0 \leq PF_s \leq 1$

3.4.1 Construction

Similar to the induction generator, the synchronous generator is mainly composed of a stator and a rotor. The construction of the stator of both wound-rotor and permanent-magnet synchronous generators is essentially the same as that of an induction generator and, therefore, is not repeated here. This subsection provides an overview of the rotor configuration for the WRSG and PMSG.

Wound-Rotor Synchronous Generators. As the name indicates, the wound-rotor synchronous generator has a wound-rotor configuration to generate the rotor magnetic flux. Figure 3-16 illustrates a typical salient-pole WRSG, where only twelve poles are shown for better appreciation of the rotor structure. The field winding is wound around pole shoes, which are placed symmetrically on the perimeter of the rotor in a radial configuration around the shaft to accommodate large number of poles. The generator has an uneven air gap flux distribution due to the salient structure of the rotor. A practical example of this type of generator is shown in Figure 2-7e in Chapter 2. The synchronous generators with a high number of poles (e.g., 72 poles) operating at low rotational speeds can be used in direct-driven megawatt wind energy systems where there is no need for a gearbox. This leads to a reduction in power losses and maintenance cost.

The rotor-field winding of the synchronous generator requires DC excitation. The rotor current can be supplied directly by brushes in contact with slip rings attached to

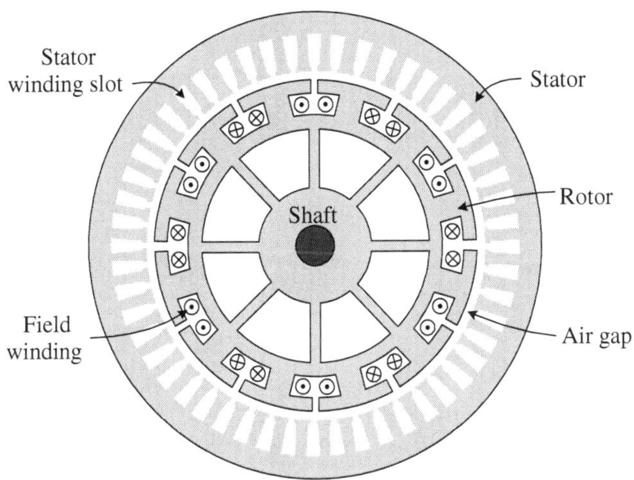

Figure 3-16. Salient-pole, wound-rotor synchronous generator (twelve-pole configuration).

the shaft and electrically connected to the rotor winding. Alternatively, a brushless exciter physically attached to the shaft can be used. The exciter generates AC currents that are rectified to DC using a diode bridge for the rotor winding. The first option is simple but requires regular maintenance of the brushes and slip rings, whereas the second option is more expensive and complex but needs little maintenance.

Permanent-Magnet Synchronous Generators. In the PMSG, the rotor magnetic flux is generated by permanent magnets, and these generators are, therefore, brushless. Because of the absence of the rotor windings, a high power density can be achieved, reducing the size and weight of the generator. In addition, there are no rotor winding losses, reducing the thermal stress on the rotor. The drawbacks of these generators lie in the fact that permanent magnets are more expensive and prone to demagnetization. Depending on how the permanent magnets are mounted on the rotor, the PMSG can be classified into surface-mounted and inset PM generators.

SURFACE-MOUNTED PMSG. In the surface-mounted PMSG, the permanent magnets are placed on the rotor surface. Figure 3-17 shows such a generator, where 16 magnets are evenly mounted on the surface of the rotor core, separated by nonferrite materials between two adjacent magnets. Since the permeability of the magnets is very close to that of the nonferrite materials, the effective air gap between the rotor core and stator is uniformly distributed around the surface of the rotor. This type of configuration is known as a nonsalient-pole PMSG.

The main advantage of the surface-mounted SG is its simplicity and low construction cost in comparison to the inset PMSG. However, the magnets are subject to centrifugal forces that can cause their detachment from the rotor and, therefore, the surface-mounted PMSGs are mainly used in low-speed applications. In a direct-driven

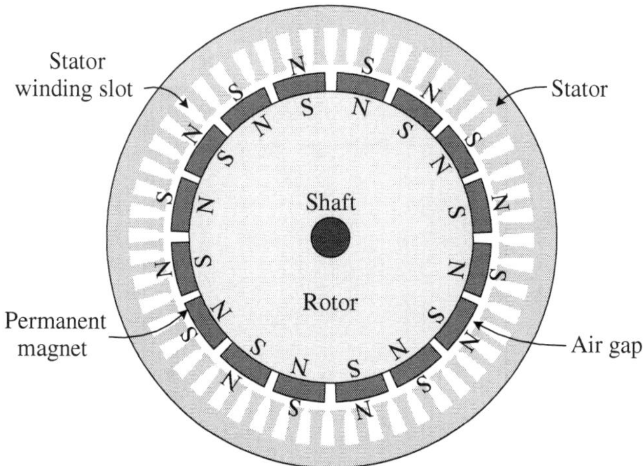

Figure 3-17. Surface-mounted nonsalient PMSG (sixteen-pole configuration).

WECS, the synchronous generator with a high number of poles is used, like the one shown in Figure 2-7d in Chapter 2. The surface-mounted PMSG can have an external rotor in which the permanent magnets are attached to the inner surface of the rotor [7]. In this case, the centrifugal forces help to keep the magnets attached to the rotor core.

INSET PMSG. In the inset PMSG, the permanent magnets are inset into the rotor surface as shown in Figure 3-18. The saliency is created by the different permeability of the rotor core material and magnets. This configuration also reduces rotational

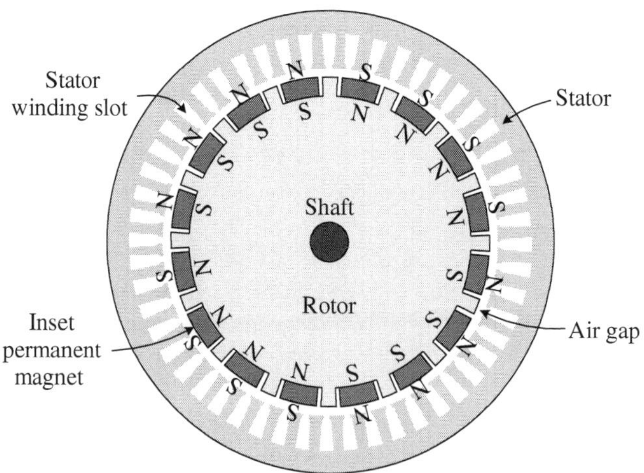

Figure 3-18. Inset PMSG with salient poles (four-pole configuration).

stress associated with centrifugal forces in comparison to the surface-mounted PMSG and, therefore, this type of generator can operate at higher rotor speeds. A practical low-pole-number PMSG used in WECS is shown in Figure 2-7e in Chapter 2.

3.4.2 Dynamic Model of SG

Figure 3-19 shows a general *dq*-axis model of a synchronous generator. To simplify the analysis, the SG is normally modeled in the rotor field synchronous reference frame. The stator circuit of the *dq*-axis model is essentially the same as that of the induction generator shown in Figure 3-8 except that

- The speed of the arbitrary reference frame ω in the IG model is replaced by the rotor speed ω_r in the synchronous frame
- The magnetizing inductance L_m is replaced by the *dq*-axis magnetizing inductances L_{dm} and L_{qm} of the synchronous generator. In a nonsalient SG, the *d*- and *q*-axis magnetizing inductances are equal ($L_{dm} = L_{qm}$), whereas in the salient-pole generators, *d*-axis magnetizing inductance is normally lower than the *q*-axis magnetizing inductance ($L_{dm} < L_{qm}$).
- The *dq*-axis stator currents, i_{ds} and i_{qs}, flow out of the stator. This is based on the generator convention since most synchronous machines are used as generators.

To model the rotor circuit, the field current in the rotor winding is represented by a constant current source I_f in the *d*-axis circuit [5]. In the PMSG, the permanent magnet that replaces the field winding can be modeled by an equivalent current source I_f with a fixed magnitude.

To simplify the SG model of Figure 3-19, the following mathematical manipulations can be performed. The voltage equations for the synchronous generator are given by

$$\begin{cases} v_{ds} = -R_s i_{ds} - \omega_r \lambda_{qs} + p\lambda_{ds} \\ v_{qs} = -R_s i_{qs} + \omega_r \lambda_{ds} + p\lambda_{qs} \end{cases} \tag{3.50}$$

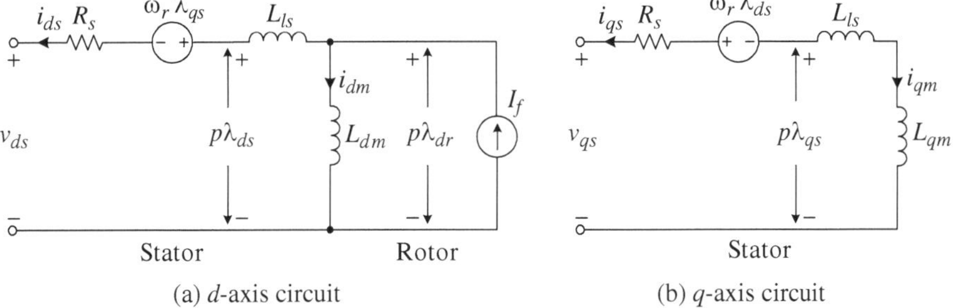

Figure 3-19. General *dq*-axis model of SG in the rotor field synchronous reference frame.

where λ_{ds} and λ_{qs} are the d- and q-axis stator flux linkages, given by

$$
\begin{cases}
\lambda_{ds} = -L_{ls}i_{ds} + L_{dm}\left(I_f - i_{ds}\right) = -(L_{ls} + L_{dm})i_{ds} + L_{dm}I_f = -L_d i_{ds} + \lambda_r \\
\lambda_{qs} = -(L_{ls} + L_{qm})i_{qs} = -L_q i_{qs}
\end{cases}
\tag{3.51}
$$

where λ_r is the rotor flux, and L_d and L_q are the stator dq-axis self-inductances, defined by

$$
\begin{cases}
\lambda_r = L_{dm}I_f \\
L_d = L_{ls} + L_{dm} \\
L_q = L_{ls} + L_{qm}
\end{cases}
\tag{3.52}
$$

Substituting (3.51) into (3.50), and considering $d\lambda_r/dt = 0$ for constant field current I_f in the WRSG and constant λ_r in the PMSG, we have

$$
\begin{cases}
v_{ds} = -R_s i_{ds} + \omega_r L_q i_{qs} - L_d p i_{ds} \\
v_{qs} = -R_s i_{qs} - \omega_r L_d i_{ds} + \omega_r \lambda_r - L_q p i_{qs}
\end{cases}
\tag{3.53}
$$

Figure 3-20 shows a simplified model for the synchronous generators, which is derived based on Equation (3.53). It should be pointed out that

- The simplified mode is as accurate as the general model of Figure 3-19 since no assumption was made during the derivation of the simplified model. The performance analysis based on the general and simplified models should give identical results.
- The SG model is valid for both wound-rotor and permanent-magnet synchronous generators. For a given field current I_f in the WRSG, the rotor flux can be calculated by $\lambda_r = L_{dm} I_f$. For the PMSG, the rotor flux λ_r is produced by permanent magnets and its rated value can be obtained from the nameplate data and generator parameters.
- The model is also valid for both salient- and nonsalient-pole synchronous generators. For a nonsalient generator, the dq-axis synchronous inductances, L_d and L_q,

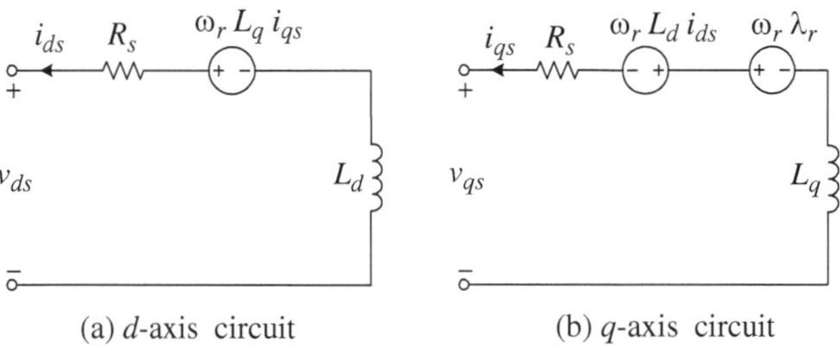

(a) d-axis circuit   (b) q-axis circuit

Figure 3-20. Simplified dq-axis model of SG in the rotor-field synchronous reference frame.

are equal, whereas they are different for a salient-pole generator. The d-axis synchronous inductance of PMSG is usually lower than that of the q-axis ($L_d < L_q$).

The electromagnetic torque produced by the SG can be calculated by the same equation for the IG given in Equation (3.16a), that is,

$$T_e = \frac{3P}{2}\left(i_{qs}\lambda_{ds} - i_{ds}\lambda_{qs}\right) \tag{3.54}$$

Substituting (3.51) into (3.54), we have

$$T_e = \frac{3P}{2}\left[\lambda_r i_{qs} - \left(L_d - L_q\right)i_{ds}i_{qs}\right] \tag{3.55}$$

The rotor speed ω_r is governed by motion equation

$$\omega_r = \frac{P}{JS}\left(T_e - T_m\right) \tag{3.56}$$

To derive the SG model for dynamic simulation of synchronous generators, Equation (3.53) is rearranged as

$$\begin{cases} i_{ds} = \dfrac{1}{S}\left(-v_{ds} - R_s i_{ds} + \omega_r L_q i_{qs}\right)/L_d \\[2mm] i_{qs} = \dfrac{1}{S}\left(-v_{qs} - R_s i_{qs} - \omega_r L_d i_{ds} + \omega_r \lambda_r\right)/L_q \end{cases} \tag{3.57}$$

Based on the above three equations, the block diagram for computer simulation of the SG is derived and shown in Figure 3-21. The input variables of the SG model are the dq-axis stator voltages v_{ds} and v_{qs}, the rotor flux linkage λ_r, and the mechanical torque T_m, whereas the output variables are the dq-axis stator currents i_{ds} and i_{qs}, the rotor mechanical speed ω_m, and the electromagnetic torque T_e.

Case Study 3-3—Analysis of Synchronous Generator in Standalone Operation. The main purpose of this case study is to

- Investigate the operation of a stand-alone SG wind energy system feeding a three-phase resistive load
- Illustrate how to effectively use the simulation model of Figure 3-21 for the simulation of synchronous generators
- Reveal the relationship between the three-phase abc variables in the stationary frame and the dq variables in the synchronous frame

The generator used in the study is a 2.45 MW, 4000 V, 53.33 Hz, 400 rpm nonsalient-pole PMSG, whose parameters are given in Table B-10 in Appendix B. The generator

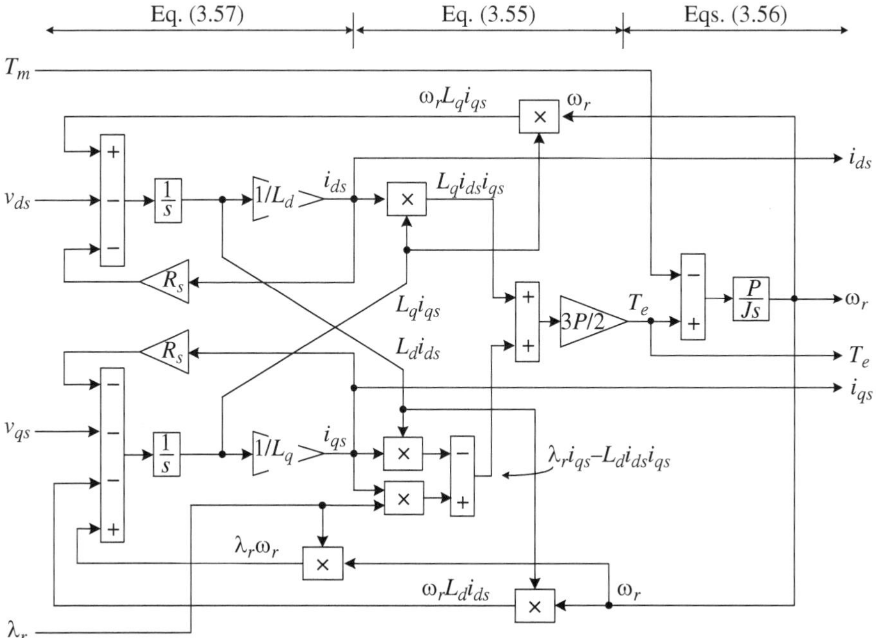

Figure 3-21. Block diagram for dynamic simulation of synchronous generators.

is loaded with a three-phase balanced resistive load R_L and operates at 320 rpm (0.8 pu) at a given wind speed. The loading of the generator can be changed by switch S. When S is closed, the load resistance is reduced to $R_L/2$ per phase.

It is assumed that the combined moment of inertia of the blades, rotor hub, and generator are very large such that the rotor speed is kept constant at 320 rpm during the transients caused by the changes in load resistance. The motion equation (3.56) is then not needed. Since the rotor speed ω_r is known, it becomes the system input variable. The SG simulation algorithm given in Figure 3-21 should be slightly modified accordingly.

The block diagram for the simulation of the SG stand-alone operation is shown in Figure 3-22. The dq-axis stator currents, i_{ds} and i_{qs}, in the synchronous frame rotating at the synchronous speed of ω_r are calculated by the SG model. They are then transformed into the abc-axis stator currents, i_{as}, i_{bs}, and i_{cs}, in the stationary frame through the dq/abc transformation. The calculated load voltages, v_{as}, v_{bs}, and v_{cs}, which are also the stator voltages, are transformed to the dq-axis voltages v_{ds} and v_{qs} in the synchronous frame, which are then fed back to the SG model.

A simple test for the synchronous generator in stand-alone operation is carried out, and results are given in Figure 3-23. The generator initially operates in steady state with a resistive load of R_L. The load resistance is reduced to $R_L/2$ by closing switch S at $t = 0.015$ sec. After a short transient period, the system reaches a new steady-state operating point. The dq-axis stator currents, i_{ds} and i_{qs}, in the synchro-

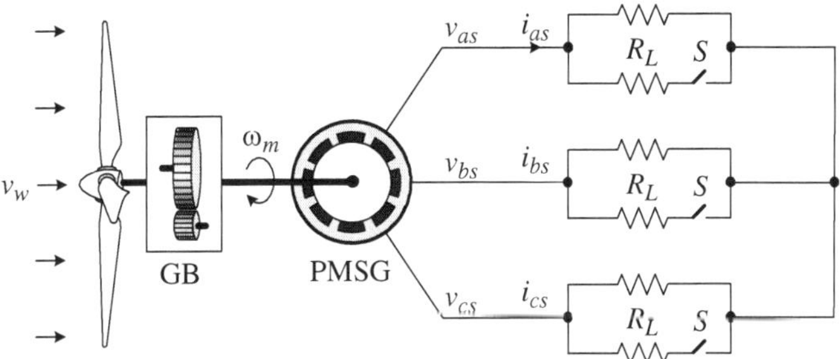

(a) SG with a three-phase resistive load

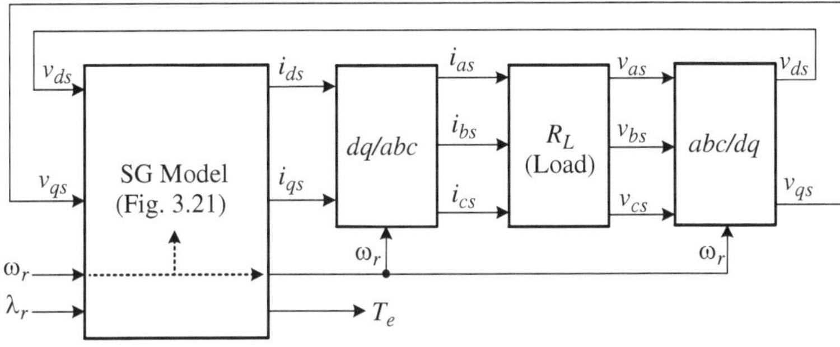

(b) Block diagram for simulation

<u>Figure 3-22.</u> Block diagram of a stand-alone SG configuration with a three-phase resistive load.

nous frame are DC variables, whereas the *abc*-axis stator currents, i_{as}, i_{bs}, and i_{cs}, in the stationary frame are sinusoids in steady state. The magnitude of the stator current i_s, given by $i_s = \sqrt{i_{qs}^2 + i_{ds}^2}$, represents the peak value of i_{as}, i_{bs}, and i_{cs}. A similar phenomenon can be observed for the stator voltages. A decrease in the load resistance results in an increase in the stator currents, but the stator voltages are reduced mainly due to the voltage drop across the stator inductances. The electromagnetic torque T_e and stator active power P_s are increased accordingly when the system operates at the new operating point.

For the nonsalient synchronous generator, its *dq*-axis inductances are equal ($L_q = L_d$). The torque equation (3.55) can then be simplified to

$$T_e = \frac{3P}{2}\left(\lambda_r\, i_{qs}\right) \tag{3.58}$$

The above equation indicates that the *d*-axis current i_{ds} in the nonsalient SG does not contribute to the torque production and, therefore, the electromagnetic torque is proportional to the *q*-axis stator current for a given rotor flux λ_r. This can be confirmed by comparing the waveforms of i_{qs} and T_e given in Figure 3-23.

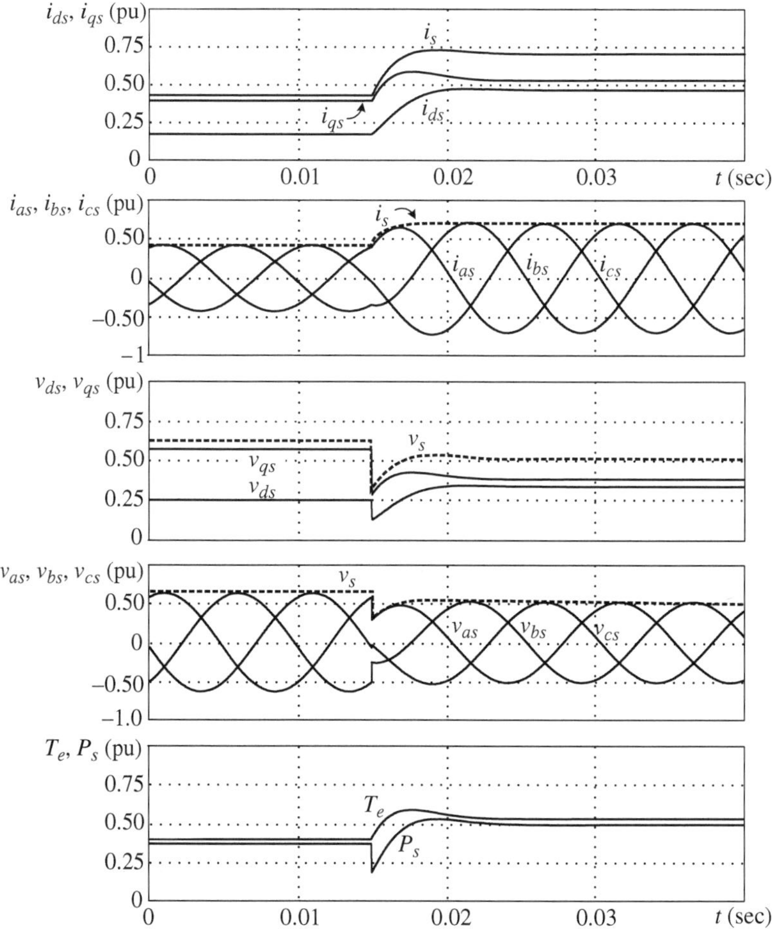

Figure 3-23. Simulated waveforms for a stand-alone PMSG system with resistive load.

3.4.3 Steady-State Equivalent Circuits

The steady-state model of a synchronous generator provides a useful tool for the analysis of the generator's steady-state performance. The SG steady-state model can be developed from its dynamic model shown in Figure 3-20. Considering that the *dq*-axis stator currents, i_{ds} and i_{qs}, in the rotor flux synchronous reference frame are DC values in steady state, their derivatives in Equation (3.53), pi_{ds} and pi_{qs}, become zero. Therefore, the equations that describe the steady-state characteristics of the synchronous generator are given by

$$\begin{cases} v_{ds} = -R_s i_{ds} + \omega_r L_q i_{qs} \\ v_{qs} = -R_s i_{qs} - \omega_r L_d i_{ds} + \omega_r \lambda_r \end{cases} \tag{3.59}$$

Since the *dq*-axis voltages and currents in the synchronous frames are all constant D values in steady state, it is convenient to use them directly for steady-state analysis. Based on Equation (3.59), the steady-state equivalent circuit for the synchronous generator is derived and shown in Figure 3-24.

Case Study 3-4—Steady-State Analysis of Stand-Alone SG with RL Load. In this case study, the steady-state performance of a stand-alone salient-pole synchronous generator with an *RL* load is analyzed using the *dq*-axis steady-state equivalent circuit of Figure 3-24. Consider a 2.5 MW, 4000 V, 40 Hz, 400 rpm six-pole salient PMSG, whose parameters are given in Table B-12 in Appendix B. The generator operates at the rotor speed of 400 rpm and supplies a three-phase *RL* load of $R_L = 4.2855\ \Omega$ and $L_L = 8.258$ mH, as shown in Figure 3-25a.

Since the *q*-axis leads the *d*-axis by 90°, the generator *dq*-axis stator voltage, which is also the load voltage, can be calculated by

$$v_{ds} + jv_{qs} = (i_{ds} + ji_{qs})(R_L + j\omega_r L_L)$$
$$= (R_L i_{ds} - \omega_r L_L i_{qs}) + j(R_L i_{qs} + \omega_r L_L i_{ds}) \tag{3.60}$$

where ω_r is the rotor electrical speed, which is also the speed of the *dq* synchronous reference frame, in which the *dq*-axis SG model is established. Equation (3.60) can be rearranged as

$$\begin{cases} v_{ds} = R_L i_{ds} - \omega_r L_L i_{qs} = R_L i_{ds} - X_L i_{qs} \\ v_{qs} = R_L i_{qs} + \omega_r L_L i_{ds} = R_L i_{qs} + X_L i_{ds} \end{cases} \tag{3.61}$$

where $X_L i_{qs} = \omega_r L_L i_{qs}$ and $X_L i_{ds} = \omega_r L_L i_{ds}$. These two terms are referred to as speed voltages, which are due to the transformation of the three-phase load inductance in the *abc* stationary frame into the *dq* synchronous frame. Other methods for transforming a three-phase inductive or capacitive circuit in the *abc* stationary frame into the *dq* rotating frame are also available [3], and the same results can be obtained. Based on Equation (3.61), the *dq*-axis equivalent circuit of the synchronous generator with an *RL* load is shown in Figure 3-25b.

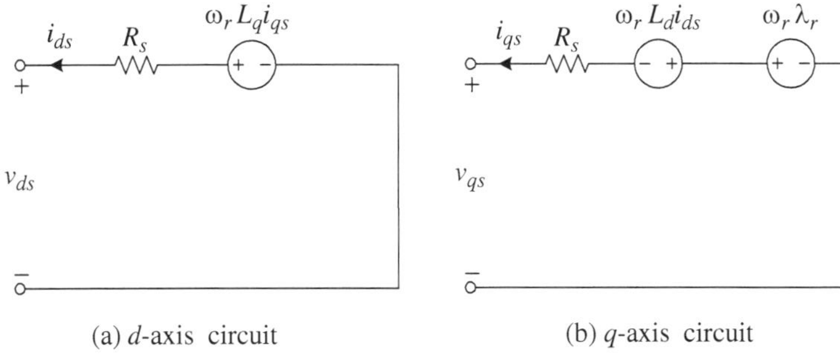

(a) *d*-axis circuit   (b) *q*-axis circuit

Figure 3-24. Steady-state model of synchronous generator.

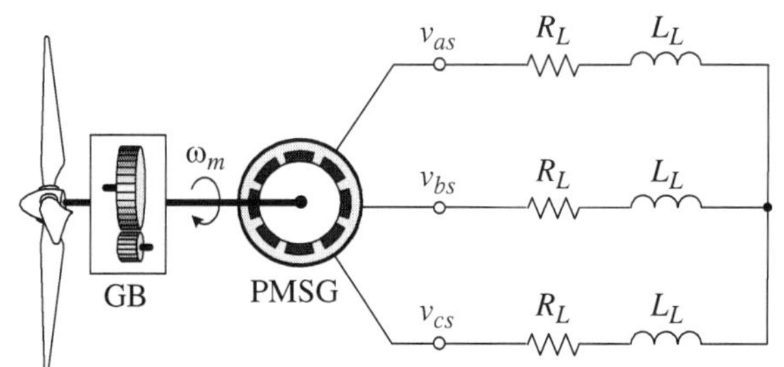

(a) SG with a three-phase *RL* load

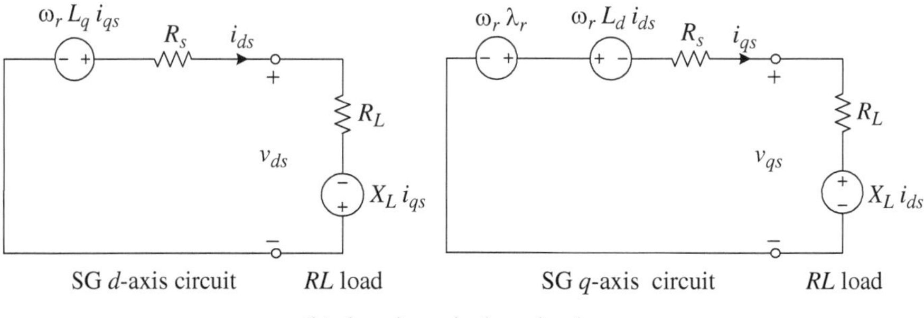

(b) *dq*-axis equivalent circuits

Figure 3-25. Steady-state analysis of PMSG with an *RL* load.

Substituting (3.59) into (3.61) yields

$$\begin{cases} -R_s i_{ds} + \omega_r L_q i_{qs} = R_L i_{ds} - \omega_r L_L i_{qs} & \text{(a)} \\ -R_s i_{qs} - \omega_r L_d i_{ds} + \omega_r \lambda_r = R_L i_{qs} + \omega_r L_L i_{ds} & \text{(b)} \end{cases} \tag{3.62}$$

From Equation (3.62a), the *d*-axis stator current i_{ds} can be expressed as

$$i_{ds} = \frac{\omega_r (L_L + L_q)}{R_L + R_s} i_{qs} \tag{3.63}$$

Substituting (3.63) into (3.62b), the *q*-axis current can be calculated by

$$i_{qs} = \frac{\omega_r \lambda_r (R_L + R_s)}{(R_L + R_s)^2 + \omega_r^2 (L_L + L_d)(L_L + L_q)} = 141.85 \text{ A} \tag{3.64}$$

where $\omega_r = n_r P 2\pi/60 = 400 \times 6 \times 2\omega/60 = 251.33$ rad/sec and $\lambda_r = 6.7302$ Wb (rated peak value; refer to Table B-12).

The *d*-axis stator current i_{ds} is then

$$i_{ds} = \frac{\omega_r (L_L + L_q)}{R_L + R_s} i_{qs} = 249.0 \text{ A} \tag{3.65}$$

The rms stator current is evaluated by

$$I_s = \sqrt{i_{ds}^2 + i_{qs}^2} / \sqrt{2} = 202.7\text{A} \tag{3.66}$$

which is the rms value of the three-phase stator currents I_{as}, I_{bs}, and I_{cs}.

With the stator currents calculated, the dq-axis stator voltages are found from

$$\begin{cases} v_{ds} = -R_s i_{ds} + \omega_r L_q i_{qs} = 772.9 \text{ V} \\ v_{qs} = -R_s i_{qs} - \omega_r L_d i_{ds} + \omega_r \lambda_r = 1124.7 \text{ V} \end{cases} \tag{3.67}$$

The rms stator voltage is then calculated by

$$V_s = \sqrt{v_{ds}^2 + v_{qs}^2} / \sqrt{2} = 965.0 \text{ V} \tag{3.68}$$

which is also the rms value of the three-phase stator voltages V_{as}, V_{bs}, and V_{cs}.

The electromagnetic torque of the generator is obtained by

$$T_e = \frac{3}{2} P(\lambda_r i_{qs} - (L_d - L_q) i_{ds} i_{qs}) = 12.7 \text{ kN} \cdot \text{m} \tag{3.69}$$

The mechanical power of the generator is

$$P_m = T_m \omega_m = T_e \omega_r / P = 531.0 \text{ kW} \tag{3.70}$$

where the mechanical torque T_m is equal to the electromagnetic torque T_e when the generator operates in steady state. The stator winding loss is

$$P_{cu,s} = 3 I_s^2 R_s = 3.0 \text{ kW} \tag{3.71}$$

The active power delivered to the load is calculated by subtracting the stator winding losses from the mechanical power:

$$P_L = P_m - P_{cu,s} = 528.0 \text{ kW} \tag{3.72}$$

The load power factor angle is

$$\varphi_L = \tan^{-1}\left(\frac{\omega_r L_L}{R_L}\right) = 25.8° \tag{3.73}$$

from which the load power factor is

$$PF_L = \cos(\varphi_L) = 0.9 \tag{3.74}$$

Alternatively, the load active and reactive power can be obtained by

$$\begin{cases} P_L = 1.5(v_{ds}i_{ds} + v_{qs}i_{qs}) = 528.0\,\text{kW} \\ Q_L = 1.5(v_{qs}i_{ds} - v_{ds}i_{qs}) = 255.7\,\text{kVA} \end{cases} \tag{3.75}$$

from which the load power factor is

$$PF_L = \frac{P_L}{\sqrt{P_L^2 + Q_L^2}} = 0.9 \tag{3.76}$$

The dq-axis stator voltages can also be calculated from the load conditions

$$\begin{cases} v_{ds} = R_L i_{ds} - \omega_r L_L i_{qs} = R_L i_{ds} - X_L i_{qs} = 772.9\,\text{V} \\ v_{qs} = R_L i_{qs} + \omega_r L_L i_{ds} = R_L i_{qs} + X_L i_{ds} = 1124.7\,\text{V} \end{cases} \tag{3.77}$$

The phase angles of the stator voltage and current are given by

$$\begin{cases} \theta_v = \tan^{-1}(v_{qs}/v_{ds}) = 55.5° \\ \theta_i = \tan^{-1}(i_{qs}/i_{ds}) = 29.7° \end{cases} \tag{3.78}$$

from which the stator voltage and current phasors can be defined by

$$\begin{cases} \overline{V}_s = V_s \angle \theta_v = 965.0 \angle 55.5°\,\text{V} \\ \overline{I}_s = I_s \angle \theta_i = 202.7 \angle 29.7°\,\text{A} \end{cases} \tag{3.79}$$

The load power factor angle and power factor are

$$\begin{cases} \phi_L = \theta_v - \theta_i = 25.8° \\ PF_L = \cos \phi_L = 0.9 \end{cases} \tag{3.80}$$

Finally, the active load power P_L can be calculated by

$$P_L = 3V_s I_s \cos \phi_L = 528.0\,\text{kW} \tag{3.81}$$

which is equal to that given in Equations (3.72) and (3.74).

Assuming that the rotational losses P_{rot} of the generator account for 0.5% (12.5 kW) of the rated generator power, the efficiency of the generator is

$$\eta = \frac{P_L}{P_m + P_{rot}} = \frac{528.0}{531.0 + 12.5} = 0.972 \tag{3.82}$$

The permanent-magnet synchronous generators usually have higher efficiency than other types of generators due to the use of permanent magnets for rotor flux generation.

3.5 SUMMARY

The construction, operating principle, and dynamic and steady-state models of induction and synchronous generators used in wind energy systems are introduced in this chapter. The induction generator (IG) models are derived in the arbitrary reference frame, which can be easily transformed to any other reference frame. These models are suitable for both squirrel-cage induction generators (SCIGs) and doubly fed induction generator (DFIG). The synchronous generator (SG) models are developed in the synchronous reference frame, which is the most the convenient reference frame for the analysis of synchronous generators. These models are applicable to both wound rotor synchronous generators (WRSGs) and permanent-magnet synchronous generators (PMSGs). Case studies are provided to investigate the dynamic and steady-state behavior of the generators. The generator models developed in this chapter will be used for the analysis of different WECS configurations with different control strategies in later chapters.

REFERENCES

1. I. Boldea, *The Electric Generators Handbook: Variable Speed Generators,* CRC Press, 2005.
2. J. M. Carrasco, L. G. Franquelo, J. Bialasiewicz, E. Galvan, R. Guisado, M. Prats, J. Leon, and N. Moreno-Alfonso, Power-Electronic Systems for the Grid Integration of Renewable Energy Sources: A Survey, *IEEE Transactions on Industrial Electronics,* Vol. 53, No. 4, 1002–1016, 2006.
3. P. Krause, O. Wasynczuk, and S. Sudhoff, *Analysis of Electric Machinery and Drive Systems,* 2nd Edition, Wiley-IEEE Press, 2002.
4. B. Wu, *High-Power Converters and AC Drives,* Wiley-IEEE Press, 2006.
5. B. K. Bose, *Power Electronics and Motor Drives: Advances and Trends,* Academic Press, 2006.
6. P. C. Sen, *Principle of Electric Machines and Power Electronics,* 2nd Edition, Wiley, 1997.
7. S. Jöckel, New Gearless Wind Turbines in the 1.5 MW Class, in *European Wind Energy Conference and Exhibition,* EWEC, 2006.

4

POWER CONVERTERS IN WIND ENERGY CONVERSION SYSTEMS

4.1 INTRODUCTION

Power converters are widely used in wind energy conversion systems (WECS). In fixed-speed WECS, the converters are used to reduce inrush current and torque oscillations during the system start-up, whereas in variable-speed WECS they are employed to control the speed/torque of the generator and also the active/reactive power to the grid [1]. According to the system power ratings and type of wind turbines, a variety of power converter configurations are available for the optimal control of wind energy systems.

Figure 4-1 illustrates three practical wind energy conversion systems using different power converter configurations. Figure 4-1a shows the fixed-speed, induction-generator-based WECS, in which a soft starter is employed to reduce inrush current caused by electromagnetic transients that take place at the moment the generator is connected to the grid. The soft starter is essentially an AC voltage controller using SCR devices, whose output voltage is adjusted such that it increases slowly with time during the system start-up. Figure 4-1b shows a variable-speed WECS using squirrel cage induction generators (SCIGs) or synchronous generators (SGs), where a back-to-back converter configuration with two identical PWM converters is used. The converters can be either voltage source converters (VSCs) or current source converters (CSCs). Figure 4-1c is also a variable-speed wind energy system only for synchronous

Power Conversion and Control of Wind Energy Systems. By B. Wu, Y. Lang, N. Zargari, and S. Kouro
© 2011 the Institute of Electrical and Electronics Engineers, Inc. Published 2011 by John Wiley & Sons, Inc.

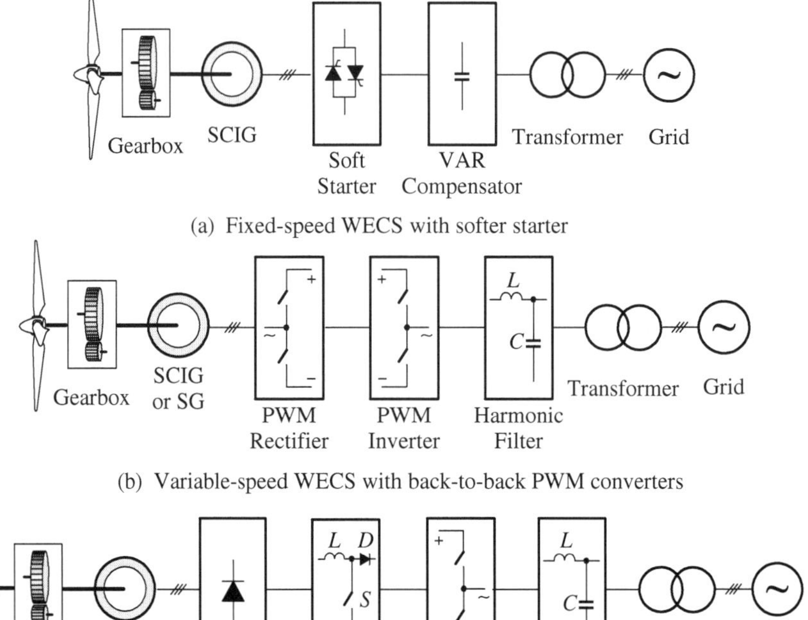

(a) Fixed-speed WECS with softer starter

(b) Variable-speed WECS with back-to-back PWM converters

(c) Variable-speed WECS with DC/DC boost converter

Figure 4-1. Three typical WECS using different power converter topologies.

generators, where a low-cost diode rectifier with a DC/DC boost converter can be used instead of the PWM rectifier.

In this chapter, different power converter topologies for wind energy systems are introduced, their operating principles are discussed, and switching schemes are elaborated. These converters include AC voltage controllers, DC/DC boost converters, two-level voltage source converters, three-level neutral point clamped (NPC) converters, and PWM current source converters [2]. Finally, the control of grid-connected converters is presented. The equations and tables derived in this chapter will be used to assist in the analysis of wind energy systems in the subsequent chapters.

4.2 AC VOLTAGE CONTROLLERS (SOFT STARTERS)

The AC voltage controller is often referred to as soft starter in WECS since its main function is to help the wind turbine to start smoothly with reduced inrush current and mechanical stress. After the system is started, the AC voltage controller is usually bypassed (short circuited) by a bypass switch, which eliminates the power losses of the controller. The AC voltage controllers in WECS normally use a SCR (thyristor) as switching device. Through the delay (firing) angle control for the SCRs, the output voltage of the controller can be adjusted from zero all the way up to its supply voltage, which effectively reduces the starting current of the system. This section starts with an introduction to the single-phase AC voltage controller, followed by detailed studies for three-phase AC voltage controllers.

4.2.1 Single-Phase AC Voltage Controller

The simplified circuit for a single-phase AC voltage controller is shown in Figure 4-2. It is composed of a pair of SCR thyristors, connected in antiparallel between the power supply and the load. The operating principle of the voltage controller, the gating arrangement for the thyristors, and the resultant output voltage and current waveforms are illustrated in Figure 4-3.

Assuming a resistive load, the waveforms for the gate signals i_{g1} and i_{g2}, output current i_o, and output voltage v_o of the controller with a delay angle of $\alpha = \pi/3$ are given in Figure 4-3a. During the positive half-cycle of the power supply, thyristor T_1 is turned on at $\omega t = \alpha = \pi/3$ by i_{g1} and is turned off at π when its current falls to zero. During the negative half-cycle, thyristor T_2 is triggered on at $\omega t = (\alpha + \pi) = 4\pi/3$ and is switched off at 2π.

With a resistive load, the rms value of the output voltage V_o can be found from

$$V_o = \left(\frac{1}{\pi} \int_{\alpha}^{\pi} (\sqrt{2}V_s \sin \omega t)^2 \, d(\omega t) \right)^{1/2} = V_s \left(1 - \frac{\alpha}{\pi} + \frac{\sin 2\alpha}{2\pi} \right)^{1/2} \tag{4.1}$$

Figure 4-3b illustrates the operation of the voltage controller with an RL load and $\alpha = \pi/3$. Thyristor T_1 is turned on at $\omega t = \pi/3$, but will not be turned off when the supply voltage v_s falls to zero at $\omega t = \pi$. This is due to the lagging inductive load current flowing through T_1, which is not yet zero at $\omega t = \pi$. Thyristor T_1 remains on until its current becomes zero, at which the energy stored in the load inductance is fully released and, thus, T_1 is turned off.

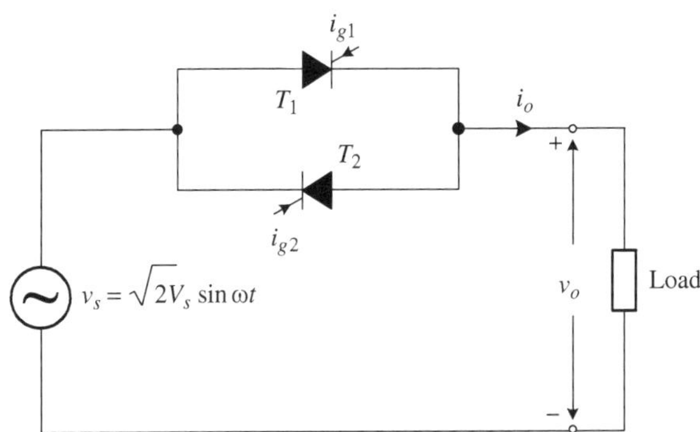

Figure 4-2. Single-phase AC voltage controller.

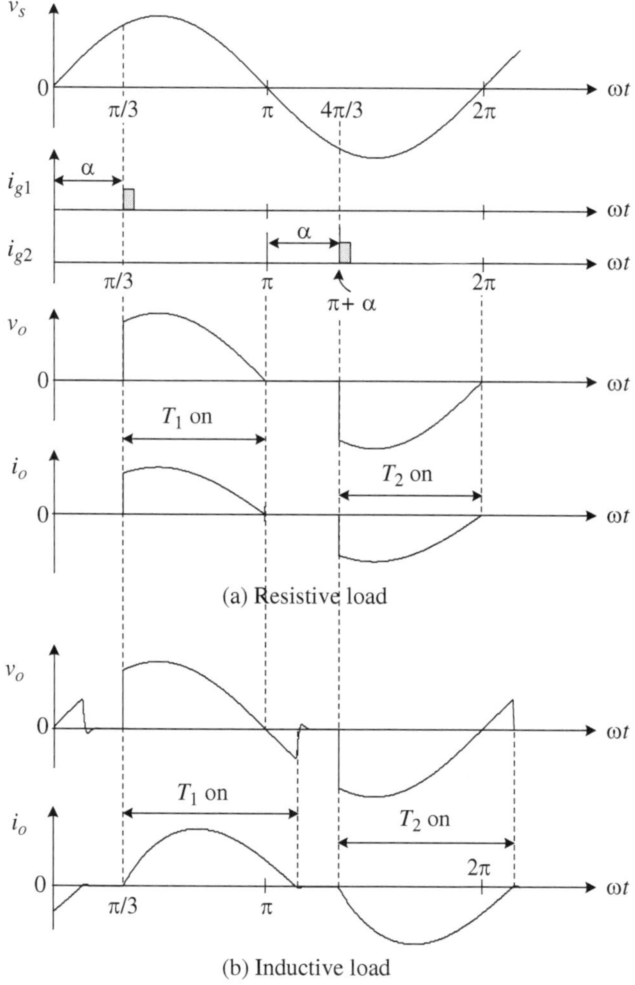

Figure 4-3. Waveforms for the single-phase AC voltage controller.

It is noted that when the delay angle α is smaller than the load power factor angle φ, defined by $\varphi = \tan^{-1}(\omega L/R)$, the output voltage v_o of the controller will be equal to its input supply voltage v_s and, thus, v_o is no longer adjustable. Take Figure 4-4 as an example, where the load power factor angle φ is $\pi/3$ and delay angle a is $\pi/6$. During the positive half-cycle of the supply voltage, thyristor T_1 conducts for a certain period of time. When the gate signal for T_2 arrives at $\omega t = \pi + \alpha$, T_2 will not be turned on since the load current i_o is still positive due to the inductive load and, thus, T_1 continues to conduct. T_2 will be turned on only when i_o falls to zero and becomes negative, provided that the gate current i_{g2} for T_2 is still there. When T_2 is turned on, T_1 is reverse biased and, thus, turned off. Both T_1 and T_2 conduct $180°$ alternatively per the fundamental-frequency cycle and, thus, the output voltage v_o is equal to the supply voltage v_s.

It is also noted that with an inductive load, continuous gating with extended duration, such as i_{g1} and i_{g2} in Figure 4-4, should be used. If the gate signals are of short duration, the controller will not operate properly. For instance, with a short gating pulse i_{g2}, like the one shown with a solid block in the figure, T_2 will not be turned on during the negative cycle of the supply voltage.

Assuming a pure inductive load, the rms value of the output voltage v_o of the controller can be calculated by

$$V_o = \begin{cases} V_s & \text{for } 0 \le \alpha < \pi/2 \\ V_s\left(2 - \dfrac{2\alpha}{\pi} + \dfrac{\sin 2\alpha}{\pi}\right)^{1/2} & \text{for } \pi/2 \le \alpha \le \pi \end{cases} \tag{4.2}$$

where V_o is equal to V_s for $0 \le \alpha < \pi/2$ due to the reasons discussed earlier.

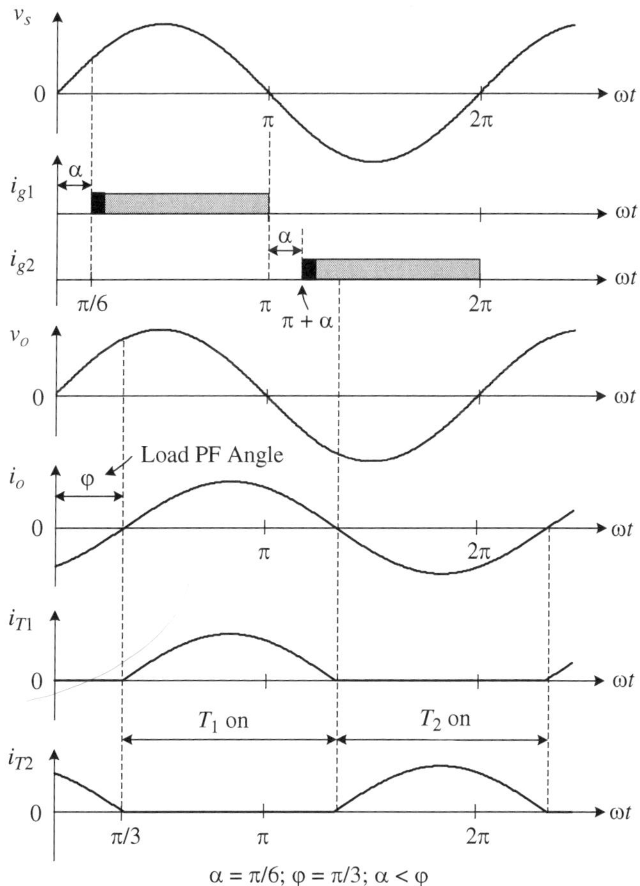

Figure 4-4. Waveforms for single-phase AC voltage controller with an RL load.

Based on Equations (4.1) and (4.2), the relationship between the voltage ratio V_o/V_s and delay angle α with a pure resistive ($\varphi = 0$) and pure inductive ($\varphi = 90°$) load is drawn in Figure 4-5. The other curves in the figure for load power factor angle of φ 45°, 60°, and 75° are obtained by computer simulation.

4.2.2 Three-Phase AC Voltage Controller

The configuration of a three-phase AC voltage controller with a three-phase Y-connected load is shown in Figure 4-6. It is composed of three pairs of SCR thyristors connected between the three-phase power supply and the load. The operating principle of the controller is illustrated in Figure 4-7, where the waveforms for the supply voltages, thyristor gating currents, and phase-a load voltage v_{an} are illustrated.

Consider a three-phase, balanced, Y-connected resistive load for the controller. During period I, thyristors T_6 and T_1 are turned on, and the line-to-line supply voltage v_{AB} is applied to the phase-a and b load resistors. Since T_5 and T_2 in phase-c are both off, the phase-a load voltage v_{an} is equal to $v_{AB}/2$ as shown in Figure 4-7. For period II, thyristors T_1 and T_2 conduct, leading to $v_{an} = v_{AC}/2$. During period III, thyristors T_2 and T_3 are turned on, but none of the phase-a thyristors is on, resulting in $v_{an} = 0$. Following the same procedure, the load voltage v_{an} during the negative half-cycle can be drawn. Similarly, the load voltage waveforms for the other two phases, v_{bn} and v_{cn}, can be determined.

The waveforms for the load line-to-line voltages can be obtained by $v_{ab} = v_{an} - v_{bn}$, $v_{bc} = v_{bn} - v_{cn}$, and $v_{ca} = v_{cn} - v_{an}$, respectively. Figure 4-8 shows the waveforms for v_{an} and v_{ab} with the delay angle changes from $2\pi/3$ to zero in steps. It can be observed that

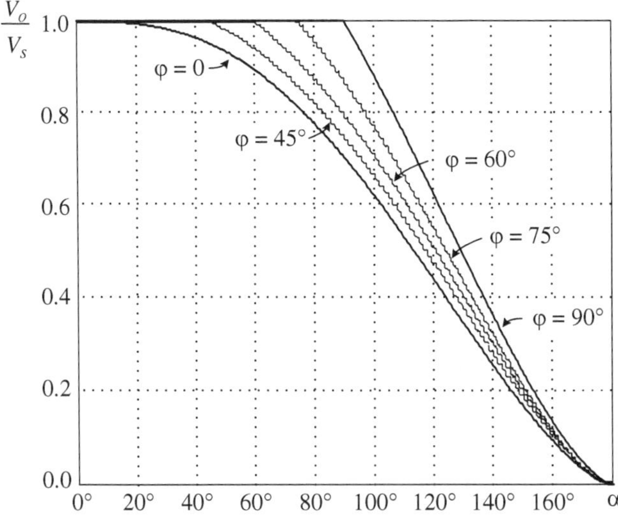

Figure 4-5. Output voltage to supply voltage ratio V_o/V_s versus delay angle α for single-phase AC voltage controller.

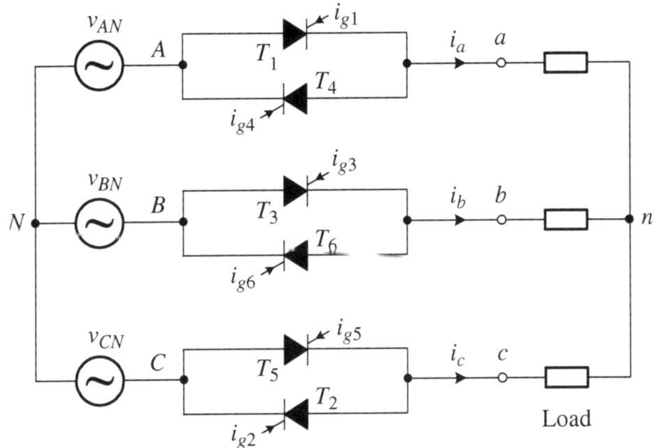

Figure 4-6. Three-phase AC voltage controller with Y-connected load.

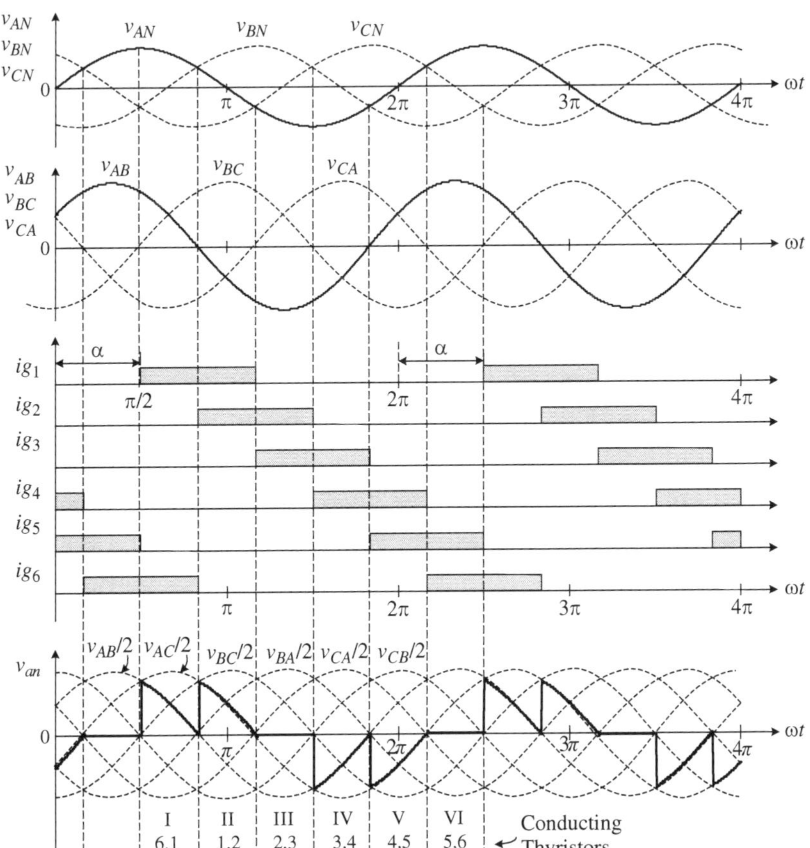

Figure 4-7. Waveforms of three-phase AC voltage controller with a Y-connected resistive load.

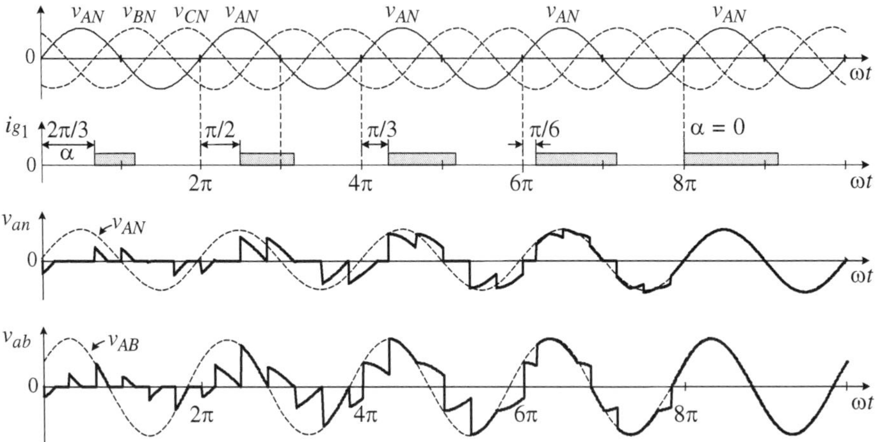

Figure 4-8. Waveforms of three-phase AC voltage controller with a resistive load and various delay angles. Delay angle: $\alpha = 2\pi/3$ (first cycle), $\pi/2$ (second cycle), $\pi/3$ (3rd cycle), $\pi/6$ (fourth cycle) and 0 (fifth cycle).

with the decreasing of delay angle, the load phase voltage v_{an} and line-to-line voltage v_{ab} increase accordingly, and, finally, are equal to the phase voltage v_{AN} and line-to-line voltage v_{AB} of the power supply at $\alpha = 0$, respectively.

Depending on the delay angle α, the operation of the three-phase AC voltage controller can be classified into three operating modes: Mode I for $\pi/2 \leq \alpha < 5\pi/6$, during which there are periods when none or two thyristors in each phase conduct; Mode II for $\pi/3 \leq \alpha < \pi/2$, during which two thyristors in each phase are turned on; and Mode III for $0 \leq \alpha < \pi/3$, during which three thyristors or two thyristors conduct simultaneously [4]. Unlike the single-phase AC voltage controller, in which the delay angle α is in the range of zero to π, the range of the three-phase AC voltage controller is from zero to $5\pi/6$ (150°), beyond which ($5\pi/6 < \alpha \leq \pi$) the output voltage of the controller is kept to zero. Therefore, there is no need to extend the delay angle beyond $5\pi/6$.

The typical waveforms of v_{an} for the voltage controller operating in the these three modes are shown in Figure 4-9, where the delay angle α is $2\pi/3$ (120°) in Mode I, $5\pi/12$ (75°) in Mode II, and $\pi/6$ (30°) in Mode III, respectively.

Based on the waveforms given in Figure 4-9, the rms value of the load phase voltage can be calculated by

$$
\begin{aligned}
V_{an} &= \left(\frac{1}{\pi} \left(\int_{\alpha}^{5\pi/6} \left(\frac{\sqrt{6}}{2} V_s \sin(\omega t + \pi/6) \right)^2 d(\omega t) + \int_{\alpha+\pi/3}^{7\pi/6} \left(\frac{\sqrt{6}}{2} V_s \sin(\omega t - \pi/6) \right)^2 d(\omega t) \right) \right)^{1/2} \\
&= V_s \left(\frac{5}{4} - \frac{3\alpha}{2\pi} + \frac{3\sin(2\alpha + \pi/3)}{4\pi} \right)^{1/2} \qquad \text{for Mode I } (\pi/2 \leq \alpha < 5\pi/6)
\end{aligned}
$$

$$(4.3)$$

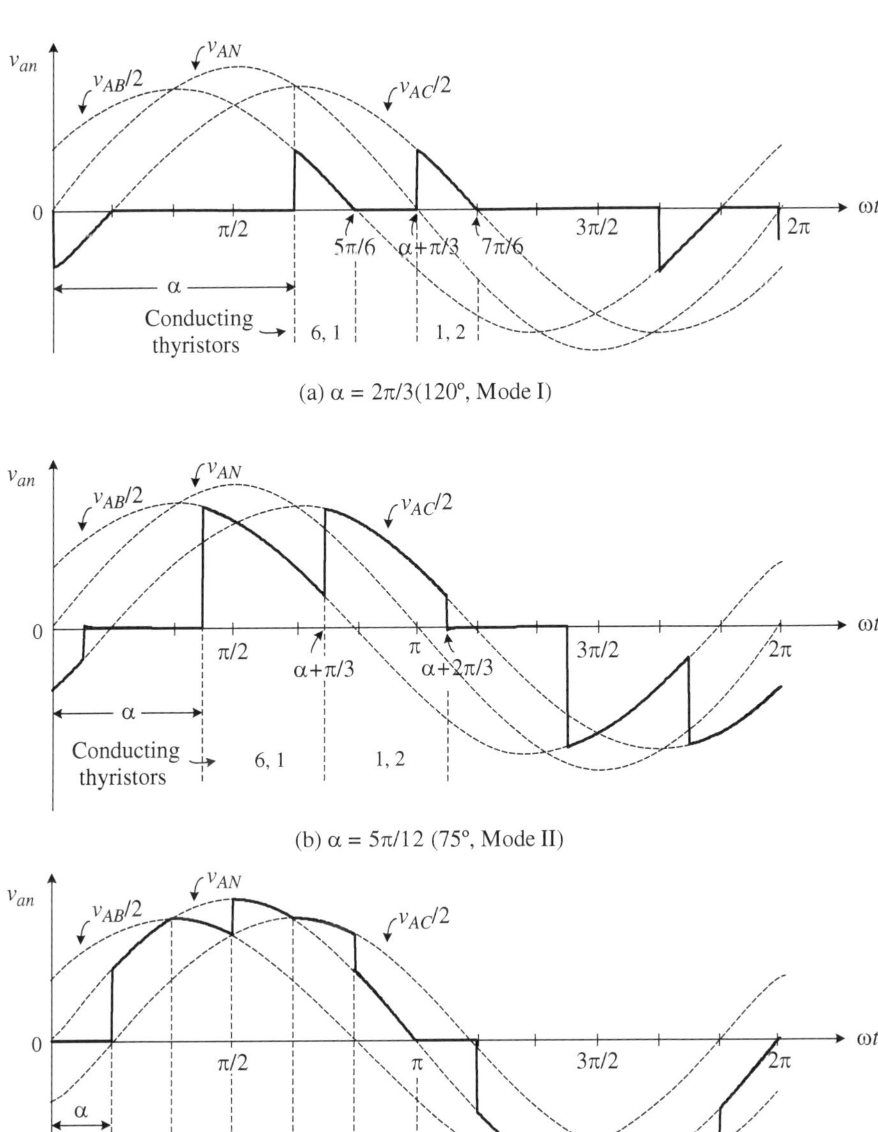

(a) $\alpha = 2\pi/3\,(120°,\ \text{Mode I})$

(b) $\alpha = 5\pi/12\ (75°,\ \text{Mode II})$

(c) $\alpha = \pi/6\,(30°,\ \text{Mode III})$

Figure 4-9. Typical waveforms of v_{an} when the three-phase controller operates in three different modes with a resistive load.

$$V_{an} = \left(\frac{1}{\pi} \left[\int_{\alpha}^{\alpha+\pi/3} \left(\frac{\sqrt{6}}{2} V_s \sin(\omega t + \pi/6) \right)^2 d(\omega t) + \int_{\alpha+\pi/3}^{\alpha+2\pi/3} \left(\frac{\sqrt{6}}{2} V_s \sin(\omega t - \pi/6) \right)^2 d(\omega t) \right] \right)^{1/2}$$

$$= V_s \left(\frac{1}{2} + \frac{3\sqrt{3}}{4\pi} \sin\left(2\alpha + \pi/6 \right) \right)^{1/2} \qquad \text{for Mod e II } (\pi/3 \le \alpha < \pi/2)$$

$$(4.4)$$

and

$$V_{an} = \left(\frac{1}{\pi} \left(\int_{\alpha}^{\pi/3} (\sqrt{2}V_s \sin \omega t)^2 d(\omega t) + \int_{\pi/3}^{\alpha+\pi/3} \left(\frac{\sqrt{6}}{2} V_s \sin(\omega t + \pi/6) \right)^2 d(\omega t) \right. \right.$$

$$+ \int_{\alpha+\pi/3}^{2\pi/3} (\sqrt{2}V_s \sin \omega t)^2 d(\omega t) +$$

$$\left. \left. \int_{2\pi/3}^{\alpha+2\pi/3} \left(\frac{\sqrt{6}}{2} V_s \sin(\omega t - \pi/6) \right)^2 d(\omega t) + \int_{\alpha+2\pi/3}^{\pi} (\sqrt{2}V_s \sin \omega t)^2 d(\omega t) \right) \right)^{1/2}$$

$$= V_s \left(1 - \frac{3\alpha}{2\pi} + \frac{3\sin 2\alpha}{4\pi} \right)^{1/2} \qquad \text{for Mode III } (0 \le \alpha < \pi/3)$$

$$(4.5)$$

where V_s is the rms value of the phase voltage of the power supply given by

$$v_{AN} = \sqrt{2}V_s \sin \omega t; \quad v_{BN} = \sqrt{2}V_s \sin \left(\omega t - 2\pi/3 \right); \text{ and } v_{CN} = \sqrt{2}V_s \sin \left(\omega t + 2\pi/3 \right) \quad (4.6)$$

The analysis of the three-phase AC voltage controller with inductive load is quite complex since the thyristors do not cease conducting when the supply voltage falls down to zero and becomes negative, the same phenomenon as discussed in the single-phase AC voltage controller. Computer simulation provides an effective means of obtaining the load voltage and current waveforms. Figure 4-10 shows simulated waveforms for phase-*a* load voltage v_{an}, line-to-line voltage v_{ab}, and load current i_a when the voltage controller operates with a three-phase, Y-connected RL load having a power factor of 0.9 at different delay angles. The waveform for the phase-*a* load current i_a is much smoother than its phase voltage v_{an} due to the filtering effect of the load inductance. The load power factor angle φ is equal to 25.8° as indicated in the figure.

With the pure inductive load ($\varphi = \pi/2$) the rms value of the load voltage v_{an} of the three-phase AC voltage controller is given by [4]

$$V_{an} = \begin{cases} V_s & \text{for } 0 \le \alpha < \pi/2 \\[2ex] V_s \left(\frac{5}{2} - \frac{3\alpha}{\pi} + \frac{3\sin(2\alpha)}{2\pi} \right)^{1/2} & \text{for } \pi/2 \le \alpha < 2\pi/3 \\[2ex] V_s \left(\frac{5}{2} - \frac{3\alpha}{\pi} + \frac{3\sin(2\alpha + \pi/3)}{2\pi} \right)^{1/2} & \text{for } 2\pi/3 \le \alpha < 5\pi/6 \end{cases} \qquad (4.7)$$

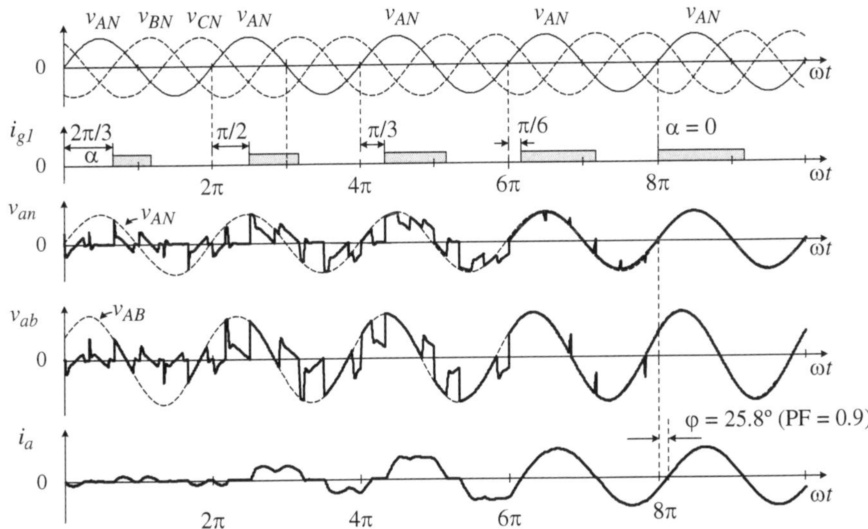

Figure 4-10. Waveforms of three-phase AC voltage controller with an RL load (cos $\varphi = 0.9$) and various delay angles. $\alpha = 2\pi/3$ (first cycle), $\pi/2$ (second cycle), $\pi/3$ (third cycle), $\pi/6$ (fourth cycle), and 0 (fifth cycle).

Based on Equations (4.3) to (4.7), the relationship between V_{an}/V_s and α is given in Figure 4-11 with the load power factor angle φ as a parameter. The other curves for $\varphi = 45°$, 60°, and 75° are obtained by computer simulations. It is noted that when the delay angle α is smaller than the load power factor angle φ, the load voltage V_{an} is equal to the supply voltage V_s and, therefore, is no longer adjustable, the same phenomenon as discussed in the single-phase AC voltage controller.

4.3 INTERLEAVED BOOST CONVERTERS

The DC/DC boost converter is one of the converter topologies often used in synchronous generator (SG) based wind energy conversion systems. As shown in Figure 4-1c, the converter is placed between the diode rectifier and the inverter of the power conversion system. The boost converter serves two main functions: tracking maximum power from the wind and boosting DC voltage to an appropriate value for the inverter. The second function facilitates the capture of maximum power from the wind at all wind speeds. For low- and medium-power wind energy systems of a few kilowatts to hundreds of kilowatts, a single-channel boost converter is often used.

In high-power megawatt wind energy systems, the current and voltage ratings can easily go beyond the range that one switching device can handle. Multiple switching devices connected in parallel or series can be a solution. However, extra measures

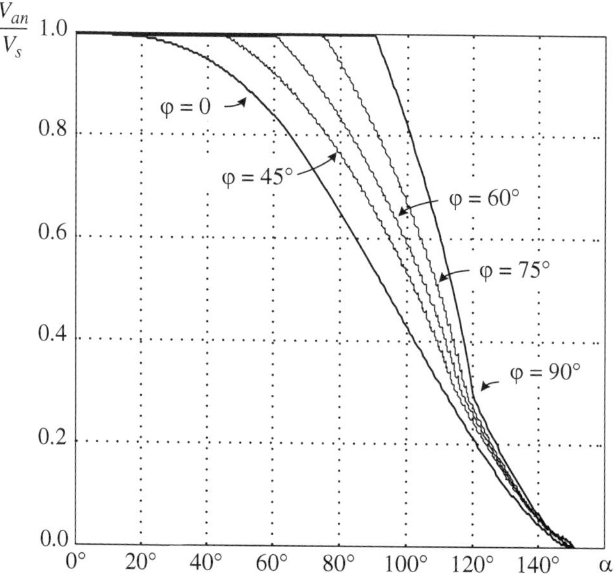

__Figure 4-11.__ Load voltage to supply voltage ratio V_{an}/V_s versus delay angle α for three-phase AC voltage controller.

should be taken for equal sharing of the current or voltage among the parallel or series devices. Instead of connecting the switching devices in parallel or series, paralleling or cascading power converters is another valid solution.

In the low-voltage (e.g., 690 V) megawatt wind energy systems, multichannel interleaved boost converters are often used to handle high currents in the system. An interleaved boost converter can be realized by interleaving (phase shifting) the gating signals for each of the parallel converters. One of the main benefits of the interleaved converter over the single converter is that the equivalent switching frequency of converter is increased. The equivalent switching frequency of the converter can be twice of the device switching frequency for a two-channel converter, or three times for a three-channel converter. The increase in the equivalent switching frequency in the interleaved converter offers a number of advantages over the single-channel converter, such as lower input current ripple and output voltage ripple, faster dynamic response, and better power handling capability.

The interleaved boost converter in the WECS normally employs an IGBT as a switching device instead of the MOSFET that is often used in low-power switch-mode power supplies. To reduce the switching losses, the IGBT operates at low switching frequencies of a few hundred hertz to a few kilohertz, whereas the MOSFET often operates at much higher switching frequencies.

In this section, the single-channel and multichannel interleaved boost converters are introduced. The operating principles of these converters are discussed, and their input current and output voltage ripples are analyzed.

4.3.1 Single-Channel Boost Converter

A boost converter is a power converter with an output DC voltage greater than its input DC voltage. A typical circuit diagram for the single-channel boost converter is shown in Figure 4-12. It is composed of a switch S_1, a diode D_1, a DC inductor L_1, and a filter capacitor C. It is assumed in the following analysis that (1) all the components in the converter are ideal (no power or voltage losses) and (2) the output filter capacitor C is very large and the output voltage of the converter is ripple free.

When switch S_1 is turned on, diode D_1 is reverse biased, and the output is isolated from the input. The input supplies energy to the inductor L_1. When the switch is turned off, diode D_1 is forward biased, and the energy stored in L_1 is released to the load through the diode. In this case, the output voltage v_o is the sum of the input voltage v_i and the inductor voltage v_{L1}, making the converter output voltage v_o higher than its input voltage v_i.

Depending on the continuity of the DC inductor current i_{L1}, the operation of the converter can be divided into two operating modes: continuous-current mode (CCM) and discontinuous-current mode (DCM). When a boost converter operates in CCM, the inductor current i_{L1} never falls to zero. Figure 4-13 shows the typical waveforms of currents and voltages in the boost converter operating in this mode.

In steady-state operation of the converter, the integral of the inductor voltage v_{L1} over time period T_s must be zero. This implies that the average voltage across the inductor L_1 over T_s is zero. Its graphical interpretation is that the area A_1 in Figure 4-13 must equal area A_2, that is,

$$V_i t_{\mathrm{on}} = (V_o - V_i)t_{\mathrm{off}} \tag{4.8}$$

from which

$$\frac{V_o}{V_i} = \frac{1}{1-D} \quad \text{for} \ \ 0 \le D < 1 \tag{4.9}$$

where D is duty cycle of the converter, defined by $D = t_{\mathrm{on}}/T_s$; T_s is the switching period; and t_{on} and t_{off} are the turn-on and turn-off times of the switch S, respectively. The

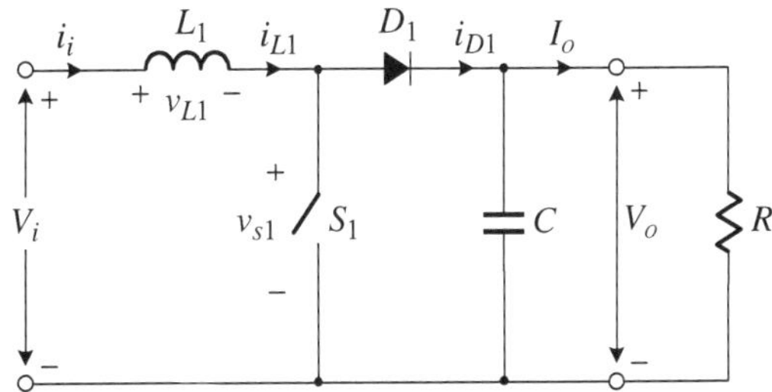

Figure 4-12. A simplified circuit for single-channel boost converter.

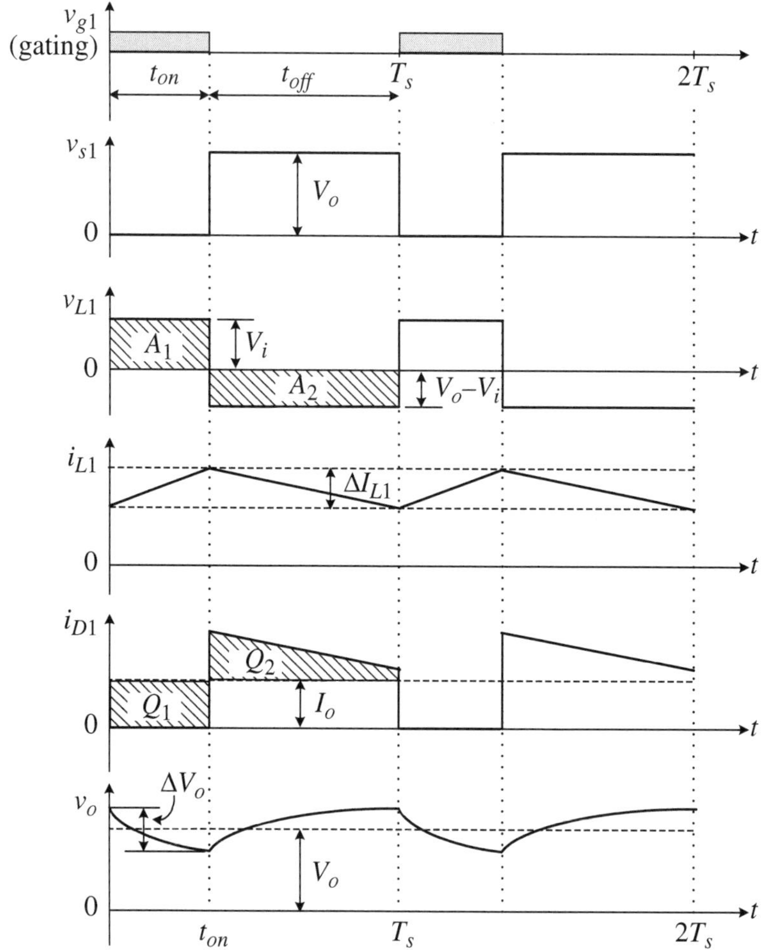

Figure 4-13. Waveforms of single-boost converter operating in a continuous current mode.

above expression indicates that the output voltage of the converter is always higher than its input voltage.

The relationship between the converter input current I_i and output current I_o can be derived from $V_i I_i = V_o I_o$, from which

$$\frac{I_o}{I_i} = 1 - D \quad \text{for} \quad 0 \le D < 1 \tag{4.10}$$

To calculate the ripple current in the inductor, differential equation $v_{L1} = L_1(di_{L1}/dt)$ can be replaced by a difference equation $\Delta v_{L1} = L_1(\Delta i_{L1}/\Delta t)$ since the inductor current changes linearly with time. For the t_{off} period in Figure 4-13, the inductor ripple current can be expressed by

$$\Delta i_{L1} = \frac{\Delta v_{L1}}{L_1}\Delta t = \frac{\left(V_o - V_i\right)}{L_1}t_{\text{off}} = D\left(1-D\right)\frac{V_o T_s}{L_1} \tag{4.11}$$

The maximum current ripple $\Delta I_{L1,\text{max}}$ for the single-channel boost converter occurs when the duty cycle D is 0.5, at which

$$\Delta I_{L1,\text{max}} = \frac{V_o T_s}{4L_1} \tag{4.12}$$

When the converter operates under light load conditions, the load current is low, and so is the current in the inductor L_1. The energy stored in the inductor during the t_{on} period may not be sufficient to maintain its current during the t_{off} period. Consequently, the inductor current i_{L1} reaches zero before the end of the t_{off} period and, therefore, becomes discontinuous. The converter thus operates in the discontinuous current mode. The inductor current at the boundary between the CCM and DCM is given by [5]

$$I_{LB} = D(1-D)\frac{V_o T_s}{2L_1} = \frac{\Delta i_{L1}}{2} \tag{4.13}$$

The maximum inductor boundary current occurs at $D = 0.5$, which can be calculated by

$$I_{LB,\text{max}} = \frac{V_o T_s}{8L_1} \tag{4.14}$$

The boundary output current can be found from

$$I_{oB} = D(1-D)^2 \frac{V_o T_s}{2L_1} = (1-D)\frac{\Delta i_{L1}}{2} \tag{4.15}$$

and its maximum value occurs at $D = 1/3$, and can be determined by [5]

$$I_{oB,\text{max}} = \frac{2}{27}\frac{V_o T_s}{L_1} \tag{4.16}$$

To calculate the output voltage ripple in the single-channel boost converter, we can look into the waveform of the current i_{D1} in the diode D_1 as shown in Figure 4-13. Assuming that all the ripple current component in D_1 is absorbed by the large output capacitor C, the capacitor C is discharged to the load during the t_{on} period when the diode is turned off, and charged during the t_{off} period when D_1 is on. The amount of charges, Q_1 during t_{on} and Q_2 during t_{off}, represented by the shaded areas should be equal. The peak-to-peak ripple voltage can then be calculated by

$$\Delta V_o = \frac{Q_1}{C} = \frac{I_o t_{on}}{C} = \frac{V_o D T_s}{RC} \tag{4.17}$$

from which

$$\frac{\Delta V_o}{V_o} = \frac{DT_s}{RC} \tag{4.18}$$

For a given load resistance R and filter capacitor C, the ripple voltage ΔV_o increases with the duty cycle D.

Typical waveforms of the converter operating in a discontinuous current mode are shown in Figure 4-14.

During the t_{on} period, the operation of the converter in the DCM is the same as that in the CCM. The current i_{L1} increases over time and energy is stored in L_1. During the t_{off} period, the inductor current i_{L1} falls to zero at the end of $K_1 T_s$ period, at which all the energy stored in L_1 during the t_{on} period is completely released. Since the average voltage across the inductor over switching period T_s is equal to zero, area A_1 in Figure 4-14 must equal A_2, that is,

$$V_i t_{on} = (V_o - V_i) K_1 T_s \tag{4.19}$$

from which

$$\frac{V_o}{V_i} = \frac{K_1 + D}{K_1} \qquad \text{for} \quad 0 \le D < 1 \tag{4.20}$$

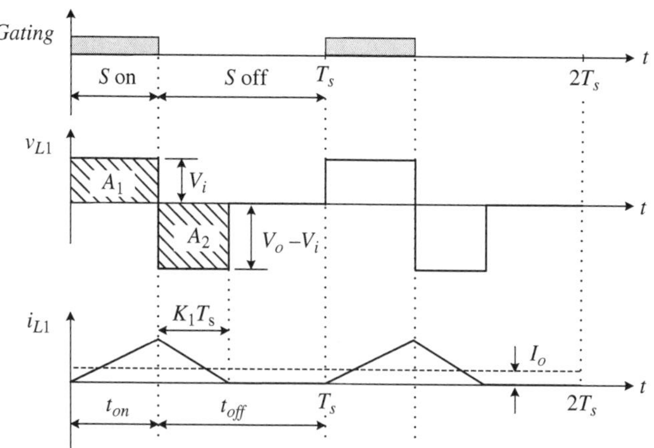

Figure 4-14. Waveforms of single-channel boost converter operating in a discontinuous current mode.

where K_1 can be calculated by [5]

$$K_1 = \frac{2L_1}{K_1 T_s D} I_o \tag{4.21}$$

4.3.2 Two-Channel Interleaved Boost Converter

The converter topology for a two-channel interleaved boost converter is shown in Figure 4-15 [6]. There are two parallel converter channels in the circuit. The first channel is composed of inductor L_1, switch S_1, and diode D_1, whereas the second channel consists of L_2, S_2, and D_2. The two converter channels are essentially connected in parallel but operate in an interleaved mode. They share the same filter capacitor C at the output. It is assumed that the parameters of the two channels are identical.

The gating arrangement and the inductor current waveforms of the converter are shown in Figure 4-16. With the interleaving design, the gating signals v_{g1} and v_{g2} for S_1 and S_2 are identical but shifted by $360°/N = 180°$, where N is the number of parallel converter channels. The operation and waveforms of individual converter channels are the same as those for the single-channel converter and, therefore, are not repeated here. Attention should be paid to the total input current i_i, which is the sum of the two inductor currents i_{L1} and i_{L2}. The input current i_i has the following characteristics:

- The average DC component of the input current (I_i) is twice that of the individual inductor ($I_i = I_{L1} + I_{L2}$). Since the input and output voltages for the parallel converters are the same, each channel handles only half of the total power of the load.

- The peak-to-peak input current ripple ΔI_i is smaller than that in the individual channels due to the use of the interleaved technique. This helps to reduce the volume of the input filter (not shown in the figure).

- The frequency of the input ripple current is twice that of the individual channels. In other words, the equivalent switching frequency of the interleaved converter is twice of that of each channel. For a given output voltage ripple, the capacitance of the output capacitor C can be reduced.

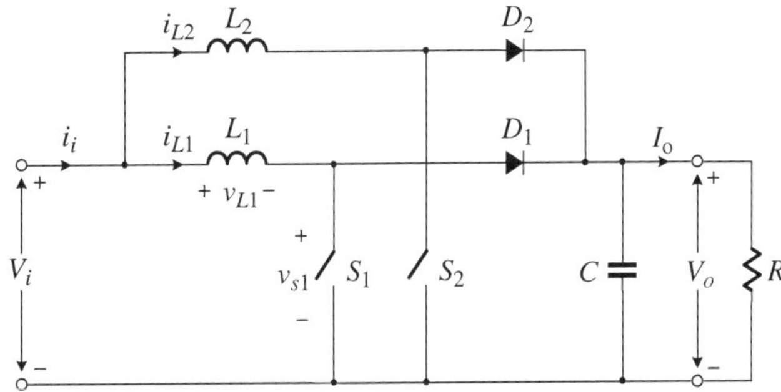

Figure 4-15. Two-channel interleaved boost converter.

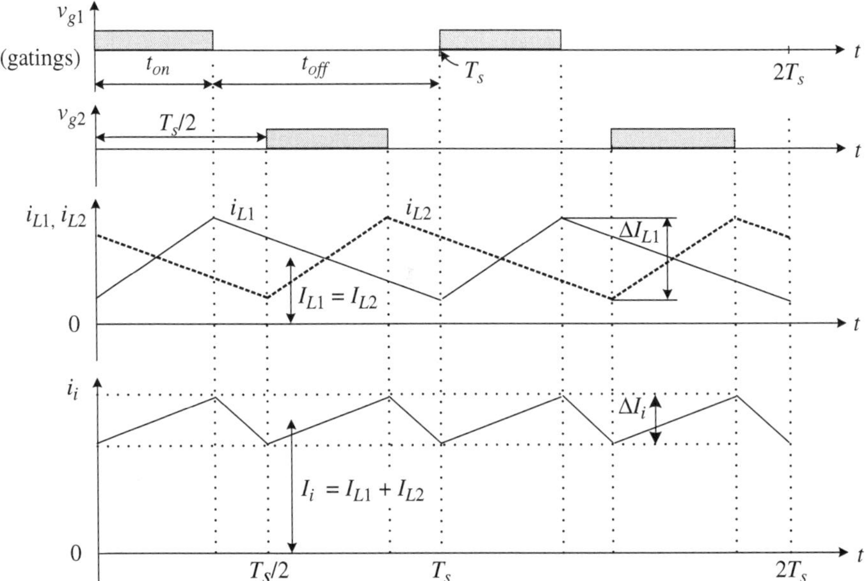

<u>Figure 4-16.</u> Waveforms for the analysis of input current ripple in a two-channel interleaved converter $(D < 0.5)$.

Input Current Ripple. It is interesting to note that when the duty cycle D increases from about 0.35 in Figure 4-16 to 0.5, the ripples in the two inductor currents i_{L1} and i_{L2} cancel each other and do not appear in i_i, that is, $\Delta I_i = 0$. The ripple current starts to increase when $D > 0.5$. Therefore, the analysis for the input current ripple can be carried out for the following two cases.

CASE 1: $0 < D \leq 0.5$. The waveforms in Figure 4-16 can be utilized to analyze the input current ripple ΔI_i in the two-channel converter. It is more convenient to perform the analysis for the t_{on} period, during which the total input current i_i increases monotonously. The input current ripple of the converter can be calculated by

$$\Delta I_i = (K_1 - K_2)t_{on} = (K_1 - K_2)DT_s \qquad (4.22)$$

where $K_1 = di_{L1}/dt$ and $K_2 = di_{L2}/dt$, which are the slopes of the inductor currents i_{L1} and i_{L2} during charging and discharging process, respectively. Thus,

$$\Delta I_i = \left(\frac{V_i}{L} - \frac{V_o - V_i}{L}\right)\left(1 - \frac{V_i}{V_o}\right)T_s = (1 - 2D)D\frac{V_o T_s}{L} \qquad (4.23)$$

where L is the inductance of the each converter channel, that is, $L = L_1 = L_2$.

The maximum input current ripple $\Delta I_{i,\max}$ can be determined by differentiating the above equation with respect to V_i:

$$\frac{\partial \Delta I_i}{\partial V_i} = \left(\frac{2V_i - V_o}{L}\right)\left(1 - \frac{V_i}{V_o}\right)T_s = 0 \tag{4.24}$$

from which

$$V_i = \frac{3}{4}V_o \text{ and } D = 0.25 \quad \text{for } \Delta I_i = \Delta I_{i,\max} \tag{4.25}$$

The maximum current ripple can be found by substituting Equation (4.25) into Equation (4.24):

$$\Delta I_{i,\max} = \frac{V_o T_s}{8L} \tag{4.26}$$

Compare the above equation with Equation (4.12), where the maximum current ripple in the two-channel boost converter is half of the single-channel converter for $D \leq 0.5$.

CASE 2: $0.5 < D < 1$. The waveforms for the analysis of the input current ripple in the two-channel converter are shown in Figure 4-17, where the duty cycle D is 0.65. It

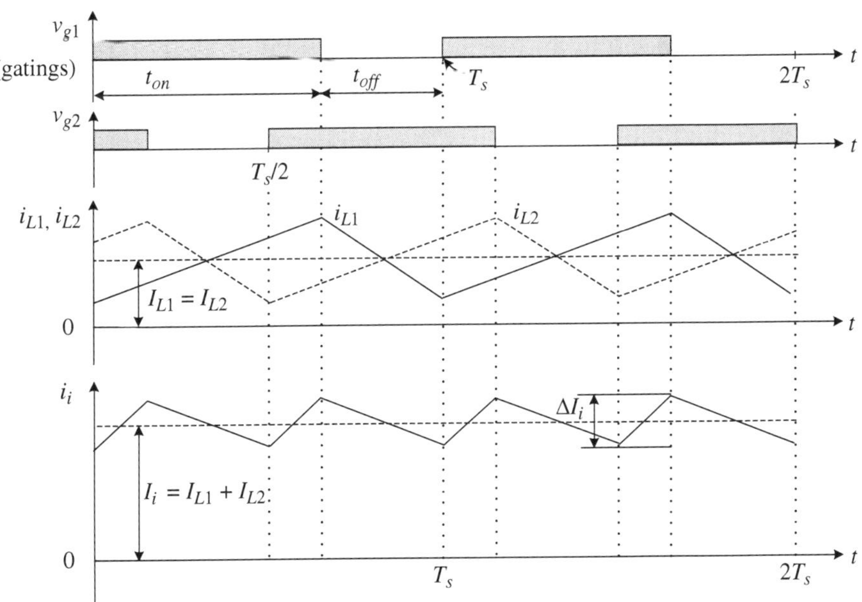

Figure 4-17. Waveforms for analysis of input ripple current in a two-channel interleaved converter ($D > 0.5$).

is more convenient to perform the analysis for the t_{off} period, during which the input current i_i decreases monotonously. The input current ripple can be determined by

$$\Delta I_i = \left(K_2 - K_1\right)t_{\text{off}} = \left(K_2 - K_1\right)\left(1 - D\right)T_s$$
$$= \left(\frac{V_o - V_i}{L} - \frac{V_i}{L}\right)\left(\frac{V_i}{V_o}\right)T_s = \left(2D - 1\right)\left(1 - D\right)\frac{V_o T_s}{L} \qquad (4.27)$$

Following the same procedure, the maximum input current ripple $\Delta I_{L1,\text{max}}$ can be determined, which is the same as that given in Equation (4.26) for the t_{off} period.

Based on Equations (4.10), (4.24), and (4.27), the input current ripple ΔI_i for the single-channel ($N = 1$) and two-channel ($N = 2$) converters versus duty cycle D is given in Figure 4-18 [6]. ΔI_i is normalized according to the maximum ripple current $\Delta I_{L1,\text{max}}$ in the single-channel converter given by Equation (4.11). It is shown that the magnitude of input ripple current for the two-channel converter is much lower than that in the single-channel converter. In particular, the ripple current for the two-channel converter becomes zero at $D = 0.5$, whereas it reaches its maximum value for the single-channel converter. Figure 4-19 shows the ratio of the input ripple current ΔI_i to the channel ripple current ΔI_L versus duty cycle D for the N-channel boost converters.

Output Voltage Ripple. As mentioned earlier, one of the benefits of the interleaved converters is the reduction of the output ripple. In the two-channel converter,

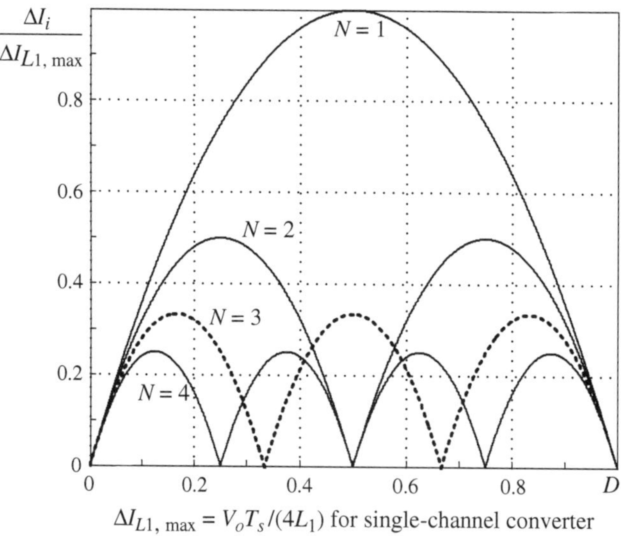

Figure 4-18. Normalized input ripple current versus duty cycle D for N-channel interleaved boost converters.

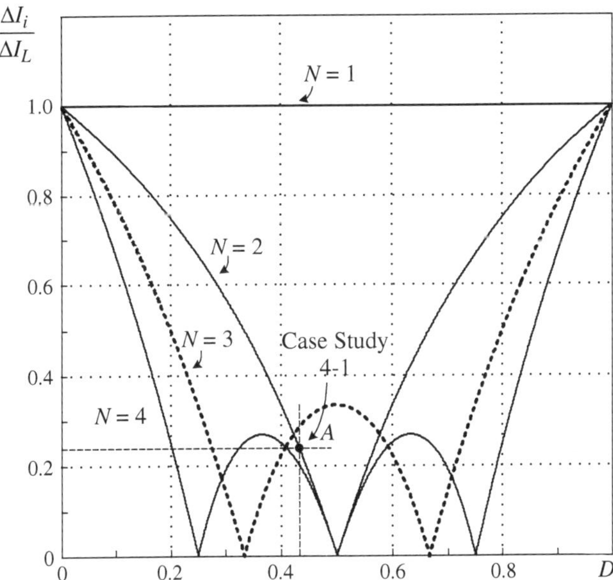

Figure 4-19. Ratio of the input ripple current ΔI_i to the channel ripple current ΔI_L versus duty cycle D for N-channel boost converters.

the two parallel converters share the same output capacitor C, which makes the analysis a little tedious. Computer simulation techniques can be used to determine the output voltage ripples. Figure 4-20 shows the results for the relative output ripple voltage for the two-channel converter under the assumption that the converter operates in a continuous-current mode [6]. The relative output ripple voltage for the single-stage converter is also given in the figure, which is calculated according to Equation (4.13). The two-channel converter produces much lower voltage ripple in comparison to the single-channel converter.

4.3.3 Multichannel Interleaved Boost Converters

Figure 4-21a shows the converter topology for a three-channel ($N = 3$) boost converter interleaved by $360°/N = 120°$. It is essentially composed of three single-channel converters connected in parallel and operating in the interleaving manner. The gate signals for the converters are identical except for a time delay of $T_s/3$ among the converters. The waveforms for the inductor currents, i_{L1}, i_{L2}, and i_{L3}, and the total input current i_L, are shown in Figure 4-21b. It can be observed that the frequency of the converter input current i_i is three times that of the individual converters.

Following the same procedure presented for the two-channel converters, the input current ripple in the three-channel converter can be derived and given by

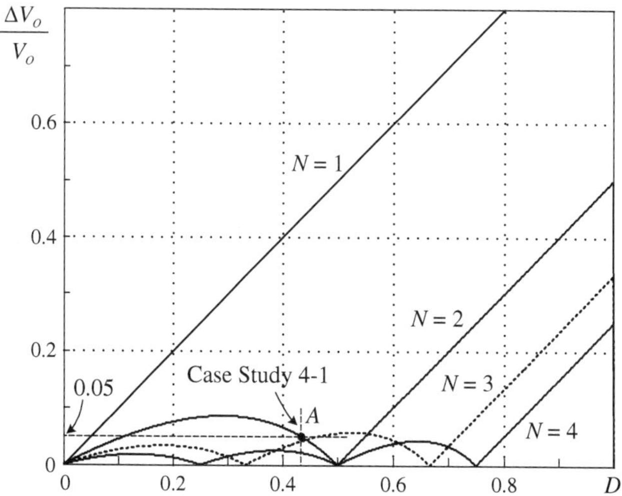

Figure 4-20. Relative output ripple voltages [$\times T_s/(RC)$] versus duty cycle D for N-channel boost converters.

$$\Delta I_i = \begin{cases} (1-3D)D\dfrac{V_o T_s}{L} & \text{for } 0 \le D < \dfrac{1}{3} \\[2ex] \left(3D(1-D)-\dfrac{2}{3}\right)\dfrac{V_o T_s}{L} & \text{for } \dfrac{1}{3} \le D < \dfrac{2}{3} \\[2ex] (3D-2)(1-D)\dfrac{V_o T_s}{L} & \text{for } \dfrac{2}{3} \le D < 1 \end{cases} \tag{4.28}$$

where L is the inductance of the each converter channel, that is, $L = L_1 = L_2 = L_3$.

The input current ripple ΔI_i and output voltage ripple ΔV_o for the three- and four-channel interleaved converters are illustrated in Figures 4-18 and 4-20, respectively. These current and voltage ripples are further reduced in comparison to those in the single- and two-channel converters. As a result, the size and cost of the input and output filters can be further reduced.

In practical wind energy conversion systems, the maximum power rating for each converter channel is around 500 kW to 600 kW. For a 1.5 MW WECS, three interleaved converters are required.

Case Study 4-1—PMSG Wind Energy System with a Two-Channel Interleaved Boost Converter.

A 1.2 MW/690 V permanent magnet synchronous-generator (PMSG) based wind energy conversion system is shown in Figure 4-22, where a two-channel interleaved boost converter and a PWM voltage source inverter are employed. The boost converter provides two main functions: (1) to boost its input DC voltage V_i to a higher DC voltage V_o at its output, and (2) to perform maximum

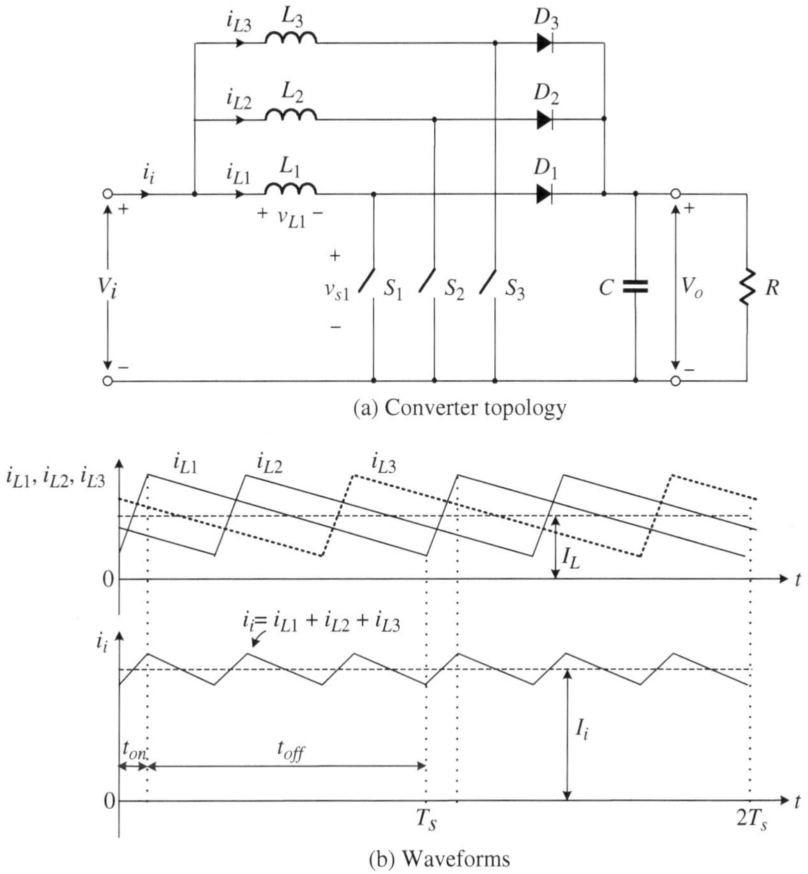

(a) Converter topology

(b) Waveforms

Figure 4-21. Converter topology and waveforms of three-channel interleaved boost converter.

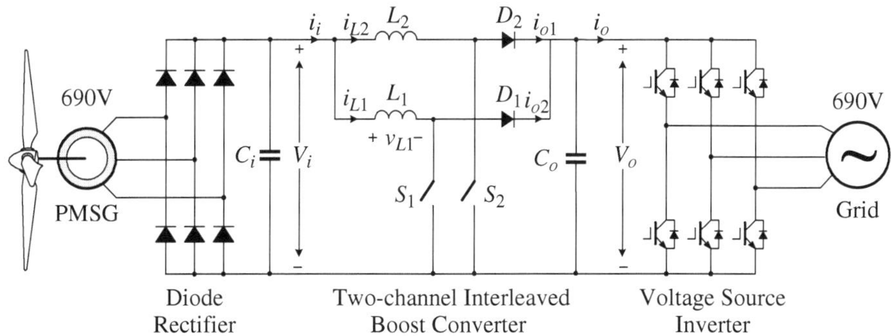

Figure 4-22. A wind energy conversion system with a two-channel interleaved boost converter.

power point tracking (MPPT) such that the system can deliver its maximum possible power captured by the turbine to the grid at any wind speed. The main function of the inverter is to keep its input DC voltage, which is the output of the boost converter, at a fixed value and also to control the reactive power to the grid.

At a given wind speed, the generator operates at a speed of 0.7 pu and delivers 412 kW (0.7^3 pu) power to the grid. The input voltage of the boost converter V_i is 680 V. The output voltage of the boost converter V_o is kept at a constant value of 1200 V by the inverter, which is the voltage required to deliver power to the grid of 690 V (refer to Section 4.7 for details). The inductance of the two-channel converter is $L = L_1 = L_2 = 270$ μH and capacitance of the output filter capacitor C_o is 300 μF. The boost converter operates at a switching frequency of 2 kHz. Assuming that all the converters are ideal without power losses, investigate the input current ripple and output voltage ripple of the boost converter.

Assuming that the boost converter operates in the continuous-current mode, the duty cycle of the each converter channel is

$$D = D_1 = D_2 = 1 - \frac{V_i}{V_o} = 1 - \frac{680}{1200} = 0.4333 \tag{4.29}$$

The ripple current in the two inductors can be calculated by

$$\Delta I_L = \Delta I_{L1} = \Delta I_{L2} = D(1-D)\frac{V_o T_s}{L} = 545.6 \text{ A} \tag{4.30}$$

The boundary inductor current relates the inductor ripple current by

$$I_{LB} = I_{LB1} = I_{LB2} = \Delta I_L / 2 = 272.8 \text{ A} \tag{4.31}$$

The boundary output current for each channel

$$I_{oB} = I_{oB1} = I_{oB2} = (1-D)I_{LB} = 154.6 \text{ A} \tag{4.32}$$

The total average output current of the interleaved converter is

$$I_o = P_o / V_o = 343.3 \text{ A} \tag{4.33}$$

from which the average output current of each channel is

$$I_{o1} = I_{o2} = I_o / 2 = 171.7 \text{ A} \tag{4.34}$$

Since $I_{o1} = I_{o2} > I_{oB}$, the converter operates in the continuous current mode.

The total input current of the boost converter is given by

$$I_i = \begin{cases} P_i / V_i = P_o / V_i = 412 \times 10^3 / 680 = 605.8 \text{ A} \\ I_o / (1-D) = 343.4 / (1-0.4333) = 605.8 \text{ A} \end{cases} \tag{4.35}$$

The percentage inductor ripple current in each channel can be found from

$$\frac{\Delta I_{L1}}{I_{L1}} = \frac{\Delta I_{L2}}{I_{L2}} = \frac{\Delta I_L}{I_i / 2} = \frac{545.6}{605.8 / 2} = 180.1\% \tag{4.36}$$

The total input current ripple

$$\Delta I_i = (1 - 2D)D \frac{V_o T_s}{L} = 128.4 \text{ A} \tag{4.37}$$

The total input current ripple can be found from

$$\frac{\Delta I_i}{I_i} = \frac{128.4}{606} = 21.2\% \tag{4.38}$$

which is much lower than the inductor ripple current of 180.1% in each of the channels.

The ratio of the total input ripple current ΔI_i to the inductor ripple current ΔI_L of each channel is

$$\frac{\Delta I_i}{\Delta I_L} = \frac{128.4}{545.6} = 23.5\% \tag{4.39}$$

which is verified by Point A in Figure 4-19.

To determine the output ripple voltage of the interleaved boost converter, it is assumed that the effect of the inverter operation on the DC voltage ripple is neglected. The load of the boost converter, which is the inverter, can be modeled by an equivalent resistance given by

$$R_{eq} = \frac{V_o}{I_o} = 3.495 \ \Omega \tag{4.40}$$

Making use of Figure 4-20, the percentage output ripple voltage can be obtained by

$$\frac{\Delta V_o}{V_o} = 0.05 \times \left(\frac{T_s}{R_{eq} C_o} \right) = 0.05 \times \frac{1/2000}{3.495 \times 300 \times 10^{-6}} = 2.38\% \tag{4.41}$$

If the operation of the two-channel converters were not interleaved, switches S_1 and S_2 would be turned on and off simultaneously. The output ripple voltage would then be

$$\frac{\Delta V_o}{V_o} = D \left(\frac{T_s}{R_{eq} C_o} \right) = 0.433 \times \frac{1/2000}{3.495 \times 300 \times 10^{-6}} = 20.6\% \tag{4.42}$$

which is around 8.6 times higher than that for the interleaved boost converter.

In summary, the multichannel interleaved converter produces much lower input current ripple and output voltage ripple in comparison to the single-channel boost converter.

4.4 TWO-LEVEL VOLTAGE SOURCE CONVERTERS

Figure 4-23a shows the simplified circuit diagram for a three-phase, two-level voltage source converter. The converter is composed of six switches, S_1 to S_6, with an antiparallel free-wheeling diode for each switch. The switches can be IGBT or IGCT devices, depending on the power and voltage ratings of the converter.

The converter has been widely used in industry for many different applications. When the converter transforms a fixed DC voltage to a three-phase AC voltage with variable magnitude and frequency for an AC load, as shown in Figure 4-23b, it is often called an inverter. When the converter transforms an AC grid voltage with fixed magnitude and frequency to an adjustable DC voltage for a DC load, it is normally known as an active rectifier or PWM rectifier. Whether it serves as an inverter or a rectifier, the power flow in the converter circuit is bidirectional: the power can flow from its DC side to the AC side, and vice versa.

In wind energy conversion systems, the converter is often connected to an electric grid, and delivers the power generated from the generator to the grid as shown in Figure 4-23d. The converter in this application is referred to as a grid-connected or grid-tied converter. It is also called an inverter since the converter normally delivers power from its DC side to the AC side.

This section focuses on pulse-width modulation (PWM) schemes for two-level voltage source converters. Since the modulation schemes are applicable to the converter that may operate as an inverter or a rectifier, the inverter is used as an example for the discussion. The section starts with an introduction to carrier-based sinusoidal PWM (SPWM) schemes, followed by a detailed analysis of space vector modulation (SVM) algorithms.

4.4.1 Sinusoidal PWM

The principle of the sinusoidal PWM scheme for the two-level converter is illustrated in Figure 4-24, where v_{ma}, v_{mb}, and v_{mc} are the three-phase sinusoidal modulating waveforms and v_{cr} is the triangular carrier signal. The fundamental-frequency component in the inverter output voltage can be controlled by the amplitude-modulation index:

$$m_a = \frac{\hat{V}_m}{\hat{V}_{cr}} \tag{4.43}$$

where $\hat{V}_m$ and $\hat{V}_{cr}$ are the peak values of the modulating and carrier waves, respectively. The amplitude-modulation index m_a is usually adjusted by varying $\hat{V}_m$ while keeping $\hat{V}_{cr}$ fixed. The frequency-modulation index is defined by

$$m_f = \frac{f_{cr}}{f_m} \tag{4.44}$$

where f_m and f_{cr} are the frequencies of the modulating and carrier waves, respectively.

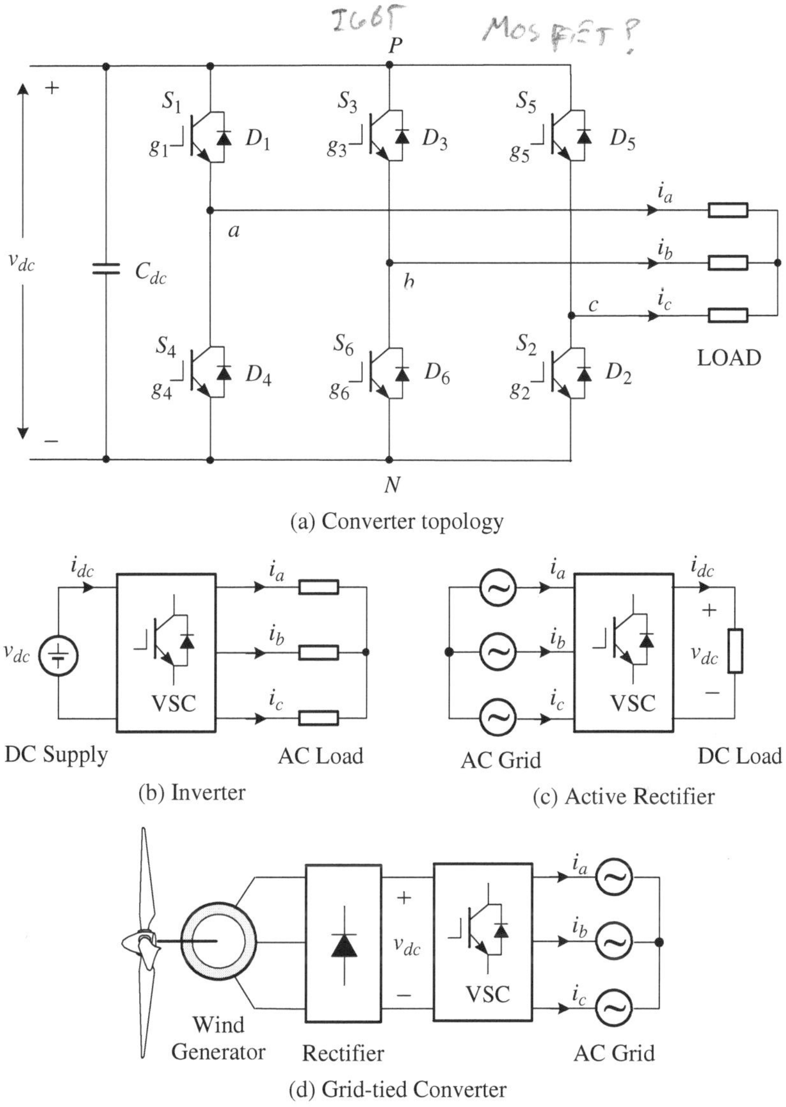

Figure 4-23. Simplified two-level voltage-source converter (VSC).

The operation of switches S_1 to S_6 is determined by comparing the modulating waves with the carrier wave. When $v_{ma} > v_{cr}$, the upper switch S_1 in inverter leg a is turned on. The lower switch S_4 operates in a complementary manner and thus is switched off. The resultant inverter terminal voltage v_{aN}, which is the voltage at the phase-a terminal with respect to the negative DC bus N, is equal to the DC voltage V_{dc}. When $v_{ma} < v_{cr}$, S_4 is on and S_1 is off, leading to $v_{aN} = 0$ as shown in Figure 4-24. Since the waveform of v_{aN} has only two levels, V_{dc} and 0, the inverter is often referred to as a two-level inverter. It is noted that to avoid possible short-circuiting during switching

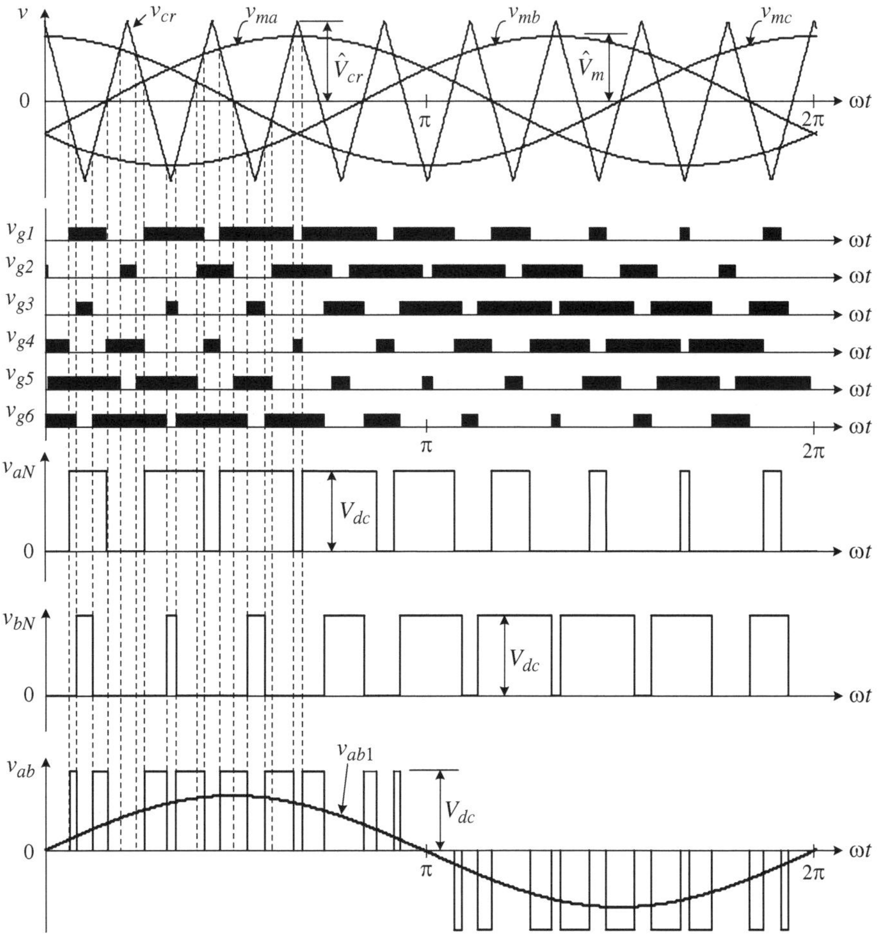

Figure 4-24. Sinusoidal pulse-width modulation (SPWM).

transients of the upper and lower devices in an inverter leg, a blanking time (or dead time) should be implemented, during which both switches are turned off.

The inverter line-to-line voltage v_{ab} can be determined by $v_{ab} = v_{aN} - v_{bN}$. The waveform of its fundamental-frequency component v_{ab1} is also given in the figure. The magnitude and frequency of v_{ab1} can be independently controlled by m_a and f_m, respectively.

The switching frequency of the active switches in the two-level inverter can be found from $f_{sw} = f_{cr} = f_m \times m_f$. For instance, v_{aN} in Figure 4-24 contains nine pulses per cycle of the fundamental frequency. Each pulse is produced by turning S_1 on and off once. With the fundamental frequency of 60 Hz, the resultant switching frequency for S_1 is $f_{sw} = 60 \times 9 = 540$ Hz, which is also the carrier frequency f_{cr}. It is worth noting that the device switching frequency may not always be equal to the carrier frequency in multilevel inverters. This issue will be addressed in the later sections.

When the carrier wave is synchronized with the modulating wave (m_f is an integer), the modulation scheme is known as synchronous PWM, in contrast to asynchronous PWM, whose carrier frequency f_{cr} is usually fixed and independent of f_m. The asynchronous PWM features a fixed switching frequency and easy implementation with analog circuits. However, it may generate noncharacteristic harmonics, whose frequency is not a multiple of the fundamental frequency. The synchronous PWM scheme is more suitable for implementation with a digital processor.

Case Study 4-2—Harmonic Analysis of Two-Level VSI with Carrier-Based Sinusoidal Modulation. The purpose of this case study is to investigate the harmonic characteristics of the two-level voltage source inverter with carrier-based sinusoidal modulation. Figure 4-25 shows a set of simulated waveforms for the two-level inverter, where v_{ab} is the inverter line-to-line voltage, v_a is the phase-*a* load voltage, and i_a is the phase-*a* load current. The inverter operates under the condition of $m_a =$

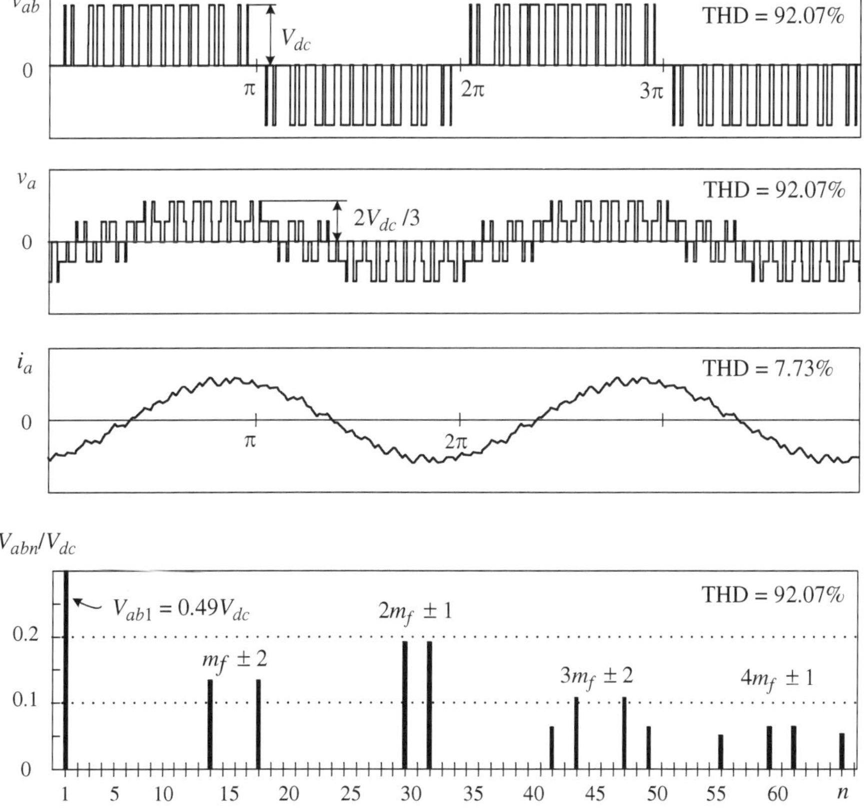

Figure 4-25. Simulated waveforms for the two-level inverter operating at $m_a = 0.8$ and $m_f = 15$, and $f_m = 60$ Hz and $f_{sw} = 900$ Hz.

0.8, $m_f = 15$, $f_m = 60$ Hz, and $f_{sw} = 900$ Hz with a rated three-phase inductive load. The load power factor is 0.9 per phase. It can be observed that

- All the harmonics in v_{ab} with order lower than $(m_f - 2)$ are eliminated.
- The harmonics are centered around m_f and its multiples, such as $2m_f$ and $3m_f$.

The above statements are valid for $m_f \geq 9$ with a m_f multiple of 3.

The waveform of the load current i_a is close to sinusoidal with a total harmonic distortion (THD) of 7.73%. The low amount of harmonic distortion is due to the elimination of low-order harmonics by the modulation scheme and the filtering effect of the load inductance.

Figure 4-26 shows the harmonic content of the inverter line-to-line voltage v_{ab} normalized to its DC voltage V_{dc} as a function of m_a, where V_{abn} is the nth-order harmonic voltage (rms). The fundamental-frequency component V_{ab1} increases linearly with m_a, whose maximum value can be found from

$$V_{ab1,\max} = 0.612\,V_d \qquad \text{for } m_a = 1 \tag{4.45}$$

The THD curve for v_{ab} is also given in the figure.

4.4.2 Space Vector Modulation

Space vector modulation (SVM) is one of the real-time modulation techniques and is widely used for digital control of voltage source inverters. This section presents the principle and implementation of the space vector modulation for the two-level inverter.

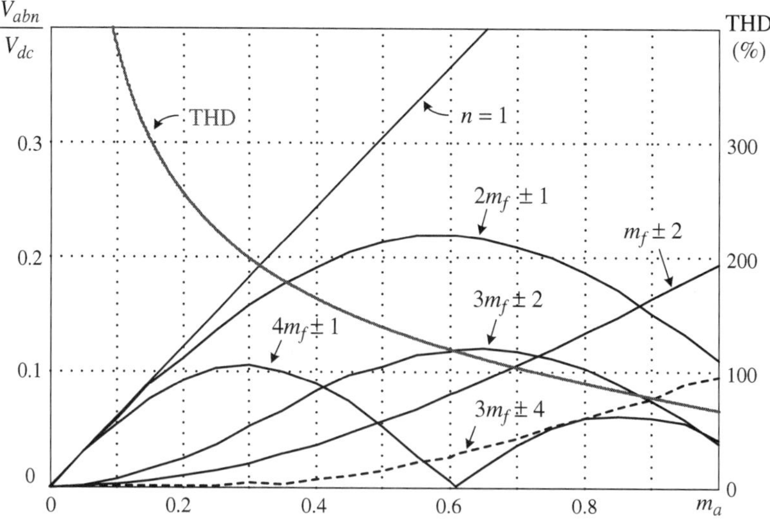

Figure 4-26. Harmonic content of v_{ab} in Figure 4-25.

Switching States. The operating status of the switches in the two-level inverter in Figure 4-23a can be represented by switching states. As indicated in Table 4-1, switching state P denotes that the upper switch in an inverter leg is on and the inverter terminal voltage (v_{aN}, v_{bN}, or v_{cN}) is positive ($+V_{dc}$), whereas O indicates that the inverter terminal voltage is zero due to the conduction of the lower switch.

There are eight possible combinations of switching states in the two-level inverter as listed in Table 4-2. The switching state [POO], for example, corresponds to the conduction of S_1, S_6, and S_2 in the inverter legs a, b, and c, respectively. Among the eight switching states, [PPP] and [OOO] are zero states and the others are active states.

Space Vectors. The active and zero switching states can be represented by active and zero space vectors, respectively. A typical space vector diagram for the two-level inverter is shown in Fig. 4-27, where the six active vectors $\vec{V}_1$ to $\vec{V}_6$ form a regular hexagon with six equal sectors (I to VI). The zero vector $\vec{V}_0$ lies at the center of the hexagon.

To derive the relationship between the space vectors and switching states, refer to the two-level inverter in Figure 4-23a. Assuming that the operation of the inverter is three-phase balanced, we have

$$v_a(t) + v_b(t) + v_c(t) = 0 \tag{4.46}$$

where v_a, v_b, and v_c are the instantaneous load phase voltages. From a mathematical point of view, one of the phase voltages is redundant since, given any two phase voltages, the third one can be readily calculated. Therefore, it is possible to transform the three-phase variables to two-phase variables through the $abc/\alpha\beta$ transformation presented in Chapter 3:

$$\begin{bmatrix} v_\alpha(t) \\ v_\beta(t) \end{bmatrix} = \frac{2}{3} \begin{bmatrix} 1 & -\dfrac{1}{2} & -\dfrac{1}{2} \\ 0 & \dfrac{\sqrt{3}}{2} & -\dfrac{\sqrt{3}}{2} \end{bmatrix} \begin{bmatrix} v_a(t) \\ v_b(t) \\ v_c(t) \end{bmatrix} \tag{4.47}$$

A space vector can be generally expressed in terms of the two-phase voltages in the α-β frame:

$$\vec{v}(t) = v_\alpha(t) + j v_\beta(t) \tag{4.48}$$

Table 4-1. Definition of switching states

Switching state	Leg a			Leg b			Leg c		
	S_1	S_4	v_{aN}	S_3	S_6	v_{bN}	S_5	S_2	v_{cN}
P	On	Off	V_{dc}	On	Off	V_{dc}	On	Off	V_{dc}
O	Off	On	0	Off	On	0	Off	On	0

Table 4-2. Space vectors, switching states, and on-state switches

Space vector		Switching state (three phases)	on-state switch	Vector definition
Zero vector	$\vec{V}_0$	[PPP]	S_1, S_3, S_5	$\vec{V}_0 = 0$
		[OOO]	S_4, S_6, S_2	
Active vector	$\vec{V}_1$	[POO]	S_1, S_6, S_2	$\vec{V}_1 = \dfrac{2}{3} V_{dc}\, e^{j0}$
	$\vec{V}_2$	[PPO]	S_1, S_3, S_2	$\vec{V}_2 = \dfrac{2}{3} V_{dc}\, e^{j\frac{\pi}{3}}$
	$\vec{V}_3$	[OPO]	S_4, S_3, S_2	$\vec{V}_3 = \dfrac{2}{3} V_{dc}\, e^{j\frac{2\pi}{3}}$
	$\vec{V}_4$	[OPP]	S_4, S_3, S_5	$\vec{V}_4 = \dfrac{2}{3} V_{dc}\, e^{j\frac{3\pi}{3}}$
	$\vec{V}_5$	[OOP]	S_4, S_6, S_5	$\vec{V}_5 = \dfrac{2}{3} V_{dc}\, e^{j\frac{4\pi}{3}}$
	$\vec{V}_6$	[POP]	S_1, S_6, S_5	$\vec{V}_6 = \dfrac{2}{3} V_{dc}\, e^{j\frac{5\pi}{3}}$

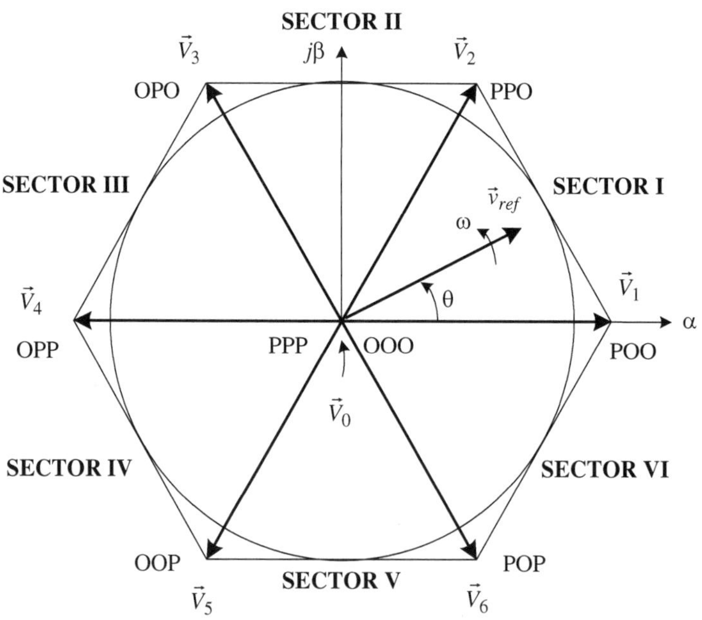

Figure 4-27. Space-vector diagram for the two-level inverter.

Substituting (4.47) into (4.48), we have

$$\vec{v}(t) = \frac{2}{3}\left[v_a(t)e^{j0} + v_b(t)e^{j2\pi/3} + v_c(t)e^{j4\pi/3} \right] \tag{4.49}$$

where $e^{jx} = \cos x + j \sin x$ and $x = 0$, $2\pi/3$ or $4\pi/3$. For the active switching state [POO], the generated load phase voltages are

$$v_a(t) = \frac{2}{3}V_{dc}, \; v_b(t) = -\frac{1}{3}V_{dc}, \text{ and } \; v_c(t) = -\frac{1}{3}V_{dc} \tag{4.50}$$

The corresponding space vector, denoted as $\vec{V}_1$, can be obtained by substituting (4.50) into (4.49):

$$\vec{V}_1 = \frac{2}{3}V_{dc}\,e^{j0} \tag{4.51}$$

Following the same procedure, all six active vectors can be derived:

$$\vec{V}_k = \frac{2}{3}V_{dc}\,e^{j(k-1)\frac{\pi}{3}} \qquad k = 1, 2, \dots , 6 \tag{4.52}$$

The zero vector $\vec{V}_0$ has two switching states [PPP] and [OOO], one of which is redundant. As will be seen later, the redundant switching state can be utilized to minimize the switching frequency of the inverter or perform other useful functions. The relationship between the space vectors and their corresponding switching states is given in Table 4-2.

Note that the zero and active vectors do not move in space and, thus, are referred to as stationary vectors. On the contrary, the reference vector $\vec{v}_{ref}$ in Figure 4-27 rotates in space at an angular velocity

$$\omega = 2\pi f \tag{4.53}$$

where f is the fundamental frequency of the inverter output voltage. The angular displacement between $\vec{v}_{ref}$ and the α-axis of the α-β frame can be obtained by

$$\theta(t) = \int_0^t \omega(t)\, dt + \theta_0 \tag{4.54}$$

For a given magnitude (length) and position, $\vec{v}_{ref}$ can be synthesized by three nearby stationary vectors, based on which the switching states of the inverter can be selected and gate signals for the active switches can be generated. When $\vec{v}_{ref}$ passes through the sectors one by one, different sets of switches will be turned on or off. As a result, when $\vec{v}_{ref}$ rotates one revolution in space, the inverter output voltage varies one cycle over time. The inverter output frequency corresponds to the rotating speed of $\vec{v}_{ref}$, whereas its output voltage can be adjusted by the magnitude of $\vec{v}_{ref}$.

Dwell Time Calculation. As mentioned earlier, the reference $\vec{v}_{ref}$ can be synthesized by three stationary vectors. The dwell time for the stationary vectors essentially represents the duty-cycle time (on-state or off-state time) of the chosen switches during a sampling period T_s. The dwell time calculation is based on the volt-second balancing principle, that is, the product of the reference voltage $\vec{v}_{ref}$ and sampling period T_s equals the sum of the voltage multiplied by the time interval of chosen space vectors.

Assuming that the sampling period T_s is sufficiently small, the reference vector $\vec{v}_{ref}$ can be considered constant during T_s. Under this assumption, $\vec{v}_{ref}$ can be approximated by two adjacent active vectors and one zero vector. For example, when $\vec{v}_{ref}$ falls into sector I, as shown in Figure 4-28, it can be synthesized by $\vec{V}_1$, $\vec{V}_2$, and $\vec{V}_0$. The volt-second balancing equation is

$$\begin{cases} \vec{v}_{ref}\, T_s = \vec{V}_1\, T_a + \vec{V}_2\, T_b + \vec{V}_0\, T_0 \\ T_s \;=\; T_a + T_b + T_0 \end{cases} \tag{4.55}$$

where T_a, T_b, and T_0 are the dwell times for the vectors $\vec{V}_1$, $\vec{V}_2$, and $\vec{V}_0$, respectively. The space vectors in Equation (4.52) can be expressed as

$$\vec{v}_{ref} = v_{ref}\, e^{j\theta}, \;\; \vec{V}_1 = \frac{2}{3} V_{dc}, \;\; \vec{V}_2 = \frac{2}{3} V_{dc}\, e^{j\frac{\pi}{3}}, \text{ and } \vec{V}_0 = 0 \tag{4.56}$$

where v_{ref} represents the magnitude of the reference vector and θ is the angle between $\vec{v}_{ref}$ and the α-axis of the α-β frame as shown in Figure 4-27.

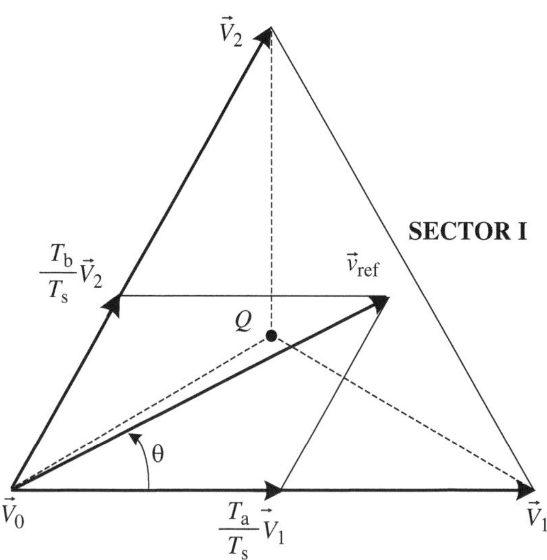

Figure 4-28. $\vec{v}_{ref}$ synthesized by $\vec{V}_1, \vec{V}_2$, and $\vec{V}_0$.

Substituting (4.56) into (4.55) and then splitting the resultant equation into the real (α-axis) and imaginary (β-axis) components in the α-β frame, we have

$$
\begin{cases}
\text{Re}: & v_{ref}(\cos\theta)T_s = \dfrac{2}{3}V_{dc}T_a + \dfrac{1}{3}V_{dc}T_b \\[3mm]
\text{Im}: & v_{ref}(\sin\theta)T_s = \dfrac{1}{\sqrt{3}}V_{dc}T_b
\end{cases}
\tag{4.57}
$$

Solving (4.57) together with $T_s = T_a + T_b + T_0$ yields

$$
\begin{cases}
T_a = \dfrac{\sqrt{3}\,T_s\,v_{ref}}{V_{dc}}\sin\left(\dfrac{\pi}{3}-\theta\right) \\[4mm]
T_b = \dfrac{\sqrt{3}\,T_s\,v_{ref}}{V_{dc}}\sin\theta \qquad\qquad \text{for } 0 \le \theta < \pi/3 \\[4mm]
T_0 = T_s - T_a - T_b
\end{cases}
\tag{4.58}
$$

To visualize the relationship between the location of $\vec{v}_{ref}$ and the dwell times, let us examine some special cases. If $\vec{v}_{ref}$ lies exactly in the middle between $\vec{V}_1$ and $\vec{V}_2$ (i.e., $\theta = \pi/6$), the dwell time T_a of $\vec{V}_1$ will be equal to T_b of $\vec{V}_2$. When $\vec{v}_{ref}$ is closer to $\vec{V}_2$, T_b will be greater than T_a. If $\vec{v}_{ref}$ is coincident with $\vec{V}_2$, T_a will be zero. With the head of $\vec{v}_{ref}$ located right on the central point Q, $T_a = T_b = T_0$. The relationship between the $\vec{v}_{ref}$ location and dwell times is summarized in Table 4-3.

Note that although Equation (4.58) is derived when $\vec{v}_{ref}$ is in sector I, it can also be used when $\vec{v}_{ref}$ is in other sectors, provided that a multiple of $\pi/3$ is subtracted from the actual angular displacement θ such that the modified angle θ' falls into the range between zero and $\pi/3$ for use in the equation, that is,

$$
\theta' = \theta - (k-1)\pi/3 \qquad \text{for } 0 \le \theta' < \pi/3
\tag{4.59}
$$

where $k = 1, 2, \ldots, 6$ for sectors I, II, $\ldots$, VI, respectively. For example, when $\vec{v}_{ref}$ is in sector II, the calculated dwell times T_a, T_b, and T_0 based on Equation (4.58) are for vectors $\vec{V}_2$, $\vec{V}_3$, and $\vec{V}_0$, respectively.

Modulation Index. Equation (4.58) can also be expressed in terms of modulation index m_a:

Table 4-3. $\vec{v}_{ref}$ location and dwell times

$\vec{v}_{ref}$ Location	$\theta = 0$	$0 < \theta < \dfrac{\pi}{6}$	$\theta = \dfrac{\pi}{6}$	$\dfrac{\pi}{6} < \theta < \dfrac{\pi}{3}$	$\theta = \dfrac{\pi}{3}$
Dwell times	$T_a > 0$ $T_b = 0$	$T_a > T_b$	$T_a = T_b$	$T_a < T_b$	$T_a = 0$ $T_b > 0$

$$\begin{cases} T_a = T_s m_a \sin\left(\dfrac{\pi}{3} - \theta\right) \\[2mm] T_b = T_s m_a \sin\theta \\[2mm] T_0 = T_s - T_a - T_b \end{cases} \tag{4.60}$$

where

$$m_a = \frac{\sqrt{3}\, v_{ref}}{V_{dc}} \tag{4.61}$$

The length of the reference vector $\vec{v}_{ref}$ represents the peak value of the fundamental-frequency component in the inverter output phase voltage, that is,

$$v_{ref} = \hat{V}_{a1} = \sqrt{2} V_{a1} \tag{4.62}$$

where V_{a1} is the rms value of the fundamental component in the inverter output phase (phase-a) voltage.

Substituting (4.62) into (4.61), one can find the relationship between m_a and V_{a1}:

$$m_a = \frac{\sqrt{3}\, v_{ref}}{V_{dc}} = \frac{\sqrt{6}\, V_{a1}}{V_{dc}} \tag{4.63}$$

For a given DC voltage V_{dc}, the inverter output voltage V_{a1} is proportional to modulation index m_a.

The maximum length of the reference vector, $v_{ref,\max}$, corresponds to the radius of the largest circle that can be inscribed within the hexagon, as shown in Figure 4-27. Since the hexagon is formed by six active vectors having a length of $2V_{dc}/3$, $v_{ref,\max}$ can be found from

$$v_{ref,\max} = \frac{2}{3} V_{dc} \times \frac{\sqrt{3}}{2} = \frac{V_{dc}}{\sqrt{3}} \tag{4.64}$$

Substituting (4.64) into (4.61) gives the maximum modulation index

$$m_{a,\max} = 1 \tag{4.65}$$

from which the modulation index for the SVM scheme is in the range of

$$0 \leq m_a \leq 1 \tag{4.66}$$

Switching Sequence. With the space vectors selected and their dwell times calculated, the next step is to arrange the switching sequence. In general, the switching sequence design for a given $\vec{v}_{ref}$ is not unique, but it should satisfy the following two requirements for the minimization of the device switching frequency:

1. The transition from one switching state to the next involves only two switches in the same inverter leg, one being switched on and the other switched off.
2. The transition for $\vec{v}_{ref}$ moving from one sector in the space vector diagram to the next requires no or a minimum number of switchings.

Figure 4-29 shows a typical seven-segment switching sequence and inverter output voltage waveforms for $\vec{v}_{ref}$ in sector I, where $\vec{v}_{ref}$ is synthesized by $\vec{V}_1$, $\vec{V}_2$, and $\vec{V}_0$. The sampling period T_s is divided into seven segments for the selected vectors. It can be observed that

- The dwell times for the seven segments add up to the sampling period ($T_s = T_a + T_b + T_0$).
- Design requirement (1) is satisfied. For instance, the transition from [OOO] to [POO] is accomplished by turning S_1 on and S_4 off, which involves only two switches.
- The redundant switching sates for $\vec{V}_0$ are utilized to reduce the number of switchings per sampling period. For the $T_0/2$ segment in the center of the sampling period, the switching state [PPP] is selected, whereas for the $T_0/4$ segments on both sides, the state [OOO] is used.
- Each of the switches in the inverter turns on and off once per sampling period. The switching frequency f_{sw} of the devices is thus equal to the sampling frequency f_{sp}, that is, $f_{sw} = f_{sp} = 1/T_s$.

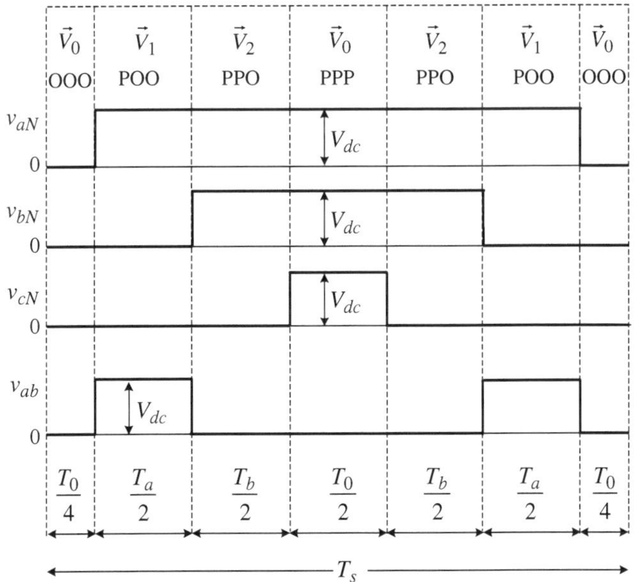

Figure 4-29. Seven-segment switching sequence for $\vec{v}_{ref}$ in sector I.

4.4.3 Harmonic Analysis

The harmonic performance of the SVM scheme for the two-level voltage source inverter is analyzed through the following case study.

Case Study 4-3—Two-Level Voltage Source Inverter with Space Vector Modulation. In this case study, the procedure for computer simulation and real-time digital implementation of the SVM scheme is introduced. Simulation for the two-level voltage source inverter is performance-based, and harmonic performance of the SVM scheme is analyzed.

Figure 4-30 shows a block diagram for computer simulation and digital implementation of the SVM scheme. The input variables, v_a^*, v_b^*, and v_c^*, are the three-phase reference voltages, which are also the required output-phase voltages of the inverter. The reference voltages are normally generated by the controller in a wind energy conversion system. Through the $abc/\alpha\beta$ transformation, the three-phase reference voltages in the abc stationary frame are transformed into two-phase variables, v_β and v_β, in the α-β stationary frame, from which the reference vector for the SVM scheme is established:

$$\vec{v}_{ref} = v_{ref} e^{j\theta} \tag{4.67}$$

where

$$\begin{cases} v_{ref} = \sqrt{(v_\alpha)^2 + (v_\beta)^2} \\[2ex] \theta = \tan^{-1}\dfrac{v_\beta}{v_\alpha} \end{cases} \tag{4.68}$$

With the reference vector in place, the modulation index m_a and sector number can be calculated by Equations (4.61) and (4.59), the dwell times can be determined by Equation (4.60), and the switching sequence can be designed according to Table 4-4. Finally, the gate signals for the six switches in the inverter can be generated. With the SVM

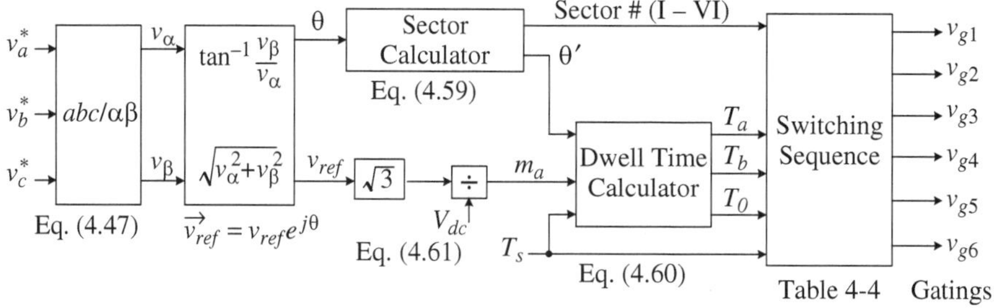

Figure 4-30. Block diagram for computer simulation and real-time digital implementation of the SVM algorithm.

Table 4-4. Seven-segment switching sequence

Sector	Switching segment						
	1	2	3	4	5	6	7
I	$\vec{V}_0$	$\vec{V}_1$	$\vec{V}_2$	$\vec{V}_0$	$\vec{V}_2$	$\vec{V}_1$	$\vec{V}_0$
	OOO	POO	PPO	PPP	PPO	POO	OOO
II	$\vec{V}_0$	$\vec{V}_3$	$\vec{V}_2$	$\vec{V}_0$	$\vec{V}_2$	$\vec{V}_3$	$\vec{V}_0$
	OOO	OPO	PPO	PPP	PPO	OPO	OOO
III	$\vec{V}_0$	$\vec{V}_3$	$\vec{V}_4$	$\vec{V}_0$	$\vec{V}_4$	$\vec{V}_3$	$\vec{V}_0$
	OOO	OPO	OPP	PPP	OPP	OPO	OOO
IV	$\vec{V}_0$	$\vec{V}_5$	$\vec{V}_4$	$\vec{V}_0$	$\vec{V}_4$	$\vec{V}_5$	$\vec{V}_0$
	OOO	OOP	OPP	PPP	OPP	OOP	OOO
V	$\vec{V}_0$	$\vec{V}_5$	$\vec{V}_6$	$\vec{V}_0$	$\vec{V}_6$	$\vec{V}_5$	$\vec{V}_0$
	OOO	OOP	POP	PPP	POP	OOP	OOO
VI	$\vec{V}_0$	$\vec{V}_1$	$\vec{V}_6$	$\vec{V}_0$	$\vec{V}_6$	$\vec{V}_1$	$\vec{V}_0$
	OOO	POO	POP	PPP	POP	POO	OOO

scheme, the fundamental frequency and the magnitude of the inverter output voltages v_a, v_b, and v_c are equal to those of the three-phase reference voltage v_a^*, v_b^*, and v_c^*. As result, the inverter output voltage is fully controllable by its references.

The simulated waveforms for the inverter output voltages and load current are shown in Figure 4-31. The inverter operates under the condition of $f = 60$ Hz, $f_{sw} = 720$ Hz, $T_s = 1/720$ sec, and $m_a = 0.8$ with a rated three-phase inductive load. The load power factor is 0.9 per phase. It can be observed that the waveform of the inverter line-to-line voltage v_{ab} is not half-wave symmetrical, that is, $v_{ab}(\omega t) \neq -v_{ab}(\omega t + \pi)$. Therefore, it contains even-order harmonics, such as 2nd, 4th, 8th, and 10th, in addition to odd-order harmonics. The THDs of v_{ab} and i_a are 80.2% and 8.37%, respectively.

The two-level voltage source converter technology is widely accepted in the variable-speed WECS, including doubly fed induction generator (DFIG), squirrel cage induction generator (SCIG), and synchronous generator (SG) based wind energy systems. The converter can be used either as a rectifier that converts three-phase AC voltage produced by the generator to a DC voltage or as a grid-tied inverter that delivers the active power from the generator and rectifier to the grid. The modulation schemes presented above are applicable to the converter that may operate as a rectifier or an inverter.

4.5 THREE-LEVEL NEUTRAL POINT CLAMPED CONVERTERS

The diode-clamped multilevel inverter employs clamping diodes and cascaded DC capacitors to produce AC voltage waveforms with multiple levels [2]. The inverter can be generally configured as a three-, four-, or five-level topology, and the three-level inverter, often known as neutral point clamped (NPC) inverter, has found wide practical application, especially in medium-voltage (MV) variable-speed drives [3]. The NPC

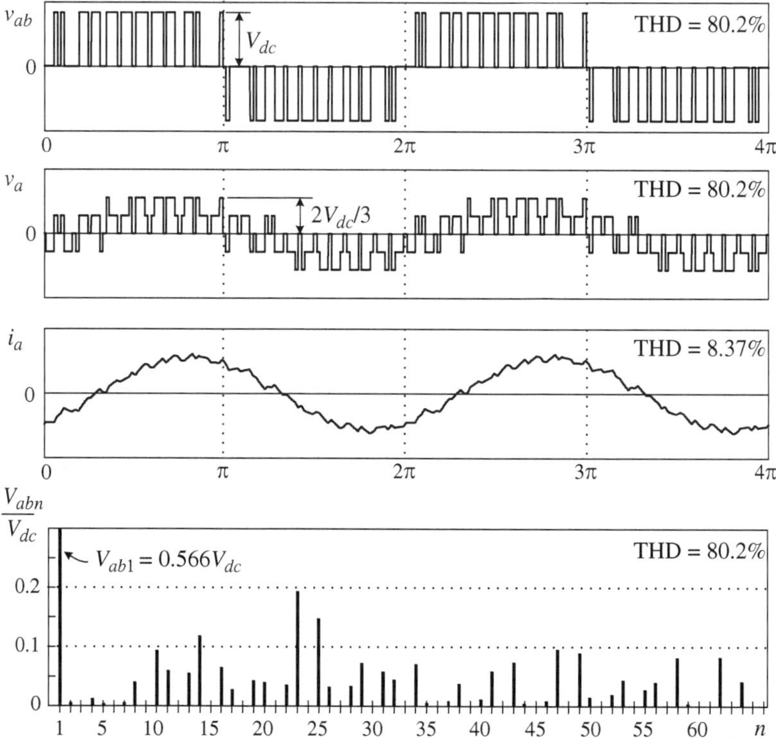

Figure 4-31. Inverter output waveforms produced by SVM scheme with $f_1 = 60$ Hz, $f_{sw} = 720$ Hz, and $m_a = 0.8$.

inverter is also a good candidate for MV (3 kV–4 kV) wind energy systems. The main features of the NPC inverter include reduced dv/dt and THD in its AC output voltages in comparison to the two-level inverter discussed earlier. More importantly, the inverter can be used in the MV wind energy systems without switching devices in series. For instance, the NPC inverter using 6 kV IGBT or IGCT devices is suitable for the 4 kV WECS, for which there is no need to connect the switches in series.

4.5.1 Converter Configuration

Figure 4.32 shows the simplified circuit diagram for the three-level NPC inverter. The inverter leg a is composed of four active switches S_1 to S_4 with four antiparallel diodes D_1 to D_4. In practice, either an IGBT or IGCT can be employed as a switching device.

On the DC side of the inverter, the DC bus capacitor is split into two, providing a neutral point Z. The diodes connected to the neutral point, D_{Z1} and D_{Z2}, are the clamping diodes. When switches S_2 and S_3 are turned on, the inverter output terminal a is connected to the neutral point through one of the clamping diodes. The voltage across each of the DC capacitors is E, which is normally equal to half of the total DC voltage V_{dc}.

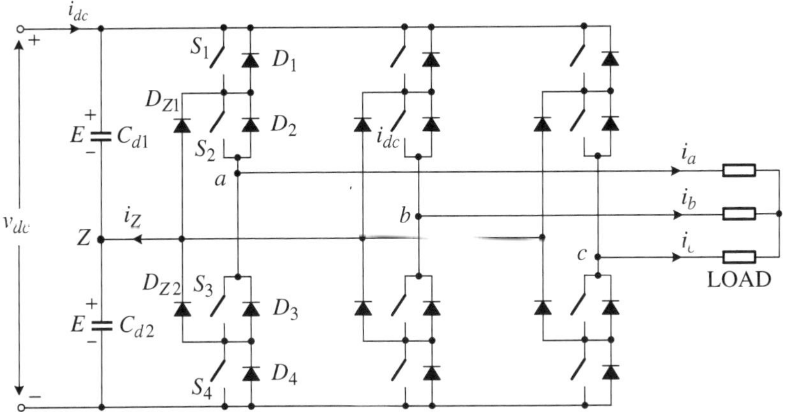

Figure 4-32. Three-level NPC inverter.

The operating status of the switches in the NPC inverter can be represented by switching states as shown in Table 4-5. Switching state P denotes that the upper two switches in leg a are on and the inverter terminal voltage v_{aZ}, which is the voltage at terminal a with respect to the neutral point Z, is $+E$, whereas N indicates that the lower two switches conduct, leading to $v_{aZ} = -E$.

Switching state O signifies that the inner two switches S_2 and S_3 are on and v_{aZ} is clamped to zero through the clamping diodes. Depending on the direction of load current i_a, one of the two clamping diodes is turned on. For instance, a positive load current ($i_a > 0$) forces D_{Z1} to turn on, and the terminal a is connected to the neutral point Z through the conduction of D_{Z1} and S_2.

It can be observed from Table 4.5 that switches S_1 and S_3 operate in a complementary manner. With one switched on, the other must be off. Similarly, S_2 and S_4 are a complementary pair as well.

Figure 4-33 shows an example of switching states and gate signal arrangements, where v_{g1} to v_{g4} are the gate signals for S_1 to S_4, respectively. The gate signals can be generated by carrier-based modulation, space vector modulation, or selective harmonic elimination schemes. The waveform for v_{aZ} has three voltage levels, $+E$, 0, and $-E$, based on which the inverter is referred to as a three-level inverter.

Table 4-5. Definition of switching states

Switching state	Device switching states (Phase a)				Inverter terminal voltage V_{aZ}
	S_1	S_2	S_3	S_4	
P	On	On	Off	Off	E
O	Off	On	On	Off	0
N	Off	Off	On	On	$-E$

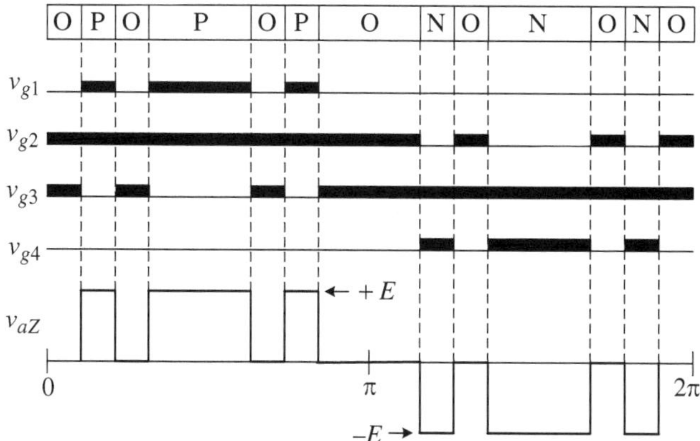

Figure 4-33. Switching states, gate signals, and inverter terminal voltage v_{aZ}.

Figure 4-34 shows how the line-to-line voltage waveform is obtained. The inverter terminal voltages v_{aZ}, v_{bZ}, and v_{cZ} are three-phase balanced with a phase shift of $2\pi/3$ between each other. The line-to-line voltage v_{ab} can be found from $v_{ab} = v_{aZ} - v_{bZ}$, which contains five voltage levels ($+2E$, $+E$, 0, $-E$, and $-2E$).

4.5.2 Space Vector Modulation

As indicated earlier, the operation of each inverter phase leg can be represented by three switching states P, O, and N. Taking all three phases into account, the inverter

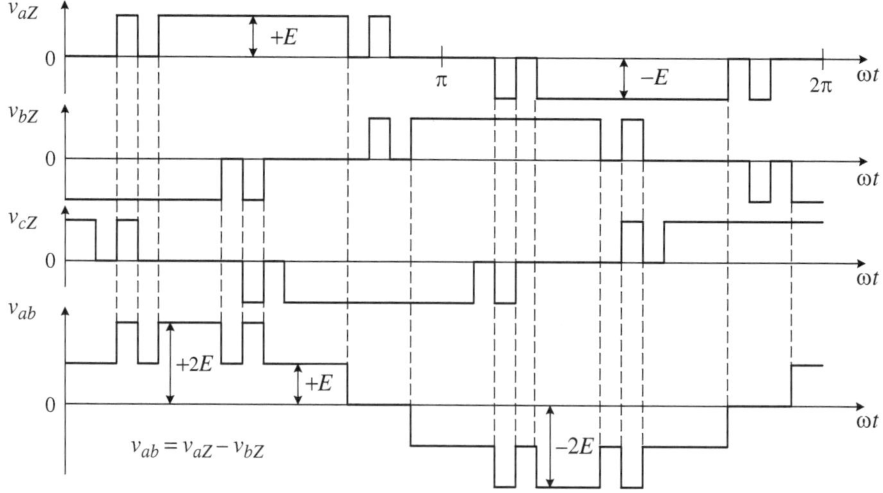

Figure 4-34. Inverter terminal and line-to-line voltage waveforms.

has a total of 27 possible combinations of switching states and 19 different stationary space vectors ($\vec{V}_0 \sim \vec{V}_{18}$) as shown in Figure 4-35. The principle of the space vector modulation for the NPC inverter is the same as that for the two-level inverter, but the implementation is more complicated. For a given position in space, the reference vector $\vec{v}_{ref}$ can be synthesized by three adjacent stationary vectors, based on which the dwell times for each of the selected stationary vectors can be calculated. According to the calculated dwell times, the turn-on times of the switches during a sampling period T_s can be determined, from which the switching sequence can be designed. The gate signals for the switches in the inverter can then be generated, based on which of the inverter output voltages are generated. When the reference vector $\vec{v}_{ref}$ rotates in space for one cycle, the inverter output voltage varies by one cycle at the fundamental frequency [3].

Case Study 4-4—Three-Level NPC Inverter with Space Vector Modulation. The main purpose of this case study is to investigate the harmonic profile of the three-level NPC inverter with space vector modulation, and then compare its THD profile with that of the two-level inverter.

Consider a 3 MW/4000 V three-level NPC inverter modulated by the SVM scheme. The inverter is loaded with a three-phase inductive load with a power factor of 0.9. Figure 4-36 shows the simulated waveforms for the inverter operating at f = 60 Hz, T_s = 1/1080 sec, $f_{sw,dev}$ = 1080/2 + 60/2 = 570 Hz, and m_a = 0.8. The gate signals v_{g1} and v_{g4}

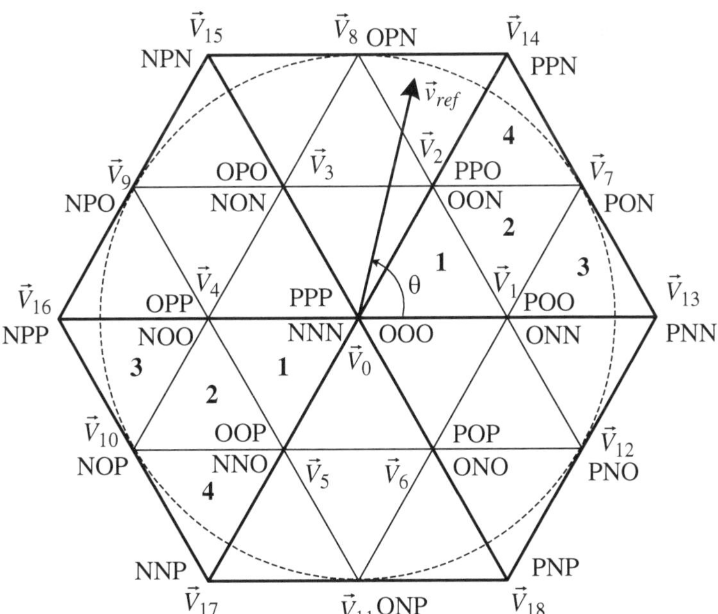

Figure 4-35. Space-vector diagram of the NPC inverter.

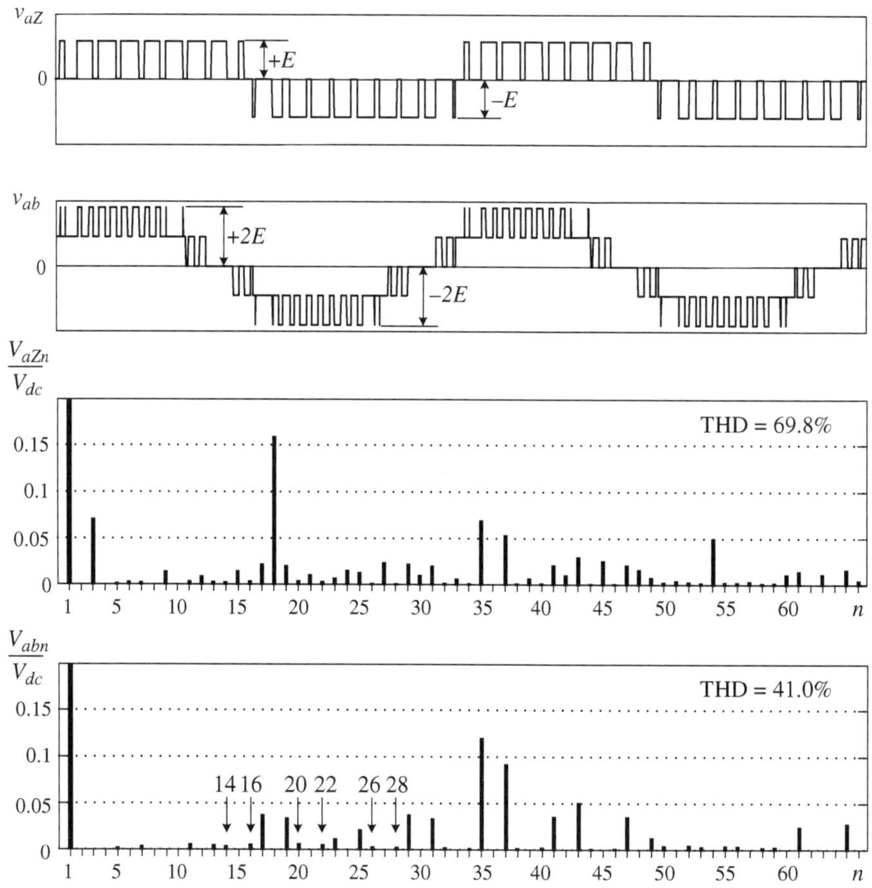

Figure 4-36. Simulated voltage waveforms of the NPC inverter (f_1 = 60 Hz, T_s = 1/1080 sec, $f_{sw,dev}$ = 570 Hz, and m_a = 0.8).

are for the switches S_1 and S_4 of the inverter circuit in Figure 4-32. Since the inner switches S_2 and S_3 operate complementarily with S_4 and S_1, their gatings are not shown.

The waveform of the inverter terminal voltage v_{aZ} is composed of three voltage levels, whereas the inverter line-to-line voltage v_{ab} has five voltage levels. The waveform for v_{aZ} contains triplen harmonics with the third and eighteenth being dominant. Since the triplen harmonics are of zero sequence, they do not appear in the line-to-line voltage v_{ab}. However, v_{ab} contains even-order harmonics such as the fourteenth and sixteenth in addition to odd-order harmonics. This is due to the fact that waveform of v_{ab} produced by the SVM scheme is not half-wave symmetrical.

The dominant harmonics in v_{ab} are the seventeenth and nineteenth, centered around the eighteenth harmonic, whose frequency is 1080 Hz. This frequency can be considered as the equivalent inverter switching frequency $f_{sw,inv}$, which is approximately twice the device switching frequency $f_{sw,dev}$.

Figure 4-37 shows the THD profile of the output line-to-line voltage produced by the two- and three-level inverters. Both inverters are modulated by the SVM schemes. The device switching frequencies for the two- and three-level inverters are 720 Hz and 570 Hz, respectively. It can be observed that the THD produced by the three-level inverter is much lower than that of the two-level inverter, which is one of the main advantages of the three-level converters.

The two-level inverter is often used in low-voltage (e.g., 690 V) WECS, whereas the NPC inverter is normally used in medium-voltage (e.g , 4000 V) WECS. For a given power rating, the NPC inverter has much lower output current than the two-level inverter. For example, the rated output current of a 3 MW/4000 V NPC inverter with unity power factor operation is 433 A in comparison to 2510 A for a 3 MW/690 V two-level inverter. The power cable that carries 2510 A current and runs from the top of the wind turbine tower to ground is much heavier and more costly than the 433 A cable and also has higher I^2R losses.

4.6 PWM CURRENT SOURCE CONVERTERS

Solid-state converters can be generally classified into voltage source converters (VSCs) and current source converters (CSCs). The voltage source converter produces a defined three-phase PWM output voltage waveform, whereas the current source converter outputs a defined PWM current waveform. The PWM current source converter features simple converter topology, nearly sinusoidal waveforms, and reliable short-circuit protection. It is particularly suitable for high-power applications such as megawatt variable-speed drives and wind energy conversion systems.

This section mainly deals with the modulation schemes for the current source inverter. Two modulation techniques for the inverter are discussed: selective harmonic

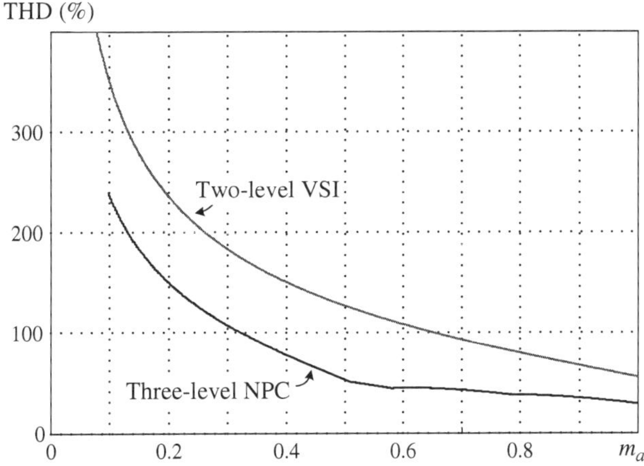

Figure 4-37. THD profile of the output voltage produced by the two-level and three-level NPC inverters.

elimination (SHE) and space vector modulation (SVM). These modulation schemes are developed for high-power inverters operating with a switching frequency below 1 kHz for reduction in switching losses.

4.6.1 Current Source Inverter Topology

A typical three-phase PWM current source inverter (CSI) circuit is shown in Figure 4-38. The inverter is composed of six IGCT devices of symmetrical type or reverse blocking IGBTs. The inverter requires a DC current source i_{dc} at its DC input and produces a defined PWM output current i_{aw}.

The current source inverter normally requires a three-phase capacitor C_i at its output to assist in the commutation of the switching devices. For instance, at the turn-off of switch S_1, the inverter PWM current i_{aw} falls to zero within a very short period of time. The capacitor provides a current path for the energy trapped in the phase-a load inductance. Otherwise, a high-voltage spike would be induced, causing damages to the switching devices. The capacitor also acts as a harmonic filter, improving the load current and voltage waveforms. The value of the capacitor is normally in the range of 0.3 to 0.6 per unit for an inverter with a switching frequency of 200 Hz to 400 Hz.

4.6.2 Selective Harmonic Elimination

The switching pattern design for the CSI should generally satisfy two conditions: (1) the DC current i_{dc} should be continuous and (2) the inverter PWM current i_{aw} should be defined. The two conditions can be translated into a switching constraint: at any instant of time (excluding commutation intervals) there are only two switches conducting: one in the top half of the bridge and the other in the bottom half. With only one

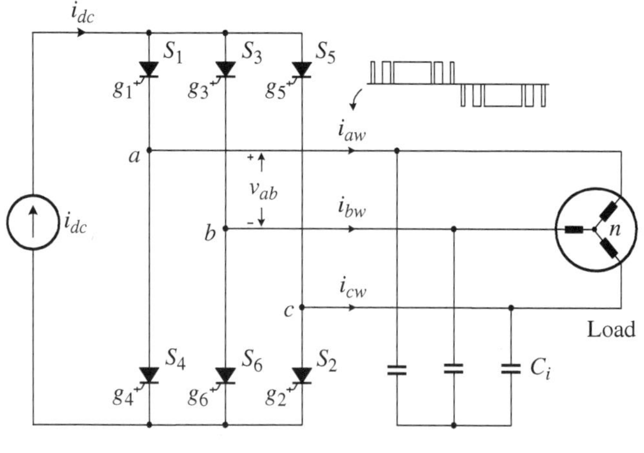

Figure 4-38. PWM current-source inverter.

switch turned on, the continuity of the DC current is lost. A very high voltage will be induced by the constant DC current, causing damage to the switching devices. If more than two devices are on simultaneously, the PWM current i_{aw} is not defined by the switching pattern. For instance, with S_1, S_2, and S_3 conducting at the same time, the currents in S_1 and S_3, which are the PWM currents in the inverter phases a and b, are load-dependent although the sum of the two currents is equal to i_{dc}.

Selective harmonic elimination is an offline modulation scheme, which is able to eliminate a number of low-order unwanted harmonics in the inverter PWM current i_{aw}. The switching angles are precalculated and then imported into a digital controller for implementation. Figure 4-39 shows a typical SHE waveform that satisfies the switching constraint for the CSI. There are five pulses per half-cycle ($N_p = 5$) with five switching angles in the first $\pi/2$ period. However, only two out of the five angles, θ_1 and θ_2, are independent. Given these two angles, all other switching angles can be calculated.

The two switching angles provide two degrees of freedom, which can be used to either eliminate two harmonics in i_{aw} without modulation index control or eliminate one harmonic and provide an adjustable modulation index m_a. The first option is preferred since the adjustment of i_{aw} is normally done by varying i_{dc}. The number of harmonics to be eliminated is then given by $k = (N_p - 1)/2$.

To determine switching angles such as θ_1 and θ_2 in Figure 4-39 for harmonic elimination, Fourier analysis can be performed, from which a set of nonlinear equations can be formulated [3]. These equations can be solved by numerical methods. Table 4-6 gives a list of switching angles for the elimination of up to four harmonics in i_{aw}. It is noted that the nonlinear equations for harmonic elimination may not always have a valid solution. If this happens, optimization techniques can be used to find optimal switching angles that minimize the magnitude of low-order harmonics.

4.6.3 Space Vector Modulation

In addition to the SHE scheme, the current source inverter can also be controlled by space vector modulation (SVM). This section presents the principle of the SVM scheme for current source inverters.

Switching States. As stated earlier, the PWM switching pattern for the CSI shown in Figure 4-38 must satisfy the constraint that only two switches in the inverter conduct at any time instant, one in the top half of the CSI bridge and the other in the

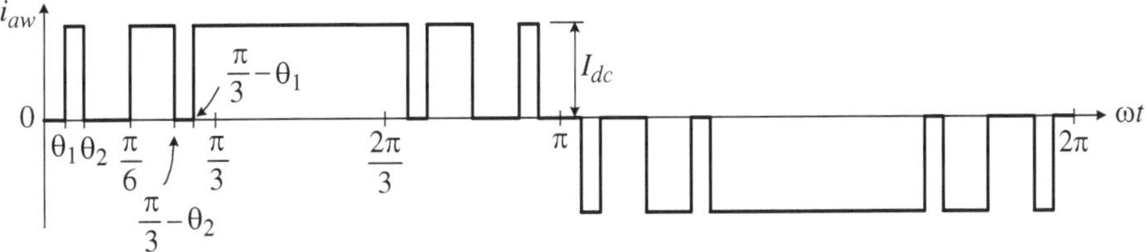

Figure 4-39. Selective harmonic elimination (SHE) scheme.

Table 4-6. SHE switching angles

Harmonics eliminated	Switching angles			
	θ_1	θ_2	θ_3	θ_4
5	18.00	—	—	—
7	21.43	—	—	—
11	24.55	—	—	—
5,7	7.93	13.75	—	—
5,11	12.96	19.14	—	—
5,13	14.48	21.12	—	—
5,7,11	2.24	5.60	21.26	—
5,7,13	4.21	8.04	22.45	—
5,7,17	6.91	11.96	25.57	—
5,7,11,13*	0.00	1.60	15.14	20.26
5,7,11,17	0.07	2.63	16.57	21.80
5,7,11,19	1.11	4.01	18.26	23.60

*Harmonics not completely eliminated but their magnitude minimized.

bottom half. Under this constraint, the three-phase inverter has a total of nine switching states, as listed in Table 4-7. These switching states can be classified as zero switching states and active switching states.

There are three zero switching states: (1,4), (3,6), and (5,2). The zero state (1,4) signifies that switches S_1 and S_4 in inverter-phase leg a conduct simultaneously and the other four switches in the inverter are off. The DC current source i_{dc} is bypassed, leading to $i_{aw} = i_{bw} = i_{cw} = 0$. This operating mode is often referred to as bypass operation.

There exist six active switching states. State (1,2) indicates that switch S_1 in leg a and S_2 in leg c are on. The DC current flows through S_1, the load, S_2, and then back to the DC source, resulting in $i_{aw} = I_{dc}$ and $i_{cw} = -I_{dc}$. The definition of the other five active states is also given in the table.

Table 4-7. Switching states and space vectors

Type	Switching state	On-state switch	Inverter PWM current			Space vector
			i_{aw}	i_{bw}	i_{cw}	
Zero states	(1,4)	S_1, S_4	0	0	0	$\vec{I}_0$
	(3,6)	S_3, S_6				
	(5,2)	S_5, S_2				
Active states	(6,1)	S_6, S_1	I_{dc}	$-I_{dc}$	0	$\vec{I}_1$
	(1,2)	S_1, S_2	I_{dc}	0	$-I_{dc}$	$\vec{I}_2$
	(2,3)	S_2, S_3	0	I_{dc}	$-I_{dc}$	$\vec{I}_3$
	(3,4)	S_3, S_4	$-I_{dc}$	I_{dc}	0	$\vec{I}_4$
	(4,5)	S_4, S_5	$-I_{dc}$	0	I_{dc}	$\vec{I}_5$
	(5,6)	S_5, S_6	0	$-I_{dc}$	I_{dc}	$\vec{I}_6$

Space Vectors. The active and zero switching states can be represented by active and zero space vectors, respectively. A space vector diagram for the CSI is shown in Figure 4-40, where $\vec{I}_1$ to $\vec{I}_6$ are the active vectors and $\vec{I}_0$ is the zero vector. The active vectors form a regular hexagon with six equal sectors, whereas the zero vector $\vec{I}_0$ lies at the center of the hexagon.

Assuming that the operation of the inverter in Figure 4-38 is three-phase balanced, that is,

$$i_{aw}(t) + i_{bw}(t) + i_{cw}(t) = 0 \tag{4.69}$$

where i_{aw}, i_{bw}, and i_{cw} are the instantaneous PWM output currents in the inverter phases a, b, and c, respectively, the three-phase currents can be transformed into two-phase currents in the α-β frame:

$$\begin{bmatrix} i_\alpha(t) \\ i_\beta(t) \end{bmatrix} = \frac{2}{3} \begin{bmatrix} 1 & -\dfrac{1}{2} & -\dfrac{1}{2} \\ 0 & \dfrac{\sqrt{3}}{2} & -\dfrac{\sqrt{3}}{2} \end{bmatrix} \begin{bmatrix} i_{aw}(t) \\ i_{bw}(t) \\ i_{cw}(t) \end{bmatrix} \tag{4.70}$$

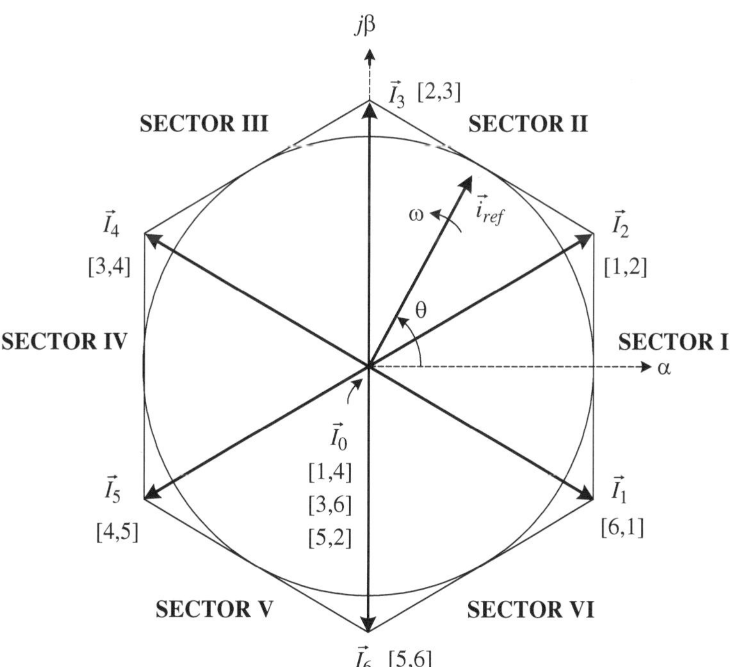

Figure 4-40. Space-vector diagram for the current-source inverter.

A current-space vector can be generally expressed in terms of the two-phase currents as

$$\vec{i}(t) = i_\alpha(t) + j\,i_\beta(t) \tag{4.71}$$

Substituting (4.70) into (4.71), $\vec{i}(t)$ can be expressed in terms of i_{aw}, i_{bw}, and i_{cw}:

$$\vec{i}(t) = \frac{2}{3}\left[i_{aw}(t)\,e^{j0} + i_{bw}(t)\,e^{j2\pi/3} + i_{cw}(t)\,e^{j4\pi/3} \right] \tag{4.72}$$

For the active state $(6,1)$, S_1 and S_6 are turned on, and the inverter PWM currents are

$$i_{aw}(t) = I_{dc}, \;\; i_{bw}(t) = -I_{dc} \;\; \text{and} \;\; i_{cw}(t) = 0 \tag{4.73}$$

Substituting (4.73) into (4.72) yields

$$\vec{I}_1 = \frac{2}{\sqrt{3}} I_{dc}\, e^{j(-\pi/6)} \tag{4.74}$$

Similarly, the other five active vectors can be derived. The active vectors can be expressed as

$$\vec{I}_k = \frac{2}{\sqrt{3}} I_{dc}\, e^{j\left((k-1)\frac{\pi}{3} - \frac{\pi}{6} \right)} \qquad \text{for } k = 1, 2, ..., 6. \tag{4.75}$$

Note that the active and zero vectors do not move in space, and thus are referred to as stationary vectors. On the contrary, the current reference vector $\vec{i}_{ref}$ in Figure 4-40 rotates in space at an angular velocity

$$\omega = 2\pi f \tag{4.76}$$

where f is the fundamental frequency of the inverter output current i_{aw}. The angular displacement between $\vec{i}_{ref}$ and the α-axis of the α-β frame can be obtained by

$$\theta(t) = \int_0^t \omega(t)\, dt + \theta_0 \tag{4.77}$$

For a given length and position, $\vec{i}_{ref}$ can be synthesized by three nearby stationary vectors, based on which the switching states of the inverter can be selected and gate signals for the active switches can be generated. When $\vec{i}_{ref}$ passes through sectors one by one, different sets of switches are turned on or off. As a result, when $\vec{i}_{ref}$ rotates one revolution in space, the inverter output current varies one cycle over time. The frequency and magnitude of the inverter output current correspond to the rotating speed and length of $\vec{i}_{ref}$, respectively.

Dwell Time Calculation. As indicated above, the reference $\vec{i}_{ref}$ can be synthesized by three stationary vectors. The dwell time for the stationary vectors essentially represents the duty-cycle time (on-state or off-state time) of the chosen switches during a sampling period T_s. The dwell time calculation is based on the ampere-second balancing principle, that is, the product of the reference vector $\vec{i}_{ref}$ and sampling period T_s equals the sum of the current vectors multiplied by the time interval of chosen space vectors. Assuming that the sampling period T_s is sufficiently small, the reference vector $\vec{i}_{ref}$ can be considered constant during T_s. Under this assumption, $\vec{i}_{ref}$ can be approximated by two adjacent active vectors and a zero vector. For example, with $\vec{i}_{ref}$ falling into Sector I as shown in Figure 4-41, it can be synthesized by $\vec{I}_1$, $\vec{I}_2$, and $\vec{I}_0$. The ampere-second balancing equation is thus given by

$$\begin{cases} \vec{i}_{ref}\, T_s = \vec{I}_1\, T_1 + \vec{I}_2\, T_2 + \vec{I}_0\, T_0 \\ T_s = T_1 + T_2 + T_0 \end{cases} \qquad (4.78)$$

where T_1, T_2 and T_0 are the dwell times for the vectors $\vec{I}_1$, $\vec{I}_2$, and $\vec{I}_0$, respectively. Substituting

$$\vec{i}_{ref} = i_{ref}\, e^{j\theta}, \ \ \vec{I}_1 = \frac{2}{\sqrt{3}} I_{dc}\, e^{-j\frac{\pi}{6}}, \ \vec{I}_2 = \frac{2}{\sqrt{3}} I_{dc}\, e^{j\frac{\pi}{6}}, \text{ and } \vec{I}_0 = 0 \qquad (4.79)$$

into (4.78) and then splitting the resultant equation into the real (α-axis) and imaginary (β-axis) components leads to

$$\begin{cases} \text{Re}: \ \ i_{ref}(\cos\theta)T_s \ = \ I_{dc}\,(T_1 + T_2) \\ \text{Im}: \ \ i_{ref}(\sin\theta)\, T_s \ = \ \dfrac{1}{\sqrt{3}} I_{dc}\,(-T_1 + T_2) \end{cases} \qquad (4.80)$$

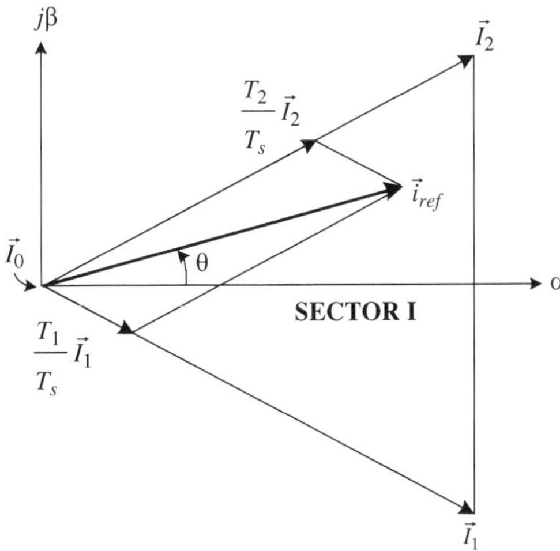

Figure 4-41. Synthesis of $\vec{i}_{ref}$ by $\vec{I}_1$, $\vec{I}_2$, and $\vec{I}_0$.

Solving (4.80) together with $T_s = T_1 + T_2 + T_0$ gives

$$\begin{cases} T_1 = m_a \sin\left(\pi/6 - \theta\right) T_s \\ T_2 = m_a \sin\left(\pi/6 + \theta\right) T_s & \text{for } -\pi/6 \le \theta < \pi/6 \\ T_0 = T_s - T_1 - T_2 \end{cases} \tag{4.81}$$

where m_a is the modulation index, given by

$$m_a = \frac{i_{ref}}{I_{dc}} = \frac{\hat{I}_{aw1}}{I_{dc}} \tag{4.82}$$

in which $\hat{I}_{aw1}$ is the peak value of the fundamental-frequency component in i_{aw}.

Note that although Equation (4.81) is derived when $\vec{i}_{ref}$ is in sector I, it can also be used when $\vec{i}_{ref}$ is in other sectors provided that a multiple of $\pi/3$ is subtracted from the actual angular displacement θ such that the modified angle θ' falls into the range of $-\pi/6 \le \theta' < \pi/6$ for use in the equation, that is,

$$\theta' = \theta - (k-1)\pi/3 \qquad \text{for } -\pi/6 \le \theta' < \pi/6 \tag{4.83}$$

where $k = 1, 2, \ldots, 6$ for sectors I, II, $\ldots$, VI, respectively.

The maximum length of the reference vector, $i_{ref,\max}$, corresponds to the radius of the largest circle that can be inscribed within the hexagon as shown in Figure 4-40. Since the hexagon is formed by the six active vectors having a length of $2I_{dc}/\sqrt{3}$, $i_{ref,\max}$ can be found from

$$i_{ref,\max} = \frac{2I_{dc}}{\sqrt{3}} \times \frac{\sqrt{3}}{2} = I_{dc} \tag{4.84}$$

Substituting (4.84) into (4.82) gives the maximum modulation index

$$m_{a,\max} = 1 \tag{4.85}$$

from which the modulation index is in the range of

$$0 \le m_a \le 1 \tag{4.86}$$

Switching Sequence. Similar to the space vector modulation for the two-level VSI, the switching sequence design for the CSI should satisfy the following two requirements for the minimization of switching frequencies:

1. The transition from one switching state to the next involves only two switches, one being switched on and the other switched off.
2. The transition for $\vec{i}_{ref}$ moving from one sector to the next requires the minimum number of switchings.

Figure 4-42 shows a typical three-segment sequence for the reference vector $\vec{i}_{ref}$ residing in sector I, where v_{g1} to v_{g6} are the gate signals for switches S_1 to S_6, respectively. The reference vector $\vec{i}_{ref}$ is synthesized by $\vec{I}_1$, $\vec{I}_2$, and $\vec{I}_0$. The sampling period T_s is divided into three segments composed of T_1, T_2, and T_0. The switching states for vectors $\vec{I}_1$ and $\vec{I}_2$ are (6,1) and (1,2), and their corresponding on-state switch pairs are (S_6, S_1) and (S_1, S_2). The zero state (1,4) is selected for $\vec{I}_0$ such that design requirement 1 above is satisfied.

Case Study 4-5—Current Source Inverter with Space Vector Modulation.

Consider a 1 MW/4000 V CSI using the space vector modulation. The inverter operates at $f = 60$ Hz, $f_{sp} = 1080$ Hz, and $f_{sw} = 540$ Hz with $m_a = 1$. The filter capacitor C_i is 0.3 pu per phase. The inverter is loaded with a three-phase balanced inductive load having a resistance of 1.0 pu and an inductance of 0.1 pu per phase. The DC current of the inverter is adjusted such that the fundamental-frequency inverter output current i_{aw1} is rated. The simulated waveforms for the converter are shown in Figure 4-43, where i_{aw} is the inverter PWM current and v_{ab} is the line-to-line output voltage of the inverter.

The spectra for i_{aw} and v_{ab} are also shown in the figure, where I_{awn} is the rms value of the nth-order harmonic current in i_{aw} and $I_{aw1,max}$ is the maximum rms fundamental-frequency current that can be found from Equation (4.82):

$$I_{aw1,max} = \frac{m_{a,max} \times I_{dc}}{\sqrt{2}} = 0.707 I_{dc} \qquad \text{for } m_{a,max} = 1 \qquad (4.87)$$

The PWM current i_{aw} contains no even-order harmonics and its THD is 45.7%. Similar to the two-level inverter, the SVM current source inverter produces low-order harmon-

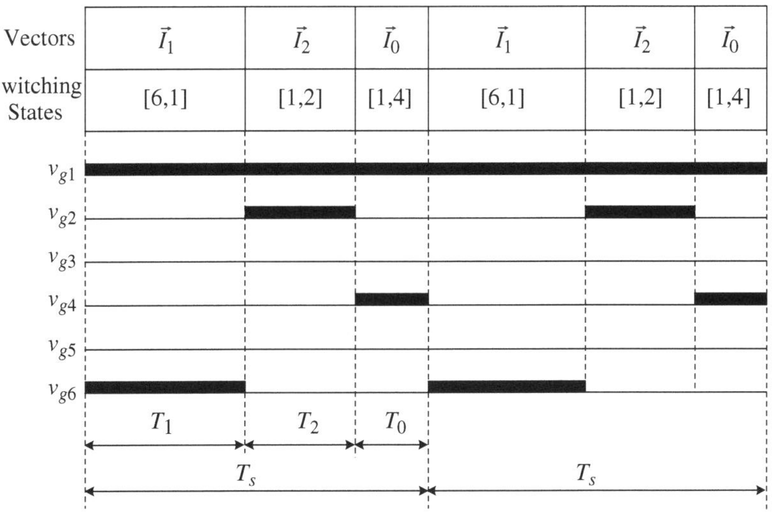

Figure 4-42. Switching sequence for $\vec{i}_{ref}$ in sector I.

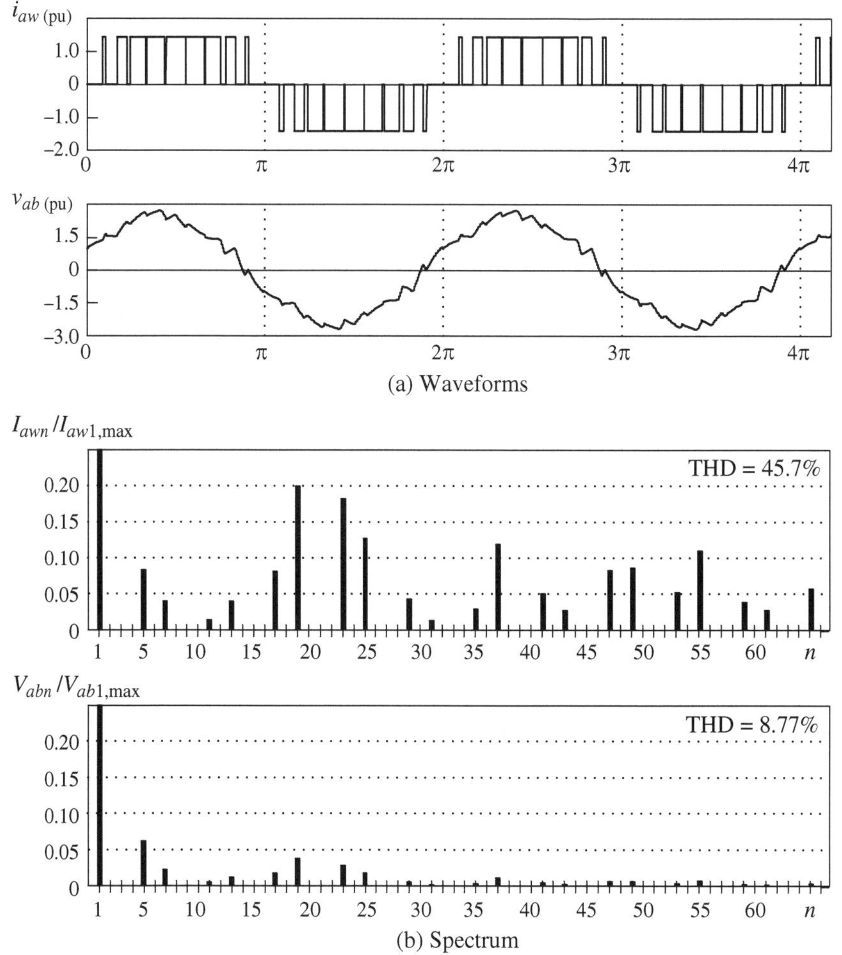

Figure 4-43. Waveforms in a 1 MVA CSI using the SVM scheme (f_1 = 60 Hz, f_{sw} = 540 Hz, m_a = 1, and C_i = 0.3 pu).

ics such as the fifth and seventh. The magnitude of these harmonics can be substantially reduced by using multisampling techniques [7]. Unlike the voltage source inverters, the current source inverter does not have high dv/dt in its output voltage waveform, and the THD of its line-to-line voltage v_{ab} is only 8.77%.

4.6.4 PWM Current Source Rectifier

Figure 4.44 shows a typical configuration of a PWM current source rectifier (CSR) in a wind energy conversion system. Like the current source inverter (CSI), the PWM rectifier requires a filter capacitor C_r to assist the commutation of switching devices

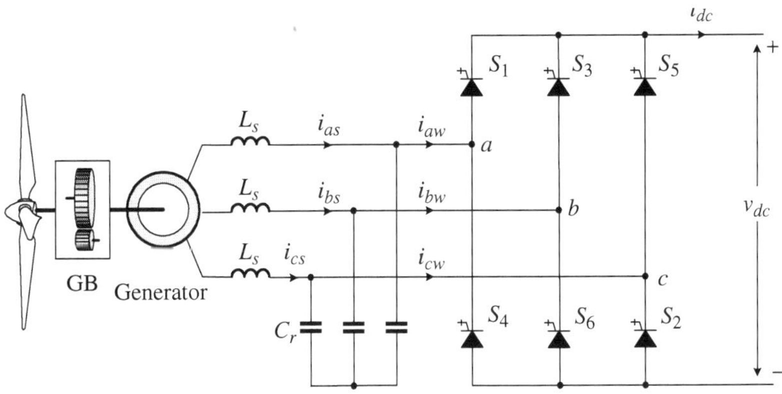

Figure 4-44. Typical configuration of a PWM current-source rectifier in a wind energy system.

and filter out current harmonics. The capacitor size is dependent on a number of factors such as the rectifier switching frequency, LC resonant mode, required line current THD, and type of generator. It is normally in the range of 0.1 to 0.3 pu for WECS with a large synchronous generator and 0.3 to 0.6 pu for induction-generator-based WECS with a switching frequency of a few hundred hertz. Both the SHE and SVM schemes developed for the CSI can be used for the CSR.

The DC output voltage v_{dc} of the rectifier can be adjusted by two methods: modulation index (m_a) control and delay angle (α) control. The operating principle of delay angle control is the same as that of phase-controlled SCR rectifiers.

The input active power of the rectifier can be expressed as

$$P_{ac} = 3\,V_{a1}\,I_{aw1}\cos\alpha = \sqrt{3}\,V_{ab1}\,I_{aw1}\cos\alpha \qquad (4.88)$$

where V_{a1}, V_{ab1}, and I_{aw1} are the fundamental-frequency rms phase input voltage, line-to-line input voltage, and PWM input current of the rectifier, respectively. The delay angle α is defined as the angle between V_{a1} and I_{aw1}, respectively. The DC output power is given by

$$P_{dc} = V_{dc}\,I_{dc} \qquad (4.89)$$

where V_{dc} and I_{dc} are the average DC output voltage and current, respectively. Neglecting the power losses in the rectifier, the AC input power is equal to the DC output power:

$$\sqrt{3}\,V_{ab1}I_{aw1}\cos\alpha = V_{dc}I_{dc} \qquad (4.90)$$

from which

$$V_{dc} = \sqrt{3/2}\,V_{ab1}\,m_a\cos\alpha \qquad (4.91)$$

where

$$m_a = \hat{I}_{aw1} / I_{dc} = \sqrt{2} I_{aw1} / I_{dc} \qquad (4.92)$$

Equation (4.91) illustrates that for a given line-to-line input voltage V_{ab1}, the average DC voltage of the rectifier can be controlled by both modulation index m_a and delay angle α. The above results will be used for the analysis of wind energy systems discussed in Chapters 7 and 9.

4.7 CONTROL OF GRID-CONNECTED INVERTER

Most commercial wind turbines deliver the generated power to the electric grid through power converters. A typical grid-connected (grid-tied) inverter for wind energy applications is shown in Figure 4-45, where a two-level voltage source inverter is used as an example. The inverter is connected to the grid through a line inductance L_g, which represents the leakage inductance of the transformer, if any, and the line reactor of 0.05 to 0.1 per unit, which is normally added to the system for the reduction of line current distortion. The line resistance is negligibly small and has little impact on the system performance. It is, therefore, omitted in the analysis.

The grid-tied inverter can be modulated by the PWM schemes presented in the previous sections, such as the space vector modulation scheme. The inverter is a boost converter by nature, and its average DC voltage V_{dc} can be derived from Equation (4.63) and given by

$$V_{dc} = \frac{\sqrt{6}\, V_{ai1}}{m_a} \qquad \text{for } 0 < m_a \leq 1 \qquad (4.93)$$

where m_a is the modulation index and V_{ai1} is the rms value of the fundamental-frequency component of the inverter phase (phase-a) voltage v_{ai1}. Assuming that V_{ai1}

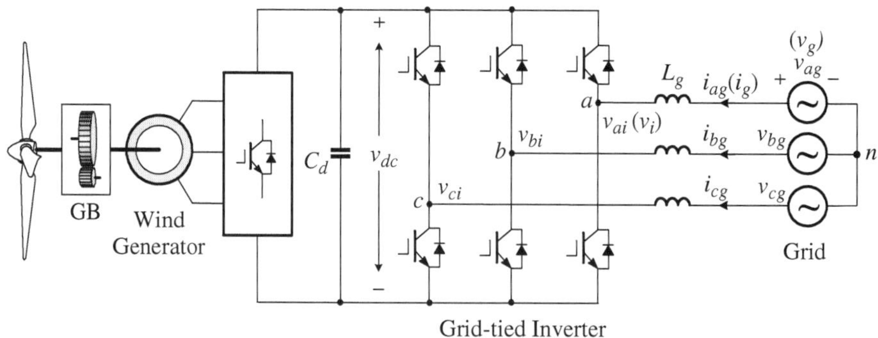

Grid-tied Inverter

Figure 4-45. Grid-connected inverter in a wind energy system.

is equal to the rms value of grid phase voltage V_g, which can be considered constant, the DC voltage can be boosted to a high value by a small m_a.

Figure 4-46a shows a simplified diagram for a wind energy system, in which the wind turbine, generator, and rectifier are replaced by a battery in series with a small resistance that represents the power losses in the system. The power flow between the inverter and grid is bidirectional. Power can be transferred from the grid to the DC circuit of the inverter, or vice versa. For wind energy applications, the power is normally delivered from the inverter to the grid. The active power of the system delivered to the grid can be calculated by

$$P_g = 3\ V_g I_g \cos \varphi_g \tag{4.94}$$

where φ_g is the grid power factor angle, defined by

$$\varphi_g = \angle \overline{V}_g - \angle \overline{I}_g \tag{4.95}$$

The grid power factor can be unity, leading, or lagging, as shown in Figure 4-46b. It is often required by the grid operator that a wind energy system provide a controllable reactive power to the grid to support the grid voltage in addition to the active power production. Therefore, a wind energy system can operate with the power factor angle in the range of $90° \le \varphi_g < 270°$.

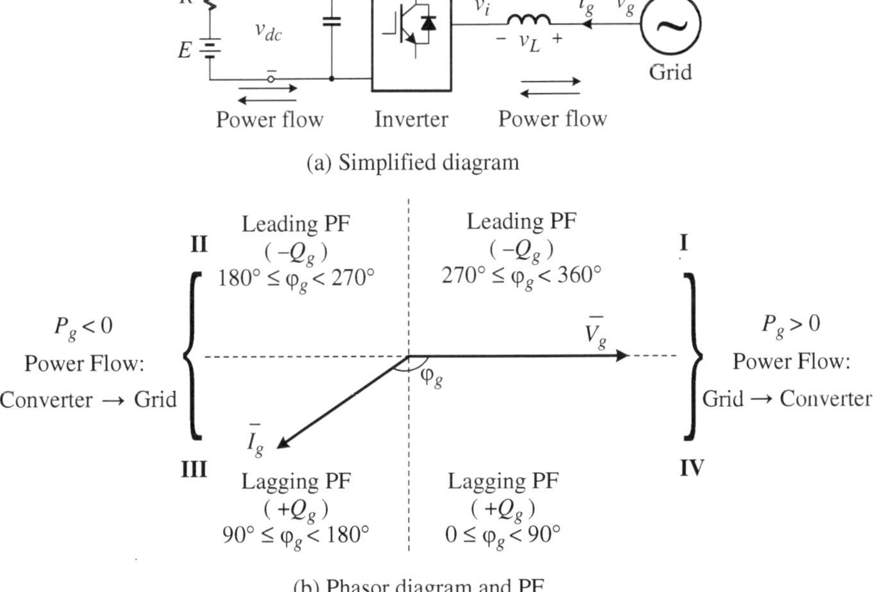

(a) Simplified diagram

(b) Phasor diagram and PF

Figure 4-46. Simplified system diagram and definition of power factor.

4.7.1 Voltage Oriented Control (VOC)

The grid-connected inverter can be controlled with various schemes. One of the schemes is known as voltage oriented control (VOC), as shown in Figure 4-47. This scheme is based on transformation between the *abc* stationary reference frame and *dq* synchronous frame as introduced in Chapter 3. The control algorithm is implemented in the grid-voltage synchronous reference frame, where all the variables are of DC components in steady state. This facilitates the design and control of the inverter.

To realize the VOC, the grid voltage is measured and its angle θ_g is detected for the voltage orientation. This angle is used for the transformation of variables from the *abc* stationary frame to the *dq* synchronous frame through the *abc/dq* transformation or from the synchronous frame back to the stationary frame through the *dq/abc* transformation, as shown in Figure 4-47. Various methods are available to detect the grid voltage angle θ_g. Assuming that the grid voltages, v_{ag}, v_{bg}, and v_{cg}, are three-phase balanced sinusoidal waveforms, θ_g can be obtained by

$$\theta_g = \tan^{-1}\frac{v_\beta}{v_\alpha} \tag{4.96}$$

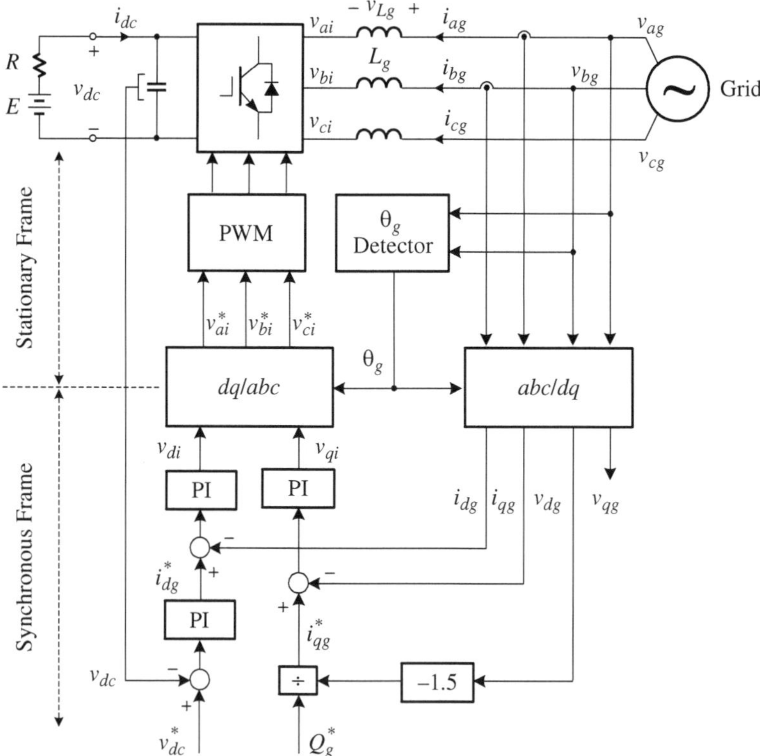

Figure 4-47. Block diagram of voltage-oriented control (VOC).

where v_β and v_β can be obtained by the *abc*/$\alpha\beta$ transformation:

$$\begin{cases} v_\alpha = \dfrac{2}{3}\left(v_{ag} - \dfrac{1}{2}v_{bg} - \dfrac{1}{2}v_{cg} \right) = v_{ag} \\ v_\beta = \dfrac{2}{3}\left(\dfrac{\sqrt{3}}{2}v_{bg} - \dfrac{\sqrt{3}}{2}v_{cg} \right) = \dfrac{\sqrt{3}}{3}\left(v_{ag} + 2v_{bg} \right) \end{cases} \quad \text{for} \quad v_{ag} + v_{bg} + v_{cg} = 0 \quad (4.97)$$

The above equation indicates that there is no need to measure the phase-*c* grid voltage v_{cg} as shown in Figure 4-47. In practice, the grid voltage may contain harmonics and be distorted, so digital filters or phase-locked loops (PLLs) may be used for the detection of the grid voltage angle θ_g.

There are three feedback control loops in the system: two inner current loops for the accurate control of the *dq*-axis currents i_{dg} and i_{qg}, and one outer DC voltage feedback loop for the control of DC voltage v_{dc}. With the VOC scheme, the three-phase line currents in the *abc* stationary frame i_{ag}, i_{bg}, and i_{cg} are transformed to the two-phase currents i_{dg} and i_{qg} in the *dq* synchronous frame, which are the active and reactive components of the three-phase line currents, respectively. The independent control of these · two components provides an effective means for the independent control of system active and reactive power.

To achieve the VOC control scheme, the *d*-axis of the synchronous frame is aligned with the grid voltage vector, therefore the *d*-axis grid voltage is equal to its magnitude ($v_{dg} = v_g$), and the resultant *q*-axis voltage v_{qg} is then equal to zero ($v_{qg} = \sqrt{v_g^2 - v_{dg}^2} = 0$), from which the active and reactive power of the system can be calculated by

$$\begin{cases} P_g = \dfrac{3}{2}(v_{dg}i_{dg} + v_{qg}i_{qg}) = \dfrac{3}{2}v_{dg}i_{dg} \\ Q_g = \dfrac{3}{2}(v_{qg}i_{dg} - v_{dg}i_{qg}) = -\dfrac{3}{2}v_{dg}i_{qg} \end{cases} \quad \text{for} \quad v_{qg} = 0 \quad (4.98)$$

The *q*-axis current reference i_{qg}^* can then be obtained from

$$i_{qg}^* = \frac{Q_g^*}{-1.5v_{dg}} \quad (4.99)$$

where Q_g^* is the reference for the reactive power, which can be set to zero for unity power factor operation, a negative value for leading power factor operation, or a positive value for lagging power factor operation.

The *d*-axis current reference i_{dg}^*, which represents the active power of the system, is generated by the PI controller for DC voltage control. When the inverter operates in steady state, the DC voltage v_{dc} of the inverter is kept constant at a value set by its reference voltage v_{dc}^*. The PI controller generates the reference current i_{dg}^* according to the operating conditions. Neglecting the losses in the inverter, the active power on the AC side of the inverter is equal to the DC-side power, that is,

$$P_g = \frac{3}{2}v_{dg}i_{dg} = v_{dc}i_{dc} \quad (4.100)$$

As mentioned earlier, the power flow of the inverter system is bidirectional. When the active power is delivered from the grid to the DC circuit, the inverter operates in a rectifying mode ($P_g > 0$), whereas when the power is transferred from the DC circuit to the grid ($P_g < 0$), the inverter is in an inverting mode. The control system will automatically switch between the two operating modes and, therefore, no extra measures should be taken for the controller. To study the bidirectional power flow, the DC load of the inverter can be modeled by a resistor R in series with a battery E, as shown in Figure 4-47. Since the average DC voltage V_{dc} of the inverter is set by its reference v^*_{dc} and is kept constant by the PI controller, the direction of the power flow is set by the difference between E and V_{dc} according to the following conditions

$$\begin{cases} E < V_{dc} & \rightarrow & I_{dc} > 0 & \rightarrow & P_g > 0 & \rightarrow \text{Power from grid to load (rectifying mode)} \\ E > V_{dc} & \rightarrow & I_{dc} < 0 & \rightarrow & P_g < 0 & \rightarrow \text{Power from load to grid (inverting mode)} \\ E = V_{dc} & \rightarrow & I_{dc} = 0 & \rightarrow & P_g = 0 & \rightarrow \text{No power flow between the DC circuit and the grid} \end{cases}$$

$$(4.101)$$

To determine an appropriate DC voltage reference reference v^*_{dc}, one should take system transients and possible grid voltage variations into account. Assume that when the inverter operates under the rated conditions, the modulation index m_a is 0.8. The DC reference voltage can then be set by

$$V^*_{dc} = \frac{\sqrt{6}V_{ai1}}{m_a} = \frac{\sqrt{6}}{0.8} = 3.06 \text{ pu} \quad (V_{ai1} = 1 \text{ pu}) \tag{4.102}$$

which gives around a 20% voltage margin for adjustment during the transients and grid voltage variations.

4.7.2 VOC with Decoupled Controller

To further investigate the VOC scheme, the state equation for the grid-side circuit of the inverter in the *abc* stationary reference frame can be expressed as

$$\begin{cases} \dfrac{di_{ag}}{dt} = \left(v_{ag} - v_{ai}\right)/L_g \\[2mm] \dfrac{di_{bg}}{dt} = \left(v_{bg} - v_{bi}\right)/L_g \\[2mm] \dfrac{di_{cg}}{dt} = \left(v_{cg} - v_{ci}\right)/L_g \end{cases} \tag{4.103}$$

The above equations can be transformed into the *dq* synchronous reference frame

$$\begin{cases} \dfrac{di_{dg}}{dt} = \left(v_{dg} - v_{di} + \omega_g L_g\, i_{qg} \right)/ L_g \\[3mm] \dfrac{di_{qg}}{dt} = \left(v_{qg} - v_{qi} - \omega_g L_g\, i_{dg} \right)/ L_g \end{cases} \tag{4.104}$$

where ω_g is the speed of the synchronous reference frame, which is also the angular frequency of the grid, and $\omega_g L_g i_{qg}$ and $\omega_g L_g i_{dg}$ are the induced speed voltages due to the transformation of the three-phase inductance L_g from the stationary reference frame to the synchronous frame, as discussed in Chapter 3.

Equation (4.104) illustrates that the derivative of the d-axis line current i_{dg} is related to both d- and q-axis variables, as is the q-axis current i_{qg}. This indicates that the system control is cross-coupled, which may lead to difficulties in controller design and unsatisfactory dynamic performance. To solve the problem, a decoupled controller shown in Figure 4-48 can be implemented.

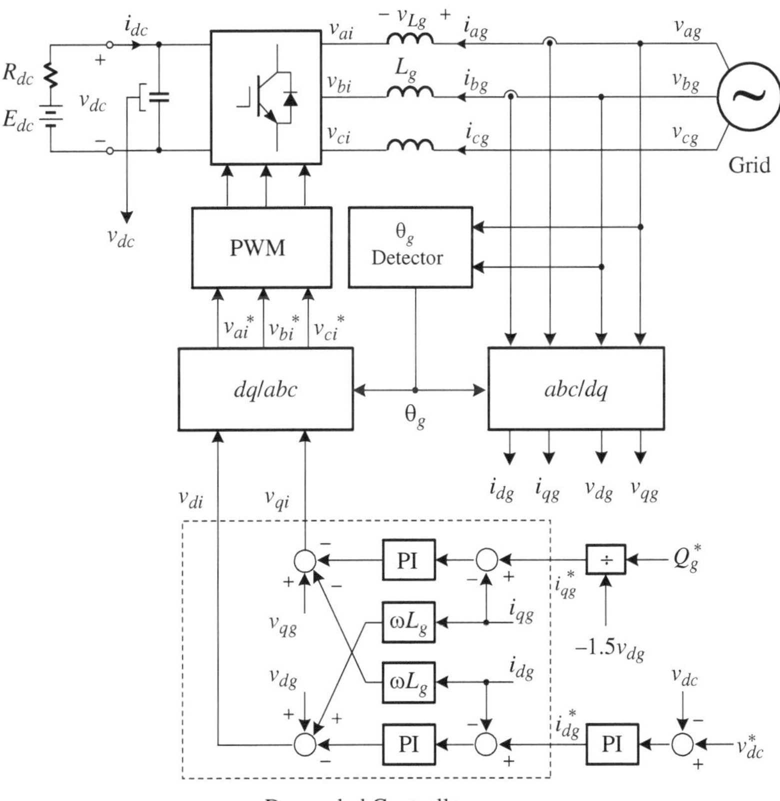

Figure 4-48. Voltage-oriented control (VOC) with a decoupled controller.

Assuming that the controllers for the dq-axis currents in Figure 4-48 are of the PI type, the output of the decoupled controller can be expressed as

$$\begin{cases} v_{di} = -\left(k_1 + k_2 / S\right)\left(i_{dg}^* - i_{dg}\right) + \omega_g L_g i_{qg} + v_{dg} \\ v_{qi} = -\left(k_1 + k_2 / S\right)\left(i_{qg}^* - i_{qg}\right) - \omega_g L_g i_{dg} + v_{qg} \end{cases} \quad (4.105)$$

where $(k_1 + k_2/S)$ is the transfer function of the PI controller.

Substituting (4.105) into (4.104) yields

$$\begin{cases} \dfrac{di_{dg}}{dt} = \left(k_1 + k_2 / S\right)\left(i_{dg}^* - i_{dg}\right) / L_g \\ \dfrac{di_{qg}}{dt} = \left(k_1 + k_2 / S\right)\left(i_{qg}^* - i_{qg}\right) / L_g \end{cases} \quad (4.106)$$

The above equation indicates that the control of the d-axis grid current i_{dg} is decoupled, involving only d-axis components, as is the q-axis current i_{qg}. The decoupled control makes the design of the PI controllers more convenient, and the system is more easily stabilized.

4.7.3 Operation of Grid-Connected Inverter with VOC and Reactive Power Control

The operation of the grid-tied inverter with VOC and reactive power control is analyzed through a case study below.

Case Study 4-6—Operation and Analysis of Grid-Connected Inverter.
Consider a 2.3 MW/690 V grid-connected inverter. This inverter is controlled by the VOC scheme with a decoupled PI controller as shown in Figure 4-48. The DC reference is set to 1220 V, which is 3.06 pu as specified by Equation (4.102). The system parameters and operating conditions are given in Table 4.8.

Figure 4-49a illustrates the space vector diagram for the grid voltage vector $\vec{v}_g$. With the VOC scheme, $\vec{v}_g$ is aligned with the d-axis of the synchronous frame, and rotates in space at the synchronous speed of ω_g, which is also the grid angular frequency given by

$$\omega_g = 2\pi f_g \quad (4.107)$$

where f_g is the frequency of the grid voltage.

The q-axis voltage v_{qg} of the space vector $\vec{v}_g$ is zero, and d-axis voltage v_{dg} is equal to v_g, which is the magnitude (peak value) of $\vec{v}_g$. The angle θ_g of the vector is referenced to the a-axis of the stationary frame. Based on $\vec{v}_g$ and θ_g in Figure 4-49a, the three-phase grid voltage in the stationary frame is reconstructed by

Table 4-8. Parameters and operating condition the grid-connected inverter

Inverter ratings	2.3 MW, 690 V, 1924.5 A	
Control scheme	VOC with decoupled PI controller	Figure 4-48
System input references	v_{dc}^*	1220V (3.062 pu)
	Q_g^*	Adjustable
Inverter	Converter type	Two-level VSC
	Modulation scheme	Space-vector modulation
	Switching frequency	2.04 kHz
DC link circuit	Resistance R	0.0207 ohms (0.1 pu)
	Battery E	1259 V (3.16 pu)
Electric grid	Grid voltage/frequency	690 V/60 Hz
	Line inductance	0.1098 mH (0.2 pu)
Reference frame transformation	*abc*/*dq* and *dq*/*abc* transformation	Equations 3.1 and 3.2, Chapter 3

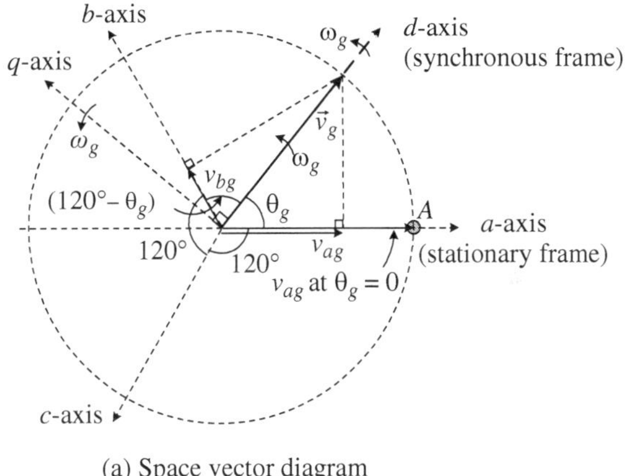

(a) Space vector diagram

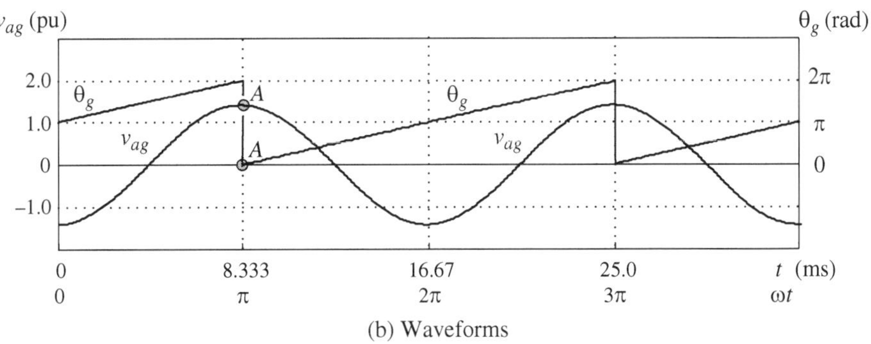

(b) Waveforms

Figure 4-49. Angle of grid-voltage vector for the VOC scheme.

$$\begin{cases} v_{ag} = v_g \cos \theta_g = v_g \cos \omega_g t \\ v_{bg} = v_g \cos \left(\theta_g - 120°\right) = v_g \cos \left(\omega_g t - 120°\right) \\ v_{cg} = v_g \cos \left(\theta_g - 240°\right) = v_g \cos \left(\omega_g t - 240°\right) \end{cases} \qquad (4.108)$$

Figure 4-49b shows the waveforms of the phase-*a* grid voltage v_{ag} and the space angle θ_g. When $\vec{v}_g$ rotates in space, θ_g and v_{ag} varies from zero to 2π periodically. When θ_g is equal to zero, v_{dg} reaches its peak value, shown at point A in Figure 4-49.

The transient waveforms of the inverter are illustrated in Figure 4-50, where the inverter initially delivers the rated active power ($P_g = -1$ pu) and zero reactive power ($Q_g = 0$) to the grid. Ignoring all the ripples (produced by current harmonics), the *d*-axis current i_{dg} is -1.41 pu (rated) and the zero *q*-axis current i_{qg} is zero. The corresponding waveforms of phase-*a* grid voltage and current during the transient are also given in the figure.

At $t = 0.05$ sec, the battery voltage E starts to reduce such that the active power to the grid is reduced to 0.8 pu around $t = 0.075$ sec, which leads the reduction of the *d*-axis current from its rated value to -1.13pu ($\sqrt{2} \times 0.8$). The *q*-axis current remains unchanged during the transients due to the decoupled control of the active and reactive power. The magnitude of the phase-*a* grid current i_{ag} is reduced, but kept out of phase with its voltage.

At $t = 0.15$ sec, the reference for the reactive power Q_g^* starts to vary from zero to -0.5 pu, demanding a leading power factor operation. The *q*-axis current i_{qg} reaches 0.707 pu at $t = 0.20$ sec, which is half of the rated value. The *d*-axis current is almost kept constant during the transients.

Figure 4-51a shows the simulated waveforms of the inverter operating in steady state I defined in Figure 4-50. The peak value of the phase-*a* grid current i_{ag} is 1.41 pu

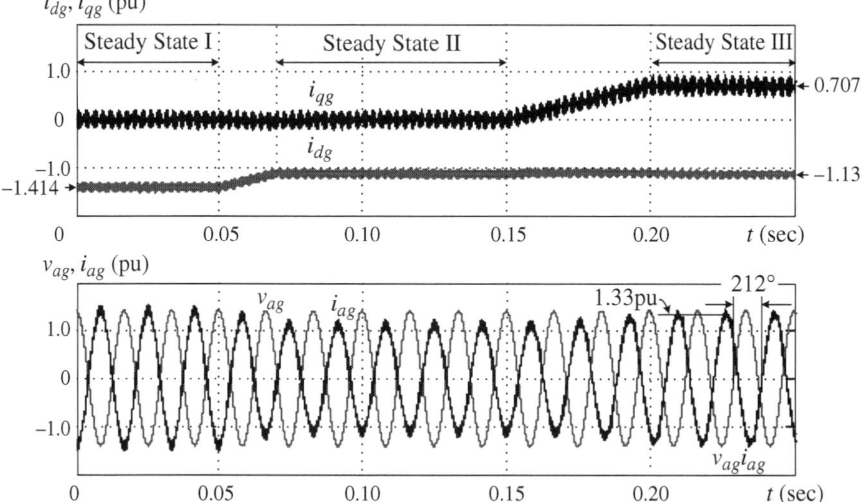

Figure 4-50. Transient waveforms of the grid-tied inverter with voltage-oriented control.

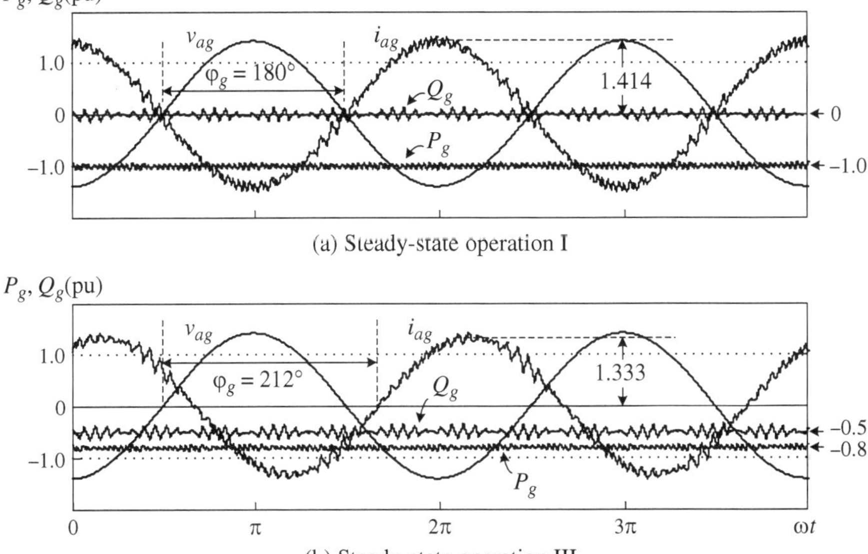

Figure 4-51. Steady-state waveforms of the grid-tied inverter with VOC scheme.

(rated). The grid current i_{ag} is out of phase with its voltage v_{ag}. The active power delivered to the grid is

$$P_g = V_{ag}I_{ag}\cos\varphi_g = \frac{i_{ag}}{\sqrt{2}} \times \frac{v_{ag}}{\sqrt{2}} \times \cos 180° = -1\,\text{pu} \qquad (4.109)$$

The negative value in the above equation indicates that the inverter delivers active power to the grid.

Figure 4-51b shows the simulated waveforms when the system reaches steady state III in Figure 4-50. The measured phase-*a* current i_{ag} is 1.33 pu, which lags phase-*a* voltage v_{ag} by 212°. The active and reactive power to the grid can be calculated by

$$P_g = V_{ag}I_{ag}\cos\varphi_g = \frac{1.333}{\sqrt{2}} \times \frac{1.414}{\sqrt{2}} \times \cos 212° = -0.8\,\text{pu}$$

$$Q_g = V_{ag}I_{ag}\cos\varphi_g = \frac{1.333}{\sqrt{2}} \times \frac{1.414}{\sqrt{2}} \times \sin 212° = -0.5\,\text{pu}$$

$$(4.110)$$

The negative reactive power indicates that the inverter operates with a leading (capacitive) power factor, which corresponds to operation in quadrant II of Figure 4-46. In practical WECS, the capacitive leading power operation is often required to support the grid voltage.

4.8 SUMMARY

A variety of power converter topologies used in wind energy conversion systems (WECS) were analyzed in this chapter, including AC voltage controllers, DC/DC boost converters, two-level voltage source converters (VSC), three-level neutral point clamped (NPC) converters, and PWM current source converters (CSC). The operating principles and switching schemes of these converters were discussed in detail. For wind energy systems with grid-tied converters, a voltage oriented control (VOC) scheme with decoupled PI controllers was elaborated. Case studies were provided for the in-depth analysis of the converter systems. The equations and tables derived in this chapter will be used for the analysis and design of wind energy conversion systems in the subsequent chapters.

REFERENCES

1. F. Blaabjerg and Z. Chen, *Power Electronics for Modern Wind Turbines,* Morgan & Claypool, 2006.
2. B. K. Bose, *Power Electronics and Motor Drives: Advances and Trends,* Academic Press, 2006.
3. B. Wu, *High-Power Converters and AC Drives,* Wiley-IEEE Press, 2006.
4. M. H. Rashid, *Power Electronics Handbook,* 2nd Edition, Academic Press, 2007.
5. N. Mohan, T. M. Undeland, and W. P. Bobbins, *Power Electronics—Converters, Applications and Design,* 2nd Edition, Wiley, 1998.
6. M. O'Loughlin, 350W Two-Phase Interleaved PFC Pre-regulator Design Review, Texas Instruments, 2005.
7. J. Dai, Y. Lang, B. Wu, D. Xu, and N. R. Zargari, A Multisampling SVM Scheme for Current Source Converter with Superior Harmonic Performance, *IEEE Transactions on Power Electronics,* Vol. 24, No. 11, 2436–2445, 2009.

5

WIND ENERGY SYSTEM CONFIGURATIONS

5.1 INTRODUCTION

In a continued effort to reduce the cost, increase the reliability, and improve the efficiency of wind energy conversion systems (WECS), a variety of WECS configurations have been developed. A classification of the most common configurations is given in Figure 5-1, where the wind turbines can be generally classified into the fixed- and variable-speed turbines. The fixed-speed turbines employ a squirrel-cage induction generator (SCIG) connected directly to the grid and, thus, do not need any power converter during normal operation.

The variable-speed wind turbines can be divided into direct- and indirect-drive turbines. In the direct-drive turbines, low-speed synchronous generators (SGs) with a large number of poles are employed. The speed of the synchronous generators is designed to match the turbine speed such that the gearbox normally required in other configurations is eliminated. Both wound-rotor synchronous generators (WRSGs) and permanent-magnet synchronous generators (PMSGs) are suitable for the direct-drive turbines, for which a full-capacity power converter system is required. The converter system serves as an interface between the generator and the power grid.

The indirect-drive turbines require a gearbox to match the low turbine speed to the high generator speed. The WRSG, PMSG, and SCIG equipped with full-capacity power converters have all been used in practical wind energy systems. In addition, doubly fed induction generators (DFIGs) with reduced-capacity converters and wound-rotor

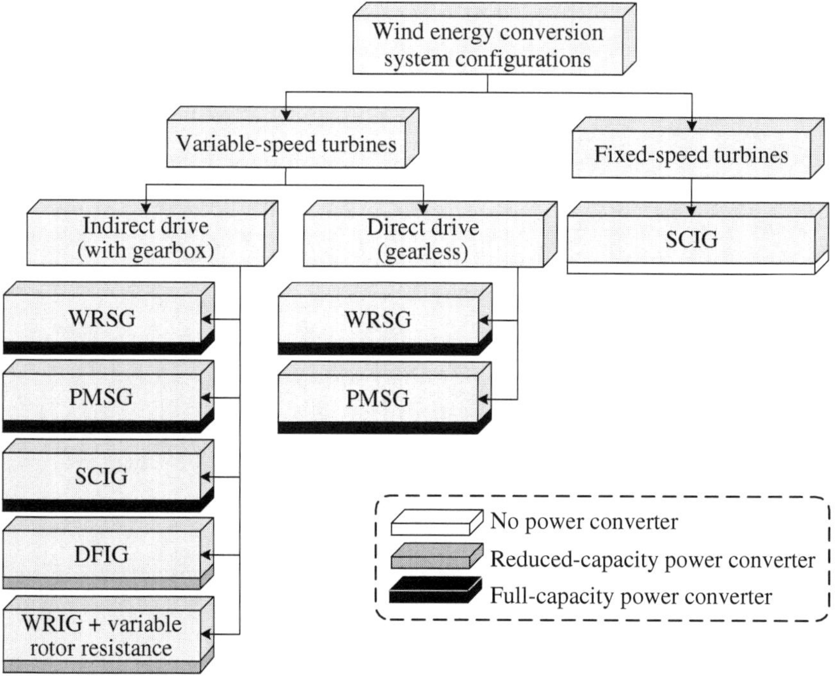

Figure 5-1. Classification of wind energy system configurations.

induction generators (WRIGs) with converter-controlled variable rotor resistance have also found practical applications.

This chapter provides an overview of a variety of WECS configurations, including fixed-speed SCIG wind turbines without power converters, variable-speed WRIG and DFIG turbines with reduced-capacity converters, and variable-speed SCIG and SG turbines with full-capacity power converters. Most of these configurations have found practical applications, and some have been proposed with promising features for further development. The main features and drawbacks of each configuration are discussed.

5.2 FIXED-SPEED WECS

The fixed-speed wind energy systems can be divided into (1) single-speed WECS, in which the generator operates at only one fixed speed; and (2) two-speed WECS, in which the generator can operate at two fixed speeds.

5.2.1 Single-Speed WECS

A typical configuration for a high-power (megawatts), fixed-speed wind energy system is shown in Figure 5-2. The turbine is normally of horizontal-axis type with three

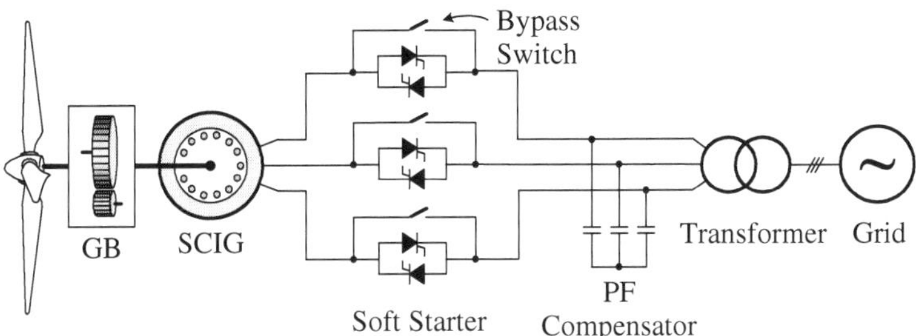

Figure 5-2. Fixed-speed wind energy system configuration.

rotor blades rotating at low speeds, for example, 15 rpm as the rated speed. Squirrel-cage induction generators are exclusively used in the system. Assuming that a four-pole generator is connected to a 50 Hz grid, its speed is slightly higher than 1500 rpm, for which a gear ratio of about 100:1 is required. An example of a commercial fixed-speed WECS is given in Table 1-8 in Chapter 1.

To assist the start-up of the turbine, a soft starter is used to limit the inrush current in the generator winding. The soft starter is essentially a three-phase AC voltage controller (presented in Chapter 4). It is composed of three pairs of bidirectional thyristor switches. To start the system, the firing angle of the thyristors is gradually adjusted such that the voltage applied to the generator is increased gradually from zero to the grid voltage level. As a result, the stator current is effectively limited. Once the start-up process is over, the soft starter is bypassed by a switch, and the WECS is then connected to the grid through a transformer. Since the system does not need a power converter interface during normal operation, it is classified as a WECS without power converters.

To compensate for the inductive reactive power consumed by the induction generator, a capacitor-based power-factor (PF) compensator is normally used. In practice, the compensator is composed of multiple capacitor banks, which can be switched into or out of the system individually to provide an optimal compensation according to the operating conditions of the generator.

Due to the use of a cost-effective and robust squirrel-cage induction generator with inexpensive soft starter, the fixed-speed WECS features simple structure, low cost, and reliable operation. However, compared to the variable-speed WECS, the fixed-speed system has a lower energy conversion efficiency since it can achieve the maximum efficiency only at one given wind speed.

5.2.2 Two-Speed WECS

Two-Speed Operation by Changing Number of Poles. To improve the energy conversion efficiency, two-speed SCIG wind energy systems have been developed. The speed of the generator changes with the number of stator poles. Switching from a four-pole to a six- or eight-pole configuration can introduce a speed reduction of one-third or one-half, respectively. The number of poles can be changed by reconfiguring the stator winding through appropriate parallel and series connection of the stator coils [1]. With the number of poles switched from four to six, for example, a generator connected to a 50 Hz grid can operate at slightly higher than 1500 rpm and 1200 rpm, so the system can capture the maximum power at two different wind speeds, leading to improvements in energy efficiency. The detailed analysis of the efficiency improvements is given in Chapter 6.

Two-Speed Operation by Two Generators. The two-speed operation can also be obtained by having two separate generators mechanically coupled to a single shaft: one is a fully rated high-speed generator (normally four poles) and the other is a partially rated low-speed generator (six or eight poles), as illustrated in Figure 5-3a. The selection of the generators is done through switch S according to the wind speeds. At high wind speeds, switch S is in Position 1, connecting the high-speed generator to the grid. When the wind speed reduces to a certain level, S is switched to Position 2. The low-speed generator is selected and delivers power to the grid. This WECS configuration uses two off-the-shelf generators and, therefore, does not need a customized generator to achieve the two-speed operation. However, this approach requires two separate generators and also a long drive train that needs special consideration for the coupling of both generators.

The two-speed operation can also be obtained by using a split gearbox with two shafts as shown in Figure 5-3b. The two shafts have the same gear ratio, and each shaft is connected to a separate SCIG. Similar to the single-shaft configuration, a fully rated four-pole generator is selected at high wind speeds, whereas a partially rated six- or eight-pole generator is switched on at low wind speeds. This configuration requires a special gearbox, but off-the-shelf generators may be used.

The single- and dual-shaft WECS configurations require two generators, which increases the cost and weight of the system in addition to the added complexity in the mechanical components. Therefore, they have found limited practical applications.

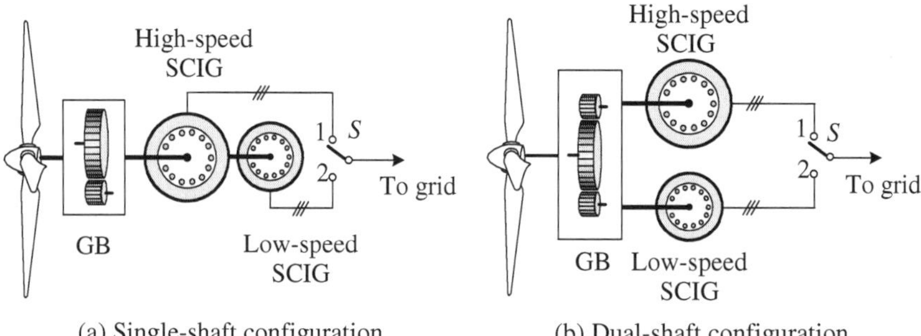

(a) Single-shaft configuration   (b) Dual-shaft configuration

Figure 5-3. Configurations of two-speed WECS with two generators.

5.3 VARIABLE-SPEED INDUCTION GENERATOR WECS

The variable-speed operation of a wind energy system can be realized by a WRIG, in which the rotor is connected to an external variable resistance, or by a DFIG with a reduced-capacity power converter in the rotor circuit. The former is one of the earliest commercial variable-speed WECS, and latter is one of the most widely used WECS to date.

5.3.1 Wound-Rotor Induction Generator with External Rotor Resistance

Figure 5-4 shows the simplified configuration for a variable-speed WRIG wind energy system with a converter-controlled external rotor resistance. The system configuration is the same as that of the fixed-speed wind energy system except that the SCIG is replaced with the WRIG. The external rotor resistance R_{ext} is made adjustable by a converter composed of a diode bridge and an IGBT chopper. The equivalent value of R_{ext} seen by the rotor varies with the duty cycle of the chopper.

The torque-slip characteristics of the generator vary with the external rotor resistance R_{ext}, as shown in Figure 5-5. With different values of R_{ext}, the generator can operate at different operating points. This introduces a moderate speed range, usually less than 10% of the rated speed [2]. Slip rings and brushes of the WRIG can be avoided in some practical WECS by mounting the external rotor resistance circuit on the rotor shaft. This reduces maintenance needs, but introduces additional heat dissipation inside the generator.

The main advantage of this configuration compared to the variable-speed WECS is the low cost and simplicity. The major drawbacks include limited speed range, inability to control grid-side reactive power, and reduced efficiency due to the resistive loss-

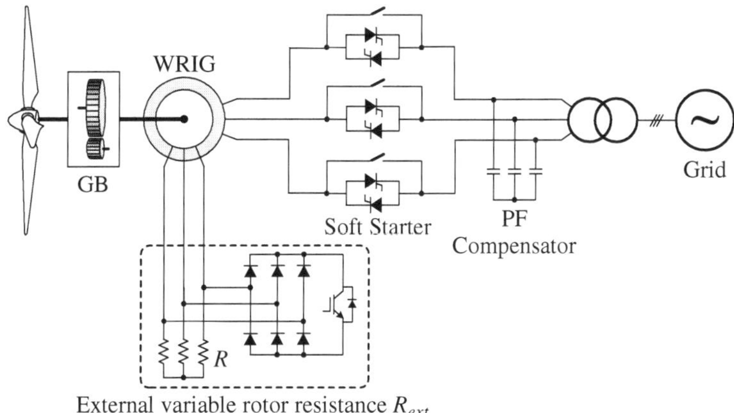

Figure 5-4. WRIG wind energy system with a converter-controlled external rotor resistance.

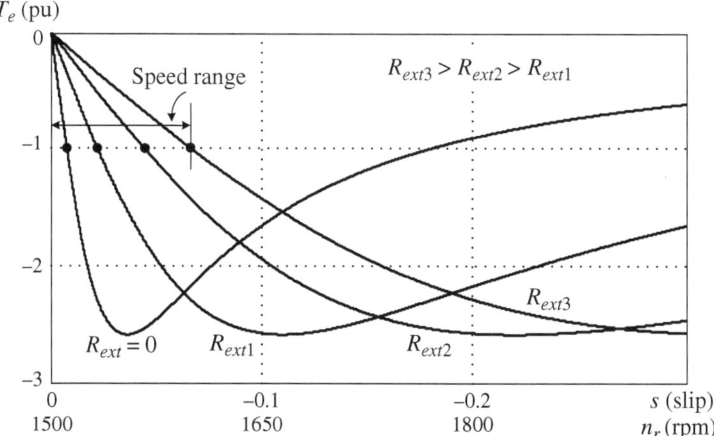

Figure 5-5. Torque-slip characteristics of WRIG with external rotor resistance R_{ext}.

es. An example of commercial WECS with variable rotor resistance is given in Table 1-9 in Chapter 1.

5.3.2 Doubly Fed Induction Generator WECS with Reduced-Capacity Power Converter

The variable-speed DFIG wind energy system is one of the main WECS configurations in today's wind power industry. As shown in Figure 5-6, the stator is connected to the grid directly, whereas the rotor is connected to the grid via reduced-capacity power converters [3]. A two-level IGBT voltage source converter (VSC) system in a back-to-back configuration is normally used. Since both stator and rotor can feed energy to the grid, the generator is known as a doubly fed generator. The typical stator voltage for the commercial DFIG is 690 V and power rating is from a few hundred kilowatts to several megawatts.

The rotor-side converter (RSC) controls the torque or active/reactive power of the generator while the grid-side converter (GSC) controls the DC-link voltage and its AC-side reactive power. Since the system has the capability to control the reactive power, external reactive power compensation is not needed.

The speed range of the DFIG wind energy system is around ±30%, which is 30% above and 30% below synchronous speed [3]. The speed range of 60% can normally meet all the wind conditions and, therefore, it is sufficient for the variable-speed operation of the wind turbine. The maximum slip determines the maximum power to be processed by the rotor circuit, which is around 30% of the rated power. Therefore, the power flow in the rotor circuit is bidirectional: it can flow from the grid to the rotor or vice versa. This requires a four-quadrant converter system as shown in Figure 5-6. However, the converter system needs to process only around 30% of the rated power. The use of reduced-capacity converters results in reduction in cost, weight, and physi-

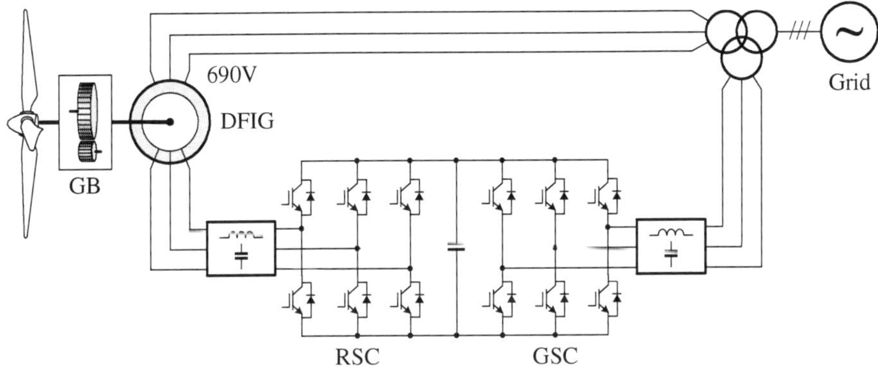

Figure 5-6. Configuration of DFIG wind energy system.

cal size as well. Compared with the fixed-speed systems, the energy conversion efficiency of the DFIG wind turbine is greatly enhanced. Examples of commercial DFIG wind turbines are provided in Table 1-10 in Chapter 1.

Power converters normally generate switching harmonics. To solve the problems caused by the harmonics, different types of harmonic filters are used in practical wind energy conversion systems. The first three filters in Figure 5-7, parts a to c, are normally used in voltage source converters. To simplify the diagrams of various WECS configurations in this chapter, these filters are represented by a general block diagram in part d. The last filter (part e) in the figure is suitable for current source converters (CSC).

The L filter is often used in the generator-side converters to reduce the harmonic distortion of the generator current and voltage, which leads to a reduction of harmonic losses in generator's magnetic core and winding. LC filters may also be used to achieve better results. The LCL filter is often employed in the grid-side converters to

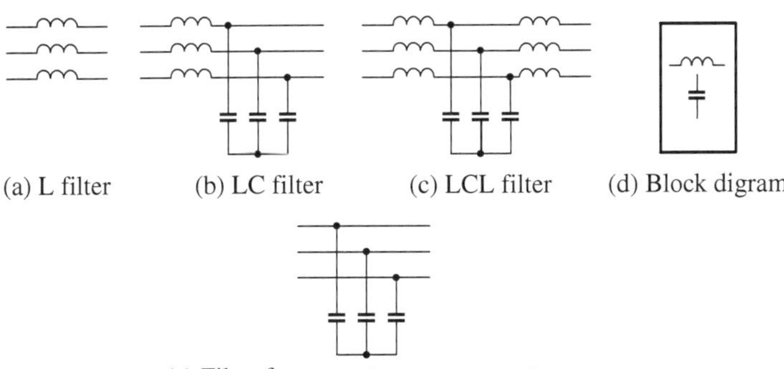

Figure 5-7. Harmonic filters in wind energy conversion systems.

meet stringent harmonic requirements specified by various grid codes. LC filters are also found in practical WECS, but they are not as effective as the LCL filters. An added benefit of using these filters is that they can effectively mitigate high dv/dt problems caused by fast switching of semiconductor switches. However, both LC and LCL filters may cause LC resonances. The filter parameters and resonant modes should be carefully designed to avoid possible LC oscillations.

The filter shown in Figure 5-7e is essentially a three-phase capacitor for current source converters. In addition to the filter function, the capacitor is required by the CSC to assist in the commutation of the semiconductor switches [4]. Therefore, this filter capacitor is indispensable in current source converters.

5.3.3 SCIG Wind Energy Systems with Full-Capacity Power Converters

With Two-Level Voltage Source Converters. A typical voltage source converter configuration for SCIG wind energy systems is shown in Figure 5-8, where a two-level voltage source rectifier (VSR) and voltage source inverter (VSI) using IGBT devices are employed. The two converters are identical in topology and linked by a DC-link capacitive filter. The generator and converters are typically rated for 690 V, and each converter can handle up to 0.75 MW.

For wind turbines larger than 0.75 MW, the power rating of the converter can be increased by paralleling IGBT modules. Measures should be taken to ensure minimum circulating current among the parallel modules. To minimize the circulating current, issues such as dynamic and static characteristics of IGBTs, design and arrangement of gate driver circuits, and physical layout of IGBT modules and DC bus should be considered. Some semiconductor manufacturers provide IGBT modules for parallel operation to achieve a power rating of several megawatts.

An alternative approach to the paralleled converter channels is illustrated in Figure 5-9, where three converter channels are in parallel for a megawatt IG wind turbine. Each converter channel is mainly composed of two-level voltage source converters in a back-to-back configuration with harmonic filters. An additional benefit of the paralleled converter channels is the improvement of energy efficiency [5]. For example, when the system delivers a small amount of power to the grid at low wind speeds, one

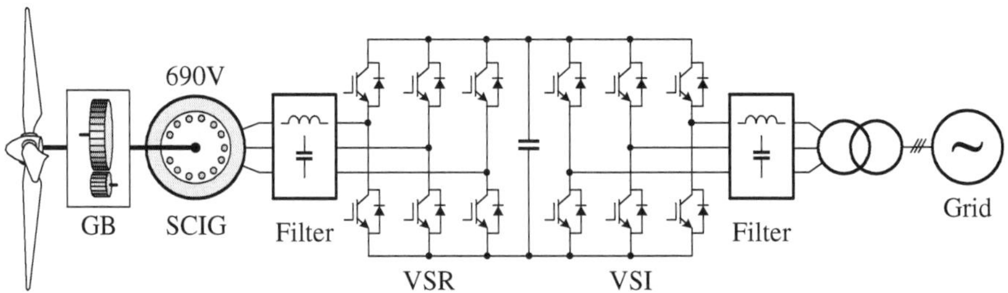

Figure 5-8. Configuration for SCIG wind energy system with two-level VSCs.

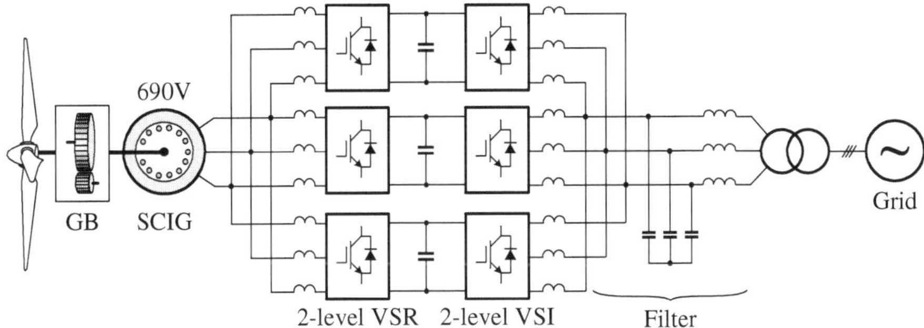

Figure 5-9. Configuration of paralleled converter channels for low-voltage megawatt SCIG wind energy system.

or two converter channels out of three can be turned off, leading to higher system efficiency. This configuration provides redundancy as well, due to the paralleled converter channels. If one channel fails, the other two channels can continue to operate under certain conditions. However, similar to the paralleled IGBT modules, measures should be taken to minimize circulating current among the paralleled converter channels.

With Three-Level NPC Converters. The low-voltage converters discussed in the previous sections are cost-effective at low power levels. As the power rating of wind turbines increases to several megawatts, medium-voltage (MV) wind energy systems of 3 kV or 4 kV become competitive [6]. For example, the rated current of the generator and inverter in a 4 kV/3 MW wind turbine is around 433 A, which is much lower than 2510 A for a 690 V system. The cable cost and losses are reduced in the MV wind energy systems.

Figure 5-10 shows a MV wind turbine that employs a full-capacity converter system with a MV generator. Two back-to-back connected three-level neutral point clamp (NPC) converters are used in the system, where the converter power rating can reach

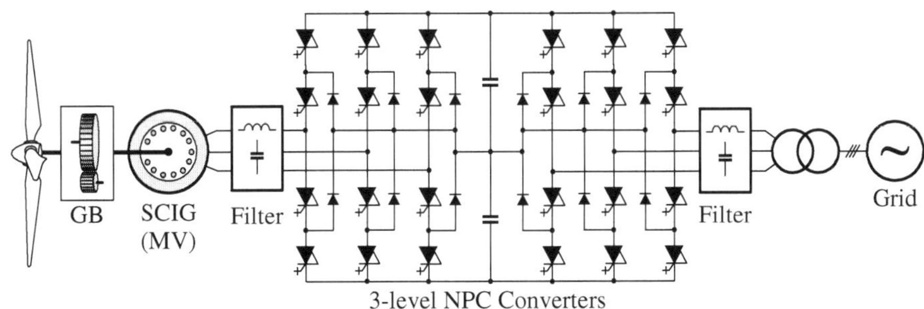

Figure 5-10. Configuration for medium-voltage SCIG wind energy systems with three-level NPC converters.

6 MVA without any series or parallel switching devices or converters [7]. High-voltage switching devices, such as HV-IGBT and IGCTs of 4.5 kV to 6.5 kV, can be employed in the converters. To minimize switching losses, the device switching frequency is normally around a few hundred hertz. Although the NPC converter has found application in commercial medium-voltage SG WECS, commercial medium-voltage SCIG wind turbines have not been reported yet.

All the power converters presented in previous configurations have been of the voltage source type. The current source converter (CSC) technology is also suitable for use in multimegawatt wind energy systems. The CSC technology has been successfully used in high-power applications such as large industrial drives [8], but the application of this technology to MV wind energy systems is yet to be explored.

Figure 5-11 shows a typical current source converter configuration for variable-speed wind energy systems [9]. Two identical converters are employed, one operating as a PWM current source rectifier (CSR) on the generator side and the other as a PWM current source inverter (CSI) on the grid side. As discussed in Chapter 4, these converters require a three-phase capacitor on their respective AC sides to assist commutation of switching devices and mitigate switching harmonics. The rectifier and inverter are linked by a DC choke L_{dc}, which smoothes the DC current and also decouples the generator from the grid.

The current source converter features simple converter structure with low switch count, low switching dv/dt, and reliable short-circuit protection compared to the voltage-ource converter. Although the dynamic response of the CSC may not be as fast as the VSC, it is a promising converter configuration for use in medium-voltage wind energy systems.

5.4 VARIABLE-SPEED SYNCHRONOUS GENERATOR WECS

Synchronous generator wind energy systems have many more configurations than the induction generator WECS. This is mainly due to the fact that (1) the synchronous generator provides the rotor flux by itself through permanent magnets or rotor field winding and, thus, diode rectifiers can be used as generator-side converters, which is

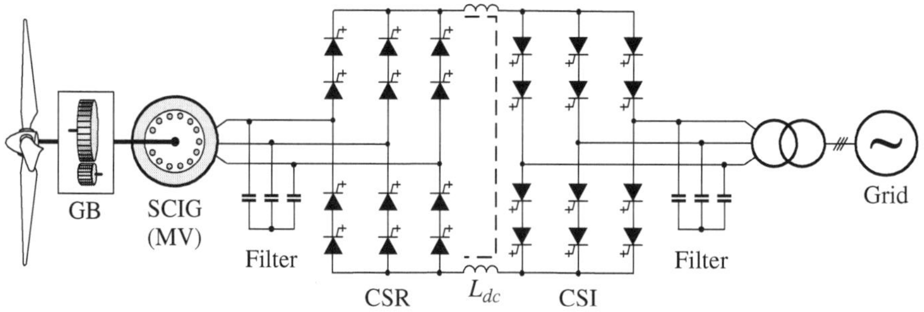

Figure 5-11. Configuration for SCIG with medium-voltage current source converters.

impossible in the induction generator WECS, and (2) it is easier and more cost-effective for the synchronous generator to have multiple-pole (e.g., 72 poles) and multiple-phase (e.g., six phases) configurations than its counterpart.

5.4.1 Configuration with Full-Capacity Back-to-Back Power Converters

With Two-Level VSC and Three-Level NPC Converters. A typical configuration for SG wind energy systems with full-capacity power converters is shown in Figure 5-12, where back-to-back two-level voltage source converters are employed in low-voltage wind energy systems and three-level NPC converters are used in medium-voltage wind turbines. Similar to the SCIG system presented earlier, parallel modules or converter channels are required in the LV systems for generators of more than 0.75 MW, whereas in the MV systems a single NPC converter can handle power up to a few megawatts.

Not all the SG wind turbines need a gearbox. When a low-speed generator with high number of poles is employed, the gearbox can be eliminated. The gearless wind turbine is attractive due to the reduction in cost, weight, and maintenance. Examples of practical SG wind turbine configurations are given in Table 1-11 in Chapter 1.

With PWM Current Source Converters. Figure 5-13 shows a typical configuration for a medium-voltage SG wind energy system using current source converter

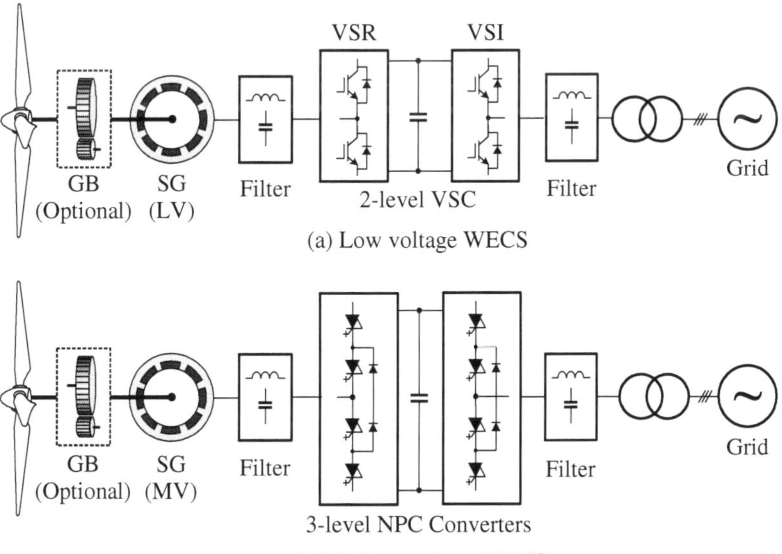

Figure 5-12. SG wind energy system with two-level VSC and three-level NPC converters.

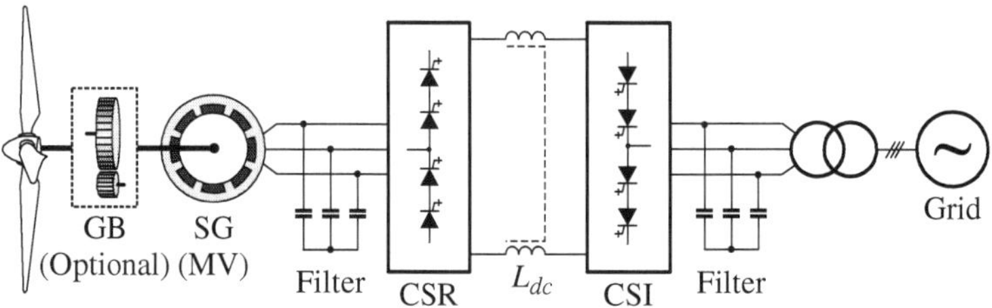

Figure 5-13. Configuration of medium-voltage SG wind energy system using CSC technology.

technology. As discussed earlier, the current source converter has a number of advantages over its counterpart. It is a promising converter topology for large SG WECS at the medium-voltage level of 3 kV or 4 kV.

5.4.2　Configuration with Diode Rectifier and DC/DC Converters

With Diode Rectifier and Multichannel Boost Converters. To reduce the cost of the wind energy systems, the two-level voltage source rectifier in Figure 5-12a can be replaced by a diode rectifier and a boost converter as shown in Figure 5-14a. This converter configuration cannot be used for SCIG wind turbines since the diode rectifier cannot provide the magnetizing current needed for the induction generator. The diode rectifier converts variable generator voltage to a DC voltage, which is boosted to a higher DC voltage by the boost converter. It is important that the generator voltage at low wind speeds be boosted to a sufficiently high level for the inverters, which ensures the delivery of the maximum captured power to the grid in the full wind speed range.

The two-level inverter controls the DC link voltage and grid-side reactive power. The power rating of the system is in the range of a few kilowatts to several hundred kilowatts, and can be further increased to the megawatt level by using a two-channel or three-channel interleaved boost converter as shown in Figure 5-14b.

Compared with the PWM voltage source rectifier, the diode rectifier and boost converter are simpler and more cost-effective. However, the stator current waveform is distorted due to the use of the diode rectifier, which increases the losses in the generator and causes torque ripple as well. Both system configurations illustrated in Figure 5-14 are used in practical systems.

An alternative WECS configuration using a six-phase generator with a multichannel boost converter is shown in Figure 5-15, where the output of the generator is rectified by two diode bridge rectifiers. To increase the power rating, a three-channel interleaved boost converter and two paralleled three-phase inverters are used [10]. This topology provides a low-cost alternative as compared to the full-capacity back-to-back VSC solution.

With Diode Rectifier and Multilevel Boost Converters. Another configuration for the diode boost inverter interface is illustrated in Figure 5-16, where a three-

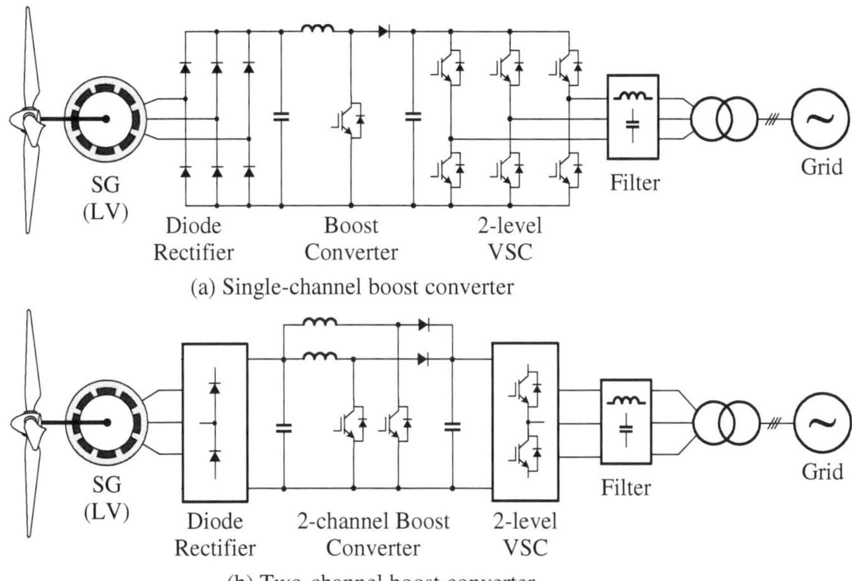

Figure 5-14. Configurations of SG wind energy systems with diode rectifier and boost converters.

level boost converter is used. The three-level boost converter is composed of two single-boost converters connected in cascade. This alternative has found practical application with a power rating up to 1.2 MW.

A variant of this configuration for medium-voltage wind turbines is shown in Figure 5-17, where the operation at the medium voltage of 3 or 4 kV is fulfilled by a three-level boost converter and a three-level NPC inverter. Both IGCT and IGBT can be used in the converters. The added benefit of this configuration is that the neutral point voltage can be effectively controlled by the boost converters, and there is no need to develop a complex control scheme for the NPC inverter to balance the neutral point

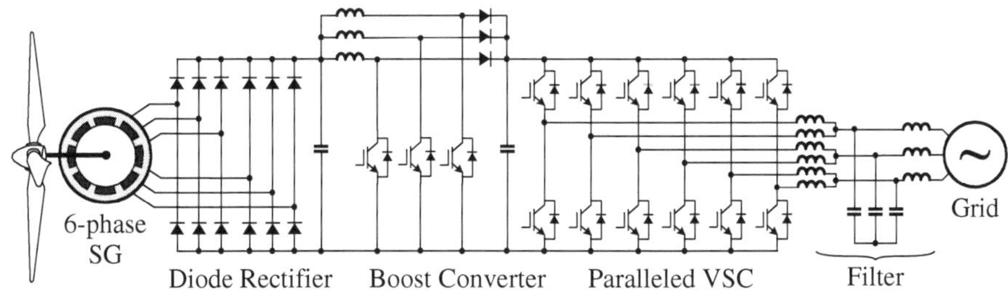

Figure 5-15. Configuration of SG WECS with six-phase generator and three-channel boost converter.

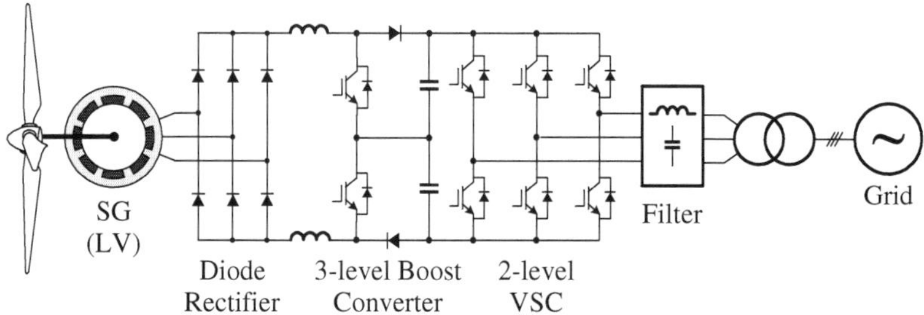

Figure 5-16. Low-voltage wind turbine with diode rectifier and three-level boost converter.

voltage. Although this configuration is not commercialized yet, it is a promising topology for medium-voltage wind turbines.

With Diode Rectifier and Buck Converter for CSC WECS. Considering the concept of duality for voltage- and current source converters, a CSC configuration with diode rectifier and buck converter can be deduced from the VSC configurations presented in the previous section. The boost converter in the VSC topology that boosts the DC output voltage can be replaced by a buck converter that boosts the DC output current [11]. This enables the use of the simple diode rectifier for the CSC configurations, as shown in Figure 5-18.

The buck converter is the natural choice for this topology as it needs an output inductor, which can also serve as the DC-link inductor needed by the CSC. This is contrast to the VSC topology in Figure 5-14a, where the boost converter shares the DC capacitor with the inverter. By controlling the duty cycle of the buck converter and modulation index and delay angle of the inverter, the generator-side active power (or generator torque), DC link current, and grid-side reactive power can be tightly controlled.

Compared to the back-to-back CSC configuration introduced earlier in this chapter, the buck converter based WECS represents a reliable, simple, and cost-effective solu-

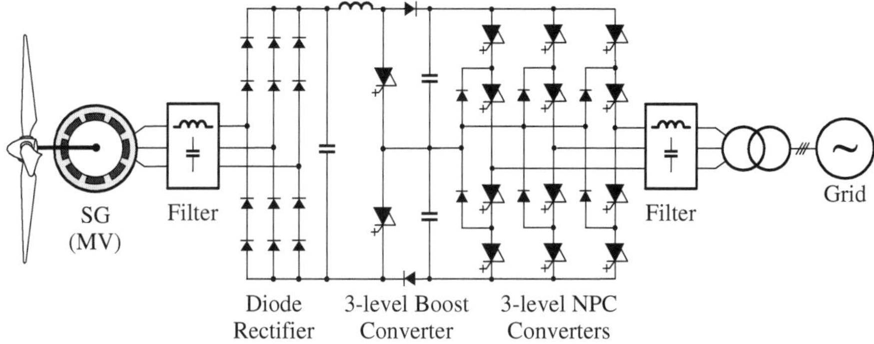

Figure 5-17. Medium-voltage WECS with three-level boost converter and NPC converter.

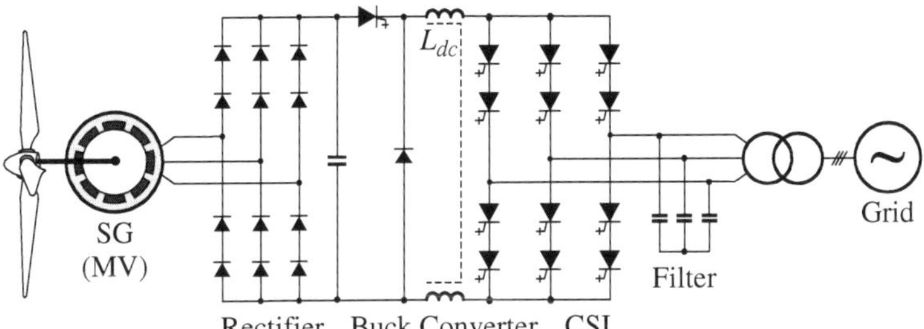

<u>Figure 5-18.</u> Configuration of MV wind turbine with buck converter and current source inverter.

tion. However, the stator current contains higher THD due to the use of the diode rectifier, causing harmonic losses and torque ripples.

5.4.3 Configurations with Distributed Converters for Multiwinding Generators

In addition to paralleling devices or converters as discussed previously, it is possible to increase the power rating of wind energy systems by using distributed converters for a generator with multiple windings or for multiple generators. Correspondingly, the grid-side transformer can also be designed with multiple windings. This configuration has a number of advantages, including

- Low-power converters for megawatt wind turbines. The total generated power can be delivered to the grid through a number of standard two-level voltage source converters. These converters can be mass-produced with low cost and improved reliability.
- No circulating current or power derating. The distributed converters are insulated from each other. There are no circulating currents among the converters, which also leads to no power derating for the converters.
- Low torque ripple and harmonic distortion. In six-phase synchronous generators, the stator voltages across the two stator windings are phase-shifted such that low-order harmonic currents produced by the generator-side converters can be cancelled, leading to the reduction in torque ripples. On the grid side, phase-shifting transformers can be used, which can cancel low-order harmonic currents produced by the grid-side converters. Consequently, smaller size filters can be used with reduced costs and losses.

Configuration with Multiwinding Generators. The multiwinding generator approach is illustrated in Figure 5-19, where a six-phase generator is used and the power is delivered to the grid through two distributed converter channels. Each converter channel is composed of two-level voltage source converters and filters. Since

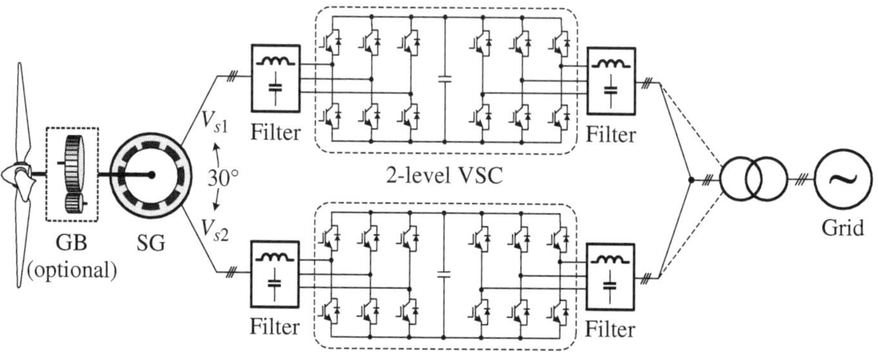

Figure 5-19. Converter configuration for six-phase generator wind energy system.

the two sets of the stator windings are insulated, there is no circulating current between the two converter channels. Therefore, the outputs of the two converter channels can be connected to the same transformer winding. Alternatively, a phase-shifting transformer can be used, as shown in dashed lines of Figure 5-19. With a proper design of switching schemes of the two inverters, the grid-side harmonic performance of the system can be further improved by the phase-shifting transformer.

Another example for the multiwinding generator is shown in Figure 5-20, where the generator has six sets of three-phase windings, and each winding feeds the wind power to the grid through a power converter channel. The system configuration is the same as that with the six-phase generator except that there are no phase displacements between the stator voltages of different windings. Although the multiwinding generator is specially designed at higher cost, this configuration has found practical application due to the advantages discussed earlier [12].

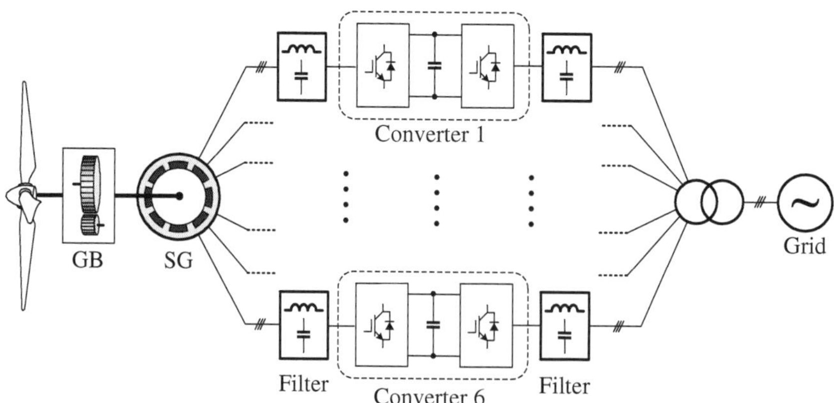

Figure 5-20. Generator with six stator windings for multimegawatt WECS.

Configuration with Multiple Generators. A recent configuration with four synchronous generators and distributed power converter stages is shown in Figure 5-21 [13]. The system uses a distributed gearbox with multiple high-speed shafts that drive four independent generators. Each generator is interfaced to the grid via a partially rated converter channel, composed of a diode bridge rectifier and two-level voltage source converters. Since the power is divided among the four distributed converters, the wind turbine can reach multimegawatt power range without using paralleled switching devices or converters.

The main advantage of this configuration is the high power density achieved by the distributed gearbox and multiple generator system. This leads to a light and small nacelle for a multimegawatt WECS and thus reduces transportation and installation costs. The use of diode rectifier and standard two-level converter makes this a cost-effective solution. This configuration can provide redundancy for possible fault tolerant operation. If one converter channel has a fault, it can be taken out of service, and the power can be easily distributed among the other channels. If the wind speed is high, the blades can be pitched to reduce captured power to compensate for the faulty converter channel. The main disadvantage of the system is that it requires a complex gearbox. This system is commercially available on the market. An example of this configuration is given in Table 1-11 in Chapter 1.

5.5 SUMMARY

With the combination of different types of generators and power converters, a variety of WECS configurations have been developed. This chapter presented the most common configurations for practical and emerging wind turbines. Various technical issues related to these configurations were discussed, including generator types, power converter topologies, active power control, energy conversion efficiency, and grid-side reactive power compensation. Features and drawbacks for each of the configurations

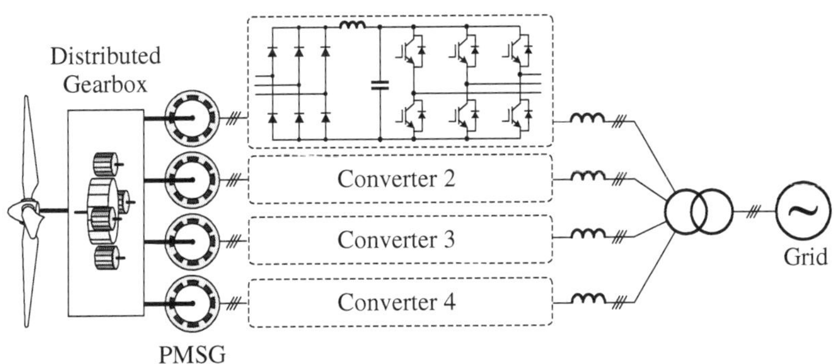

Figure 5-21. Configuration with four generators interfaced by diode rectifiers and two-level VSCs.

Table 5-1. Summary of large WECS configurations

Wine turbine type	Fixed speed		Variable speed			
Generator	SCIG	WRIG + rotor resistance	DFIG	SCIG	WRSG	PMSG
Power converter topologies	No	Diode rectifier + chopper	2-level VSC	2- and 3-level VSC, PWM CSC	2- and 3-level VSC, diode rectifier + boost, PWM CSC	
Converter capacity	Not applicable	Small	Reduced	Full	Full	Full
Speed range	< 1%	< 10%	± 30%	Full	Full	Full
Soft starter	Yes	Yes	No	No	No	No
Gearbox	Yes	Yes	Yes	Yes	Optional	Optional
Aerodynamic power control	Stall, active stall, pitch	Pitch	Pitch	Pitch	Pitch	Pitch
External reactive power compensation	Needed	Needed	No	No	No	No
Active power control and MPPT	Not applicable	Limited range	Yes	Yes	Yes	Yes

were analyzed. A summary of the WECS configurations presented in this chapter is given in Table 5-1.

REFERENCES

1. N. Ammasaigounden, M. Subbiah, and M. Krishnamurthy, Wind-Driven Self-Excited Pole-Changing Induction Generators, *IEE Proceedings Electric Power Applications,* Vol. 133, No. 5, 315–321, 1986.

2. D. Burnham, S. Santoso, and E. Muljadi, Variable Rotor-Resistance Control of Wind Turbine Generators, in *IEEE Power and Energy Society General Meeting* (PES), 2009.

3. S. Muller, M. Deicke, and R. W. de Doncker, Doubly Fed Induction Generator Systems for Wind Turbines, *IEEE Industry Applications Magazine,* Vol. 8, No. 3, 26–33, 2002.

4. B. Wu, *High-Power Converters and AC Drives,* Wiley-IEEE Press, 2006.

5. D. Ehlert, and H. Wrede, Wind Turbines with Doubly-Fed Induction Generator Systems with Improved Performance due to Grid Requirements, in *IEEE Power Engineering Society General Meeting,* pp. 1–7, 2007.

6. S. Kouro, M. Malinowski, K. Gopakumar, J. Pou, L. G. Franquelo, B. Wu, J. Rodríguez, M. Pérez, and J. I. León, Recent Advances and Industrial Applications of Multilevel Converters, *IEEE Transactions on Industrial Electronics,* Vol. 57, No. 8, 2553–2580, 2010.

7. E. J. Bueno, S. Cóbreces, F. J. Rodríguez, A. Hernández, and F. Espinosa, Design of a Back-to-Back NPC Converter Interface for Wind Turbines with Squirrel-Cage Induction Generator, *IEEE Transactions on Energy Conversion,* Vol. 23, No. 3, 932–945, 2008.

8. B. Wu, J. Pontt, J. Rodriguez, S. Bernet, and S. Kouro, Current-Source Converter and Cycloconverter Topologies for Industrial Medium-Voltage Drives, *IEEE Transactions on Industrial Electronics,* Vol. 55, No.7, 2786–2797, 2008.

9. Jingya Dai, Dewei Xu, and Bin Wu, A Novel Control System for Current Source Converter Based Variable Speed PM Wind Power Generators, in *IEEE Power Electronics Specialists Conference* (PESC), pp. 1852–1857, 2007.

10. X. Xiong, and H. Xin, "Research on Multiple Boost Converter Based on MW-Level Wind Energy Conversion System, in *Proceedings of the Eighth International Conference on Electrical Machines and Systems* (ICEMS), Vol. 2, 1046–1049, 2005.

11. J. Dai, *Current Source Converters for Megawatt Wind Energy Conversion Systems,* PhD Thesis, Ryerson University, 2010.

12. J. Birk, and B. Andresen, Parallel-Connected Converters for Optimizing Efficiency, Reliability and Grid Harmonics in a Wind Turbine, in *European Conference on Power Electronics and Applications* (EPE), pp. 1–7, 2007.

13. A. Mikhail, K. Cousineau, L. Howes, W. Erdman, and W. Holley, Variable Speed Distributed Drive Train Wind Turbine System, United States Patent, US 7,042,110 B2, 2006.

6

FIXED-SPEED INDUCTION GENERATOR WECS

6.1 INTRODUCTION

As introduced in Chapters 1 and 5, there are several types of wind energy conversion systems (WECS). One of the earliest systems adopted by the industry and the simplest in terms of structure and control is the fixed-speed WECS [1]. Its simplicity and low costs compared to the variable-speed WECS made this configuration popular during the early years of the wind energy industry, particularly during the 1990s. This configuration is also known as the Danish concept since it was initially developed and successfully marketed by Danish companies.

In the fixed-speed WECS, a squirrel-cage induction generator (SCIG) is normally used. The generator shaft is driven by the wind turbine and its stator is directly connected to the grid. Under normal operating conditions, the stator frequency is fixed to that of the grid and the slip is very low (less than 1% for megawatt generators). The rotor speed, determined by the difference between the stator and slip frequencies, varies little with wind speed. Such a system is, therefore, referred to as a fixed-speed WECS.

The fixed-speed WECS features simple configuration, reliable operation, and low costs for manufacturing, installation, and maintenance. However, the fixed-speed operation causes higher mechanical stress and higher power fluctuation compared with the variable-speed wind energy systems. The system has a lower overall energy conversion efficiency since the generator speed cannot be adjusted to achieve maximum

Power Conversion and Control of Wind Energy Systems. By B. Wu, Y. Lang, N. Zargari, and S. Kouro **173**
© 2011 the Institute of Electrical and Electronics Engineers, Inc. Published 2011 by John Wiley & Sons, Inc.

power operation at different wind speeds. Furthermore, the fixed-speed wind energy system is unable to control the reactive power to the grid without using additional devices. Nevertheless, the fixed-speed WECS is one of the viable technologies in the wind industry, especially in offshore wind farms where the wind is steadier and maintenance is more costly than inland wind farms.

This chapter presents the operating principle and main characteristics of the fixed-speed WECS. The single-speed and two-speed wind energy systems are introduced. The system start-up with both direct grid connection and soft starter are investigated. The grid-side reactive power compensation using capacitor banks is analyzed. The system analysis is assisted by computer simulations and steady-state calculations through case studies.

6.2 CONFIGURATION OF FIXED-SPEED WIND ENERGY SYSTEMS

Figure 6-1 shows a typical fixed-speed wind energy system, which is mainly composed of a three-blade wind turbine, a gearbox, a squirrel-cage induction generator, a three-phase soft starter, a power-factor (PF) compensator, and a transformer for grid connection. Examples of commercial fixed-speed WECS are given in Table 1-8 of Chapter 1. The functions of the main components are introduced below.

6.2.1 Wind Turbine

The wind power is captured by the blades, which convert the wind kinetic energy into rotational mechanical energy. Due to aerodynamic and mechanical design considerations, the three-blade wind turbine rotating at low speeds has been the mainstream choice for large (megawatt) fixed-speed wind energy systems.

The fixed-speed WECS starts to produce power when the wind speed is higher than the cut-in speed, which is normally around 3–4 m/s. When the wind speed is higher than the rated speed of around 12–15 m/sec, the power captured by the turbine is limited either by passive aerodynamic stall, active stall, or pitch control of the blades, as

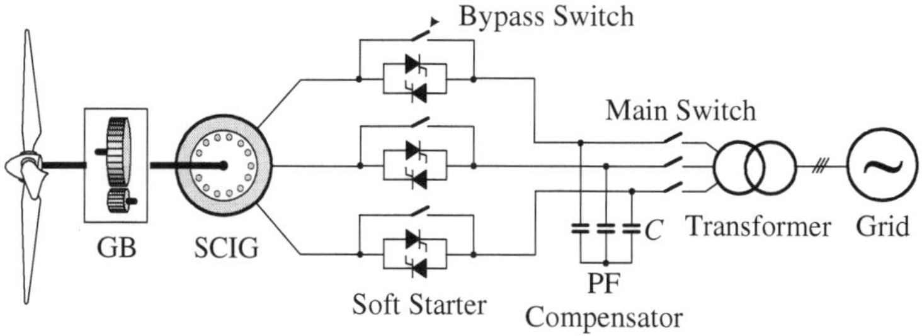

Figure 6-1. Configuration of fixed-speed SCIG wind energy conversion system.

discussed in Chapter 2. When the wind speed exceeds the cut-out speed of around 25 m/s, the turbine is stopped by either full stall or full pitch of the blades to protect the turbine and generator from possible damage.

6.2.2 Gearbox

The rotating speed of large fixed-speed wind turbines is normally in the 6 to 25 rpm range, whereas the induction generator operates at much higher rotating speeds. The generator operating speed is determined by the grid frequency and the number of poles. Since a commonly used large SCIG has four, six, or eight poles, the synchronous speed of the generator is in the range of 750 rpm to 1800 rpm for a grid frequency of 50 or 60 Hz. For instance, for a four-pole generator connected to a grid of 50 Hz, the synchronous speed is 1500 rpm. Therefore, a gearbox with a high gear ratio is required to match the low speed of the turbine to the high speed of the generator. A multiple-stage gearbox is usually required to achieve the high gear ratio. Examples of gear ratios for fixed-speed WECS are given in Table 6-1.

6.2.3 Generator

Squirrel-cage induction generators are exclusively used in the fixed-speed WECS. The generator power rating is in the range of a few kilowatts to a few megawatts. For large wind farms, megawatt generators are widely employed. The squirrel-cage induction generator is simple, reliable, cost-effective, and maintenance-free compared to other types of wind generators. This is achieved by using a robust squirrel-cage rotor structure, in which the rotor winding is made of copper bars embedded in the rotor magnetic core. As a result, slip rings and brushes that are required in the wound-rotor induction and synchronous generators are eliminated.

With the generator connected to the grid of 50 Hz or 60 Hz, the synchronous speed of the generator is fixed, and the rotor speed varies slightly with the system operating conditions. Figure 6-2 shows a torque-versus-speed curve of a 2.3 MW, 690 V, 50 Hz SCIG. When this four-pole generator delivers the rated power to the grid of 50 Hz, its rated rotor speed is 1512 rpm, which is only 0.8% higher than the synchronous speed of 1500 rpm. Therefore, the speed range for the generator is from 1500 rpm to 1512

Table 6-1. Examples of gear ratios for fixed-speed WECS (rated slip = −1%)

| Rated turbine speed (rpm) | Gear ratio | | | | | |
| | 50 Hz grid | | | 60 Hz grid | | |
	4-pole	6-pole	8-pole	4-pole	6-pole	8-pole
12	126	84	63	152	101	76
14	108	72	54	130	87	65
16	94	63	47	114	76	57

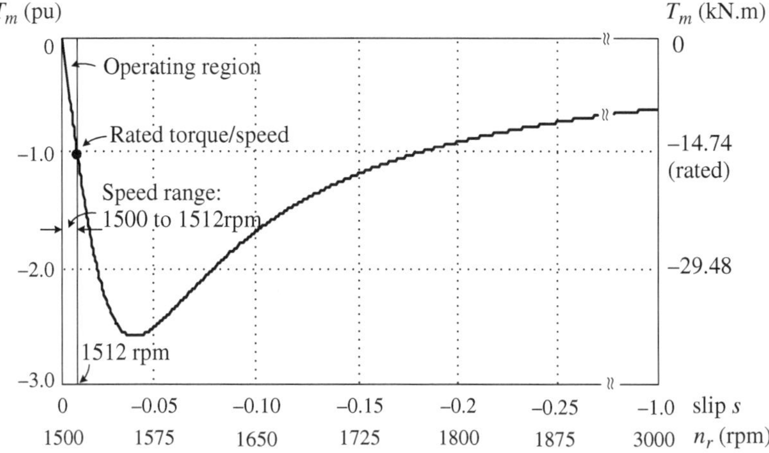

Figure 6-2. Torque-versus-speed curve of a 2.3 MW, 690 V, 50 Hz SCIG.

rpm under normal operation conditions. The speed range is so narrow that this type of wind energy system is known as the fixed-speed WECS.

6.2.4 Soft Starter

A soft starter is normally used between the generator and the grid to limit the high inrush current during the system start-up. The soft starter is basically an AC voltage controller that increases the stator voltage gradually by controlling the firing angles of the SCR devices. The underlying operating principle is to apply segments of the grid voltage waveform in such a way that the resulting voltage to the stator has a smaller fundamental component, which effectively limits the current drawn by the stator during start-up. The firing angles are varied to increase the stator voltage gradually from zero to the full voltage of the grid. After the full voltage of the grid is applied to the generator, the soft starter is shorted by a bypass switch to eliminate the conduction losses of the SCR devices. In-depth analyses of the soft starter and its operating principle is given in Chapter 4 and, therefore, not be repeated here.

6.2.5 Reactive Power Compensation

The squirrel-cage induction generator draws lagging (inductive) reactive power from the grid during operation. A PF compensator device is normally installed to meet the grid code for reactive power compensation. The most commonly used PF compensator for the fixed-speed wind energy system is a power capacitor, which provides the leading reactive power to the grid. To effectively compensate for the reactive power of the system over the full operating range, a number of capacitor banks can be used, which can be switched in or out of the system for optimal results.

6.2.6 Main Features and Drawbacks

The fixed-speed wind energy system has a number of features. Compared with the variable-speed WECS, the fixed-speed system is simpler and more cost-effective. The induction generator in a fixed-speed WECS is directly connected to the grid without the need of a full-capacity power converter system. A soft starter is required to assist in the system start-up, but it uses low-cost SCRs instead of expensive IGBTs or IGCTs. The starter is bypassed as soon as the start-up process is over and, therefore, does not require large-size heat sinks. The squirrel-cage induction generator is robust and inexpensive, and does not have slip rings or brushes, which makes the generator almost maintenance-free. In addition, the system does not need digital controllers or associated sensors for sophisticated controls that are implemented in the variable-speed WECS.

The fixed-speed wind energy system has some drawbacks. First of all, the system has low power conversion efficiency. The power that can be captured by the wind turbine for a given wind speed varies with the rotational speed of the turbine. In order to capture the maximum power available from the wind at different wind speeds, the speed of the turbine and generator must be adjusted according to the optimal tip speed ratio (TSR), that is, the ratio of the tip speed of the blade to wind speed [2]. Hence, when the wind speed varies, the turbine speed needs to be changed accordingly to maintain the optimal TSR and capture the maximum power from the wind. Since direct grid connection fixes the speed of the generator and turbine, the optimal TSR cannot be maintained for different wind speeds, which reduces the power conversion efficiency of the system.

The fixed-speed operation and uncontrolled generator do not allow the system to react quickly to wind gusts. This causes large power fluctuations transmitted to the grid and severe mechanical stress to the blades, the drive train, and, in particular, the gearbox, which may shorten the life of these mechanical components. The inability to control the active power combined with the inability to control reactive power makes large fixed-speed WECS unsuitable for weak grids. The main advantages and drawbacks of the fixed-speed wind conversion system are summarized in Table 6-2.

6.3 OPERATION PRINCIPLE

6.3.1 Fixed-Speed Operation of SCIG

To discuss the operating principle of fixed-speed wind energy systems, the power-versus-speed characteristics of both wind turbine and generator are illustrated in Figure 6-3, where P_m is the mechanical power applied to the shaft of the generator and ω_m is the mechanical speed of the generator. Assuming that the mechanical losses of the gearbox and drivetrain are neglected, the mechanical power produced by the turbine is equal to the mechanical power of the generator in per unit terms. In addition, the mechanical speed of the turbine is equal to that of the generator in per unit terms. Therefore, the power versus speed curves of the turbine and generator can be shown in the same plot.

Table 6-2. Summary of main characteristics of fixed-speed WECS

Advantages	Description
Simple system and low costs	No PWM power converters No closed-loop controls Cost-effective generator (SCIG)
Reliable generator and low maintenance	Squirrel-cage rotor, no rotor winding No slip rings or brushes Compact size and light weight

Disadvantages	Description
Low conversion efficiency	Fixed-speed operation, unable to implement MPPT
Low power quality	No reactive power compensation capability Fluctuation in output power No voltage ride-through capability
High mechanical stress	Caused by gusts of wind and inability to control active power Reduction in life span of mechanical components

Each turbine P_m–ω curve of the turbine corresponds to a specific wind speed. In theory, the number of such curves is infinite, but only a few curves are plotted in Figure 6-3 for analysis. On the contrary, the generator has only one P_m–ω_m curve. The speed of the generator varies slightly with its mechanical power, but the variation is so small (e.g., 0.008 pu in Figure 6-2) that the P_m–ω_m curve of the generator is virtually a vertical line. In steady state, the system operates at one of the intersections of the turbine and generator P–ω curves.

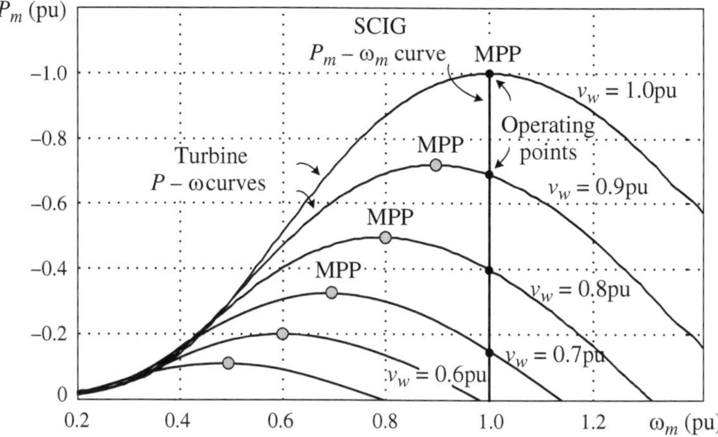

Figure 6-3. System operating points and maximum power points (MPP) at different wind speeds.

For a given wind speed, there is a corresponding maximum power point (MPP) on the turbine P–ω curves. For the fixed-speed WECS, the system can operate only at one MPP, which is at the rated wind speed of 1.0 pu in the example of Figure 6-3. At the other wind speeds, the system operates at the points that are lower than MPP and, therefore, cannot capture the maximum power available from the wind, leading to lower power-conversion efficiency. In particular, when the wind speed is below 0.6 pu, no power can be captured by the system.

6.3.2 Two-Speed Operation of Fixed-Speed WECS

In order to improve power conversion efficiency of the fixed-speed WECS, two-speed SCIG wind energy systems have found practical applications [3]. The two-speed operation is realized by changing the number of poles of the stator winding. The synchronous speed of a generator connected to a grid of 60 Hz is 1800 rpm and 1200 rpm with four-pole and six-pole configurations, respectively. By changing number of poles, the generator can operate at two different rotor speeds that are slightly higher than its synchronous speed. The generator configured with four-pole operation at the rated wind speed can be switched to a six-pole configuration to allow lower speed operation at lower wind speeds. In this way, two optimal tip speed ratios that correspond to two maximum power points can be achieved at two different wind speeds. Doing so, more energy from the wind can be harvested, improving the overall efficiency of the wind energy system.

To facilitate the analysis, a simplified stator winding with four-pole and eight-pole configurations is illustrated in Figure 6-4, where the phase-a winding is illustrated together with a schematic for three-phase winding for both configurations. The stator winding of the two-speed generator can be reconfigured through a switch. Changing from the four-pole to eight-pole configuration, and vice versa, is performed by switch S. When S is in Position 1, the stator winding produces only four magnetic poles. Coils a_1 and a_3 are in series, which are then in parallel with series-connected a_2 and a_4. When S is in Position 2, the stator winding has eight poles, and all four coils in phase-a are connected in series.

To better understand the principle of two-speed operation, consider Figure 6-5, where the generator can operate at two fixed rotor speeds: $\omega_{m1} = 1.0$ pu with a four-pole stator winding and $\omega_{m2} = 2/3$ (0.67) pu with a six-pole stator winding. With the four-pole configuration at the rated wind speed ($v_{w1} = 1.0$ pu), the generator operates at Point A, at which the maximum power capture is achieved. When the wind speed reduces to v_{w2}, the captured power is only 0.07 pu, which is substantially lower than the maximum power of 0.296 pu available for that wind speed. The power difference ΔP is 0.226 pu, which represents the power loss due to the fixed-speed operation. However, by changing the stator winding from four poles to six poles, the generator can operate at Point B with a fixed speed of 0.67 pu, at which the wind energy system can capture the maximum power of 0.296 pu available from the wind.

Figure 6-6 shows the switch between the two fixed rotor speeds. For a given rotor speed ω_{m1} or ω_{m2}, the operating points at difference wind speeds are indicated by solid dots and small circles. The switch between the two rotor speeds takes place at a specific wind speed, at which the two rotor speeds produce the same mechanical power P_m.

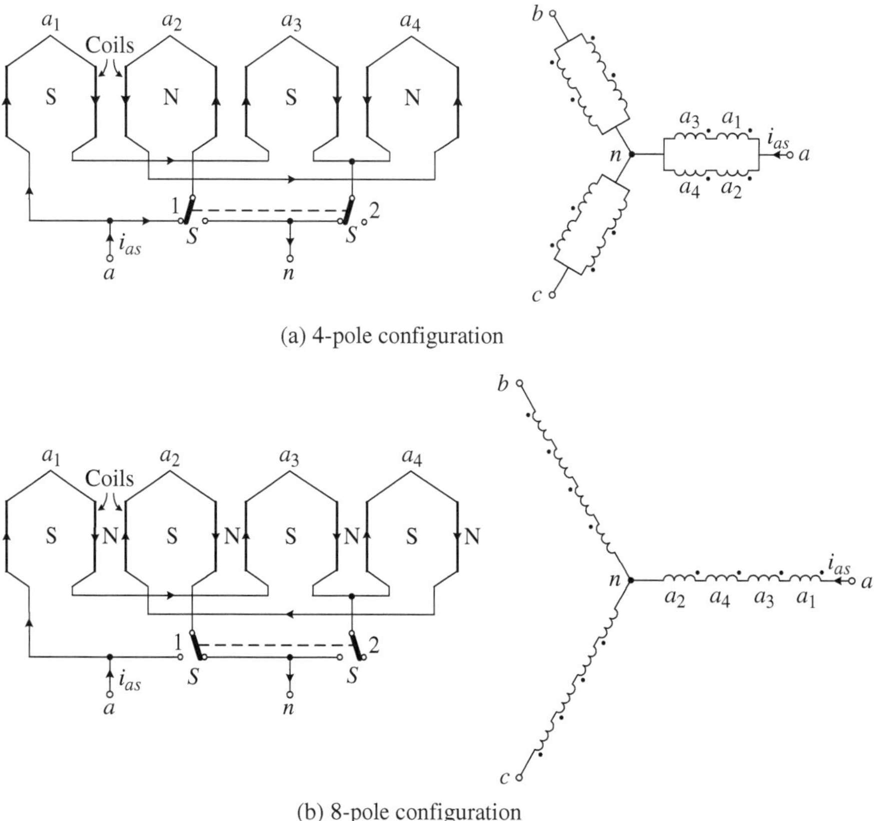

(a) 4-pole configuration

(b) 8-pole configuration

<u>Figure 6-4.</u> Induction generator with four-pole and eight-pole configurations.

In this example, the wind speed of 0.875 pu satisfies the requirement. For a given wind speed v_w, the operating point indicated by the solid dot always generates higher power than that indicated by the circle. Therefore, the power conversion efficiency of the wind energy system is improved by the two-speed operation.

6.4 GRID CONNECTION WITH SOFT STARTER

In a variable-speed WECS, the generator is connected to the grid through a converter system, and the stator current, torque, and speed of the generator are fully controllable during the system start-up. On the contrary, the generator in fixed-speed wind energy systems is directly connected to the grid without converters and, therefore, the operation of the generator is not controllable. However, a soft starter can be used during the system start-up [4], especially in large megawatt systems, to reduce the adverse impact of high inrush current on the wind energy system as well as the grid.

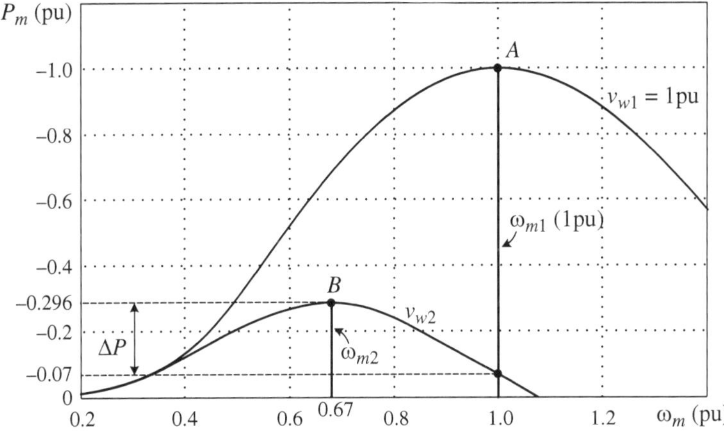

Figure 6-5. Power versus rotor speed characteristics of SCIG WECS with two fixed rotor speeds.

The dynamic performance of a fixed-speed WECS with direct grid connection was investigated in Case Study 3-1 in Chapter 3, where a 2.3 MW, 690 V, 50 Hz, 1512 rpm squirrel-cage induction generator was used. For convenience, the transient waveforms during the system start-up are rearranged and presented in Figure 6-7. The generator is directly connected to the grid by a switch at $t = 0$ when the wind brings the rotor speed close to the rated speed (0.959 pu, 1450 rpm). The direct connection causes excessive inrush currents (i_{as}, i_{bs}, and i_{cs}) in the stator with a peak value of more than 10 pu, high electromagnetic torque T_e (2 pu, peak), and, more importantly, high torque oscillations. The high inrush current may have an adverse impact to the grid, especially in the

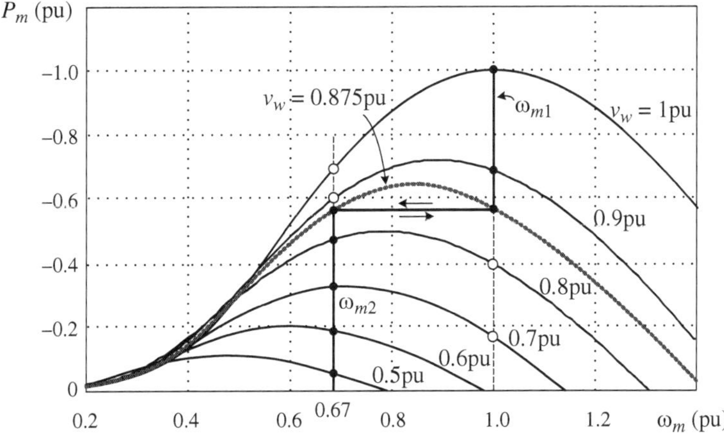

Figure 6-6. Switch between the two fixed rotor speeds ω_{m1} and ω_{m2}.

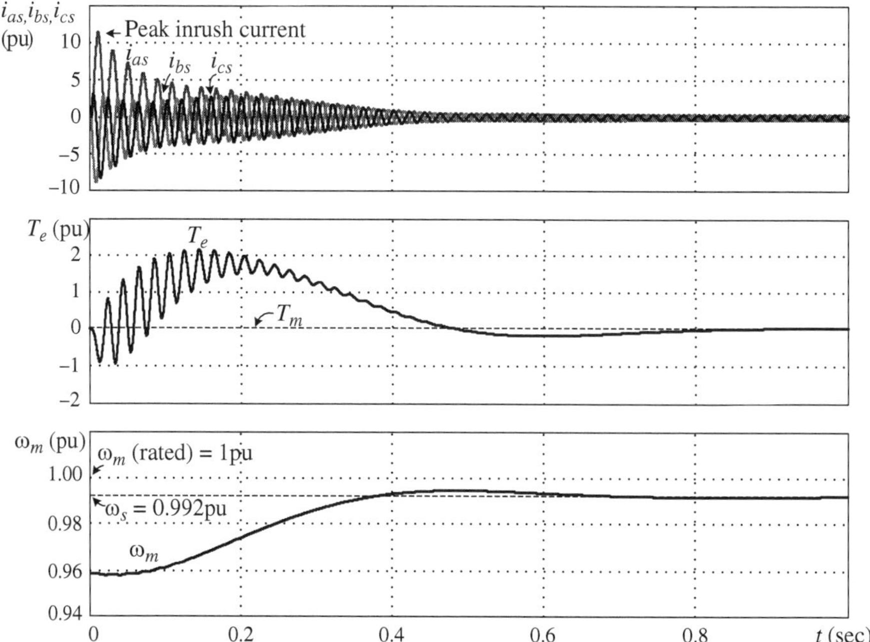

Figure 6-7. Dynamic response of SCIG with direct grid connection (without soft starter).

weak grid systems, and the high torque oscillations may cause excessive mechanical stress. Obviously, the direct grid connection of the SCIG during system start-up is not allowed in practice, especially for large turbines.

In order to reduce the inrush current and torque oscillations during start-up, a soft starter is normally employed in fixed-speed wind energy systems. The dynamic performance of the system during the soft start is investigated through the following case study.

Case Study 6-1—Soft Start of Fixed-Speed WECS

The fixed-speed wind energy system under investigation is shown in Figure 6-1. The power factor compensator is omitted for simplicity. A 2.3 MW, 690 V, 50 Hz, 1512 rpm squirrel-cage induction generator is used in the study. When the wind speed is higher than its cut-in speed, the pitch angle of the blade is slightly adjusted, and a small amount of torque is produced to accelerate the turbine and the generator. Although the generator is rotating during the acceleration period, no stator voltage is induced due to the lack of magnetizing current to produce a magnetic flux that is needed to generate the stator voltage.

When the rotor speed is close to its synchronous speed (0.959 pu, 1450 rpm), the generator is connected to the grid by the main switch shown in Figure 6-1. At the same time, the soft starter is activated with a large firing angle of 120°. A very low voltage is produced by the soft starter, and a small stator current starts to flow through the generator as shown in Figure 6-8.

The firing angle is then decreased gradually from 120° to 0° in 0.5 sec, at which the full grid voltage is applied to the generator. Compared with the transient waveforms produced by the direct grid connection in Figure 6-7, the peak inrush current is reduced from more than 10 pu to approximately 3.3 pu, which is mainly due to the slow increase of the stator voltage by the soft starter. Since the inrush current is substantially reduced, the torque oscillations are almost eliminated. The absence of torque oscillations reduces mechanical stress in the drive train, increasing the life span of the mechanical system and reducing maintenance needs as well.

During the system start-up, the mechanical torque T_m from the shaft of the generator is close to zero since a very small amount of torque is applied to the turbine. The generator operates below the synchronous speed in the motoring mode, producing a positive electromagnetic torque T_e that accelerates the generator until it reaches the synchronous speed. At this point, the bypass switch is closed and the system start-up process is completed. The system is now ready to capture and produce power from the wind.

To produce power from the wind, the pitch angle is adjusted from its start-up position to an optimal value. The mechanical torque T_m increases accordingly until it reaches its rated value of -1.0 pu, assuming that the wind speed is rated as well. Since the generator is in the generating mode and the mechanical torque $|T_m|$ from the gener-

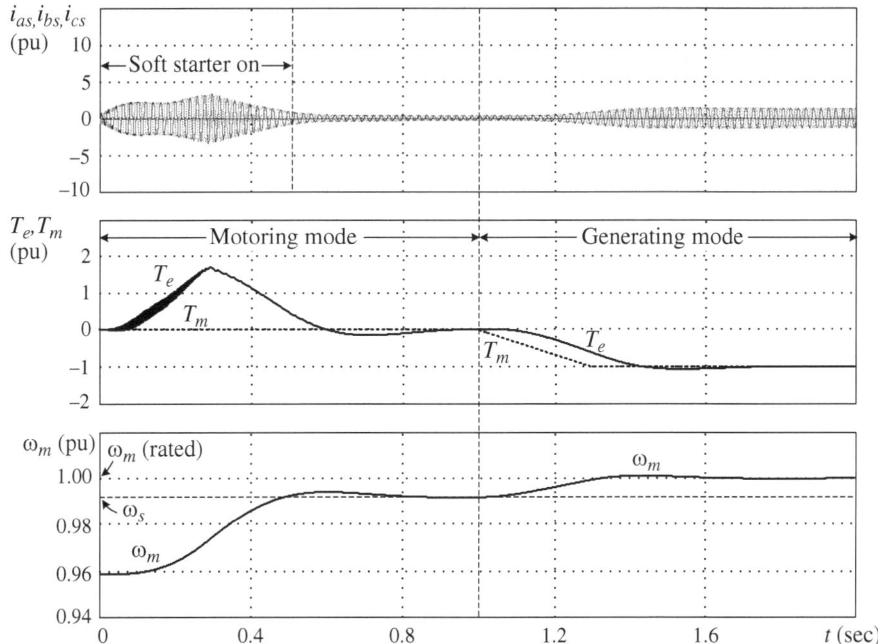

Figure 6-8. Startup transients of fixed-speed WECS with a soft starter.

ator shaft is higher than the electromagnetic torque $|T_e|$ produced by the generator, the rotor speed ω_m accelerates, and finally reaches its rated value of 1.0 pu, at which $T_e = T_m$. The system enters into its steady-state operation, delivering the rated power to the grid.

6.5 REACTIVE POWER COMPENSATION

The induction generator in a fixed-speed wind energy system draws reactive power from the grid. In a variable-speed WECS, the reactive power control is usually realized by power converters that connect the generator to grid. However, this method is not applicable in fixed-speed WECS. To meet the grid code, the reactive power of fixed-speed wind energy systems has to be compensated for [5].

The amount of reactive power Q drawn by an induction generator varies with the active stator power P_s or the slip of the generator. This can be appreciated in Figure 6-9 for a typical large induction generator, where S_R and P_R are the rated apparent power and rated stator active power of the generator, respectively. The generator draws reactive power close to a third of the total power rating when it is not delivering any active power to the grid. This reactive power is mainly associated with the magnetizing inductance L_m of the generator. The amount of reactive power increases with the stator active power delivered to the grid. This increase is mainly caused by the large rotor current flowing through the stator and rotor leakage inductances L_{ls} and L_{lr}.

To overcome this problem, three-phase capacitors are commonly used as reactive power compensators [6]. Since the amount of reactive power varies with the active power as shown in Figure 6-9, the level of compensation needs to be adjusted accordingly. Therefore, it is common to have a set of capacitors that can provide different levels of reactive power compensation by switching them in or out of the sys-

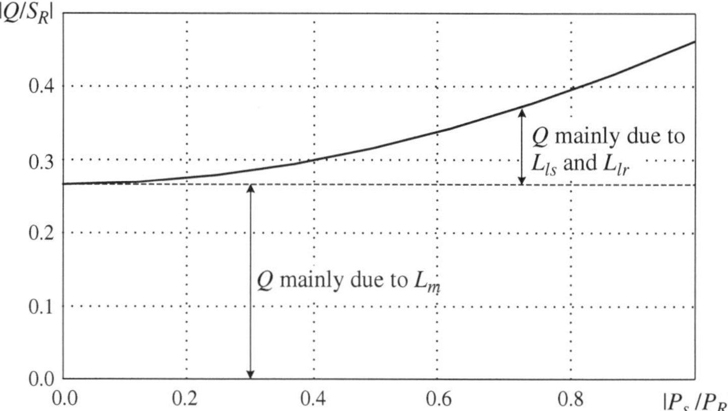

Figure 6-9. Typical characteristics of reactive versus active power of a squirrel-cage induction generator.

tem. Detailed analysis of the reactive power compensation is given in the following case study.

Case Study 6-2—Reactive Power Compensation

This case study starts with the design of a three-phase capacitor for reactive power compensation of a typical fixed-speed WECS operating under the rated conditions. The design procedure is then used to select the values of three additional capacitors to compensate for the reactive power of the generator over its full operating range.

Consider a 2.3 MW, 690 V, 50 Hz, 1512 rpm SCIG fixed-speed wind energy system. The generator parameters are given in Table B-1 of Appendix B. The system is connected to a grid of 690 V/50 Hz (assuming the transformer turns ratio of 1:1 for simplicity). The generator operates under the rated conditions with its rated speed of 1512 rpm, based on which the slip of the generator is

$$s = \frac{1500 - 1512}{1500} = -0.008 \tag{6.1}$$

Using the IG steady-state equivalent circuit of Figure 3-13b in Chapter 3, the impedance of the induction generator can be calculated by

$$\overline{Z}_s = R_s + jX_{ls} + jX_m\left(\frac{R_r}{s} + jX_{lr}\right) = 0.1837\angle152.58°\ \Omega \tag{6.2}$$

The stator current is then determined by

$$\overline{I}_s = \frac{\overline{V}_s}{\overline{Z}_s} = \frac{690/\sqrt{3}\angle0°}{0.1837\angle152.58°} = 2168\angle-152.58°\ \text{A}\ \ (1\ \text{pu}) \tag{6.3}$$

The stator power factor is obtained from

$$PF_s = \cos\varphi_s = -0.888 \tag{6.4}$$

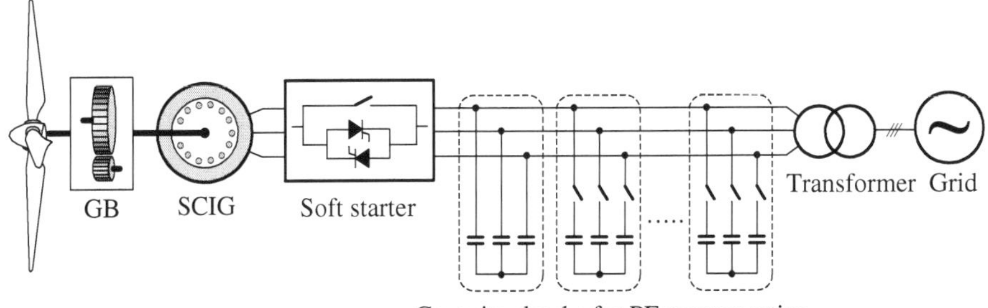

Figure 6-10. Reactive power compensation by multiple capacitor banks.

where the power factor angle is

$$\varphi_s = \angle \bar{V}_s - \angle \bar{I}_s = 152.58°$$ (6.5)

The stator apparent, active, and reactive power are calculated by

$$\begin{cases} S_s = 3V_s I_s = 3 \times 398.4 \times 2168 = 2.591 \text{ MVA} \\ P_s = S_s \cos(\phi_s) = -2.3 \text{ MW} \\ Q_s = S_s \sin(\phi_s) = 1.193 \text{ MVAR} \end{cases}$$ (6.6)

To compensate for the lagging reactive power drawn by the induction generator, a three-phase capacitor C_1 is connected to the system. The reactive power provided by the capacitor can be calculated by

$$Q_c = 3V_c I_c = 3(V_s)^2 \omega_s C_1$$ (6.7)

where I_c is the capacitor current; V_c is the capacitor voltage, which is the same as the stator voltage V_s due to the shunt connection; and ω_s is the stator/grid angular frequency.

To achieve a unity power factor, the capacitor should provide a reactive power of 1.193 MVAR, that is,

$$Q_c = 3(V_s)^2 \omega_s C_1 = 1.193 \text{ MVAR}$$ (6.8)

from which

$$C_1 = \frac{Q_c}{3(V_s)^2 \omega_s} = \frac{1.193 \times 10^6}{3 \times (398.4)^2 \times (2\pi \times 50)} = 7,975 \text{ μF} \quad (0.46 \text{ pu})$$ (6.9)

Since the voltage across each of the capacitors is relatively low (only 398.4 V, wye-connected), the three-phase capacitor can be connected in delta. The capacitance for the delta-connected capacitor can then be reduced to

$$C_\Delta = C_1 / 3 = 2,658 \text{ μF}$$ (6.10)

In this case, the line-to-line voltage of 690 V is applied to each of the capacitors.

Following the same procedure, the reactive power of the wind energy system can be calculated at different active power levels delivered to the grid. Figure 6-11 shows the relationship between the calculated reactive and active power (P–Q curve) of the system, where $|S_R| = 2.591$ MVA and $|P_R| = 2.3$ MW, which are the rated apparent and active power of the generator, respectively. With the capacitor C_1 of 0.46 pu, the reactive power drawn by the generator is fully compensated by the capacitor at the rated stator power ($P_s/P_R = 1.0$). However, with the decrease of the stator power P_s, the generator draws less reactive power from the grid, and system starts to be overcompensated.

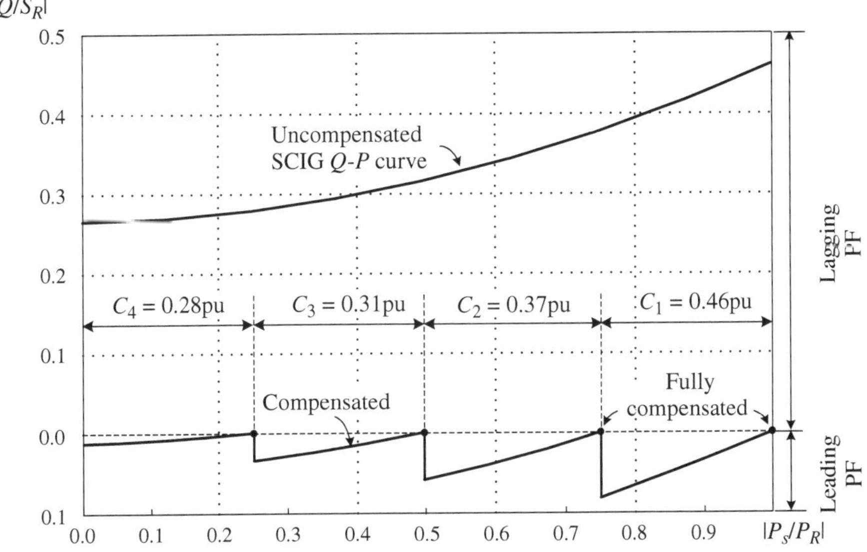

Figure 6-11. Reactive power compensation by capacitor banks.

To have an effective compensation over the full operating range of the generator, different values of the capacitor should be used. A satisfactory compensation can be made with three or four different capacitors. Figure 6-11 shows performance of the reactive power compensation with four groups of the capacitors over the full operating range of the generator. With the capacitance of $C_1 = 0.46$ pu, $C_2 = 0.37$ pu, $C_3 = 0.31$ pu, and $C_4 = 0.28$ pu, the reactive power of the generator can be fully compensated at $|P_s/P_R| = 1.0, 0.75, 0.5,$ and 0.25, respectively.

Figure 6-12 illustrates the connection of the capacitors to achieve the required capacitances for the reactive power compensation presented above. Three switches are

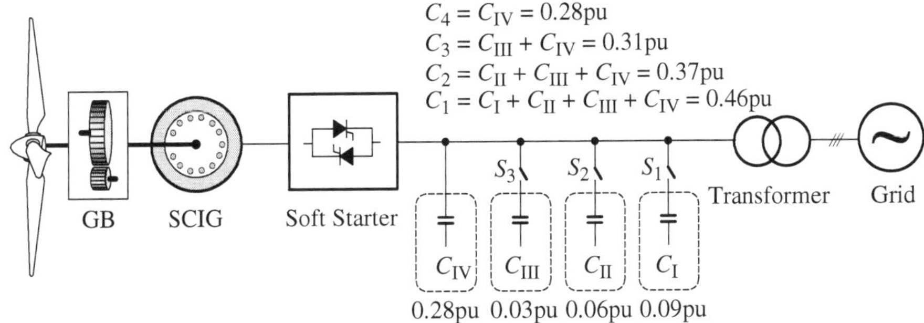

Figure 6-12. Connection of capacitors for reactive power compensation over the full operating range.

employed to connect or disconnect the capacitors C_I, C_{II}, and C_{III} to/from the system. Capacitor C_{IV} is permanently connected to the system for compensation of the reactive power drawn by the magnetizing inductance, which is always there when the system is in operation. The values for the other required capacitances, C_1, C_2, and C_3, can be obtained by selecting different combinations of switches S_1, S_2, and S_3.

Figure 6-13 shows the power factor improvement through reactive power compensation by the capacitors. The power factor of the generator is kept near to unity (> 0.99) when the turbine is delivering active power above 10% of its rated capacity to the grid. Compared with the uncompensated power factor curve, the system power factor is substantially improved, especially when the system delivers less than 50% of the rated active power to the grid.

It is worth mentioning that reactive power compensation can also be performed using shunt power converter circuits installed between the generator and grid. These circuits are also known as active VAR compensators. They can perform dynamic compensation of the reactive power over the full operating range without the need for the step changes in capacitors. However, they are more complex and expensive due to the use of power converters and controllers. These systems may be used to correct the power factor of a wind farm at the point of common coupling, where the wind farm is connected to the grid.

6.6 SUMMARY

In this chapter, the main characteristics of fixed-speed wind energy conversion systems using squirrel-cage induction generators were presented. The system operating principle was discussed, including the torque-speed curves of the generator, power

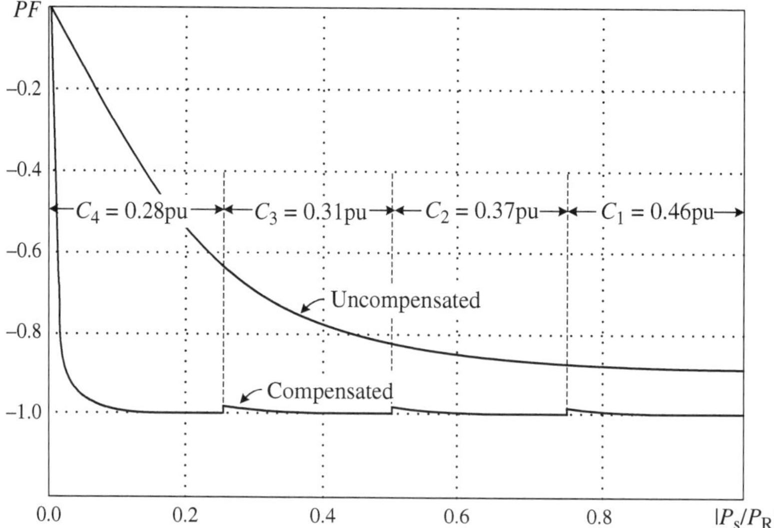

Figure 6-13. Power factor improvement through reactive power compensation by capacitor banks.

curves of the wind turbine, and reactive power compensation. The dynamic response of the fixed-speed system with hard and soft start was investigated. To improve the power conversion efficiency, a two-speed wind energy system was presented. Case studies were provided to facilitate the analysis of the fixed-speed wind energy systems.

REFERENCES

1. T. Burton, D. Sharpe, N. Jenkins, and E. Bossanyi, *Wind Energy Handbook,* Wiley, 2001.

2. Erich Hau, *Wind Turbines: Fundamentals, Technologies, Application, Economics,* 2nd Edition, Springer, 2005.

3. Ion Boldea, *Variable Speed Generators,* CRC Press, 2005.

4. F. Blaabjerg and Z. Chen, *Power Electronics for Modern Wind Turbines,* Morgan & Claypool, 2006.

5. E.ON Netz, *Grid Code: High and Extra High Voltage,* www.eon-netz.com, 2006.

6. E. Diaz-Dorado, C. Carrillo, and J. Cidras, Control Algorithm for Coordinated Reactive Power Compensation in a Wind Park, *IEEE Transactions on Energy Conversion,* Vol. 23, No. 7, 1064–1072, 2008.

7

VARIABLE-SPEED WIND ENERGY SYSTEMS WITH SQUIRREL CAGE INDUCTION GENERATORS

7.1 INTRODUCTION

In variable-speed squirrel cage induction generator (SCIG) wind energy conversion systems (WECS), full-capacity power converters are required to adjust the speed of the generator in order to harvest the maximum possible power available from the wind. A typical configuration for the SCIG wind energy system is shown in Figure 7-1, where a back-to-back power converter configuration is normally employed. The generator-side converter (rectifier) is used to control the speed or torque of the generator with a maximum power point tracking (MPPT) scheme. The grid-side converter (inverter) is employed for the control of DC link voltage and grid-side reactive power. The control of the grid-tied converter has been described in Chapter 4 and, therefore, is not discussed in this chapter.

This chapter focuses on the control schemes for the SCIG wind energy systems. The principles of direct and indirect field oriented control (FOC) and direct torque control (DTC) are discussed. The system dynamic and steady-state performance are analyzed. The analysis is assisted by computer simulations and steady-state equivalent circuits. In addition to voltage source converter (VSC) based WECS, the application of current source converters (CSCs) in SCIG wind energy systems is also presented.

Power Conversion and Control of Wind Energy Systems. By B. Wu, Y. Lang, N. Zargari, and S. Kouro **191**
© 2011 the Institute of Electrical and Electronics Engineers, Inc. Published 2011 by John Wiley & Sons, Inc.

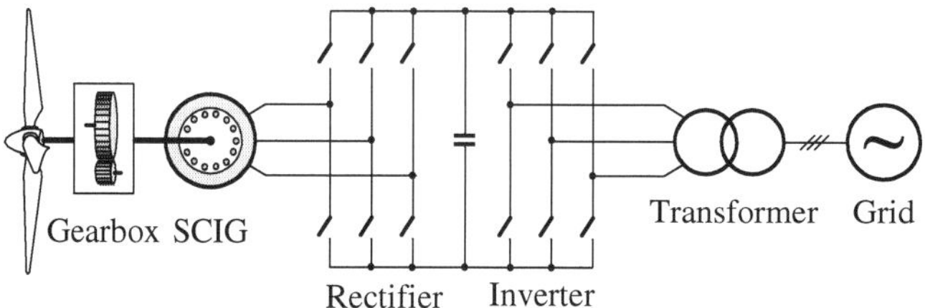

Figure 7-1. Typical configuration of variable-speed SCIG wind energy system.

7.2 DIRECT FIELD ORIENTED CONTROL

7.2.1 Field Orientation

The field orientation control can be generally classified into stator flux, air-gap flux, and rotor flux orientations. The rotor flux orientation is simple and easy to implement and is one of the more popular schemes used in both AC drives and wind energy systems [1]. The essence of field oriented control is the decoupled control of the rotor flux λ_r and electromagnetic torque T_e of the generator to achieve high dynamic performance. Using the rotor flux orientation, the stator current of the generator can be decomposed into a flux-producing component, which produces the rotor flux λ_r, and a torque-producing component, which produces the torque T_e. These two components are then controlled independently.

Rotor flux orientation is achieved by aligning the d-axis of the synchronous reference frame with the rotor flux vector $\vec{\lambda}_r$, as shown in Figure 7-2. The resultant dq-axis rotor flux components are

$$\begin{cases} \lambda_{qr} = 0 \\ \lambda_{dr} = \sqrt{\left(\lambda_r\right)^2 - \left(\lambda_{qr}\right)^2} = \lambda_r \end{cases} \tag{7.1}$$

where λ_r is the length (magnitude) of $\vec{\lambda}_r$. The rotating speed of the synchronous reference frame is given by

$$\omega_s = 2\pi f_s \tag{7.2}$$

where f_s is the stator frequency of the generator.

Substituting Equation 7.1 into the torque equation of (Equation 3.16 of Chapter 3) yields

$$T_e = K_T \, \lambda_r \, i_{qs} \tag{7.3}$$

where $K_T = (3PL_m/3L_r)$. If the rotor flux λ_r is kept constant during the generator operation, the developed torque is directly controlled by the q-axis stator current i_{qs}.

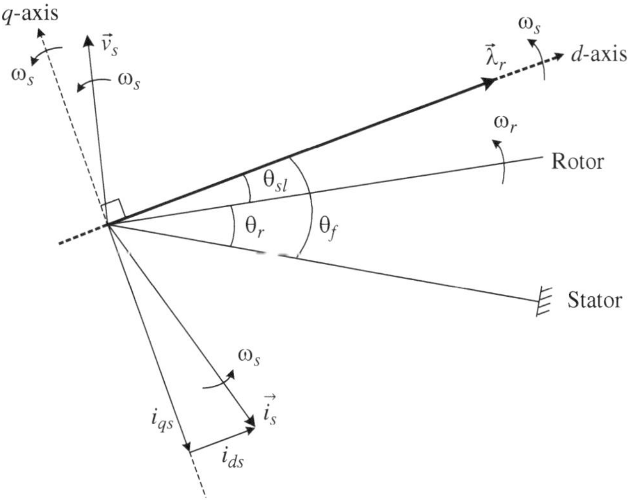

Figure 7-2. Rotor flux field oriented control (the d-axis of the synchronous frame aligned with $\vec{\lambda}_r$).

The stator current vector $\vec{i}_s$ in Figure 7-2 can be resolved into two components along the dq axes. The d-axis current i_{ds} is the flux-producing current, whereas the q-axis current i_{qs} is the torque-producing current. In field oriented control, i_{ds} is normally kept at its rated value while i_{qs} is controlled independently. With decoupled control for i_{ds} and i_{qs}, high dynamic performance can be realized. It is noted that the stator current vector $\vec{i}_s$ rotates in space at the synchronous speed and, therefore, i_{ds} and i_{qs} are DC currents in steady state.

One of the key issues associated with the rotor flux oriented control is to accurately determine the rotor flux angle θ_f for field orientation. Different methods can be used to find θ_f. If the angle is obtained through the measurement of generator terminal voltages and currents, the method is known as direct field oriented control. For the indirect field oriented control, the rotor flux angle is obtained from

$$\theta_f = \theta_r + \theta_{sl} \tag{7.4}$$

where θ_r and θ_{sl} are the measured rotor position angle and calculated slip angle, respectively.

7.2.2 Direct FOC for SCIG Wind Energy Systems

Figure 7-3 shows a typical block diagram of direct field oriented control for the induction generator [2]. To implement the FOC scheme, the rotor flux magnitude λ_r and its angle θ_f are identified by the rotor flux calculator based on the measured stator voltages (v_{as} and v_{bs}) and currents (i_{as} and i_{bs}).

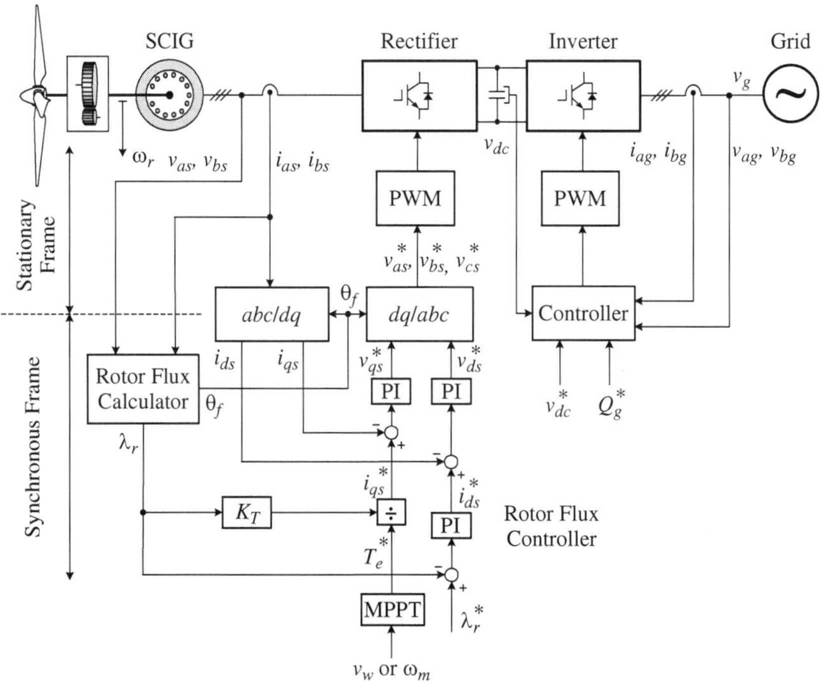

Figure 7-3. Direct field oriented control with rotor flux orientation.

There are three feedback control loops: one for the rotor flux linkage λ_r, one for the d-axis stator current i_{ds}, and another for the q-axis stator current i_{qs}. For the rotor flux control, the measured λ_r is compared with its reference λ_r^* and the error is passed through a PI controller. The output of the PI regulator yields the d-axis stator current reference i_{ds}^*. The torque reference T_e^* is generated by the maximum power point tracking scheme discussed in Chapter 2. The q-axis stator current reference can then be calculated by

$$i_{qs}^* = \frac{T_e^*}{K_T\,\lambda_r} \tag{7.5}$$

The feedback dq-axis stator currents i_{ds} and i_{qs} are compared with their references, and the errors are sent to current PI controllers to generate stator voltage references v_{ds}^* and v_{qs}^*. The dq-axis voltages in the synchronous reference frame are then transformed to the three-phase stator voltages v_{as}^*, v_{bs}^*, and v_{cs}^* in the stationary frame through the dq/abc transformation block.

Both carrier-based modulation and space vector modulation schemes presented in Chapter 4 can be used to generate gating signals for the rectifier. If a carrier-based modulation scheme is employed, v_{as}^*, v_{bs}^*, and v_{cs}^* are the modulating signals, which are compared with a triangular carrier wave to generate PWM gatings for the switching devices in the rectifier. For the space vector modulation, the reference vector, $\vec{v}_{ref}$, is generated according to the three-phase reference voltages v_{as}^*, v_{bs}^*, and v_{cs}^*, based on which the PWM gating signals are produced.

As shown in Figure 7-3, the rotor flux angle θ_f is used in the *abc/dq* and *dq/abc* transformation for field orientation. Neglecting switching harmonics produced by the rectifier, the variables above the transformation blocks (see Figure 7-3) are all in the stationary reference frame, which are sinusoidal in steady state, whereas the variables below transformation blocks are all DC signals in the rotor flux synchronous frame.

7.2.3   Rotor Flux Calculator

To calculate the rotor flux magnitude and angle, the generator model in the stationary *dq*-axis frame in Figure 7-4a can be utilized. The stator flux vector can be expressed as

$$\vec{\lambda}_s = \int \left(\vec{v}_s - R_s \vec{i}_s \right) dt \tag{7.6}$$

The rotor flux vector can be calculated by

$$\vec{\lambda}_r = L_r \vec{i}_r + L_m \vec{i}_s = L_r \frac{\vec{\lambda}_s - L_s \vec{i}_s}{L_m} + L_m \vec{i}_s = \frac{L_r}{L_m} \left(\vec{\lambda}_s - \sigma L_s \vec{i}_s \right) \tag{7.7}$$

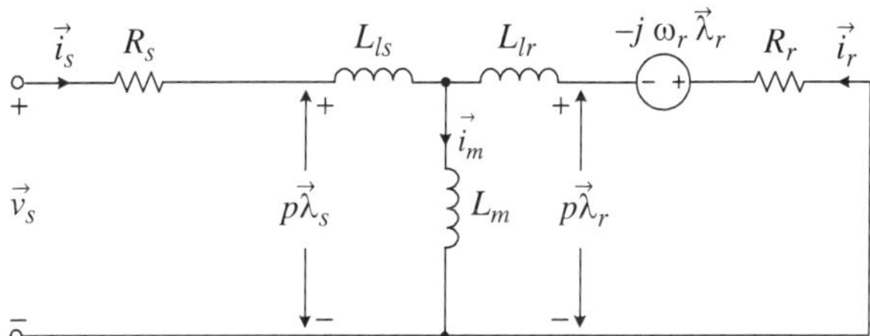

(a) SCIG dynamic model in the stationary frame

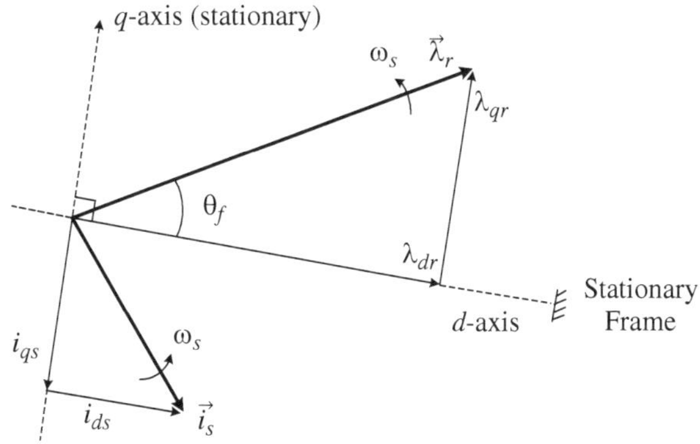

(b) Space vector diagram in the stationary frame

Figure 7-4. Dynamic model of SCIG and space vector diagram for rotor flux calculation.

where σ is the total leakage factor, defined by

$$\sigma = 1 - \frac{L_m^2}{L_s L_r} \tag{7.8}$$

Decomposing the rotor flux $\vec{\lambda}_r$ into the d- and q-axis components in the stationary frame, we have

$$\begin{cases} \lambda_{dr} = \dfrac{L_r}{L_m}\left(\lambda_{ds} - \sigma L_s i_{ds}\right) \\[2mm] \lambda_{qr} = \dfrac{L_r}{L_m}\left(\lambda_{qs} - \sigma L_s i_{qs}\right) \end{cases} \tag{7.9}$$

from which the magnitude and angle of the rotor flux are

$$\begin{cases} \lambda_r = \sqrt{\lambda_{dr}^2 + \lambda_{qr}^2} \\[2mm] \theta_f = \tan^{-1}\dfrac{\lambda_{qr}}{\lambda_{dr}} \end{cases} \tag{7.10}$$

It can be seen from the above equations that the rotor flux magnitude λ_r and its angle θ_f can be identified based on measured stator voltage, stator current, and generator parameters (L_s, L_r, L_m, and R_s). No rotor current is needed for the calculation. Figure 7-4b shows the vector diagram for the rotor flux vector $\vec{\lambda}_r$ and stator current vector $\vec{i}_s$ used in the rotor flux calculator. When the two vectors rotate one revolution in space, their dq-axis components, λ_{dr}, λ_{qr}, i_{ds}, and i_{qs}, vary one cycle over time due to the use of the stationary (stator) frame. Neglecting the switching harmonics, these variables are sinusoidal in steady state.

Figure 7-5 shows the block diagram for the digital implementation of a rotor flux calculator. Of the three stator voltages v_{as}, v_{bs}, and v_{cs}, only two need to be measured and the third can be found from $v_{as} + v_{bs} + v_{cs} = 0$. The same is true for the stator currents. The stator voltages and currents are then transformed into two-phase variables through $abc/\alpha\beta$ stationary transformation blocks. The other blocks are derived from Equations 7.6 to 7.10. The output of the rotor flux calculator is the rotor flux angle θ_f for field orientation and its amplitude λ_r for rotor flux feedback control.

It should be noted that the dq-axis voltages and currents in the rotor flux calculator are all AC signals (sinusoidal in steady state), which may cause problems for the integrators, such as undesired DC offsets during system start-up. In practice, the integrators can be replaced by first-order, low-pass filters with a low cut-off frequency up to a couple of hertz. The replacement of integrators by low-pass filters has little impact on system operation since the wind generator normally operates at much higher stator frequencies than the filter cutoff frequency.

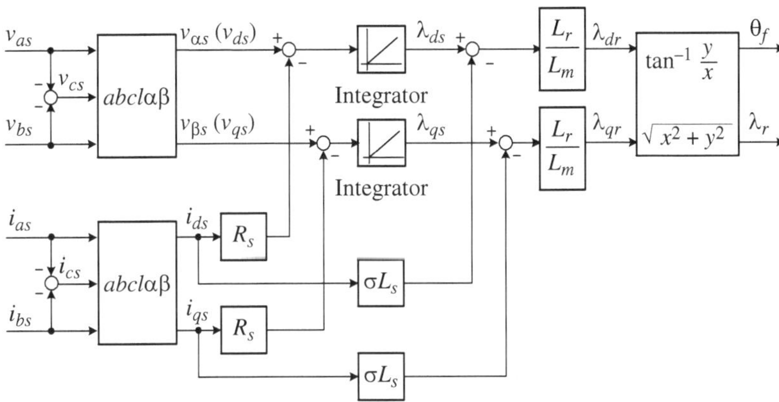

<u>Figure 7-5.</u> Block diagram of a rotor flux calculator.

7.2.4   Dynamic and Steady-State Analysis of Direct FOC WECS

To study the operation and performance of the FOC SCIG wind energy system, three case studies are provided. One investigates the dynamic behavior of the system during the system start-up, one examines the steady state operation of the system, and another analyzes the dynamic performance of the system with a step change in wind speeds.

Case Study 7-1—Start-up Transient Analysis. In this case study, the start-up process of a 2.3 MW/690 V SCIG wind energy system with direct FOC scheme is investigated. The speed of the turbine and generator is brought from zero to its rated value by wind, at which time a step rotor flux reference (rated) is applied. The rotor flux starts to build up, and when it reaches the rated value, a step torque reference (rated) is applied. The wind energy system then delivers its rated power to the grid at the rated wind speed.

It is assumed that the generator speed does not change during the electrical transients since the mechanical system of the wind turbine is much slower than the electrical system. It is also assumed that the DC link voltage is kept constant by the grid-side converter and, therefore, a constant DC voltage source instead of a grid-connected converter is used in simulation to absorb the energy from the wind turbine. The parameters and operating conditions of the SCIG wind energy system are given in Table 7-1.

Figure 7-6 shows the simulated waveforms of the wind energy system during its start-up transients. The transient process is analyzed step by step below.

1. At $t = 0$, a step rotor flux reference λ_r^* (rated) is applied. The reference for the electromagnetic torque, T_e^*, is set to zero.

2. During the period of $0 < t < 0.42$ sec, the magnitude of the phase-a rotor flux λ_{ar} increases over time t. The rotor flux PI controller is saturated due to the large er-

Table 7-1. Parameters and operating conditions of the SCIG WECS with direct FOC
scheme

Induction generator	Generator ratings: 2.3 MW, 690 V, 50 Hz, 1512 rpm 2168 A, 14.74 kN·m	Generator parameters: Table B-1 (Appendix B)
System input variables	Step input of rotor flux reference at $t = 0$	$\lambda_r^* = 1.711$ Wb (rated)
	Step input of torque reference at $t = 0.8$ sec	$T_e^* = -14.74$ kN·m (1.0 pu)
Control scheme	Direct rotor flux FOC	Figure 7-3
Generator-side converter	Converter type Modulation scheme Switching sequence Switching frequency	Two-level VSC Space vector modulation Seven segment; Table 4-4 2 kHz
Grid-side converter	Replaced by a DC voltage source with a fixed DC voltage	$V_{dc} = 1220$ V (3.06 pu)
Rotor flux calculator	Algorithm Integrator: Replaced by a first-order, low-pass filter with a cutoff frequency f_0 of 0.1 Hz	Figure 7-5 Transfer function: $$\dfrac{2\pi f_0}{2\pi f_0 s + 1}$$
Reference frame transformation	abc/dq transformation dq/abc transformation	Eq. (3.1), Chapter 3 Eq. (3.3), Chapter 3

ror between the rotor flux reference λ_r^* and measured λ_r. The output of the controller is the reference for the d-axis stator current, i_{ds}^*, which is the magnetizing
component of the stator current. The controller saturation level for i_{ds}^* is set to
the rated value of the stator current such that the rotor flux can be established at
its maximum rate. During this period, the q-axis stator current i_{qs}, which is the
torque-producing component of the stator current, is kept zero due to the zero
torque reference ($T_e^* = 0$).

3. During the period of $0.42 \leq t < 0.8$ sec, the rotor flux λ_r reaches its rated value
   (1.349 pu, peak) set by the reference, and the rotor flux controller is out of saturation. The d-axis current i_{ds} is decreased to 0.368 pu, the q-axis current is still
   zero, and the phase-a stator current i_{as} falls to 0.368 pu as well, which is, in fact,
   the magnetizing current of the generator.

4. At $t \geq 0.8$ sec, a step torque reference of the rated value (-1 pu) is applied. The
   q-axis stator current i_{qs} quickly increases to 1.365 pu, while the d-axis current i_{ds}
   is kept constant at 0.368 pu due to the independent control of the dq-axis currents. The rms stator current is given by $I_s = \sqrt{i_{ds}^2 + i_{qs}^2}/\sqrt{2} = 1.0$ pu (rated). The
   generator operates with the rated rotor flux and stator current, which leads to the
   rated electromagnetic torque T_e, as shown in Figure 7-6.

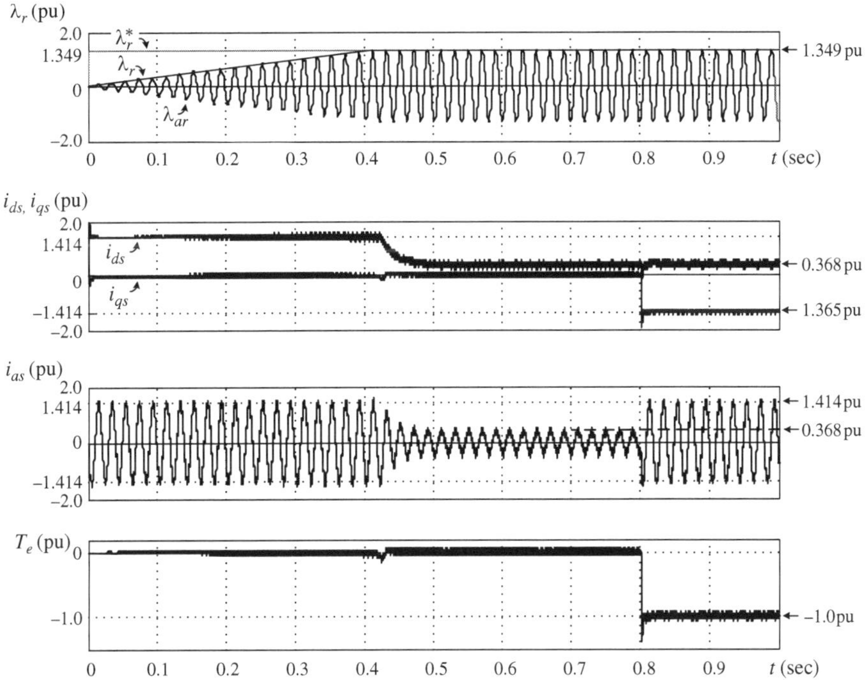

Figure 7-6. Simulated waveforms during the start-up of the SCIG wind energy system.

Case Study 7-2—Steady-State Analysis of SCIG WECS Under Rated Operating Conditions.

The steady-state waveforms of the 2.3 MW/690 V SCIG wind energy system operating under the rated conditions are shown in Figure 7-7. In fact, some of these waveforms are from Figure 7-6 in the period of $t = 0.93$ to 1.0 sec. Included are the waveforms for the phase-a PWM stator voltage v_{as}, the fundamental-frequency stator voltage v_{as1}, stator current i_{as}, phase-a rotor flux linkage λ_{ar}, rotor flux angle θ_f, and dq-axis stator currents. The steady-state analysis is performed based on the generator steady-state equivalent circuit of Figure 3-13 in Chapter 3.

With the parameters given in Table B-1 in Appendix B and the generator operating under the rated conditions, its input impendence is

$$\overline{Z}_s = R_s + jX_{ls} + jX_m \; // \left(\frac{R_r}{s} + jX_{lr} \right) = 0.1838\angle152.6° \; \Omega \;\; \text{for } s = -0.008 \;(\text{rated}) \quad (7.11)$$

The rms and peak values of the stator current are

$$\begin{cases} \overline{I}_s = \dfrac{\overline{V}_s}{\overline{Z}_s} = \dfrac{690 / \sqrt{3}\angle0°}{0.1838\angle152.6°} = 2168\angle-152.6° \; \text{A} \;\; (1.0 \text{ pu}) \\[3mm] i_s = \sqrt{2}I_s = \sqrt{2}\times2168 = 3066 \; \text{A} \;\; (1.414 \text{ pu}) \end{cases} \quad (7.12)$$

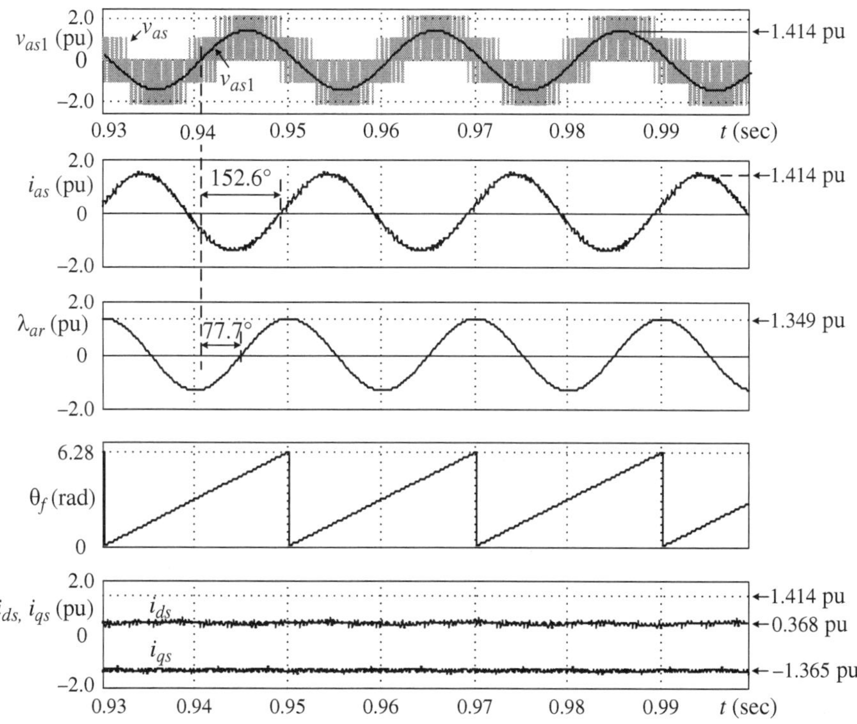

Figure 7-7. Steady-state waveforms of the SCIG WECS operating under the rated conditions.

According to the reference direction of the rotor current given in the steady-state equivalent circuit of Figure 3-13 in Chapter 3, the rotor current can be found from

$$\overline{I}_r = \frac{jX_m\overline{I}_s}{jX_m + \left(\dfrac{R_r}{s} + jX_{lr}\right)} = 2030.8\angle-167.7° \text{ A} \quad (0.937 \text{ pu}) \tag{7.13}$$

The magnetizing flux linage is calculated by

$$\overline{\Lambda}_m = (\overline{I}_s - \overline{I}_r)L_m = 1.2168\angle-83.9° \text{ Wb} \tag{7.14}$$

The rms rotor flux linkage is

$$\overline{\Lambda}_r = \overline{\Lambda}_m - L_{lr}\overline{I}_r = 1.2096\angle-77.7° \text{ Wb} \tag{7.15}$$

The peak rotor flux is then given by

$$\lambda_r = \sqrt{2}\Lambda_r = 1.7106 \text{ Wb} \quad (1.349 \text{ pu}) \tag{7.16}$$

The peak value (1.414 pu) and phase angle (–152.6°) of the stator current, and the peak value (1.349 pu) and phase angle (–77.7°) of the rotor flux obtained from the above steady-state analysis correlate well with the values obtained from the simulation results given in Figure 7-7.

A phasor diagram for the steady-state operation of the generator is given in Figure 7-8a. To find the dq-axis stator currents, i_{ds} and i_{qs}, the d and q axes are added to the phasor diagram with d-axis aligned with the rotor flux as shown in Figure 7-8b. This alignment is due to the rotor flux FOC implemented in the control of the wind energy system. Note that the dq axes in this diagram are stationary and, therefore, do not rotate in space at the synchronous speed for field orientation. They are simply added to the steady-state phasor diagram to determine the dq-axis components of the stator current. Based on this diagram, the dq-axis stator currents (rms) can be calculated by

$$\begin{cases} I_{ds} = I_s \cos(\angle \overline{I}_s - \angle \overline{\Lambda}_r) = 2168\cos(-74.9°) = 564.8 \text{ A} \quad (0.261 \text{ pu}) \\ I_{qs} = I_s \sin(\angle \overline{I}_s - \angle \overline{\Lambda}_r) = 2168\sin(-74.9°) = -2093 \text{ A} \quad (-0.965 \text{ pu}) \end{cases} \quad (7.17)$$

The current peak values are given by

$$\begin{cases} i_{ds} = \sqrt{2}I_{ds} = 798.7 \text{ A} \quad (0.368 \text{ pu}) \\ i_{qs} = \sqrt{2}I_{qs} = -2960 \text{ A} \quad (-1.365 \text{ pu}) \end{cases} \quad (7.18)$$

which confirm the values obtained by simulation in Figure 7-7.

Similarly, the dq-axis stator voltage (rms) can be calculated by

$$\begin{cases} V_{ds} = V_s \cos(\angle \overline{V}_s - \angle \overline{\Lambda}_r) = (690 / \sqrt{3})\cos(77.7°) = 84.9 \text{ V} \quad (0.2130 \text{ pu}) \\ V_{qs} = V_s \sin(\angle \overline{V}_s - \angle \overline{\Lambda}_r) = (690 / \sqrt{3})\sin(77.7°) = 389.2 \text{ V} \quad (0.977 \text{ pu}) \end{cases} \quad (7.19)$$

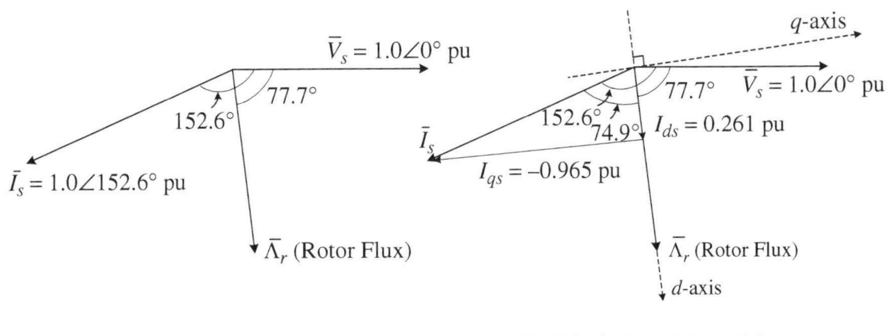

(a) Phasor diagram   (b) Calculation of I_{ds} and I_{qs}

Figure 7-8. Phasor diagram of the induction generator superimposed with dq axes.

from which their peak values are given by

$$\begin{cases} v_{ds} = \sqrt{2}V_{ds} = 120.0 \text{ V} \quad (0.301\,\text{pu}) \\ v_{qs} = \sqrt{2}V_{qs} = 550.5 \text{ V} \quad (1.382\,\text{pu}) \end{cases} \tag{7.20}$$

Case Study 7-3 SCIG WECS Transients with a Step Change in Wind Speeds.

In this case study, the transients of a 2.3 MW/690 V SCIG wind energy system caused by a step change in wind speeds is investigated. The torque and power versus rotor speed characteristics of the system at the wind speed of 8.4 m/sec and 12 m/sec are illustrated in Figure 7-9. With the implementation of the MPPT scheme, the electromagnetic torque T_e of the generator is proportional to the rotor mechanical speed square $(\omega_m)^2$, whereas its mechanical power P_m is proportional to the rotor speed cube $(\omega_m)^3$, as discussed in Chapter 2. It is noted that the turbine mechanical speed is equal to the rotor mechanical speed in per-unit terms and, therefore, we do not need to consider the gear ratio of the gearbox in the discussion.

The system initially operates with a rotor speed of 0.7 pu as indicated by Point A in Figure 7-9. The case study investigates the transients when the wind speed suddenly increases to its rated value of 12 m/sec and the operating point is moved from Point A to D. The generator and system parameters for the case study remain the same as those given in Table 7-1.

In order to reduce the simulation time, the combined moment of inertia of the wind turbine and the generator was reduced to 1.8% of its actual value. With the rotational losses of the turbine shaft neglected, the mechanical torque T_m of the generator is then equal to the mechanical torque generated by the turbine. The simulated waveforms are given in Figure 7-10.

During the period of $0 < t < 0.1$ sec, the wind energy system operates in steady state at 0.7 pu rotor speed, at which point the mechanical torque T_m from the turbine shaft is -0.49 pu (generating mode). The electromagnetic torque of the generator T_e contains certain amount of torque ripple caused by the harmonics in the stator current, but its average value (-0.49 pu) is equal to the mechanical torque. At $t = 0.1$ sec, the system operates at Point A in Figures 7-9 and 7-10.

At $t \geq 0.1$ sec, the wind speed suddenly increases to its rated value of 12 m/sec. The turbine mechanical torque T_m is instantly increased to its rated value of -1.0 pu (Point B in Figures 7-9 and 7-10). However, the mechanical speed of the generator ω_m cannot change instantaneously at $t = 0.1$ sec due to the moment of inertia. As a result, the torque reference T_e^* produced by the MPPT block remains unchanged, and the generator torque T_e remain unchanged as shown in Figure 7-10.

During the period of $0.1 < t < 0.38$ sec, the difference between the turbine mechanical torque T_m and the generator electromagnetic torque T_e accelerates the generator, and the rotor speed increases accordingly. Moving through Point C from B, the system finally reaches a new operating point D at $t = 0.38$ sec. The turbine mechanical torque T_m reaches Point D via its T_m-ω_m curve, whereas the generator torque T_e reaches the same operating point via its own T_e-ω_m curve, as shown in Figures 7-9 and 7-10.

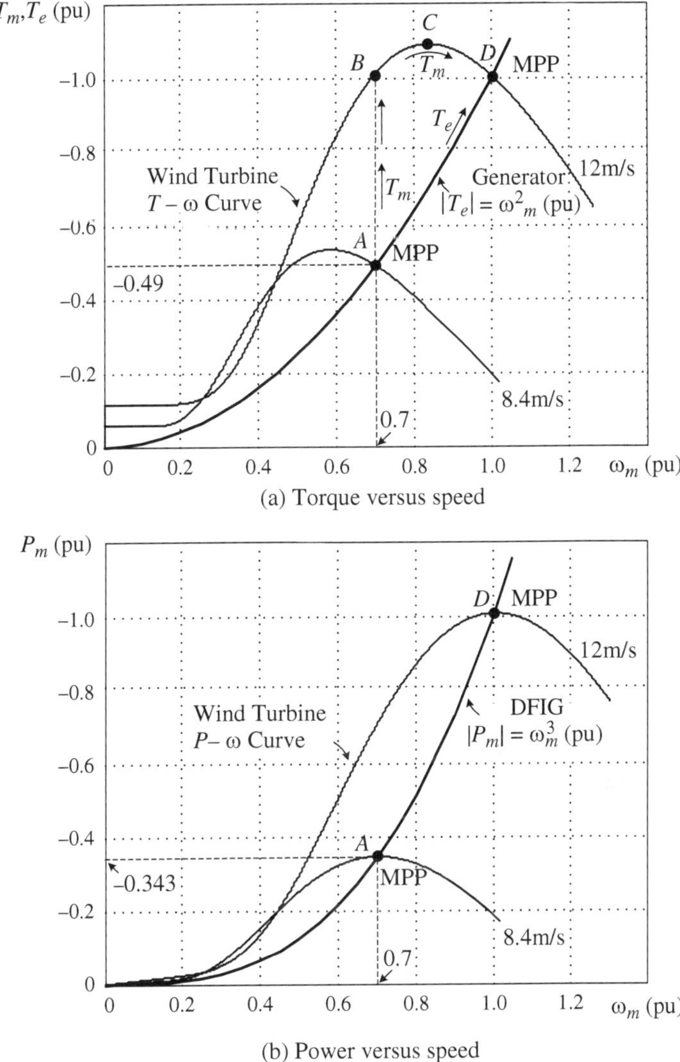

Figure 7-9. Torque and power versus rotor speed characteristics of the wind turbine and generator with MPPT control.

When $t > 0.38$ sec, the wind energy system reaches its steady state. With the rated wind speed, the generator operates under the rated conditions with rated torque and rated stator current. The waveforms of the dq-axis stator currents, i_{ds} and i_{qs}, are also given in Figure 7-10. With the rotor flux feedback control, the d-axis stator current i_{ds} is kept constant during the system transients, which in turn makes the rotor flux λ_r constant. The dq-axis stator currents ($i_{ds} = 0.368$ pu and $i_{qs} = -1.365$ pu) and stator volt-

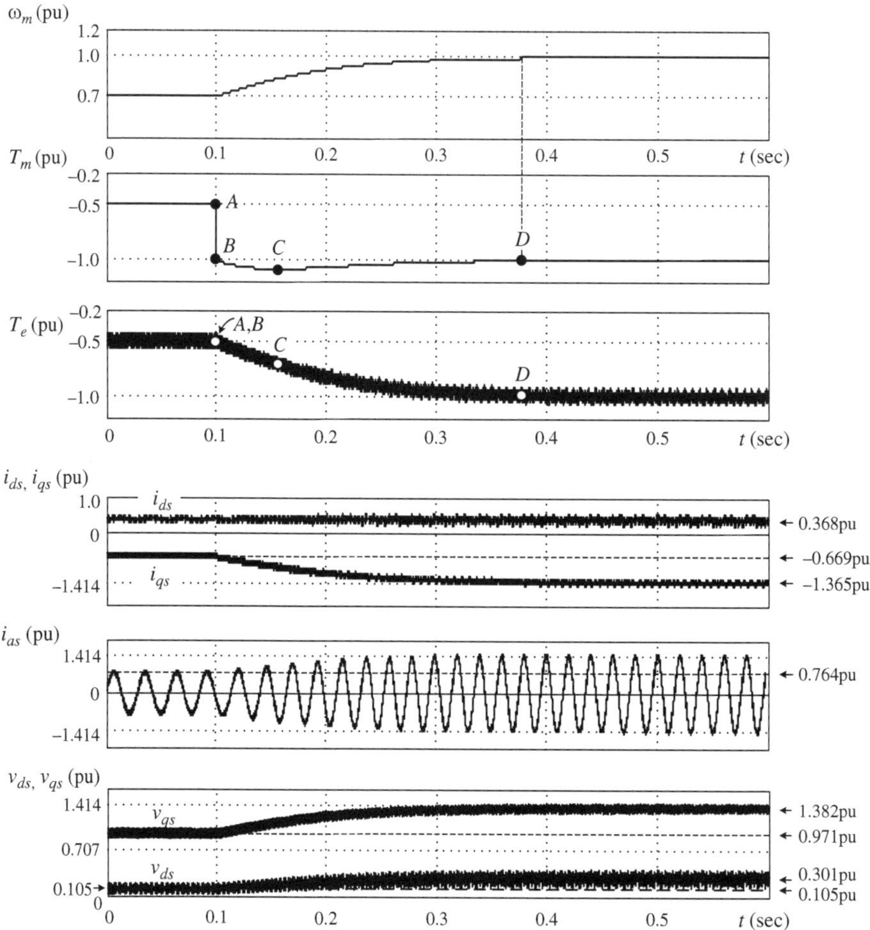

<u>Figure 7-10.</u> Transient waveforms of SCIG wind energy system with a step change in wind speed.

ages (v_{ds} = 0.301 pu and v_{qs} = 1.382 pu) correlate well with the steady-state analysis performed in Case Study 7-2.

7.3 INDIRECT FIELD ORIENTED CONTROL

7.3.1 Principle of Operation of Indirect FOC

As mentioned earlier, the indirect FOC is essentially the same as that for the direct FOC, but the method for obtaining the rotor flux angle is different [3]. The rotor flux angle in the indirect FOC can be obtained from

$$\theta_f = \int (\omega_r + \omega_{sl}) dt \qquad (7.21)$$

where ω_r is the measured rotor speed and ω_{sl} is the calculated slip frequency.

The slip frequency ω_{sl} can be derived from the generator model of Figure 3-7a in Chapter 3 in the synchronous frame, from which

$$p\vec{\lambda}_r = -R_r \vec{i}_r - j\omega_{sl} \vec{\lambda}_r \qquad (7.22)$$

Substituting the rotor current

$$\vec{i}_r = \frac{1}{L_r}\left(\vec{\lambda}_r - L_m \vec{i}_s\right) \qquad (7.23)$$

into (7.22) yields

$$p\vec{\lambda}_r = -\frac{R_r}{L_r}\left(\vec{\lambda}_r - L_m \vec{i}_s\right) - j\omega_{sl} \vec{\lambda}_r \qquad (7.24)$$

from which

$$\vec{\lambda}_r \left(1 + \tau_r (p + j\omega_{sl})\right) = L_m \vec{i}_s \qquad (7.25)$$

where τ_r is the rotor time constant, defined by

$$\tau_r = L_r / R_r \qquad (7.26)$$

Decomposing (7.25) into the dq-axis components and taking into account the rotor flux orientation ($\lambda_{qr} = 0$ and $\lambda_{dr} = \lambda_r$), we have

$$\begin{cases} \lambda_r \left(1 + p\tau_r\right) = L_m i_{ds} \\ \omega_{sl} \tau_r \lambda_r = L_m i_{qs} \end{cases} \qquad (7.27)$$

from which the slip frequency is obtained:

$$\omega_{sl} = \frac{L_m}{\tau_r \lambda_r} i_{qs} \qquad (7.28)$$

A typical block diagram of the SCIG wind energy system with indirect FOC is shown in Figure 7-11, where the rotor flux angle θ_f for field orientation is obtained by measuring the rotor speed ω_r and calculated slip frequency ω_{sl} (7.28) through an integrator. There are three feedback loops, one for the rotor flux λ_r, one for the d-axis stator current i_{ds}, and another for the q-axis stator current i_{qs}. The rotor flux reference λ_r^* is normally set at its rated value, and the actual rotor flux can be obtained from

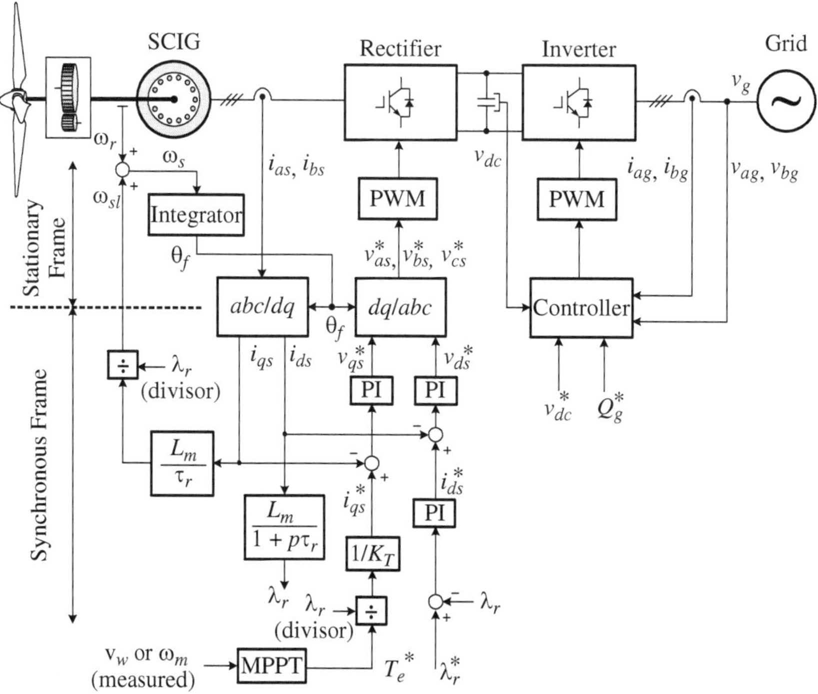

Figure 7-11. Indirect field oriented control for an SCIG wind energy system.

$$\lambda_r = \frac{L_m}{\left(1 + p\tau_r\right)} i_{ds} \qquad (7.29)$$

where i_{ds} is the measured d-axis stator current.

The reference torque T_e^* is generated by the MPPT block according to the measured wind speed or rotor speed. The reference for the q-axis current reference i_{qs}^* can be obtained from the torque equation (7.5):

$$i_{qs}^* = \frac{1}{K_T \lambda_r} T_e^* \qquad (7.30)$$

For a constant λ_r, the torque-producing current i_{qs}^* is proportional to T_e^*.

7.3.2 Steady-State Analysis of Indirect FOC SCIG Wind Energy System

The steady-state performance of an SCIG wind energy system with indirect rotor flux FOC is investigated through a case study given below.

Case Study 7-4—Steady-State Analysis of Indirect FOC SCIG Wind Energy System. For a wind energy system operating at the maximum power point, the rotor speed ω_m and turbine mechanical torque T_m can be easily determined from the T_m–ω_m curve of the system. However, the stator frequency ω_s and slip s of the generator are unknown, which may cause difficulties in analyzing the system based on the per-phase, steady-state equivalent circuit of the generator.

The problem can be solved when the induction generator is controlled by the rotor flux FOC scheme. The rotor flux λ_r is normally kept at its rated value by the FOC, which provides the starting point for the analysis. The procedure for the analysis and calculation is summarized in Table 7.2.

Let us consider the 2.3 MW/690 V SCIG wind energy system in Case Study 7-3 again. At the wind speed of 8.4 m/s, the generator operates at 0.7 pu rotor speed. The

Table 7.2. Steady-state analysis of SCIG wind energy system

Rotor flux linkage (Wb)	λ_r (Rated, controlled by FOC)
Rotor speed (rad/sec)	ω_m (Given, based on wind speed)
Electromagnetic torque (N·m)	$T_e = T_m$ (Given, from the $T_m - \omega_m$ profile)
d-axis current (A)	$i_{ds} = \dfrac{\lambda_r}{L_m}$
q-axis current (A)	$i_{qs} = \dfrac{T_e}{K_T \lambda_r} = \dfrac{2L_r}{3PL_m} \dfrac{T_e}{\lambda_r}$
Stator current (A)	$i_s = \sqrt{i_{ds}^2 + i_{qs}^2}$
Slip frequency (rad/sec)	$\omega_{sl} = \dfrac{L_m}{\tau_r \lambda_r} i_{qs} = \dfrac{R_r L_m}{L_r \lambda_r} i_{qs}$
Stator frequency (rad/sec)	$\omega_s = \omega_r + \omega_{sl}$
Slip	$s = \dfrac{\omega_{sl}}{\omega_s}$
Generator impedance (Ω)	$\bar{Z}_s = R_s + jX_{ls} + jX_m \; // \left(\dfrac{R_r}{s} + jX_{lr} \right)$
Stator current (rms) (A)	$\bar{I}_s = i_s / \sqrt{2} \angle 0°$ (Reference phasor)
Stator voltage (rms) (V)	$\bar{V}_s = \bar{I}_s \bar{Z}_s = V_s \angle \varphi$
Stator power factor angle	$\varphi_s = \angle \bar{V}_s - \angle \bar{I}_s = \varphi$

rotor flux λ_r is kept at its rated value of 1.711 Wb by the FOC scheme. The d-axis stator current can then be calculated by

$$i_{ds} = \frac{\lambda_r}{L_m} = \frac{1.7106}{2.1346 \times 10^{-3}} = 801.4 \text{ A} \quad (0.369 \text{ pu}) \tag{7.31}$$

The turbine mechanical torque is given by

$$T_m = T_{m,R} \times 0.7^2 = -14,740 \times 0.7^2 = -7.2226 \text{ kN·m} \tag{7.32}$$

The q-axis current is calculated by

$$i_{qs} = \frac{2L_r}{3PL_m} \frac{T_e}{\lambda_r} = \frac{2 \times 2.1995 \times 10^{-3}}{3 \times 2 \times 2.1346 \times 10^{-3}} \cdot \frac{-7222.6}{1.7106} = -1450.2 \text{ A} \quad (-0.669 \text{ pu}) \tag{7.33}$$

The peak stator current can then be obtained by

$$i_s = \sqrt{i_{ds}^2 + i_{qs}^2} = \sqrt{801.4^2 + 1450.2^2} = 1656.9 \text{ A} \quad (0.764 \text{ pu}) \tag{7.34}$$

The above calculated i_{ds}, i_{qs}, and i_s are essentially the same as those obtained by simulation in Figure 7-10.

The slip frequency is determined by

$$\omega_{sl} = \frac{R_r L_m}{L_r \lambda_r} i_{qs} = \frac{1.497 \times 10^{-3} \times 2.1346 \times 10^{-3}}{2.1995 \times 10^{-3} \times 1.7106} \times (-1450.2) = -1.2317 \text{ rad/sec} \tag{7.35}$$

The stator frequency can be found from

$$\omega_s = \omega_r + \omega_{sl} = (1512 \times 0.7)(2\pi / 60) P - 1.2317 = 220.4 \text{ rad/sec} \quad (35.08 \text{ Hz}) \tag{7.36}$$

from which the slip is

$$s = \frac{\omega_{sl}}{\omega_s} = -5.588 \times 10^{-3} \tag{7.37}$$

The rms value of the stator current is expressed as

$$\bar{I}_s = i_s / \sqrt{2} \angle 0° = 1171.6 \angle 0° \text{A} \tag{7.38}$$

With the generator impedance of

$$\bar{Z}_s = R_s + jX_{ls} + jX_m \ // \left(\frac{R_r}{s} + jX_{lr} \right) = 0.2349 \angle 144.9° \ \Omega \tag{7.39}$$

The rms stator voltage can be calculated by

$$\overline{V}_s = \overline{I}_s \overline{Z}_s = V_s \angle \varphi = 275.2 \angle 144.9° \text{ V} \quad (0.691 \text{ pu}) \tag{7.40}$$

Figure 7-12 shows the phasor diagram of the generator for the calculation of dq-axis stator voltages, where the stator current is the reference phasor according to Equation 7.38. The dq axes are added to the phasor diagram to assist in the calculation of the dq-axis stator voltage. According to the steady-state equivalent circuit of the generator presented in Chapter 3, the rotor current can be found from

$$\overline{I}_r = \frac{jX_m I_s}{jX_m + \left(\dfrac{R_r}{s} + jX_{lr}\right)} = 995.2 \angle -28.9° \text{ A} \tag{7.41}$$

The magnetizing flux is

$$\overline{\Lambda}_m = (\overline{I}_s - \overline{I}_r) L_m = 1.2113 \angle 58.0° \text{ Wb} \tag{7.42}$$

The rms rotor flux linkage can be obtained by

$$\overline{\Lambda}_r = \overline{\Lambda}_m - L_{lr} \overline{I}_r = \Lambda_r \angle \theta_1 = 1.2096 \angle 61.1° \text{ Wb} \tag{7.43}$$

The angle between the stator voltage and rotor flux is

$$\theta_2 = \varphi - \theta_1 = 83.8° \tag{7.44}$$

from which the dq-axis stator voltages can be calculated by

$$\begin{cases} v_{ds} = \sqrt{2} V_s \cos \theta_2 = 42.0 \text{ V} \quad (0.105 \text{ pu}) \\ v_{qs} = \sqrt{2} V_s \sin \theta_2 = 386.9 \text{ V} \quad (0.971 \text{ pu}) \end{cases} \tag{7.45}$$

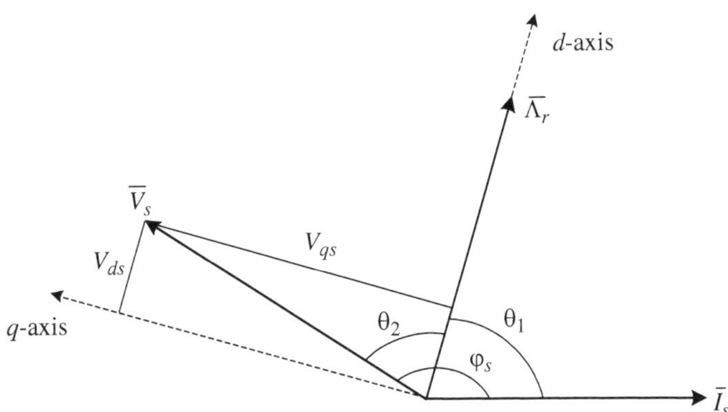

Figure 7-12. Phasor diagram for the calculation of dq-axis stator voltages.

The calculated *dq* axis stator currents (i_{ds} = 0.369 pu and i_{qs} = −0.669 pu) and stator voltages (v_{ds} = 0.105 pu and v_{qs} = 0.971 pu) at the generator speed of 0.7 pu correlate well with the simulated results given in Figure 7-10.

7.4   DIRECT TORQUE CONTROL

Direct torque control (DTC) is an advanced control scheme originally developed for motor drives. The control scheme can be easily adopted for wind energy conversion systems. It is characterized by simple control algorithm and easy digital implementation. In this section, the principles of operation of the DTC scheme are introduced and the converter switching logic is discussed [4,5]. The dynamic and steady-state analyses of a wind energy system implementing the DTC scheme are then provided.

7.4.1   Principle of Direct Torque Control

The electromagnetic torque developed by an induction generator can be expressed in a number of ways, one of which is

$$T_e = \frac{3P}{2}\frac{L_m}{\sigma L_s L_r}\lambda_s\lambda_r\sin\theta_T \tag{7.46}$$

where λ_s and λ_r are the magnitude (peak value) of the stator and rotor flux linkage vectors $\vec{\lambda}_s$ and $\vec{\lambda}_r$, and θ_T is the angle between the two vectors, often known as torque angle. The essence of direct torque control is to control the electromagnetic torque of the generator by adjusting the torque angle θ_T while keeping the magnitude of the stator flux at a constant value (normally at its rated value). Note that with the stator flux kept at its rated value, the rotor flux λ_r of the generator is almost constant, varying only a few percent around its rated value over a wide operating range. As a result, the torque can be directly controlled by θ_T.

Referring to the induction generator model of Figure 7-4a, the stator flux $\vec{\lambda}_s$ relates the stator voltage vector $\vec{v}_s$ by

$$p\vec{\lambda}_s = \vec{v}_s - R_s\vec{i}_s \tag{7.47}$$

The above equation shows that the derivative of $\vec{\lambda}_s$ reacts instantly to changes in $\vec{v}_s$. The stator voltage $\vec{v}_s$ is, in fact, the pulse-width modulated output voltage of the rectifier, which can be controlled by the reference vector $\vec{v}_{ref}$ in the space vector modulation. Since $\vec{v}_{ref}$ is synthesized by the voltage vectors (switching states) of the rectifier, a proper selection of the vectors can make the magnitude and angle of $\vec{\lambda}_s$ adjustable.

Figure 7-13 shows the direct torque control method implemented for a two-level voltage source rectifier. The *dq*-axis plane for the stator flux $\vec{\lambda}_s$ is divided into six sectors, I to VI. The stator flux $\vec{\lambda}_s$ in the figure falls into sector I, and its angle θ_s is referenced to the *d*-axis of the stationary reference frame. When the induction generator operates in the motoring mode, the stator flux vector $\vec{\lambda}_s$ leads the rotor flux vector $\vec{\lambda}_r$, and

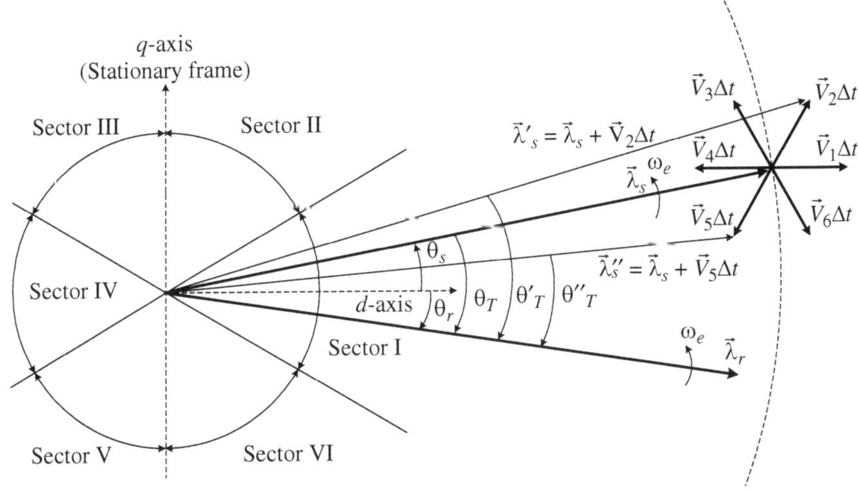

Figure 7-13. Principle of direct torque control.

the torque angle θ_T is positive. In the wind energy system, the generator normally operates in the generating mode. The torque angle θ_T is negative, and resultant torque T_e is negative, which is consistent with the induction generator model presented in Chapter 3. The torque angle can be determined from

$$\theta_T = \angle\vec{\lambda}_s - \angle\vec{\lambda}_r = \theta_s - \theta_r \tag{7.48}$$

where θ_r is the rotor flux angle referenced to the d-axis of the stationary reference frame as shown in Figure 7-13.

Let us now examine the impact of the voltage vectors of the rectifier $\vec{V}_0$ to $\vec{V}_6$ on $\vec{\lambda}_s$ and θ_T (refer to Section 4.4.2 for the definition of voltage vectors). Assume that $\vec{\lambda}_s$ and θ_T in Figure 7-13 are the initial stator flux vector and torque angle. When voltage vector $\vec{V}_2$ is selected, the stator flux vector will become $\vec{\lambda}'_s = \vec{\lambda}_s + \vec{V}_2\Delta t$ after a short time interval Δt, leading to an increase in flux magnitude ($\lambda'_s > \lambda_s$) and torque angle ($\theta'_T > \theta_T$). If voltage vector $\vec{V}_5$ is selected, $\vec{\lambda}_s$ will change to $\vec{\lambda}''_s = \vec{\lambda}_s + \vec{V}_5\Delta t$, causing a decrease in flux magnitude ($\lambda''_s < \lambda_s$) and torque angle $\theta''_T < \theta_T$. Similarly, the selection of $\vec{V}_3$ and $\vec{V}_6$ can make one variable increase and the other decrease. Therefore, λ_s and θ_T can be controlled separately by proper selection of rectifier voltage vectors.

Note that the changes in $\vec{v}_s$ have little impact on $\vec{\lambda}_r$ during the short time interval Δt due to the large rotor time constant. Therefore, in the above analysis it is assumed that the rotor flux vector $\vec{\lambda}_r$ is kept constant during Δt.

7.4.2 Switching Logic

Figure 7-14 shows a typical block diagram of a direct torque controlled induction generator WECS. The direct torque control is realized by the rectifier, whereas the DC

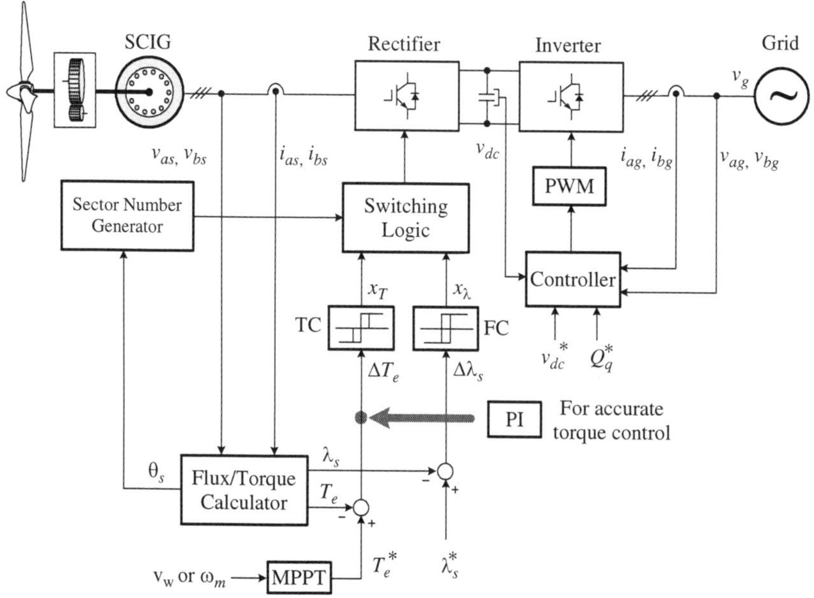

Figure 7-14. DTC scheme for an SCIG wind energy system.

link voltage control and grid-side reactive power control are achieved by the inverter. Since the grid-tied inverter is discussed in Chapter 4, it is not repeated here. Similar to the FOC scheme, the stator flux and electromagnetic torque are controlled separately to achieve good dynamic performance. The stator flux reference λ_s^* is compared with the calculated stator flux λ_s, and the error $\Delta\lambda_s$ is sent to the flux comparator (FC). The torque reference T_e^* is compared with the calculated torque T_e, and their difference ΔT_e is the input to the torque comparator (TC). The output of the flux and torque comparators (x_λ and x_T) are sent to the switching logic unit for proper selection of the voltage vectors (switching states) of the rectifier.

Both flux and torque comparators are of a hysteresis (tolerance band) type, whose transfer characteristics are shown in Figure 7-15. The flux comparator has two output levels ($x_\lambda = +1$ and -1), whereas the torque comparator has three output levels ($x_T = +1, 0$, and -1), where $+1$ requests an increase in λ_s or θ_T, -1 demands a decrease in λ_s or θ_T, and 0 signifies no changes. The tolerance bands for the flux and torque comparators are δ_λ and δ_T, respectively.

Table 7-3 gives the switching logic for the stator flux reference $\vec{\lambda}_s^*$ rotating in the counterclockwise direction. The input variables are x_λ, x_T, and the sector number, and the output variables are the rectifier voltage vectors. The output of the comparators dictates which voltage vector should be selected. For example, assuming that $\vec{\lambda}_s^*$ is in sector I, the comparator output of $x_\lambda = x_T = +1$ signifies an increase in λ_s and T_e. Voltage vector $\vec{V}_2$ can then be selected from Table 7-3. This selection will make both λ_s and θ_T increase, as shown in Figure 7-13.

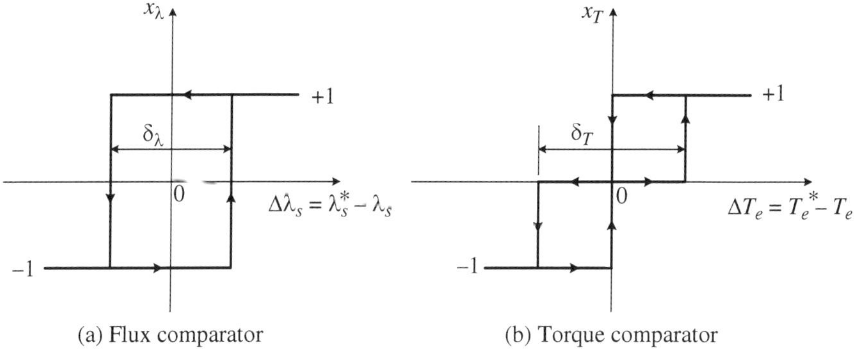

(a) Flux comparator   (b) Torque comparator

Figure 7-15. Characteristics of hysteresis comparators.

When the output of the torque comparator x_T is zero (no need to adjust T_e), the zero vector $\vec{V}_0$ can be selected. Having two alternatives for the switching states $\vec{V}_0$, that is, [OOO] and [PPP], in the switching table can help reduce the converter (semiconductor) switching frequency. For instance, when x_T changes between +1 and 0 or between 0 and −1, the zero states in the table ensure that only two switches are involved during the transition, one being turned on and the other being turned off.

It is worth noting that to achieve the MPPT control of the wind energy system in Figure 7-14, the generator torque T_e should follow the reference T_e^* accurately. This requirement can be satisfied by adding a PI controller before the torque comparator, as shown in the figure. The PI regulator ensures that the generator torque is equal to its reference given by the MPPT scheme when the system operates in steady state. It

Table 7-3. Switching logic for $\vec{\lambda}_s^*$ rotating in the counterclockwise direction

Comparator output		Sector					
x_λ	x_T	I	II	III	IV	V	VI
+1	+1	$\vec{V}_2$	$\vec{V}_3$	$\vec{V}_4$	$\vec{V}_5$	$\vec{V}_6$	$\vec{V}_1$
		[PPO]	[OPO]	[OPP]	[OOP]	[POP]	[POO]
	0	$\vec{V}_0$	$\vec{V}_0$	$\vec{V}_0$	$\vec{V}_0$	$\vec{V}_0$	$\vec{V}_0$
		[PPP]	[OOO]	[PPP]	[OOO]	[PPP]	[OOO]
	−1	$\vec{V}_6$	$\vec{V}_1$	$\vec{V}_2$	$\vec{V}_3$	$\vec{V}_4$	$\vec{V}_5$
		[POP]	[POO]	[PPO]	[OPO]	[OPP]	[OOP]
−1	+1	$\vec{V}_3$	$\vec{V}_4$	$\vec{V}_5$	$\vec{V}_6$	$\vec{V}_1$	$\vec{V}_2$
		[OPO]	[OPP]	[OOP]	[POP]	[POO]	[PPO]
	0	$\vec{V}_0$	$\vec{V}_0$	$\vec{V}_0$	$\vec{V}_0$	$\vec{V}_0$	$\vec{V}_0$
		[OOO]	[PPP]	[OOO]	[PPP]	[OOO]	[PPP]
	−1	$\vec{V}_5$	$\vec{V}_6$	$\vec{V}_1$	$\vec{V}_2$	$\vec{V}_3$	$\vec{V}_4$
		[OOP]	[POP]	[POO]	[PPO]	[OPO]	[OPP]

Table 7-4. Switching logic for the stator flux rotating in the clockwise direction

Comparator output		Sector					
x_λ	x_T	I	II	III	IV	V	VI
1	+1	$\vec{V}_6$	$\vec{V}_5$	$\vec{V}_4$	$\vec{V}_3$	$\vec{V}_2$	$\vec{V}_1$
		[POP]	[OOP]	[OPP]	[OPO]	[PPO]	[POO]
	0	$\vec{V}_0$	$\vec{V}_0$	$\vec{V}_0$	$\vec{V}_0$	$\vec{V}_0$	$\vec{V}_0$
		[PPP]	[OOO]	[PPP]	[OOO]	[PPP]	[OOO]
	−1	$\vec{V}_2$	$\vec{V}_1$	$\vec{V}_6$	$\vec{V}_5$	$\vec{V}_4$	$\vec{V}_3$
		[PPO]	[POO]	[POP]	[OOP]	[OPP]	[OPO]
−1	+1	$\vec{V}_5$	$\vec{V}_4$	$\vec{V}_3$	$\vec{V}_2$	$\vec{V}_1$	$\vec{V}_6$
		[OOP]	[OPP]	[OPO]	[PPO]	[POO]	[POP]
	0	$\vec{V}_0$	$\vec{V}_0$	$\vec{V}_0$	$\vec{V}_0$	$\vec{V}_0$	$\vec{V}_0$
		[OOO]	[PPP]	[OOO]	[PPP]	[OOO]	[PPP]
	−1	$\vec{V}_3$	$\vec{V}_2$	$\vec{V}_1$	$\vec{V}_6$	$\vec{V}_5$	$\vec{V}_4$
		[OPO]	[PPO]	[POO]	[POP]	[OOP]	[OPP]

should be pointed out that this PI controller is not required in a motor drive with DTC control, where the torque reference T_e^* is given by a speed PI controller. In that case, any errors between T_e and T_e^* will not affect the accuracy of the motor speed due to the speed PI controller, and the actual motor speed will be always equal to its reference when the drive operates in steady state.

The switching logic given in Table 7-3 is only valid for the generator rotating in the counterclockwise direction. When the generator operates in the clockwise direction, the switching logic in Table 7-4 can be used.

7.4.3 Stator Flux and Torque Calculator

The stator flux vector $\vec{\lambda}_s$ in the stationary frame can be expressed as

$$\begin{aligned}
\vec{\lambda}_s &= \lambda_{ds} + j\lambda_{qs} \\
&= \int \left(v_{ds} - R_s i_{ds} \right) dt + j \int \left(v_{qs} - R_s i_{qs} \right) dt
\end{aligned} \tag{7.49}$$

from which its magnitude and angle are

$$\begin{cases}
\lambda_s = \sqrt{\lambda_{ds}^2 + \lambda_{qs}^2} \\
\theta_s = \tan^{-1}\left(\dfrac{\lambda_{qs}}{\lambda_{ds}} \right)
\end{cases} \tag{7.50}$$

where v_{ds}, v_{qs}, i_{ds}, and i_{qs} are the measured stator voltages and currents. The developed electromagnetic torque can be calculated by

$$T_e = \frac{3P}{2}(\lambda_{ds} i_{qs} - \lambda_{qs} i_{ds}) \tag{7.51}$$

The above equations illustrate that the stator flux and developed torque can be obtained by measured stator voltages and currents. The only generator parameter required in the calculations is the stator resistance R_s. This is in contrast to the direct rotor flux FOC schemes, where most of the generator parameters are needed. The algorithm for the flux/torque calculator is represented by a block diagram of Figure 7-16.

7.4.4   Transient Analysis of SCIG WECS with DTC

The transient performance of the SCIG wind energy system with direct torque control is investigated through a case study. In order to better compare the performance of the DTC scheme with that of the FOC scheme in Case Study 7-3, the operating conditions of the two systems are kept the same.

Case Study 7-5—Transient Analysis of an SCIG WECS with DTC. Consider a 2.3 MW/690 V SCIG wind energy system controlled by the DTC scheme. The dynamic behavior of the system is investigated when the wind speed suddenly changes from 8.4 m/sec to 12 m/sec, which is the same range as that in Case Study 7-3. To ensure an accurate control of the generator torque T_e, a PI controller is added to the torque control loop as shown in Figure 7-14. The generator and system parameters for the case study are given in Table 7-5.

Figure 7-17 illustrates the simulated waveforms for the DTC wind energy system when the rotor speed changes from 0.7 pu to its rated value of 1.0 pu. The waveforms include the rotor mechanical speed ω_m, generator mechanical torque T_m, electromagnetic torque T_e, phase-*a* stator current i_{as}, and phase-*a* stator voltage v_{as} (also the rectifier PWM input voltage). Comparing the waveforms with those in Figure 7-10, one

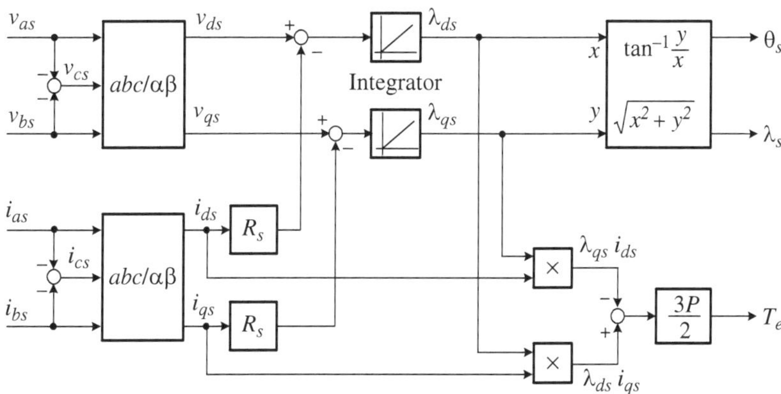

Figure 7-16. Block diagram of the flux/torque calculator for use in DTC scheme.

Table 7-3. Parameters and operating condition of the SCIG WECS with DTC

Induction generator	Generator ratings: 2.3 MW, 690 V, 50 Hz, 1512 rpm 2168 A, 14.67 kN·m	Generator parameters: Table B-1 (Appendix B)
System input variables	Stator flux reference (rated)	$\lambda_s^* = 1.803$ Wb (1.422 pu)
	Initial generator torque at wind speed of 8.4 m/s (0.7 pu)	$T_m = -7222.6$ N·m (0.49 pu)
	Step change in wind speed at $t = 0.1$ sec	$v_w = 12$ m/sec (1.0 pu)
DTC scheme	Control scheme	Figure 7-14 with the torque PI controller implemented
	Flux tolerance band	$\delta_\lambda = 0.01$ pu
	Torque tolerance band	$\delta_T = 0.08$ pu
	Switching Frequency	Around 2.7 kHz at 0.7 pu rotor speed and 2 kHz at rated rotor speed
Generator-side converter	Converter type	Two-level voltage source
	Modulation scheme	Not applicable
Grid-side converter	Replaced by a DC voltage source with a fixed DC voltage	$V_{dc} = 1220$ V (3.06 pu)
Stator flux calculator	Algorithm	Figure 7-16
PI controller	Torque control loop	Yes
	Stator flux loop	No
Reference frame transformation	Not applicable	

can observe that the dynamic performance of the DTC WECS is very close to that of the FOC wind energy system.

To further study the system performance, the waveforms of the stator flux λ_s, rotor flux λ_r, and torque angle θ_T are given in Figure 7-18. The stator flux contains some ripples, but its average value is kept at 1.422 pu set by its reference λ_s^*. During the transient process caused by the step changes in wind speed, the stator flux remains constant by the DTC scheme. The rotor flux λ_r contains little ripple due to the filtering effect of the stator and rotor leakage inductances. However, the ripple magnitude varies with the system operating conditions. At the rotor speed of 0.7 pu, the rotor flux is 1.373 pu, but it decreases to 1.349 pu when the rotor speed increases to 1.0 pu. Since the stator flux is held constant, the variation of the rotor flux is minimal, only 0.024 pu in this case. The torque angle θ_T is negative, leading to a negative torque. This indicates that the generator operates in the generating mode, delivering power from the generator to the grid through power converters. When the generator operates at 0.7 pu speed with 0.49 pu torque, the torque angle is $-5.8°$. With the rated operating conditions, the torque angle of the generator is $-12.1°$. All

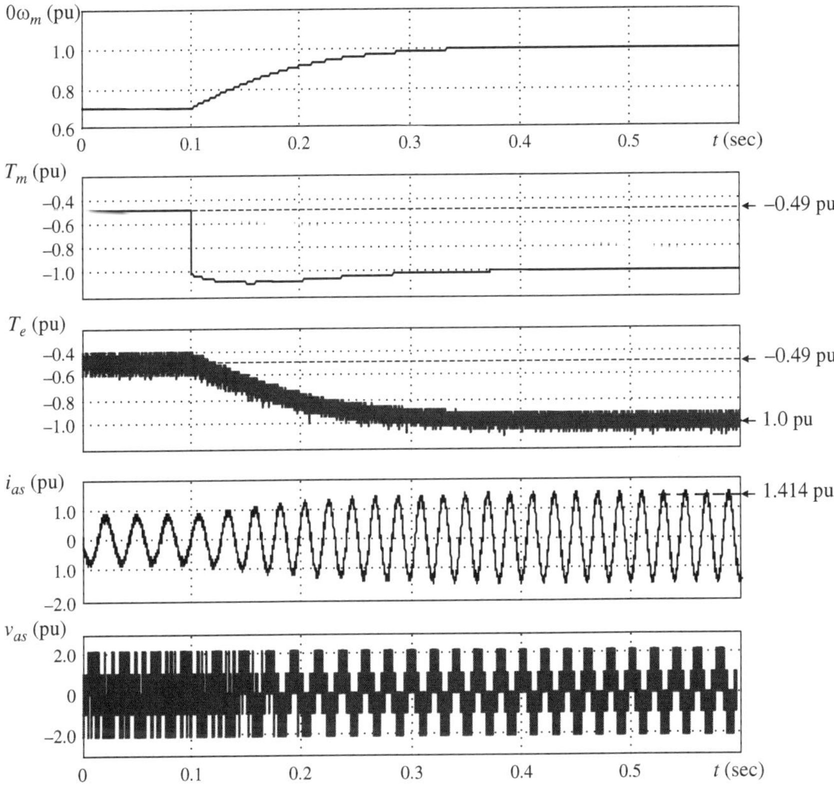

Figure 7-17. System transient waveforms with a step change in wind speed from 0.7 pu to 1.0 pu.

of the above values are verified by steady-state analysis presented in the following section.

7.4.5 Steady-State Analysis of SCIG WECS with DTC

For the direct torque controlled WECS, one of the system input variables is the stator flux reference λ_s^*. Through closed loop feedback control, the stator flux of the generator is kept to the reference $(\lambda_s = \lambda_s^*)$. The other known variables include the rotor speed ω_r and the generator mechanical torque T_m. To perform the steady-state analysis, we need to determine the slip frequency ω_{sl} and stator frequency ω_s, based on which all other system variables can be calculated. The slip and stator frequencies are determined as follows.

The rotor circuit of the induction generator model of Figure 3-7a (Chapter 3) in the synchronous frame can be described by

$$\begin{cases} R_r i_{dr} + p\lambda_{dr} - \omega_{sl}\lambda_{qr} = 0 \\ R_r i_{qr} + p\lambda_{qr} + \omega_{sl}\lambda_{dr} = 0 \end{cases} \tag{7.52}$$

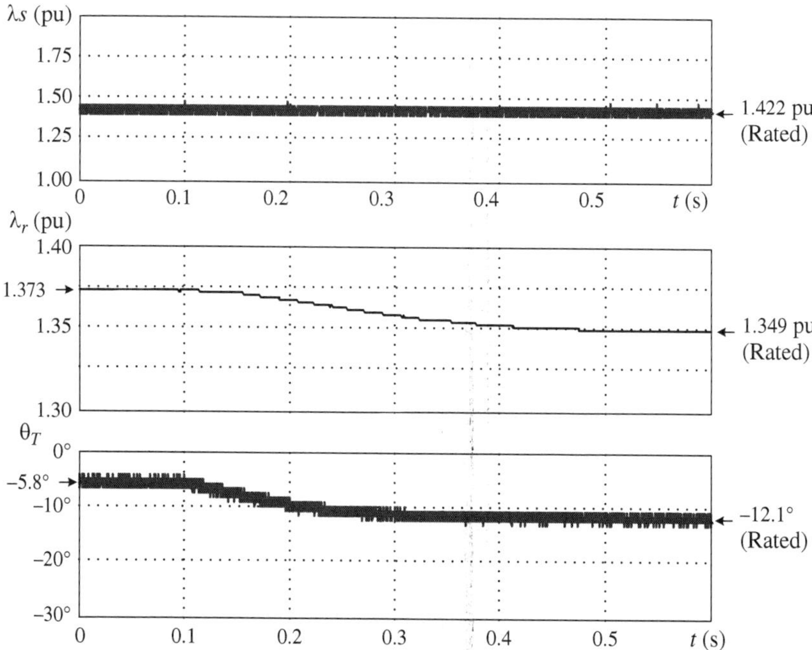

Figure 7-18. Waveforms of the rotor/stator fluxes and torque angle with a step change in wind speed from 0.7 pu to 1.0 pu.

Since all of the variables are DC quantities in the steady state, the above equation can be simplified to

$$\begin{cases} R_r i_{dr} - \omega_{sl}\lambda_{qr} = 0 \\ R_r i_{qr} + \omega_{sl}\lambda_{dr} = 0 \end{cases} \tag{7.53}$$

The dq-axis rotor currents can be expressed as

$$\begin{cases} i_{dr} = \dfrac{\lambda_{dr} - L_m i_{ds}}{L_r} \\ i_{qr} = \dfrac{\lambda_{qr} - L_m i_{qs}}{L_r} \end{cases} \tag{7.54}$$

Substituting (7.54) into (7.53) yields

$$\begin{cases} \lambda_{dr} - L_m i_{ds} - \tau_r \omega_{sl}\lambda_{qr} = 0 \\ \lambda_{qr} - L_m i_{qs} + \tau_r \omega_{sl}\lambda_{dr} = 0 \end{cases} \tag{7.55}$$

Referring to (7.9), the dq-axis rotor fluxes can be expressed in terms of stator variables:

$$\begin{cases} \lambda_{dr} = \dfrac{L_r}{L_m}(\lambda_{ds} - \sigma L_s i_{ds}) \\[4mm] \lambda_{qr} = \dfrac{L_r}{L_m}(\lambda_{qs} - \sigma L_s i_{qs}) \end{cases} \tag{7.56}$$

Substituting (7.56) into (7.55) yields

$$\begin{cases} \lambda_{ds} = L_s i_{ds} + \tau_r \omega_{sl}(\lambda_{qs} - \sigma L_s i_{qs}) \\[2mm] \lambda_{qs} = L_s i_{qs} - \tau_r \omega_{sl}(\lambda_{ds} - \sigma L_s i_{ds}) \end{cases} \tag{7.57}$$

To simplify the analysis, let us align the stator flux axis with the d-axis of the synchronous frame, that is,

$$\lambda_{qs} = 0 \text{ and } \lambda_{ds} = \lambda_s \tag{7.58}$$

Substituting (7.58) into (7.57) gives

$$\begin{cases} \lambda_s = L_s i_{ds} - \sigma L_s \tau_r \omega_{sl} i_{qs} \\[2mm] L_s i_{qs} = \tau_r \omega_{sl}(\lambda_s - \sigma L_s i_{ds}) \end{cases} \tag{7.59}$$

from which the d-axis stator current can be calculated by

$$i_{ds} = \frac{\lambda_s + \sigma L_s \tau_r \omega_{sl} i_{qs}}{L_s} \tag{7.60}$$

and the slip frequency ω_{sl} can be found by solving

$$(\tau_r \sigma)^2 \omega_{sl}^2 - \frac{(1-\sigma)\tau_r \lambda_s}{L_s i_{qs}} \omega_{sl} + 1 = 0 \tag{7.61}$$

The stator frequency can then be found from

$$\omega_s = \omega_{sl} + \omega_r \tag{7.62}$$

The q-axis stator current i_{qs} in Equations 7.60 and 7.61 can be obtained from the torque equation:

$$T_e = \frac{3P}{2}(i_{qs}\lambda_{ds} - i_{ds}\lambda_{qs})$$

With $\lambda_{qs} = 0$ and $\lambda_{ds} = \lambda_s$, the torque equation can be simplified to

$$T_e = \frac{3P}{2} i_{qs} \lambda_s \tag{7.63}$$

from which the q-axis current is

$$i_{qs} = \frac{2T_e}{3P\lambda_s} \tag{7.64}$$

where T_e is determined by the MPPT scheme at a given wind speed, and λ_s is normally equal to its rated value set by λ_s^*.

The step-by-step procedure for the analysis and calculation is summarized in Table 7-6.

Case Study 7-6—Steady-State Analysis of SCIG WECS with Direct Torque Control. In this case study, the steady-state analysis of the 2.3 MW/690 V SCIG wind energy system is performed. All system parameters remain the same as those in Case Study 7-5, and the generator operates at 0.7 pu rotor speed. With

Table 7-6. Steady-state analysis of SCIG wind energy system

Stator flux linkage (Wb)	λ_s (Rated, controlled by DTC)
Rotor speed (rad/sec)	ω_r (Given, based on wind speed)
Electromagnetic torque (N·m)	$T_e = T_m$ (Given, from the $T_m - \omega_m$ profile)
q-axis current (A)	$i_{qs} = \dfrac{2T_e}{3P\lambda_s}$
Slip frequency (rad/sec)	$(\tau_r\sigma)^2\omega_{sl}^2 - \dfrac{(1-\sigma)\tau_r\lambda_s}{L_s i_{qs}}\omega_{sl} + 1 = 0$
d-axis current (A)	$i_{ds} = \dfrac{\lambda_s + \sigma L_s \tau_r \omega_{sl} i_{qs}}{L_s}$
Stator current (A)	$i_s = \sqrt{i_{ds}^2 + i_{qs}^2}$
Stator frequency (rad/sec)	$\omega_s = \omega_r + \omega_{sl}$
Slip	$s = \dfrac{\omega_{sl}}{\omega_s}$
Generator impedance (Ω)	$\overline{Z}_s = R_s + jX_{ls} + jX_m\left(\dfrac{R_r}{s} + jX_{lr}\right)$
Stator current (rms) (A)	$\overline{I}_s = i_s/\sqrt{2}\angle 0°$ (Reference phasor)
Stator voltage (rms) (V)	$\overline{V}_s = \overline{I}_s\overline{Z}_s = V_s\angle\varphi$
Stator power factor angle	$\varphi_s = \angle\overline{V}_s - \angle\overline{I}_s = \varphi$

direct torque control, the stator flux is kept at the rated value set by its reference, that is,

$$\lambda_s = \lambda_s^* = \sqrt{2} \times 1.2748 = 1.803 \text{ Wb} \quad (1.422 \text{ pu}) \tag{7.65}$$

For a given generator speed of 0.7 pu, the generator mechanical torque is

$$T_m = T_{m,R} \times \omega_m^2 = -14{,}740 \times 0.7^2 = -7222.6 \text{ N·m} \tag{7.66}$$

With the rated stator flux, the q-axis current of the generator is

$$i_{qs} = \frac{2T_e}{3P\lambda_s} = \frac{2 \times (-7222.6)}{3 \times 2 \times 1.803} = -1335.4 \text{ A} \tag{7.67}$$

The slip frequency ω_{sl} is determined by

$$(T_r\sigma)^2 \omega_{sl}^2 - \frac{(1-\sigma)T_r\lambda_s}{L_s i_{qs}} \omega_{sl} + 1 = 0.0073 \, \omega_{sl}^2 + 0.8494 \, \omega_{sl} + 1 = 0 \tag{7.68}$$

from which

$$\omega_{sl} = \begin{cases} -1.1895 \text{ rad/sec} \quad (0.1893 \text{ Hz}) \\ -115.2 \text{ rad/sec} \quad (\text{omitted}) \end{cases} \tag{7.69}$$

The d-axis current is calculated by

$$i_{ds} = \frac{\lambda_s + \sigma L_s T_r \omega_{sl} i_{qs}}{L_s} = 955.4 \text{ A} \tag{7.70}$$

The stator frequency is

$$\omega_s = \omega_r + \omega_{sl} = \frac{1512 \times 2\pi}{60} P \times 0.7 - 1.1895 = 220.48 \text{ rad/s} \quad (35.09 \text{ Hz}) \tag{7.71}$$

from which the slip is

$$s = \frac{\omega_{sl}}{\omega_s} = -5.4 \times 10^{-3} \tag{7.72}$$

The rms stator current is then obtained by

$$\bar{I}_s = \sqrt{i_{ds}^2 + i_{qs}^2} / \sqrt{2} = \sqrt{955.4^2 + 1335.4^2} / \sqrt{2} = 1161.1\angle 0^\circ \text{ A} \tag{7.73}$$

The generator impedance is given by

$$\overline{Z}_s = R_s + jX_{ls} + jX_m \,//\left(\frac{R_r}{s} + jX_{lr}\right) = 0.2412\angle144.3^\circ\ \Omega \tag{7.74}$$

The rms stator voltage is calculated as

$$\overline{V}_s = \overline{I}_s \overline{Z}_s = 280.0\angle144.3^\circ\ \text{V} \tag{7.75}$$

The rotor current is found from

$$\overline{I}_r = \frac{jX_m \overline{I}_s}{jX_m + \left(\dfrac{R_r}{s} + jX_{lr}\right)} = 978.0\angle-29.8^\circ\,\text{A} \tag{7.76}$$

The magnetizing flux is calculated by

$$\overline{\Lambda}_m = (\overline{I}_s - \overline{I}_r)L_m = 1.2325\angle57.2^\circ\ \text{Wb} \tag{7.77}$$

The rms stator flux linkage is obtained by

$$\overline{\Lambda}_s = \overline{\Lambda}_m + L_{ls}\times\overline{I}_s = 1.2748\angle54.4^\circ\ \text{Wb} \tag{7.78}$$

and its peak value is

$$\lambda_s = \sqrt{2}\Lambda_s = \sqrt{2}\times1.2748 = 1.803\ \text{Wb}\quad(1.422\ \text{pu}) \tag{7.79}$$

which is equal to the its reference λ_s^*.

The rms rotor flux linkage is obtained by

$$\overline{\Lambda}_r = \overline{\Lambda}_m - L_{ls}\times\overline{I}_r = 1.2308\angle60.2^\circ\ \text{Wb} \tag{7.80}$$

and its peak value is

$$\lambda_r = \sqrt{2}\Lambda_r = \sqrt{2}\times1.2308 = 1.7406\ \text{Wb}\quad(1.373\ \text{pu}) \tag{7.81}$$

The torque angle is then

$$\theta_T = \angle\overline{\Lambda}_s - \angle\overline{\Lambda}_r = 54.4^\circ - 60.2^\circ = -5.8^\circ \tag{7.82}$$

The calculated values for the stator flux λ_s (1.422 pu), rotor flux λ_r (1.373 pu), and torque angle θ_T (−5.8°) at the rotor speed of 0.7 pu match well with the simulated result given in Figure 7-18. For further verification, the generator torque can be calculated by

$$T_e = \frac{3P}{2} \frac{L_m}{sL_sL_r} \lambda_s\lambda_r \sin\theta_T$$

$$= \frac{3\times2}{2} \frac{2.1346\times10^{-3}}{0.0582\times2.19952\times10^{-3}\times2.19952\times10^{-3}} \tag{7.83}$$

$$\times1.2748\times1.2308\times2\times\sin(54.4°-60.2°)$$

$$= -7222.6 \text{ N·m} \quad (-0.49 \text{ pu})$$

which is equal to the mechanical torque T_m of 0.49 pu at the wind speed of 0.7 pu.

7.5 CONTROL OF CURRENT SOURCE CONVERTER INTERFACED WECS

7.5.1 Introduction

The current source converter (CSC) is a good candidate for megawatt wind energy systems operating at the medium voltage level of a few kilovolts due to its unique features, presented in Chapter 4. Figure 7-19 shows a typical system configuration of the CSC based wind energy system, where a current source rectifier (CSR) is used to control the induction generator, whereas a current source inverter (CSI) is employed to control the DC link current and grid-side power factor. The two converters are linked by a DC choke L_{dc}, which makes the DC link current smooth and continuous. To reduce the DC link current ripple to an acceptable level (normally less than 15%), the size of the DC choke is in the range of 0.6 to 1.2 pu. As discussed in Chapter 4, the current source converter requires a capacitor (C_r for the rectifier and C_i for the inverter) at its AC terminals to assist in the commutation of the switching devices. The capacitor also acts as a harmonic filter.

To be consistent with the induction machine dynamic model presented in Chapter 3, the stator current i_s flows into the stator circuit as shown in Figure 7-19. The reference directions for the PWM rectifier current i_{wr} and the capacitor current i_{cr} are defined accordingly. As a result, the active power P and electromagnetic torque T_e produced by the induction machine when operating as a generator are negative, as summarized in Table 3-2 of Chapter 3.

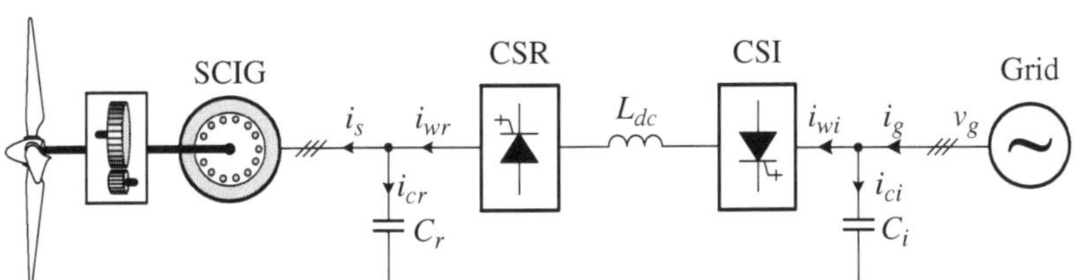

Figure 7-19. System configuration of CSC interfaced wind energy system.

Based on Equation 4.91 of Chapter 4, the average DC voltage V_{dc} of a current source rectifier can be calculated by

$$V_{dc} = \sqrt{3/2}\,V_{ab1}\,m\cos\alpha \qquad (7.84)$$

where V_{ab1} is the fundamental-frequency rms line-to-line input voltage of the rectifier, m is the modulation index of the PWM scheme, and α is the delay angle of the rectifier. Note that the subscript a associated with m_a in Equation 4.91 is omitted here for simplicity. The average DC voltage V_{dc} can be adjusted by either α or m or both. Equation 7.84 is also valid for the current source inverter since the converter topologies for the CSR and CSI are identical.

With the delay angle and modulation index control for both rectifier and inverter, the CSC wind energy system has three possible control options:

1. Delay angle control with fixed modulation index (variable α with fixed m)
2. Modulation index control with fixed delay angle (variable m with fixed α)
3. Controlling both delay angle and modulation index (variable α and variable m)

Option 1—Variable* α *with Fixed* *m. In this control option, there are two independent control variables: delay angles for the rectifier α_r and for the inverter α_i. The modulation indices for the rectifier and inverter, m_r and m_i, are fixed and kept at their maximum value $m_{\max}$, that is,

$$m_r = m_i = m_{\max} \qquad (7.85)$$

With the fixed modulation index, the most effective modulation technique is the selective harmonic elimination (SHE) scheme [6]. Compared with other modulation schemes for the current source converters presented in Chapter 4, the SHE scheme has superior harmonic performance with minimum switching frequency. Since the converter does not need DC current shoot-through (bypass) operation with the fixed $m_{\max}$, a minimum switching frequency can be achieved while eliminating a given number of harmonics. For instance, for a grid frequency of 60 Hz and a converter switching frequency of only 540 Hz, four dominant harmonics (normally the 5th, 7th, 11th, and 13th) can be eliminated. The switching losses are, therefore, minimized.

The two independent control variables, α_r and α_i, can be used to effectively control the active power set by the MPPT scheme and the DC link current of the system. However, the grid-side reactive power is not controllable, and it varies with the operating conditions of the WECS.

***Option 2—Variable* *m* *with Fixed* α.** The two independent control variables in this option are the modulation indices, m_r for the rectifier and m_i for the inverter. In principle, these two variables can be employed to control the active power and DC current of the system. However, the modulation index control requires the two converters to operate in the DC current bypass mode, which not only increases the conduction losses of switching devices due to high DC current, but also increases the switching losses. This control option can fulfill the same functionality as Option 1, but is not recommended due to higher power losses.

Option 3—Variable α **and Variable** ***m.*** With a combination of variable α and variable *m* for both rectifier and inverter, one can simultaneously control the system active power, DC link current, and grid-side reactive power, which are the three main control tasks for a wind energy system. The ability to control the system reactive power is required by various grid codes for large wind turbines.

In the following section, the control scheme for the CSC induction generator WECS with Option 1 is investigated. The control with Option 3 will be presented in Chapter 9 of this book.

7.5.2 Control of CSC WECS with Variable α and Fixed *m*

Figure 7-20 shows the space vector diagram for the CSC wind energy system with rotor flux field oriented control (FOC). The rotor flux vector $\vec{\lambda}_r$ is aligned with the *d*-axis of the synchronous reference frame. The stator voltage vector $\vec{v}_s$ leads the stator current vector $\vec{i}_s$ by an angle φ_s, which is the power factor angle of the generator, defined by $\varphi_s = \angle\vec{v}_s - \angle\vec{i}_s$. When the induction machine operates as a generator, the power factor angle is greater than $\pi/2$. The capacitor current vector $\vec{i}_{cr}$ leads $\vec{v}_s$ by $\pi/2$. The rectifier PWM current $\vec{i}_{wr}$ is a vector sum of $\vec{i}_s$ and $\vec{i}_{cr}$, and its angle with respect to the *a*-axis of the stationary frame is θ_{wr}, defined by

$$\theta_{wr} = \theta_f + \alpha_r \qquad (7.86)$$

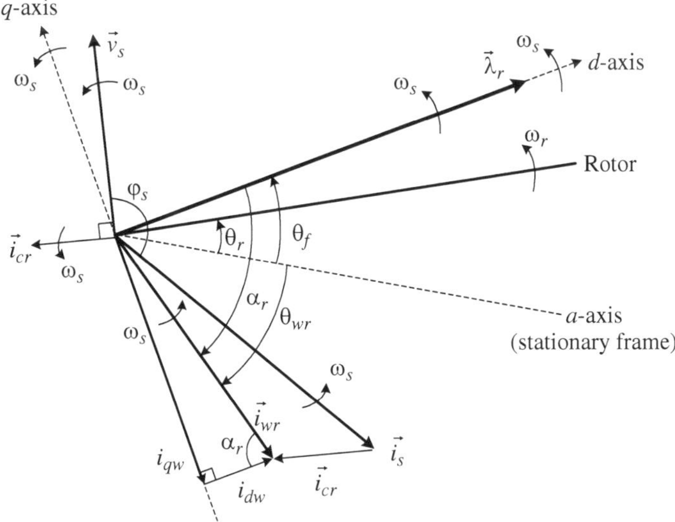

Figure 7-20. Space vector diagram for an SCIG WECS with direct FOC scheme.

where θ_f is the rotor flux angle with respect to the a-axis, and α_r is the rectifier delay angle, which is the angle of $\vec{i}_{wr}$ with respect to $\vec{\lambda}_r$. The delay angle is a constant value when the system operates in steady state since both $\vec{i}_{wr}$ and $\vec{\lambda}_r$ rotate in space at the same speed. On the contrary, θ_f and θ_{wr} vary periodically from zero and 2π over time. It is noted that for the conventional SCR rectifier, the delay angle is referenced to the input or supply voltage of the rectifier, whereas with the rotor flux FOC the delay angle is referenced to the rotor flux vector $\vec{\lambda}_r$.

Figure 7-21 shows a simplified block diagram for the CSC induction generator WECS with delay angle control. In this scheme, there are two feedback control loops, one for the rotor flux angle λ_r and the other for the DC current i_{dc}. The active power control is achieved by regulating the rectifier DC output voltage v_{dcr} and the DC current i_{dc}. The DC voltage v_{dcr} is controlled by the rectifier delay angle α_r, whereas the DC current i_{dc} is regulated by the inverter delay angle α_i.

Generator-Side Control. For the generator control, the rotor flux λ_r and rotor flux angle θ_f are calculated based on the measured stator voltages, v_{as} and v_{bs}, and currents, i_{as} and i_{bs}. The rotor flux angle θ_f is used for field orientation. The detailed algorithm for the rotor flux calculation is given in Figure 7-21 and, therefore, not repeated here. The measured stator voltages and currents in the abc stationary frame are transformed into the dq-axis synchronous frame by the abc/dq transformation block.

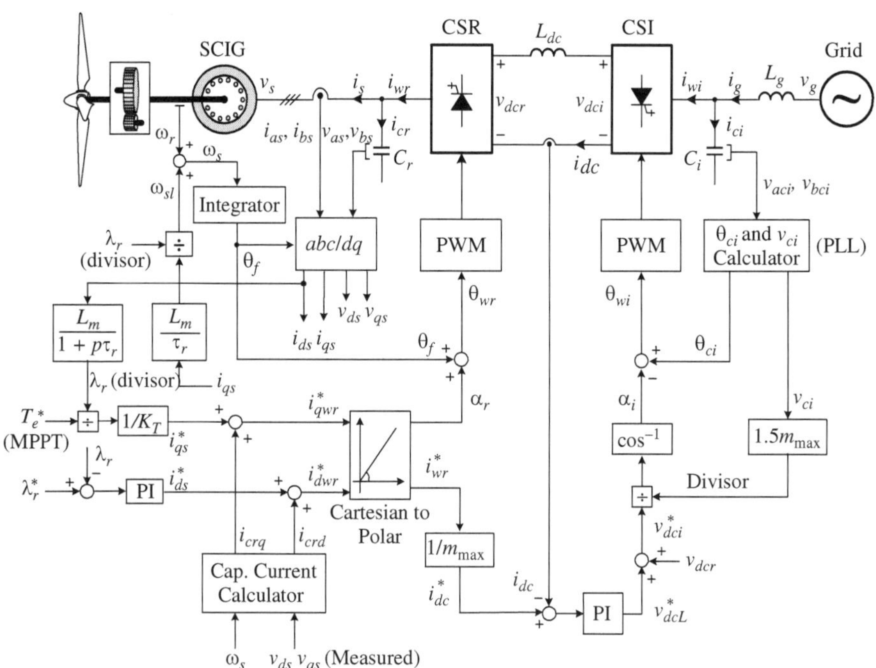

Figure 7-21. Block diagram for an CSC WECS with delay angle control.

The q-axis (torque-producing) stator current reference i_{qs}^* and d-axis (flux-producing) current reference i_{ds}^* are generated in the same manner as that shown in Figure 7-3. The dq-axis rectifier input PWM currents are calculated by

$$
\begin{cases}
i_{dwr}^* = i_{ds}^* + i_{crd} \\
i_{qwr}^* = i_{qs}^* + i_{crq}
\end{cases}
\tag{7.87}
$$

where i_{crd} and i_{crq} are the dq-axis capacitor currents, given by

$$
\begin{cases}
i_{crd} = (pv_{ds} - \omega_s v_{qs})C_r \\
i_{crq} = (pv_{qs} + \omega_s v_{ds})C_r
\end{cases}
\tag{7.88}
$$

It is noted that the subscripts crd and crq are used in the above equations instead of dcr and qcr to avoid confusion with the subscript dc for the DC link variables. The first term on the right-hand side of Equation 7.88 represents capacitor transient current and the second term is the steady-state current. To reduce the sensitivity and noise caused by the derivative terms (pv_{ds} and pv_{qs}), the effect of the capacitor transient response on the system dynamic performance may be neglected. Equation 7.88 can then be simplified to

$$
\begin{cases}
i_{crd} = -\omega_s v_{qs} C_r \\
i_{crq} = \omega_s v_{ds} C_r
\end{cases}
\tag{7.89}
$$

for use in the capacitor current calculator in Figure 7-21.

The references for the dq-axis rectifier PWM currents, i_{dwr}^* and i_{qwr}^*, are then sent to the Cartesian-to-polar transformation block for calculating the magnitude of the rectifier PWM current i_{wr}^* and the delay angle α_r:

$$
\begin{cases}
i_{wr}^* = \sqrt{\left(i_{dwr}^*\right)^2 + \left(i_{qwr}^*\right)^2} \\
\alpha_r = \tan^{-1}\left(i_{qwr}^* \big/ i_{dwr}^*\right)
\end{cases}
\tag{7.90}
$$

The sum of α_r and θ_f is the angle θ_{wr} of the rectifier PWM current i_{wr}, which is fed to the PWM generation block to produce gating pulses for the rectifier.

According to the definition of the modulation index, the DC current reference is obtained by

$$
i_{dc}^* = i_{wr}^* \big/ m_{\max}
\tag{7.91}
$$

For the SHE scheme without bypass operation, the maximum modulation index $m_{\max}$ varies slightly with the number of harmonics to be eliminated, but is close to 1.03 [1]. Equation 7.91 indicates that the use of $m_{\max}$ results in a minimum DC current that leads to the reduction of switching and conduction losses of semiconductor devices.

Grid-Side Control. The main function of the inverter is to adjust the DC current i_{dc} according to its reference i_{dc}^* through a PI controller. The phase-a and -b voltages across the filter capacitor C_i, (v_{aci}, and v_{bci}) are detected. The measured capacitor voltages are sent to the θ_{ci} and v_{ci} calculator, which is sometimes referred to as a phase-locked loop (PLL). The algorithm for the calculator is the same as that given in Section 4.7.1 and thus is not repeated here.

To improve the dynamic performance of the system, a DC voltage feed-forward control is implemented for the DC current control, in which the measured or estimated DC voltage v_{dcr} is added to the output of the PI controller. The PI controller output is the reference voltage v_{Ldc}^* for the DC choke, which determines the rate of changes of the DC current when the system is in transient. As a result, the inverter-side DC voltage reference v_{dci}^* is the sum of the rectifier-side DC voltage v_{dcr} and v_{Ldc}^*. In doing so, the changes in the rectifier output voltage v_{dcr} due to the wind velocity or other system variations will be directly reflected in v_{dci}^* for a quick adjustment.

Based on Equation 7.84, the inverter delay angle can be obtained by

$$\alpha_i = \cos^{-1}\frac{v_{dci}^*}{\sqrt{3/2}\left(V_{ab,ci}\right)m_{max}} = \cos^{-1}\left(\frac{v_{dci}^*}{1.5v_{ci}\,m_{max}}\right) \tag{7.92}$$

where $V_{ab,ci}$ is the fundamental-frequency rms line-to-line voltage of the filter capacitor C_i, and v_{ci} is the fundament peak phase voltage of the capacitor. The angle θ_{wi} of the inverter current i_{wi} for the PWM generation block is obtained by subtracting the inverter delay angle α_i from the capacitor voltage angle θ_{ci}, based on which the PWM block produces the gating signals for the inverter.

7.5.3 Steady-State Analysis of CSC WECS

The steady-state analysis of a CSC interfaced wind energy system is investigated by computer simulation and a steady-state equivalent circuit in the following case study.

Case Study 7-7—CSC Interfaced SCIG WECS. Consider a 3 MW/3000 V squirrel cage induction generator wind energy system using current source converters. The induction generator is controlled by an indirect FOC scheme and operates at its rated speed. System parameters and operating conditions are given in Table 7-7.

The simulated waveforms of the wind energy system are shown in Figure 7-22. The waveform of the phase-a stator voltage v_{as} is slightly distorted, whereas the stator current i_{as} is close to sinusoidal. The stator current lags the stator voltage by 153.3°, which is the power factor angle φ_s. Both the peak stator current and voltage are at 1.414 pu (rated), and the average DC current I_{dc} is 1.28 pu. The phase-a rotor flux λ_{ar} has a peak value of 1.351 pu, which is the rated rotor flux. The rectifier delay angle referenced to λ_{ar} is 92.7°.

To verify the simulation results in Figure 7-22, steady-state analysis based on the steady-state equivalent circuit of Figure 7-23 is performed. The procedure summarized in Table 7-2 is used to assist in the analysis.

Table 7-7. Parameters and operating condition of the CSC SCIG wind energy system

Induction generator	Generator ratings: 3 MW, 3000 V, 60 Hz, 1812 rpm 646 A, 15.8 kN·m	Generator parameters: Table B-3, Appendix B
System input variables	Rotor flux reference Mechanical torque of the generator at 1.0 pu rotor speed	$\lambda_r^* = 6.2105$ Wb (rated) $T_m = -16.02$ kN·m (1.0 pu)
Control scheme	Indirect rotor flux field oriented control	Figure 7-21

Current source rectifier
  (CSR)
	Converter type
	Modulation scheme
	
	Switching sequence
	Switching frequency
	Filter capacitor C_r
		PWM current source
		Space vector modulation
		  $m_{r,\mathrm{max}} = 1$
		Three segments, Figure 4-42
		540 Hz
		296.9 µF (0.3 pu)

Current source inverter (CSI)	Converter type Modulation scheme Harmonic elimination Switching frequency Filter capacitor C_i	PWM current source SHE $m_{i,\mathrm{max}} = 1.03$ 5th, 7th, 11th, and 13th 540 Hz 618.6 µF
DC link circuit	DC link choke L_{dc}	8.5 mH (1.196 pu) 0.0134 Ω (0.005 pu)
Electric grid	Grid voltage (no transformer) Line inductance L_g	3000 V/60 Hz 0.071 mH (0.01 pu) 0.0134 ohm (0.005 pu)

Analysis of Generator-Rectifier Operation. With the rotor flux λ_r kept at its rated value of 6.2106 Wb (peak) by the FOC scheme, the d-axis stator current can be calculated by

$$i_{ds} = \frac{\lambda_r}{L_m} = \frac{6.2106}{27.168 \times 10^{-3}} = 228.6 \text{ A} \quad (0.354 \text{ pu}) \tag{7.93}$$

With MPPT control, the mechanical torque of the generator operating at the rated speed is

$$T_m = T_{m,R} = -16{,}020 \text{ N·m} \tag{7.94}$$

The q-axis current can be calculated by

$$i_{qs} = \frac{T_e}{K_T \lambda_r} = \frac{2L_r}{3PL_m}\frac{T_e}{\lambda_r} = \frac{2 \times 27.95 \times 10^{-3}}{3 \times 2 \times 27.168 \times 10^{-3}} \cdot \frac{-16020}{6.2105} = -884.6 \text{ A} \quad (-1.369 \text{ pu}) \tag{7.95}$$

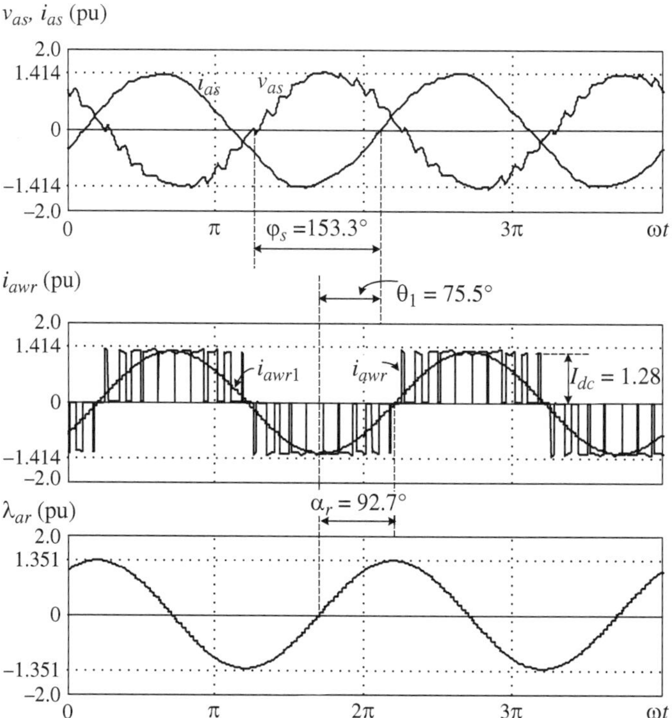

Figure 7-22. Generator-side waveforms of the CSC wind energy system under steady-state operation.

The rms stator current can then be obtained by

$$I_s = \sqrt{i_{ds}^2 + i_{qs}^2} \, / \sqrt{2} = \sqrt{228.6^2 + 884.6^2} \, / \sqrt{2} = 646.1 \, \text{A} \qquad (7.96)$$

The slip frequency can be determined by

$$\omega_{sl} = \frac{R_r L_m}{L_r \lambda_r} i_{qs} = \frac{16.623 \times 10^{-3} \times 27.168 \times 10^{-3}}{27.95 \times 10^{-3} \times 6.2105} \times (-884.6) = -2.5132 \, \text{rad/s} \qquad (7.97)$$

The stator frequency of the generator can be found from

$$\omega_s = \omega_r + \omega_{sl} = 1812 \times \left(\frac{2\pi}{60}\right) P - 2.5132 = 376.9912 \, \text{rad/s} \qquad (60 \text{ Hz}) \qquad (7.98)$$

from which the slip is

$$s = \frac{\omega_{sl}}{\omega_s} = -0.00667 \qquad (7.99)$$

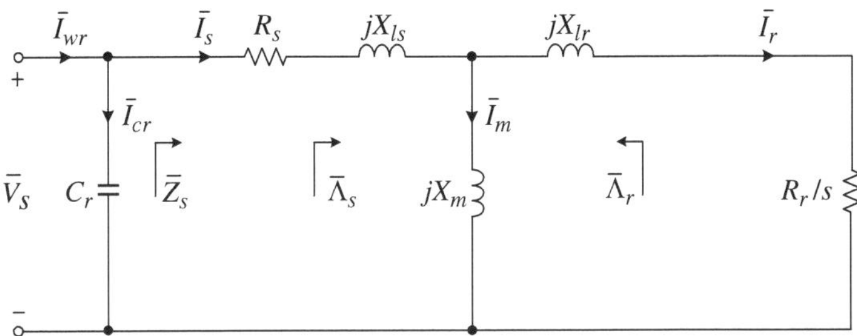

Figure 7-23. Steady-state equivalent circuit of the induction generator with rectifier-side filter capacitor C_r.

Based on the steady-state equivalent circuit of Figure 7-23, the stator impedance can be calculated by

$$\bar{Z}_s = R_s + jX_{ls} + jX_m \; // \left(\frac{R_r}{s} + jX_{lr} \right) = 2.681\angle 153.3°\,\Omega \qquad (7.100)$$

The rms stator voltage can be obtained by

$$\bar{V}_s = \bar{I}_s \bar{Z}_s = 1732.0\angle 153.3°\,\text{V} \qquad (7.101)$$

The rotor current can be found from

$$\bar{I}_r = \frac{jX_m \bar{I}_s}{jX_m + \left(\dfrac{R_r}{s} + jX_{lr} \right)} = 608.0\angle -14.5°\,\text{A} \qquad (7.102)$$

The magnetizing flux can be expressed as

$$\bar{\Lambda}_m = (\bar{I}_s - \bar{I}_r)L_m = 4.4172\angle 69.3°\,\text{Wb} \qquad (7.103)$$

The rotor flux linkage can be obtained from

$$\begin{cases} \bar{\Lambda}_r = \bar{\Lambda}_m - L_{lr}\bar{I}_r = 4.3915\angle 75.5°\,\text{Wb} \\ \lambda_r = \sqrt{2}\Lambda_r = 6.2106\,\text{Wb} \quad (1.351\,\text{pu}) \end{cases} \qquad (7.104)$$

The capacitor current can be calculated by

$$\bar{I}_{cr} = (j\omega_s C_r)\bar{V}_s = (j376.9912\times 0.0002968)(1732.0\angle 153.3°) = 193.8\angle -116.7°\,\text{A} \qquad (7.105)$$

The rectifier PWM current can be found from

$$\bar{I}_{wr} = \bar{I}_s + \bar{I}_{cr} = 646.1\angle 0° + 193.8\angle -116.7° = 585.2\angle -17.2°\,\text{A} \qquad (7.106)$$

The delay angle for the rectifier can be obtained by

$$\alpha_r = \angle \overline{\Lambda}_r - \angle \overline{I}_{wr} = 75.5° - (-17.2°) = 92.7° \tag{7.107}$$

and the stator power factor angle of the generator is

$$\varphi_s = \angle \overline{V}_s - \angle \overline{I}_s = 153.3° \tag{7.108}$$

The above calculated results validate the simulation results given in Figure 7-22.

The steady-state analysis can also be performed using the dq-axis stator currents and voltages. To obtain the dq-axis components, the d- and q-axes are added to the phasor diagram of Figure 7-24 with the d-axis aligned with the rotor flux phasor $OB\Lambda_r$. Note that the dq axes in this analysis are stationary, and do not rotate in space. These are simply added to the steady-state phasor diagram to determine the dq-axis components of the stator current and voltage. Based on the diagram, the following angles can be defined:

$$\begin{cases} \theta_1 = \angle \overline{I}_s - \angle \overline{\Lambda}_r = 0° - 75.5° = -75.5° \\ \theta_2 = \angle \overline{V}_s - \angle \overline{\Lambda}_r = 153.3° - 75.5° = 77.8° \\ \theta_3 = \angle \overline{I}_{wr} - \angle \overline{I}_s = -17.2° \end{cases} \tag{7.109}$$

The dq-axis stator current can then be determined by

$$\begin{cases} i_{ds} = \sqrt{2}I_s \cos \theta_1 = \sqrt{2}\,646.1\cos 75.5° = 228.78 \text{ A} \\ i_{qs} = \sqrt{2}I_s \sin \theta_1 = -\sqrt{2}\,646.1\sin 75.5° = -884.62 \text{ A} \end{cases} \tag{7.110}$$

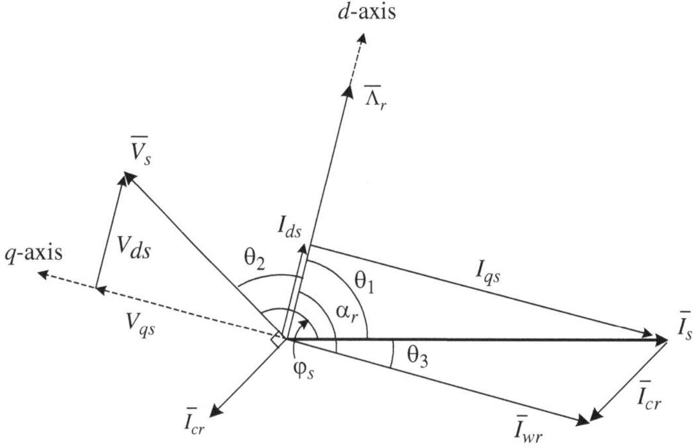

Figure 7-24. dq-axis components of stator voltage and current.

The dq-axis stator voltages can be calculated by

$$\begin{cases} v_{ds} = \sqrt{2}V_s \cos\theta_2 = \sqrt{2}\,1732.1\cos 77.8° = 517.65 \text{ V} \\ v_{qs} = \sqrt{2}V_s \sin\theta_2 = \sqrt{2}\,1732.1\sin 77.8° = 2394.24 \text{ V} \end{cases} \tag{7.111}$$

The dq-axis currents of the rectifier-side capacitor C_r can be found from

$$\begin{cases} i_{crd} = -2\pi f_s v_{qs} C_r = -2\pi \times 60 \times 2394.24 \times 296.8 \times 10^{-6} - -267.89 \text{ A} \\ i_{crq} = 2\pi f_s v_{ds} C_r = 2\pi \times 50 \times 517.65 \times 296.8 \times 10^{-6} = 57.92 \text{ A} \end{cases} \tag{7.112}$$

The dq-axis PWM currents of the rectifier can be calculated by

$$\begin{cases} i_{dwr} = i_{ds} + i_{crd} = 228.78 - 267.89 = -39.11 \text{ A} \\ i_{qwr} = i_{qs} + i_{crq} = -884.62 + 57.92 = -826.70 \text{ A} \end{cases} \tag{7.113}$$

and the rms value of the PWM current is

$$I_{wr} = \sqrt{i_{dwr}^2 + i_{qwr}^2} \,/\, \sqrt{2} = 585.2 \text{ A} \tag{7.114}$$

The rectifier delay angle is

$$\alpha_r = \tan^{-1}\frac{i_{qwr}}{i_{dwr}} = \tan^{-1}\frac{-826.70}{-39.11} = 92.7° \tag{7.115}$$

which is the same as that given in Equation 7.107.

Analysis of Inverter Operation. Figure 7-25 shows the simulated waveforms of DC link current i_{dc}, phase-a capacitor voltage v_{aci}, and phase-a inverter PWM current i_{awi}. The average DC current is 1.28 pu, and the peak of the capacitor voltage waveform is close to that of the grid voltage of $\sqrt{2}$ pu. The inverter delay angle α_i with reference to the capacitor voltage is around 165°.

For steady-state analysis, it is assumed that the DC choke is ideal without power losses. The active power delivered from the generator to the DC link circuit is

$$P_{dc} = 3V_s I_s \cos\phi_s = 3 \times 1732.1 \times 646\cos(153.3°) = -3 \times 10^6 \text{W} \tag{7.116}$$

The DC current I_{dc} is related to the fundamental-frequency component of the rectifier PWM current I_{awr1} by

$$I_{dc} = \sqrt{2}\,I_{awr1}\,/\,m_{r,\max} = \sqrt{2} \times 585.2/1.0 = 827.6 \text{ A} \quad (1.28\,\text{pu}) \tag{7.117}$$

where $m_{r,\max}$ is the maximum modulation index of the SVM scheme. The rectifier is then controlled only by its delay angle α_r. The rectifier-side DC voltage can be calculated by

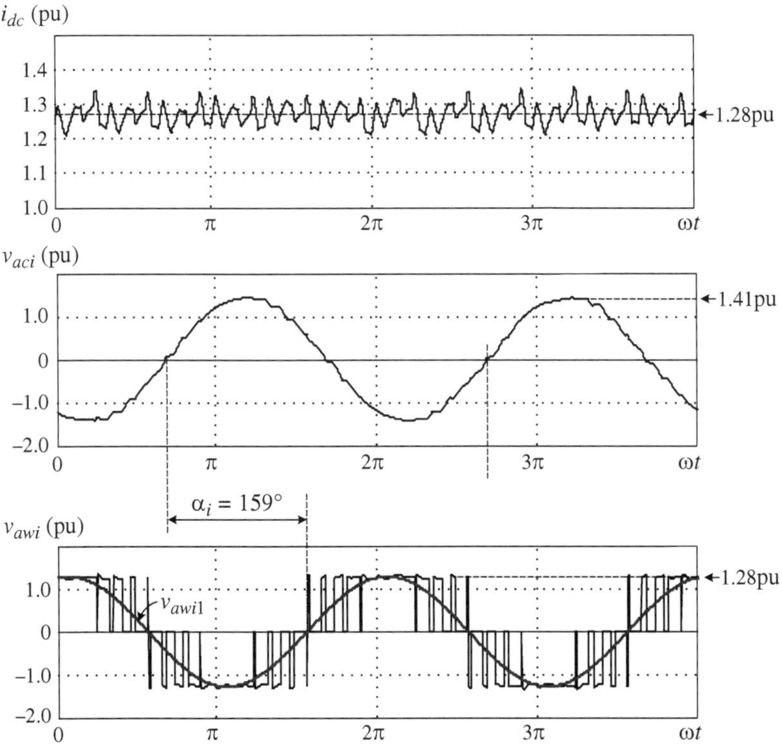

<u>Figure 7-25.</u> Grid-side waveforms of the CSC wind energy system under steady-state operation.

$$V_{dcr} = \frac{P_{dc}}{I_{dc}} = \frac{-3\times10^6}{827.6} = -3{,}624.9 \text{ V} \tag{7.118}$$

The DC voltage is negative, signifying that the rectifier delivers the active power from the generator to the DC link circuit. The DC voltage can also be calculated by using Equation 7.84:

$$V_{dcr} = \sqrt{3/2}\,V_{ab1}m_{r,\max}\cos\left(\left|\phi_s\right|+\left|\theta_3\right|\right) = \sqrt{3/2}\times3000\times1.0\times\cos\left(153.3°+17.2°\right)$$
$$= -3623.8 \text{ V} \tag{7.119}$$

Neglecting the winding resistance of the DC choke L_{dc}, the inverter-side DC voltage is

$$V_{dci} = V_{dcr} = -3623.8 \text{ V} \tag{7.120}$$

Assuming that line-to-line voltage across the inverter capacitor C_i is the same as that of grid as shown in the simulated waveforms, the inverter delay angle can be calculated by

$$\alpha_i = \cos^{-1}\left(\frac{V_{dci}}{\sqrt{3/2}\,m_{i,\max}V_{abg}}\right) = \cos^{-1}\left(\frac{-3623.8}{\sqrt{3/2}\times1.03\times3000}\right) = 163.2° \quad (7.121)$$

where $m_{i,\max}$ is the maximum modulation index of the SHE scheme as listed in Table 7-7, and V_{abg} is the line-to-line grid voltage. The inverter delay angle of 163.2° is close to that given in the simulation in Figure 7-25. The small error (2.6%) is mainly caused by using the grid voltage in the equation instead of the capacitor voltage in the simulation for simplicity.

7.6 SUMMARY

In this chapter, advanced control schemes for squirrel cage induction generator (SCIG) based wind energy conversion systems were presented. The principle of operation of direct and indirect field oriented control (FOC) schemes and direct torque control (DTC) were analyzed in detail. To help attain a deep understanding of the operation and performance of the SCIG wind energy systems, case studies for the wind energy systems using voltage and current source converter technologies were developed. The dynamic and steady-state behaviors of the systems were investigated by both theoretical analysis and computer simulation.

REFERENCES

1. B. Wu, *High-Power Converters and AC Drives,* Wiley-IEEE Press, 2006.

2. R. J. Wai and K. M. Lin, Robust Decoupled Control of Direct Field-oriented Induction Motor Drive, *IEEE Transactions on Industrial Electronics,* Vol. 52, No. 3, 837–854, 2005.

3. B. Heber, L. Xu, and Y. Tang, Fuzzy Logic Enhanced Speed Control of an Indirect Field-oriented Induction Machine Drive, *IEEE Transactions on Power Electronics,* Vol. 12, No. 5, 772–778, 1997.

4. D. Casadei, F. Profumo, G. Serra, and A. Tani, FOC and DTC: Two Viable Schemes for Induction Motors Torque Control, *IEEE Transactions on Power Electronics,* Vol. 17, No. 5, 779–787, 2002.

5. J.H. Lee, C. G. Kim, and M. J. Youn, A Dead-beat Type Digital Controller for the Direct Torque Control of an Induction Motor, *IEEE Transactions on Power Electronics,* Vol. 17, No. 5, 739–746, 2002.

6. B. Wu, S. B. Dewan, and G. R. Slemon, PWM-CSI Inverter for Induction Motor Drives, *IEEE Transactions on Industry Applications,* Vol. 28, No. 1, 64–71, 1992.

8

DOUBLY FED INDUCTION GENERATOR BASED WECS

8.1 INTRODUCTION

The doubly fed induction generator (DFIG) wind energy system is widely accepted in today's wind energy industry. The DFIG is essentially a wound rotor induction generator in which the rotor circuit can be controlled by external devices to achieve variable speed operation. A typical block diagram of the DFIG wind energy system is shown in Figure 8-1. The stator of the generator is connected to the grid through a transformer, whereas the rotor connection to the grid is done through power converters, harmonic filters, and the transformer.

The power rating for the DFIG is normally in the range of a few hundred kilowatts to several megawatts. The stator of the generator delivers power from the wind turbine to the grid and, therefore, the power flow is unidirectional. However, the power flow in the rotor circuit is bidirectional, depending on the operating conditions [1]. The power can be delivered from the rotor to the grid and vice versa through rotor-side converter (RSCs) and grid-side converters (GSCs). Since the maximum rotor power is approximately 30% of the rated stator power, the power rating of the converters is substantially reduced in comparison to the WECS with full-capacity converters.

With variable speed operation, a DFIG wind energy system can harvest more energy from the wind than a fixed-speed WECS of the same capacity when the wind speed is below its rated value. The cost of the power converters and harmonic filters is sub-

Power Conversion and Control of Wind Energy Systems. By B. Wu, Y. Lang, N. Zargari, and S. Kouro
© 2011 the Institute of Electrical and Electronics Engineers, Inc. Published 2011 by John Wiley & Sons, Inc.

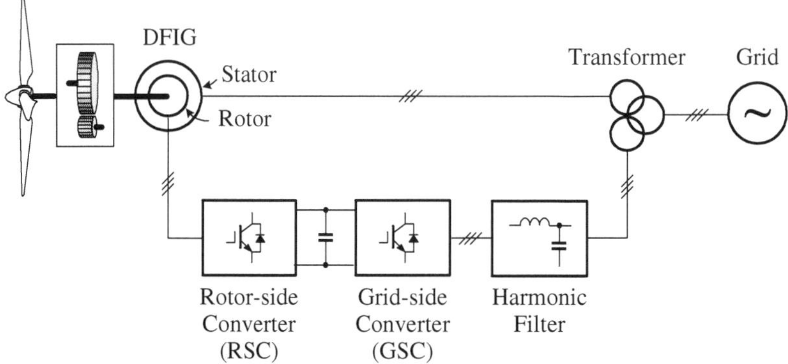

Figure 8-1. Simplified block diagram for DFIG wind energy conversion system.

stantially lower than that in the WECS with full-capacity converters. The power losses in the converters are also lower, leading to improved overall efficiency. In addition, the system can provide leading or lagging reactive power to the grid without additional devices. These features have made the DFIG wind energy system one of the preferred choices in the wind energy market.

In this chapter, the steady-state analysis for the DFIG wind energy system is presented. Dynamic behavior of the system with stator voltage oriented control (SVOC) is investigated. A number of case studies are provided to help the reader understand the principle and operation of the system.

8.2 SUPER- AND SUBSYNCHRONOUS OPERATION OF DFIG

Figure 8-2 shows an example of mechanical power P_m versus slip s characteristics of a DFIG wind energy system. The negative value of mechanical power indicates that the DFIG is in the generating mode, as discussed in Chapter 3. Since the rotor speed of the DFIG is adjustable, one of the maximum power point tracking (MPPT) schemes presented in Chapter 2 can be implemented to harvest the maximum available power from the wind turbine. When operating at the maximum power point (MPP) on the turbine power–speed curves, the generator's mechanical power from the shaft P_m is proportional to the cube of the rotor speed ω_r, as illustrated in Figure 8-2.

The rotor speed of the generator in Figure 8-2 is in the range of $0.5\omega_s$ to $1.2\omega_s$, which corresponds to about 58% of the full speed range (zero to $1.2\omega_s$). This speed range is normally sufficient for a wind energy system since the power generated at 42% of the rated speed is equal to 0.074 pu (0.42^3), only 7.4% of the rated power. The rated slip at which the rated power (1 pu) is generated in this example is –0.2, which represents the rated steady-state operating point of the system. Dynamically, the DFIG may operate up to a slip of –0.3 pu (30% above the synchronous speed ω_s). As a con-

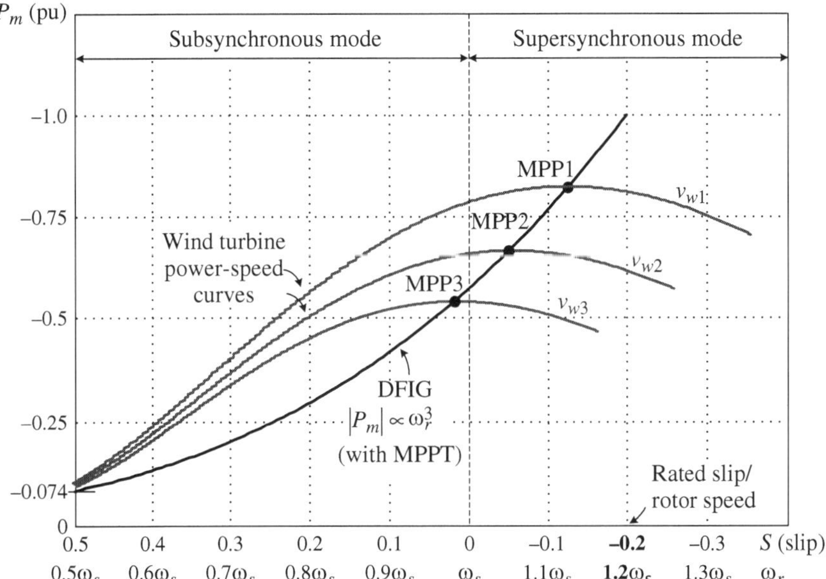

Figure 8-2. An example of power–speed characteristics in a DFIG wind energy system with MPPT control.

sequence, the power converters in the rotor circuit should be designed to handle about 30% of the rated stator power.

Depending on the rotor speed, there are two modes of operation in a DFIG WECS [2]: (1) supersynchronous mode, in which the generator operates above the synchronous speed ω_s; and (2) subsynchronous mode, in which the generator operates below the synchronous speed, as shown in Figure 8-2. The slip is negative in the supersynchronous mode and becomes positive in the subsynchronous mode.

Figure 8-3 shows the power flow in a DFIG wind energy system [3]; the harmonic filter in the rotor circuit and the grid-side transformer are omitted for simplicity. Depending on whether the slip is positive or negative, the rotor circuit can receive or deliver power from or to the grid. In the supersynchronous operation mode, the mechanical power $|P_m|$ from the shaft is delivered to the grid through both stator and rotor circuits. The rotor power $|P_r|$ is transferred to the grid by power converters in the rotor circuit, whereas the stator power $|P_s|$ is delivered to the grid directly. Neglecting the losses in the generator and converters, the power delivered to the grid $|P_g|$ is the mechanical power $|P_m|$ of the generator, as illustrated in Figure 8-3a.

For the subsynchronous operation in Figure 8-3b, the rotor receives the power from the grid. Both mechanical power $|P_m|$ and rotor power $|P_r|$ are delivered to the grid through the stator. Although the stator power $|P_s|$ is the sum of $|P_m|$ and $|P_r|$, it will not exceed its power rating since in the subsynchronous mode the mechanical power $|P_m|$ from the generator shaft is lower than that in the supersynchronous mode. As in the

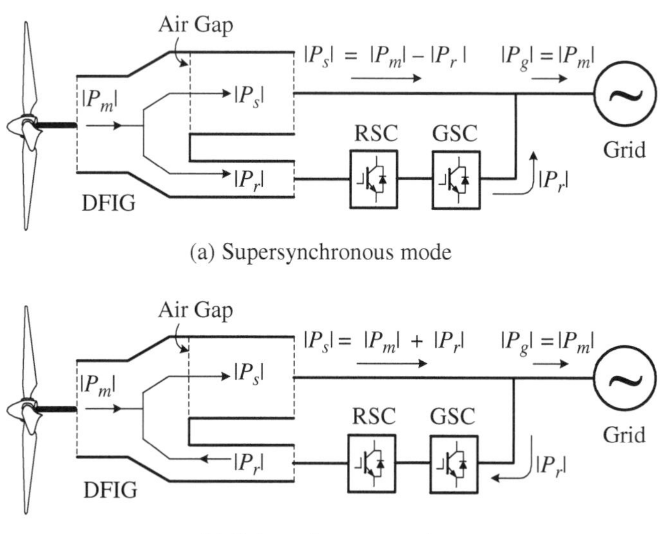

(a) Supersynchronous mode

(b) Subsynchronous mode

Figure 8-3. Power flow in DFIG wind energy conversion system.

previous case, neglecting the losses, the total power delivered to the grid $|P_g|$ is the input mechanical power $|P_m|$.

Since the DFIG generates less power when operating in the subsynchronous mode, the power rating for the converter is determined by the rated or maximum slip in the supersynchronous mode, in which the converters handle the highest rotor power. For instance, the maximum slip for a DFIG during transients caused by gusts of wind is $|-0.3|$, and the power to be processed by the converters is approximately 0.3 of the maximum stator power.

8.3   UNITY POWER FACTOR OPERATION OF DFIG

This section starts with derivation of steady-state equivalent impedance for the rotor-side converter, based on which the analysis of the DFIG wind energy system under unity power factor operation is performed. The torque-slip characteristics of the DFIG with the rotor-side converter taken into account are also developed and analyzed.

8.3.1   Steady-State Equivalent Circuit of DFIG with Rotor-Side Converter

In order to investigate the steady-state performance of the DFIG wind energy system, the rotor-side converter can be modeled by an quivalent impedance. Figure 8-4 shows a steady-state equivalent circuit of DFIG WECS. This equivalent circuit is developed

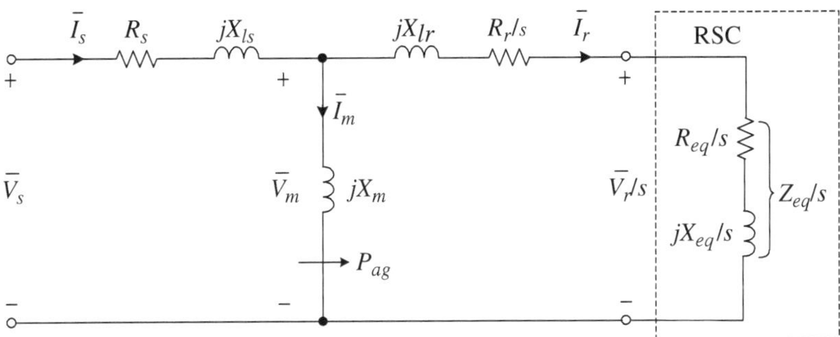

<u>Figure 8-4.</u> Steady-state equivalent circuit of DFIG with the rotor-side converter represented by R_{eq} and X_{eq}.

by adding the converter equivalent impedance to the SCIG steady-state model of Figure 3-13 (Chapter 3). The equivalent impedance of the converter is defined by

$$\overline{Z}_{eq} = R_{eq} + jX_{eq} = R_{eq} + j\omega_{sl}L_{eq} \tag{8.1}$$

where ω_{sl} is the angular slip frequency and L_{eq} is the equivalent inductance of the RSC. Note that the frequency of the rotor current in the actual rotor winding flowing into the converter is ω_{sl}, not the stator frequency ω_s. In order to integrate the converter equivalent impedance into the steady-state model with the stator frequency ω_s, the impedance $\overline{Z}_{eq}$ should be divided by slip s. The equivalent impedance referred to the stator side is then given by

$$\overline{Z}_{eq}/s = R_{eq}/s + j\omega_{sl}L_{eq}/s = R_{eq}/s + j\omega_s L_{eq} \tag{8.2}$$

where $\omega_{sl} = s\omega_s$.

Assuming that the stator operates at a unity power factor, the air-gap power of the generator can be calculated by

$$P_{ag} = 3(V_s - I_s R_s)I_s \tag{8.3}$$

From the induction machine theory, the air-gap power can also be calculated by

$$P_{ag} = \frac{\omega_s T_m}{P} \tag{8.4}$$

where T_m is the mechanical torque and P is the number of pole pairs of the generator.

Substituting (8.4) into (8.3) yields

$$\frac{\omega_s T_m}{P} = 3(V_s - I_s R_s)I_s \tag{8.5}$$

from which

$$R_s I_s^2 - V_s I_s + \frac{\omega_s T_m}{3P} = 0 \tag{8.6}$$

Solving (8.6), we have

$$I_s = \frac{V_s \pm \sqrt{V_s^2 - \dfrac{4R_s \omega_s T_m}{3P}}}{2R_s} \tag{8.7}$$

With the magnitude of the stator current calculated, we can use the equivalent circuit of Figure 8-4 to find the rotor voltage $\overline{V}_r$ and rotor current $\overline{I}_r$. The voltage across the magnetizing branch is

$$\overline{V}_m = \overline{V}_s - \overline{I}_s \left(R_s + j\omega_s L_{ls} \right) \tag{8.8}$$

where the stator voltage and current are given by

$$\overline{V}_s = V_s \angle 0° \quad \text{and} \quad \overline{I}_s = I_s \angle 180° \tag{8.9}$$

The stator voltage and current are 180° out of phase, indicating that the DFIG is in the generating mode and the stator power factor (PF_s) is unity.

The magnetizing current can be determined by

$$\overline{I}_m = \frac{\overline{V}_m}{j\omega_s L_m} \tag{8.10}$$

The rotor current is

$$\overline{I}_r = \overline{I}_s - \overline{I}_m \tag{8.11}$$

The rotor voltage can be calculated by

$$\overline{V}_r / s = \overline{V}_m - \overline{I}_r \left(\frac{R_r}{s} + j\omega_s L_{lr} \right) \tag{8.12}$$

from which

$$\overline{V}_r = s\overline{V}_m - \overline{I}_r \left(R_r + js\omega_s L_{lr} \right) \tag{8.13}$$

The rotor voltage and current relate to the equivalent resistance R_{eq} and reactance X_{eq} by

$$\frac{\overline{V}_r / s}{\overline{I}_r} = R_{eq} / s + jX_{eq} / s \tag{8.14}$$

from which

$$R_{eq} + jX_{eq} = \frac{\overline{V}_r}{\overline{I}_r} \qquad (8.15)$$

Figure 8-5 shows an example of the rotor current, rotor voltage, and equivalent impedance of the rotor-side converter when the slip of the DFIG varies from +0.3 to –0.3. The rotor current I_r increases with the slip, while the rotor voltage V_r decreases when the slip varies from 0.3 to zero, reaches near zero at $s = 0$, and increases with a positive slip. The phase angle of the rotor voltage, rotor current, and the rotor power factor angle φ_r ($\varphi_r = \angle \overline{V}_r - \varphi_r = \angle \overline{I}_r$) are also given in the figure.

When the DFIG operates in the supersynchronous mode, the equivalent resistance R_{eq} of the RSC is positive, indicating that an active power is delivered from the rotor to

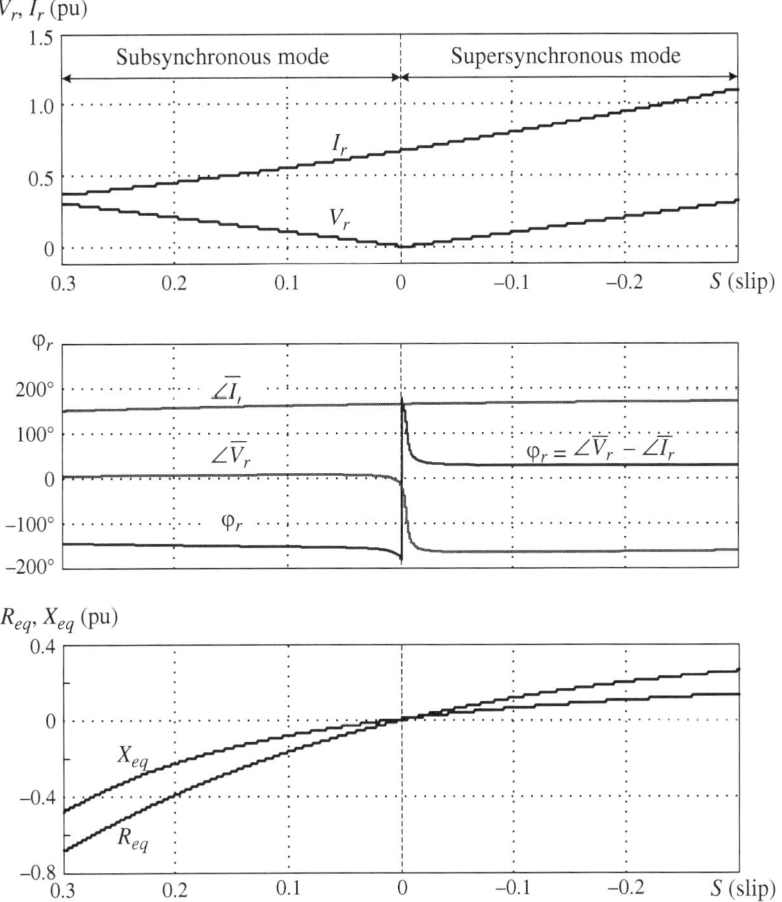

Figure 8-5. Rotor-side converter equivalent impedance ($PF_s = 1$).

the converter. When the generator is in the subsynchronous mode, R_{eq} is negative, signifying that the converter transfers active power to the rotor.

Case Study 8-1—Equivalent Impedance of Rotor-Side Converter. Consider a 1.5 MW, 690 V, 50 Hz, 1750 rpm DFIG wind energy system. The parameters of the generator are given in Table B-5 of Appendix B. The generator operates with an MPPT scheme, and its mechanical torque T_m is proportional to the square of the rotor speed. The stator power factor is unity. This case study investigates the relationship between the rotor voltage, rotor current, and the equivalent impedance of the rotor-side converter when the DFIG operates at supersynchronous, synchronous, and subsynchronous speeds.

CONVERTER EQUIVALENT IMPEDANCE AT 1750 RPM (SUPERSYNCHRONOUS MODE). From Equation (8.7), the stator current is calculated by

$$I_s = \frac{V_s \pm \sqrt{V_s^2 - \dfrac{4R_s T_m \omega_s}{3P}}}{2R_s} = -1068.2 \text{ A} \tag{8.16}$$

(the other solution, $I_s = 1.54 \times 10^5$ A, is omitted), where $V_s = 690/\sqrt{3}$ V, $T_m = -8185.1$ Nm, $\varphi_s = 2\pi \times 50$ rad/sec, $R_s = 6.25$ mΩ and $P = 2$.

Using the equivalent circuit of Figure 8-4, the voltage across the magnetizing branch is

$$\begin{aligned}
\overline{V}_m &= \overline{V}_s - \overline{I}_s \left(R_s + j\omega_s L_{ls} \right) \\
&= 690/\sqrt{3}\angle 0° - 1068.2\angle 180° \times (0.00625 + j100\pi \times 0.1687 \times 10^{-3}) \\
&= 401.2 + j56.6 = 405.2\angle 8° \text{ V}
\end{aligned} \tag{8.17}$$

The magnetizing current is calculated by

$$\overline{I}_m = \frac{\overline{V}_m}{j\omega_s L_m} = 32.92 - j233.26 = 235.6\angle -82.0° \text{ A} \tag{8.18}$$

The rotor current is

$$\overline{I}_r = \overline{I}_s - \overline{I}_m = -1101.1 + j233.26 = 1125.6\angle 168.0° \text{ A} \tag{8.19}$$

The rotor voltage is

$$\overline{V}_r = s\overline{V}_m - \overline{I}_r \left(R_r + js\omega_s L_{lr} \right) = 67.97\angle -164.9° \text{ V} \tag{8.20}$$

where $s = (\omega_s - \omega_r)/\omega_s = -0.1667$.

The equivalent impedance for the rotor-side converter is given by

$$\overline{Z}_{eq} = \overline{V}_r / \overline{I}_r = 0.05375 \text{ Ω} + j0.2751 \text{ Ω} \tag{8.21}$$

from which

$$\begin{cases} R_{eq} = 0.05375 \ \Omega \\ X_{eq} = 0.02751 \ \Omega \end{cases} \qquad (8.22)$$

CONVERTER EQUIVALENT IMPEDANCE AT 1500 RPM (SYNCHRONOUS SPEED). When the generator operates at 1500 rpm, the stator current is

$$I_s = \frac{V_s \pm \sqrt{V_s^2 - \dfrac{4R_s T_m \omega_s}{3P}}}{2R_s} = -786.3 \ \text{A} \qquad (8.23)$$

(the other solution, $I_s = 1.511 \times 10^5$ A, is omitted), where $T_m = -(1500/1750)^2 \times 8.1851 = -6.0135$ kN·m.

The voltage across the magnetizing branch is

$$\overline{V}_m = \overline{V}_s - \overline{I}_s(R_s + j\omega_s L_{ls}) = 400.5 + j41.7 = 402.6\angle 5.9° \ \text{V} \qquad (8.24)$$

The magnetizing current can be determined by

$$\overline{I}_m = \frac{\overline{V}_m}{j\omega_s L_m} = 24.23 - j232.82 = 234.1\angle -84.1° \ \text{A} \qquad (8.25)$$

The rotor current is

$$\overline{I}_r = \overline{I}_s - \overline{I}_m = -810.50 + j232.82 = 843.28\angle 164.0° \ \text{A} \qquad (8.26)$$

The rotor voltage is

$$\overline{V}_r = s\overline{V}_m - \overline{I}_r(R_r + js\omega_s L_{lr}) = -\overline{I}_r R_r = 2.2178\angle -16° \ \text{V} \qquad (8.27)$$

Thus, the equivalent resistance and reactance for the rotor-side converter are

$$\overline{Z}_{eq} = \overline{V}_r / \overline{I}_r = 0.00263 \angle -180° = -0.00263 + j0.0 \ \Omega \qquad (8.28)$$

Alternatively, the following equation can be established based on Figure 8-4:

$$\overline{V}_m - \overline{I}_r(R_r / s + j\omega_s L_{lr}) = \overline{I}_r(R_{eq} / s + jX_{eq} / s) \qquad (8.29)$$

from which

$$s\overline{V}_m - \overline{I}_r(R_r + js\omega_s L_{lr}) = \overline{I}_r(R_{eq} + j\omega_{sl} L_{eq}) \qquad (8.30)$$

With the slip s and slip frequency ω_{sl} equal to zero, the above equation is simplified to

$$-\overline{I}_r R_r = \overline{I}_r R_{eq} \qquad (8.31)$$

Thus the equivalent resistance and reactance for the rotor-side converter are

$$\begin{cases} R_{eq} = -R_r = -0.00263 \ \Omega \\ X_{eq} = \omega_{sl} L_{eq} = 0 \end{cases} \tag{8.32}$$

When the DFIG operates at the synchronous speed, both slip s and the slip frequency ω_{sl} are zero. This implies that a DC current flows through the rotor circuit and the rotor leakage reactance X_{lr} and the equivalent reactance X_{eq} are both zero. In this case, the induction generator operates just like a wound rotor synchronous generator, in which the rotor flux is produced by a DC current through a DC exciter. In the DFIG wind energy system, the DC excitation is provided by the rotor-side converter.

CONVERTER EQUIVALENT IMPEDANCE AT SUBSYNCHRONOUS SPEED. Following the same procedure, the calculated rotor voltage, rotor current, and the equivalent impedance of the rotor-side converter in the subsynchronous mode (1200 and 1350 rpm) are summarized in Table 8-1. For the convenience of comparison, the calculated results for the DFIG operating in the synchronous (1500 rpm) and supersynchronous (1650 and 1750 rpm) modes are also given in the table.

8.3.2 Torque-Slip Characteristics of DFIG WECS

To find the torque-slip characteristics of the DFIG, we can follow the same procedure as that in the Chapter 3, except that the total rotor circuit resistance is the sum of R_r and R_{eq}, and the total reactance is the sum of X_{lr} and X_{eq}. Figure 8-6 shows a set of typical torque-slip curves of the DFIG operating at the supersynchronous, synchronous, and subsynchronous speeds. With different values of R_{eq} and X_{eq}, different torque-slip characteristics can be obtained. For example, with R_{eq1} of 0.047 pu and X_{eq1} of 0.027 pu for the 1.5 MW/690 V DFIG, the torque-slip Curve 1 is obtained, whereas Curve 2 corresponds to R_{eq2} of –0.072 pu and X_{eq2} of –0.033 pu. At the synchronous speed, the torque-slip curve is a straight line (Curve 3).

In an MPPT controlled wind energy system, the mechanical torque is proportional to the square of the generator speed ($T_m \propto \omega_r^2$), as discussed in Chapter 2. This relationship is illustrated by Curve 4 in Figure 8-6. The intersections of this curve with other torque-slip curves are the steady-state operating points of the generator. For ex-

Table 8-1. Equivalent impedance of RSC in 1.5 MW/690 V DFIG WECS ($PF_s = 1$)

Rotor speed (rpm)	1200	1350	1500	1650	1750 (rated)
Slip	0.2	0.1	0	–0.1	–0.1667 (rated)
T_m (kN·m)	–3.849	–4.871	–6.014	–7.276	–8.185
$\bar{V}_r$ (V)	$83.756\angle 6.2°$	$43.068\angle 7.4°$	$2.218\angle{-16.0°}$	$39.711\angle{-165.8°}$	$67.965\angle{-164.9°}$
$\bar{I}_r$ (A)	$569.285\angle 155.9°$	$697.103\angle 160.5°$	$843.281\angle 164.0°$	$1006.991\angle 166.6°$	$1125.566\angle 168.0°$
R_{eq} (Ω)	–0.126989	–0.055113	–0.00263	0.034942	0.053751
X_{eq} (Ω)	–0.074293	–0.027918	0	0.018281	0.027513

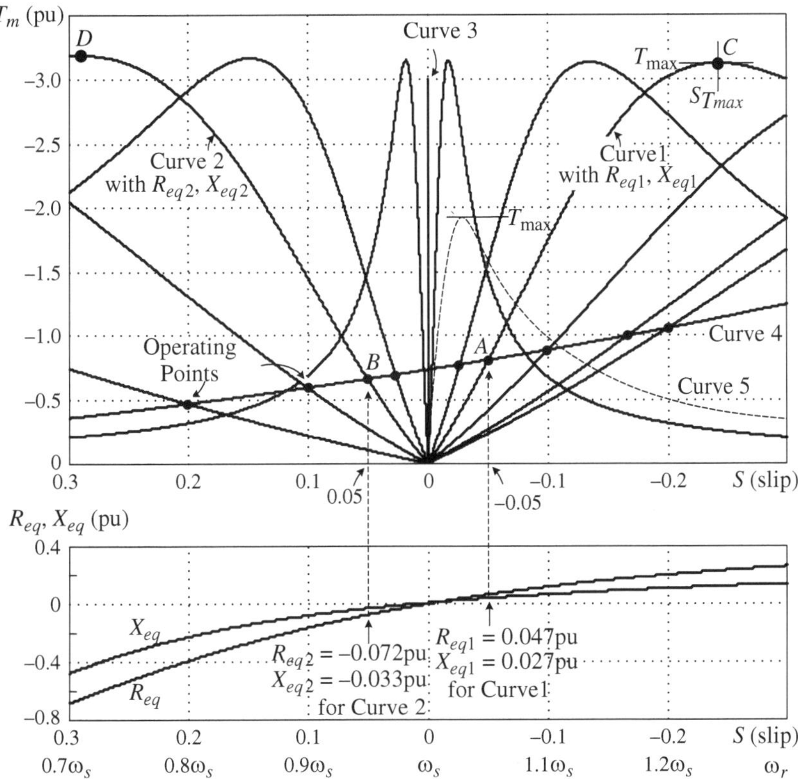

Figure 8-6. Torque-slip characteristics of DFIG wind energy system ($PF_s = 1$).

ample, at Point A, where Curve 4 intersects with Curve 1, the generator operates in the supersynchronous mode with a slip of −0.05. At a lower wind speed, the generator operates at Point B, where Curve 4 intersects with Curve 2, and the generator is in the subsynchronous operating mode with a slip of +0.05.

It should be noted that whether the slip is positive or negative, the DFIG is always in the generating mode and delivers its mechanical power $|P_m|$ to the grid. This is different from the operation of the squirrel-cage induction generator presented in Chapter 3, where the generator is in the motoring mode when the slip is positive and in the generating mode with a negative slip.

For comparison purposes, the torque-slip characteristics of the DFIG with the rotor circuit shorted ($R_{eq} = X_{eq} = 0$) is given by Curve 5 in Figure 8-6, whose maximum torque T_{max} is much lower than that of the torque-slip curves with R_{eq} and X_{eq} taken into account.

To find the maximum torque of the DFIG, a simplified steady-state equivalent circuit, shown in Figure 8-7, can be used, where the magnetizing branch is moved to the left of the stator circuit (compared with Figure 8-4). Since the magnetizing inductance is much higher than the stator resistance and inductance, such an arrangement will not

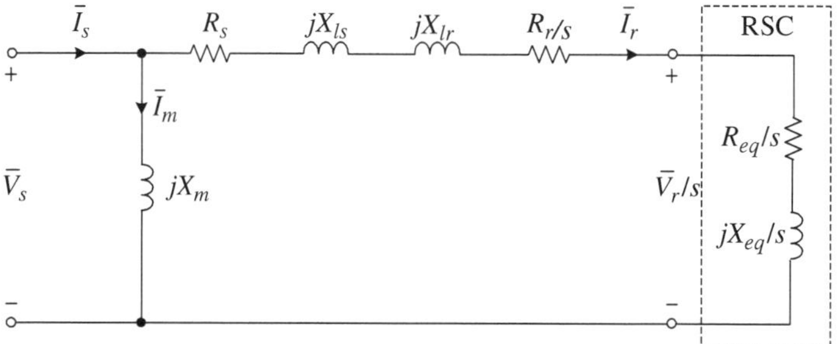

<u>Figure 8-7.</u> Simplified DFIG steady-state equivalent circuit for calculation of T_{max}.

generate large errors in calculations. The electromagnetic torque of the generator can be then expressed as

$$T_m = \frac{P_{ag}}{\omega_s / P} = \frac{1}{\omega_s / P} \times \left(3I_r^2\right) \frac{R_r + R_{eq}}{s}$$

$$= \frac{1}{\omega_s / P} \times \frac{3V_s^2}{\left(R_s + R_r / s + R_{eq} / s\right)^2 + \left(X_{ls} + X_{lr} + X_{eq} / s\right)^2} \times \frac{R_r + R_{eq}}{s} \tag{8.33}$$

The maximum torque T_{max} and the slip at the maximum torque $s_{T\text{max}}$ can be obtained by setting $dT_m/ds = 0$, from which

$$s_{T\max} = \pm \sqrt{\frac{\left(R_r + R_{eq}\right)^2 + X_{eq}^2}{R_s^2 + \left(X_{ls} + X_{lr}\right)^2}} \tag{8.34}$$

The plus and minus signs in the above equation signify the sub- and supersynchronous modes of operation. For a given value of R_{eq} and X_{eq}, the maximum torque can be found by substituting (8.34) into (8.33):

$$T_{\max} = \frac{1}{2\omega_s / P} \times \frac{3V_s^2}{R_s + \dfrac{\left(X_{ls} + X_{lr}\right)X_{eq}}{R_r + R_{eq}} - \sqrt{\left(\left(X_{ls} + X_{lr}\right)^2 + R_s^2\right) \times \left(1 + \dfrac{X_{eq}^2}{\left(R_r + R_{eq}\right)^2}\right)}} \tag{8.35}$$

The above equation is valid for both super- and subsynchronous modes of operation.

Taking the torque-slip Curve 1 with $R_{eq} = 0.017744$ Ω (0.047 pu) and $X_{eq} = 0.009985$ Ω (0.027 pu) as an example, the calculated $s_{T\text{max}}$ and T_{max} for the 1.5 MW/690 V DFIG are -0.239 and -3.27 pu, which is Point C in Figure 8-6. For Curve 2 with $R_{eq} = -0.026722$ Ω (-0.072 pu) and $X_{eq} = -0.012331$ Ω (-0.033 pu), $s_{T\text{max}}$ and T_{max} are 0.285 pu and 3.34 pu, respectively, which is Point D in the figure.

8.3.3 Steady-State Analysis of DFIG WECS with $PF_s = 1$

To facilitate the steady-state analysis of the DFIG wind energy system, the equivalent circuit of Figure 8-4 can be rearranged as shown in Figure 8-8, where the mechanical power P_m, the rotor power P_r, and stator and rotor winding losses, $P_{cu,s}$ and $P_{cu,r}$, can be easily calculated by a general equation of $P = 3I^2R$. The rotor power P_r is the power transferred from or to the rotor-side converter.

Figure 8-9 shows the power flow of the DFIG operating under the super- and subsynchronous modes. Neglecting the rotational losses P_{rot} of the turbine, the power transferred or dissipated in the generator can be calculated by

$$\begin{cases} P_m = 3I_r^2(R_r + R_{eq})(1-s)/s \\ P_r = 3I_r^2 R_{eq} \\ P_{cu,r} = 3I_r^2 R_r \\ P_{cu,s} = 3I_s^2 R_s \\ P_s = 3V_s I_s \cos\phi_s \end{cases} \tag{8.36}$$

where P_s is the stator power and φ_s is the stator power factor angle. The power delivered to the grid, P_g, is the sum of the stator and rotor power, given by

$$\left| P_g \right| = \begin{cases} \left|P_s\right| + \left|P_r\right| & \text{for supersynchronous mode} \\ \left|P_s\right| - \left|P_r\right| & \text{for subsynchronous mode} \end{cases} \tag{8.37}$$

In the supersynchronous operating mode shown in Figure 8-9a, the equivalent resistance R_{eq} of the rotor-side converter has a positive value and the rotor power P_r is positive. This implies that the resistance R_{eq} consumes power similar to the winding resistances R_r and R_s. In reality, the rotor power P_r is not dissipated in R_{eq}, but transferred from the rotor to the grid through the converters. In the subsynchronous mode, R_{eq} has a negative value and the rotor power P_r is also negative. This indicates that

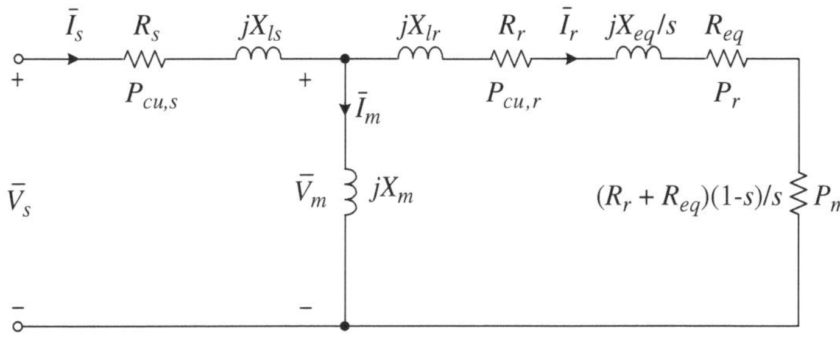

Figure 8-8. Steady-state equivalent circuit of DFIG WECS for system performance evaluation.

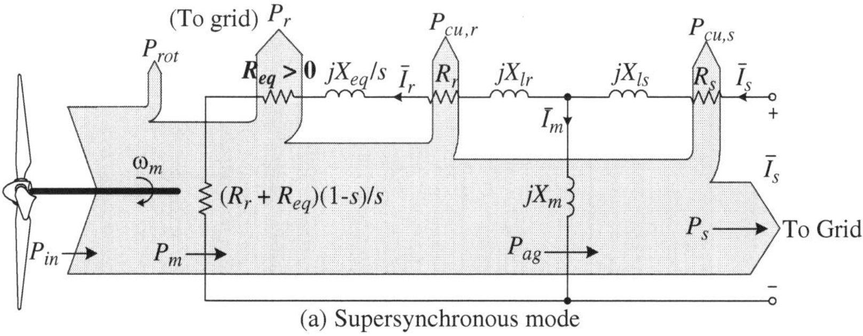

(a) Supersynchronous mode

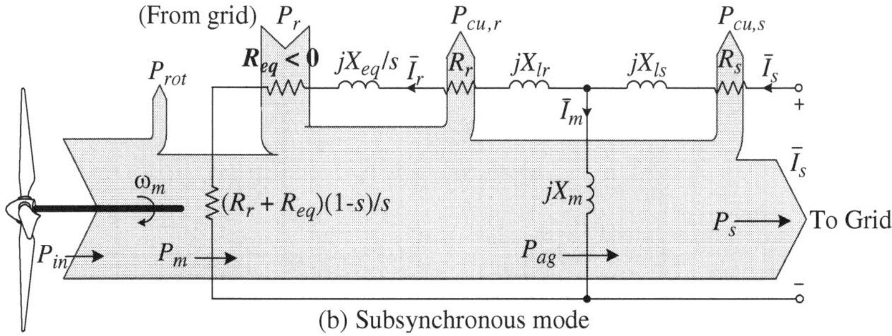

(b) Subsynchronous mode

Figure 8-9. Power flow of DFIG with rotor-side converter represented by R_{eq} and X_{eq}.

the rotor circuit receives power from the grid through the converters as shown in Figure 8-9b.

Case Study 8-2—Steady-State Analysis of DFIG WECS with PF_s = 1.

This case study is a continuation of Case Study 8-1, where the equivalent impedance for the rotor-side converter of a 1.5 MW/690 V DFIG wind energy system was developed. The steady-state operation of the above system at the supersynchronous, synchronous, and subsynchronous speeds is analyzed below.

DFIG OPERATION AT ROTOR SPEED OF 1750 RPM (SUPERSYNCHRONOUS MODE). At the rated rotor speed of 1750 rpm, the rotor current calculated in Case Study 8-1 is 1125.6 A, from which the mechanical power of the generator is calculated:

$$P_m = 3I_r^2 \left(R_{eq} + R_r \right) (1-s)/s$$
$$= 3 \times 1125.6^2 \left(0.05375 + 0.00263 \right) (1 + 0.1667)/(-0.1667) \quad (8.38)$$
$$= -1500 \ \text{kW}$$

Alternatively, the mechanical power can be calculated by

$$P_m = T_m \omega_m = -8185.1 \times 1750 \times 2\pi/60 = -1500 \ \text{kW} \quad (8.39)$$

The rotor power is

$$P_r = 3(I_r)^2 R_{eq} = 3 \times 1125.6 \times 0.05375 = 204.29 \ \text{kW} \tag{8.40}$$

The rotor and stator winding losses are

$$P_{cu,r} = 3(I_r)^2 R_r = 10.0 \ \ \text{kW} \tag{8.41}$$

$$P_{cu,s} = 3(I_r)^2 R_s = 9.07 \ \ \text{kW} \tag{8.42}$$

The stator active power is

$$P_s = 3V_s I_s \cos\varphi_s = 690/\sqrt{3} \times 1068.2 \times \cos(180°) = -1276.64 \ \text{kW} \tag{8.43}$$

where the stator power factor angle φ_s is 180° since the DFIG operates in the generating mode with a unity power factor.

The total power delivered to the grid is

$$\left|P_g\right| = \left|P_s\right| + \left|P_r\right| = 1276.64 + 204.29 = 1480.93 \ \text{kW} \tag{8.44}$$

The difference between P_m and P_g is the losses on the stator and rotor windings, that is,

$$\left|P_m\right| - \left|P_g\right| = P_{cu,r} + P_{cu,s} = 19.07 \ \ \text{kW} \tag{8.45}$$

The efficiency of the DFIG is then

$$\eta = P_g / P_m = 1480.93 / 1500 = 98.7\% \tag{8.46}$$

DFIG OPERATION AT THE SYNCHRONOUS AND SUBSYNCHRONOUS SPEEDS. Following the same procedure presented above, the operation of the DFIG in the synchronous and subsynchronous modes can be studied. Table 8-2 summarizes the calculation results. It is noted that in the subsynchronous mode the rotor circuit receives power from the grid through the converters. Therefore, the power delivered to the grid is $|P_g| = |P_s| - |P_r|$, as indicated by Equation (8.37).

Figure 8-10 shows the relationship between the stator, rotor, and mechanical power of the 1.5 MW/690 V DFIG WECS through the rotor speed range. It is clearly illustrated that the stator power $|P_s|$ is approximately equal to $|P_m| - |P_r|$ in the supersynchronous operating mode, whereas in the subsynchronous operating mode $|P_s| \approx |P_m| + |P_r|$.

8.3.4 Simplified Calculations

In large megawatt wind generators, the stator resistance of the generator is normally very small (< 0.01 pu). To simplify the analysis, the stator resistance can be neglected.

Table 8-2. Three operating modes of 1.5 MW/690 V DFIG WECS ($PF_s = 1$)

Operating mode	Subsynchronous operation	Synchronous operation	Supersynchronous operation
ω_m (rpm)	1200	1500	1750 (rated)
s Slip	0.2	0	−0.1667 (rated)
$\|T_m\|$ (kN·m)	3.849	6.014	8.1851
R_{eq} (Ω)	−0.126989	−0.002630	0.053751
X_{eq} (Ω)	−0.074293	0	0.027513
I_s (A)	504.16	786.28	1068.22
I_r (A)	569.29	843.28	1125.57
V_r (V)	83.76	2.22	67.97
$\|P_m\|$ (kW)	483.64	944.61	1500.0
$\|P_r\|$ (kW)	123.47	5.61	204.29
$P_{cu,r}$ (kW)	2.56	5.61	10.0
$P_{cu,s}$ (kW)	2.02	4.92	9.07
$\|P_s\|$ (kW)	602.53	939.69	1276.64
$\|P_g\|$ (kW)	479.06	934.08	1480.93

When the generator operates with unity stator power factor, its air-gap power can be calculated by

$$P_{ag} = 3(V_s - I_s R_s)I_s \approx 3 V_s I_s \qquad (8.47)$$

From Equations (8.47) and (8.4), the stator current can be calculated by

$$I_s = \frac{T_m \omega_s / P}{3 V_s} \qquad (8.48)$$

With the stator current I_s known, the equivalent impedance of the rotor-side converter can be calculated, and the steady-state performance of the DFIG can be analyzed. Compared with the stator current given in Equation (8.7), the calculation of the stator current by Equation (8.48) is simpler. More importantly, this method can facilitate the analysis of the DFIG wind energy systems with nonunity stator power factor, which is to be discussed in the following section.

Table 8-3 gives the calculation results for the 1.5 MW/690 V DFIG operating at 1200 rpm and 1750 rpm based on two methods: Method 1—accurate calculation of the stator current I_s based on Equation (8.7); and Method 2—simplified calculation of I_s using Equation (8.48). Results shown in Table 8-3 illustrate that the errors generated by the simplified method are minimal (less than 1.5%).

8.4 LEADING AND LAGGING POWER FACTOR OPERATION

When the generator operates with a leading or lagging power factor, the stator current can be calculated by

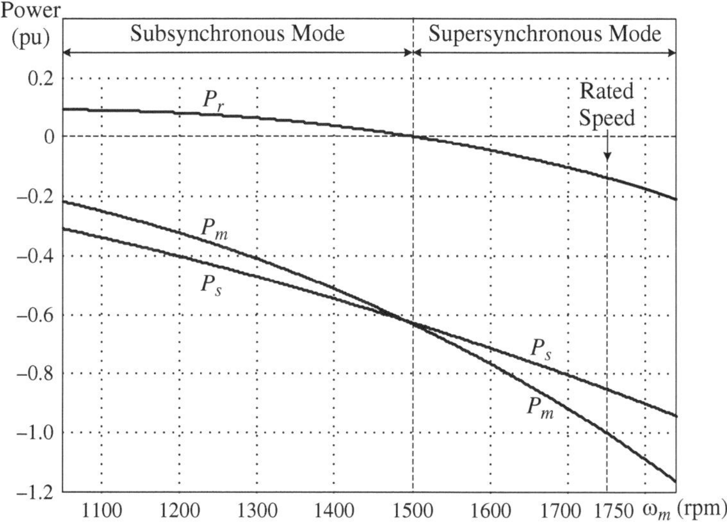

Figure 8-10. Stator, rotor, and mechanical power of the 1.5 MW/690 V DFIG operating at super- and subsynchronous speeds.

Table 8-3. Comparison of two methods for the analysis of 1.5 MW DFIG WECS (PF_s = 1)

Operating mode	Subsynchronous operation			Supersynchronous operation		
ω_m (rpm)	1200 rpm			1750 rpm		
	Method 1	Method 2	Error (%)	Method 1	Method 2	Error (%)
s, Slip	0.2	0.2	N/A	−0.1667	−0.1667	N/A
$\lvert T_m \rvert$ (kN·m)	3.849	3.849	N/A	8.185	8.185	N/A
R_{eq} (Ω)	−0.12699	−0.12671	0.22	0.05375	0.05339	0.26
X_{eq} (Ω)	−0.07429	−0.07398	0.31	0.02751	0.02735	0.57
I_s (A)	504.2	505.8	0.32	1068.2	1075.8	0.71
I_r (A)	569.3	570.9	0.28	1125.6	1133.2	0.68
V_r (V)	83.6	83.7	0.12	67.9	68.0	0.15
$\lvert P_m \rvert$ (kW)	483.6	485.3	0.35	1500	1510.7	0.71
$\lvert P_r \rvert$ (kW)	123.5	123.9	0.34	204.3	205.7	0.68
$P_{cu,r}$ (kW)	2.557	2.571	0.55	9.996	10.13	1.3
$P_{cu,s}$ (kW)	2.020	2.034	0.69	9.071	9.201	1.4
$\lvert P_s \rvert$ (kW)	602.5	604.5	0.34	1276.6	1285.7	0.71
$\lvert P_g \rvert$ (kW)	479.0	480.6	0.33	1480.9	1491.4	0.77

$$I_s = \frac{T_m \omega_s / P}{3V_s \cos\varphi_s} \tag{8.49}$$

where φ_s is the power factor angle of the stator. With the stator current calculated, the equivalent impedance of the rotor-side converter can be obtained following the same procedures given in Case Study 8-1. Table 8-4 gives the calculated converter equivalent impedance for the DFIG operating under the super- and subsynchronous modes with 0.95 leading and lagging power factor. Once the converter equivalent impedance is determined for a given operating condition, the steady-state performance of the DFIG wind energy system can be analyzed using the same procedure presented in the previous section.

Figure 8-11 shows the calculated converter equivalent impedance as well as the torque-slip curves of the DFIG with a leading power factor of 0.95. The torque-slip curves are similar to those in Figure 8-6 except that the maximum torque values differ. Figure 8-12 illustrates the torque-slip curves of generator with a lagging power factor of 0.95.

8.5 STATOR VOLTAGE ORIENTED CONTROL OF DFIG WECS

8.5.1 Principle of Stator Voltage Oriented Control (SVOC)

In DFIG wind energy systems, the stator of the generator is directly connected to the grid, and its voltage and frequency can be considered constant under the normal operating conditions. It is, therefore, convenient to use stator voltage oriented control (SVOC) for the DFIG [4]. This is in contrast to electric motor drives, where rotor- or stator-flux field oriented controls (FOC) are normally used. Figure 8-13 shows a space vector diagram for the DFIG with the stator voltage oriented control operating with unity power factor in supersynchronous mode. The stator voltage oriented control is achieved by aligning the d-axis of the synchronous reference frame with the stator voltage vector $\vec{v}_s$. The resultant d- and q-axis stator voltages are

$$v_{qs} = 0 \quad \text{and} \quad v_{ds} = v_s \tag{8.50}$$

Table 8-4. Equivalent impedance of RSC with leading and lagging stator power factor

Stator power	ω_m (rpm)	1200	1350	1500	1650	1750
factor	s (Slip)	0.2	0.1	0	−0.1	−0.1667
Leading PF	$\overline{V}_r$ (V)	86.89∠5.7°	54.02∠6.3°	2.514∠−31.6°	42.83∠−165.7°	73.67∠−165.3°
(0.95),	$\overline{I}_r$ (A)	659.3∠142.2°	798.2∠145.7°	955.9∠148.4°	1131.9∠150.5°	1259.3∠151.7°
$\varphi_s = -161.8°$	R_{eq} (Ω)	−0.0957	−0.0428	−0.00263	0.02731	0.04277
	X_{eq} (Ω)	−0.0906	−0.03667	0	0.02620	0.03992
Lagging PF	$\overline{V}_r$ [V]	80.65∠6.9°	41.16∠8.6°	2.146∠5.5°	36.59∠−165.7°	62.30∠−164.1°
(0.95),	$\overline{I}_r$ (A)	525.2∠173.3°	660.5∠178.6°	815.9∠−177.5°	990.5∠−174.7°	1117.2∠−173.3°
$\varphi_s = -161.8°$	R_{eq} (Ω)	−0.1493	−0.0614	−0.00263	0.0365	0.0550
	X_{eq} (Ω)	−0.0360	−0.0108	0	0.0058	0.0089

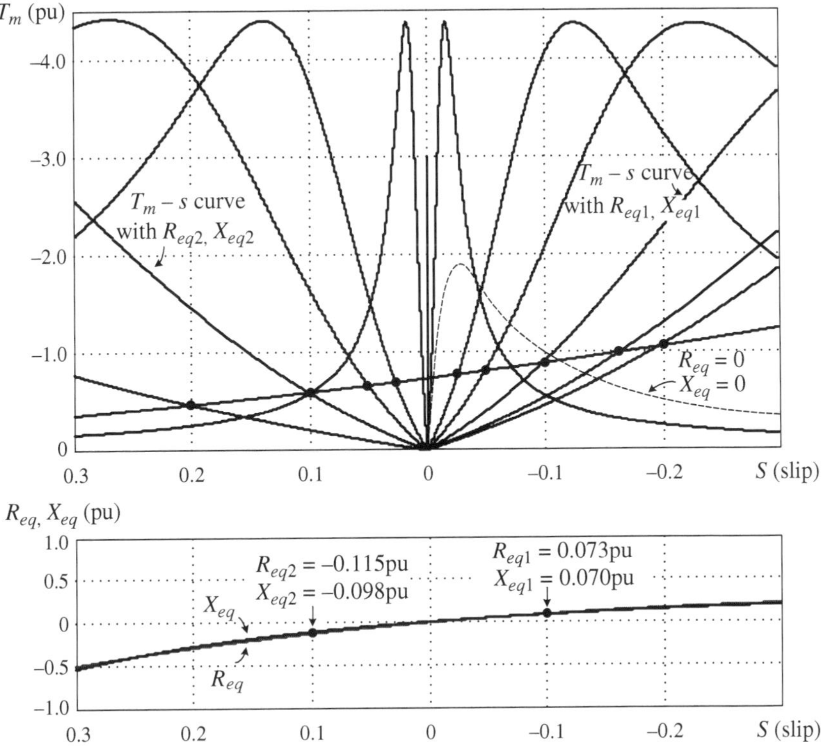

Figure 8-11. Torque-slip characteristics of DFIG operating with a leading stator power factor of 0.95.

where v_s is the magnitude of $\vec{v}_s$ (also the peak value of the three-phase stator voltage). The rotating speed of the synchronous reference frame is given by

$$\omega_s = 2\pi f_s \qquad (8.51)$$

where f_s is the stator frequency of the generator (also the frequency of the grid voltage). The stator voltage vector angle θ_s is referenced to the stator frame, which varies from zero to 2π when $\vec{v}_s$ rotates one revolution in space.

The rotor rotates at speed ω_r. The rotor position angle θ_r is also referenced to the stator frame. The angle between the stator voltage vector and the rotor is the slip angle, defined by

$$\theta_{sl} = \theta_s - \theta_r \qquad (8.52)$$

Since the DFIG operates with unity power factor, the stator current vector $\vec{i}_s$ is aligned with $\vec{v}_s$ but with opposite direction (DFIG in generating mode). The rotor voltage and current vectors, $\vec{v}_s$ and $\vec{i}_s$, which are controlled by the converters in the rotor circuit,

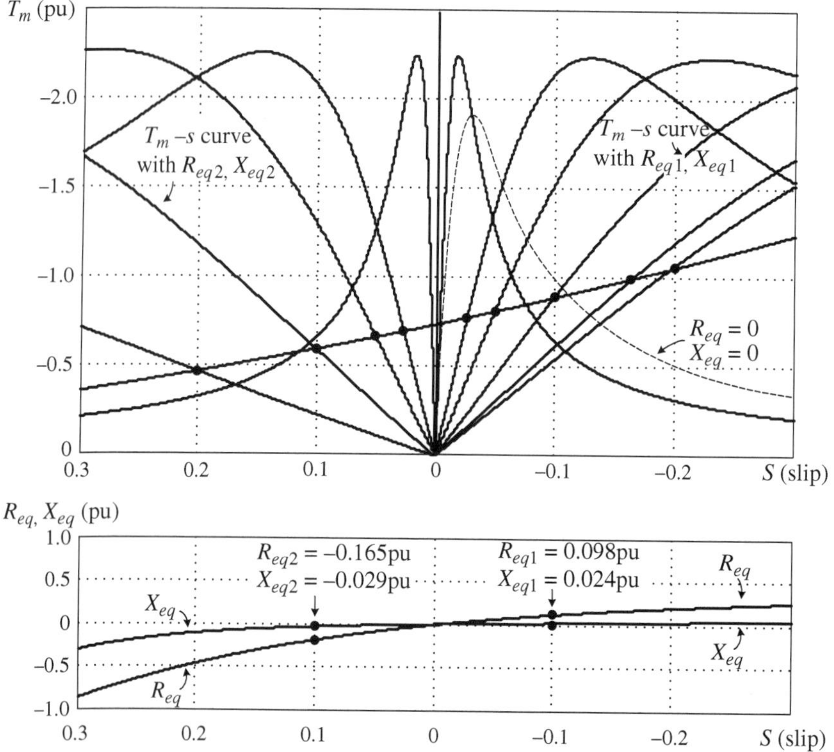

Figure 8-12. Torque-slip characteristics of DFIG operating with a lagging stator power factor of 0.95.

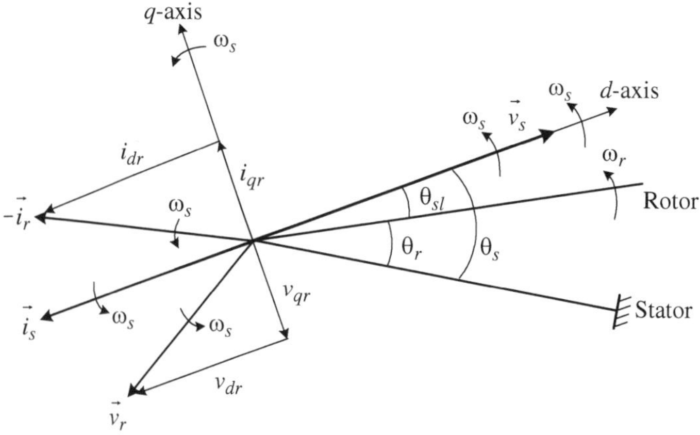

Figure 8-13. Space-vector diagram of DFIG with SVOC in the supersynchronous mode.

are also given in the diagram. The rotor voltage and current vectors can be resolved into two components along the *dq* axes: v_{dr} and v_{qr} for $\vec{v}_r$, and i_{dr} and i_{qr} for $\vec{i}_r$. These *dq*-axis components can be controlled independently by the rotor converters.

The DFIG wind energy system can be controlled by the electromagnetic torque for speed control or active power. In contrast to the other wind energy systems, the electromagnetic torque T_e of the generator, the active power P_s and the reactive power Q_s of the stator are controlled by the rotor-side converter. Therefore, it is worthwhile to investigate the controllability of T_e, P_s, and Q_s by the rotor voltage and current. The investigation will also facilitate the analysis of the stator voltage oriented control.

Referring to Equation (3.16) of Chapter 3, the electromagnetic torque of the generator can be expressed as

$$T_e = \frac{3P}{2}(i_{qs}\lambda_{ds} - i_{ds}\lambda_{qs}) \tag{8.53}$$

where λ_{ds} and λ_{qs} are the *dq*-axis stator flux linkages, given by

$$\begin{cases} \lambda_{ds} = L_s i_{ds} + L_m i_{dr} \\ \lambda_{qs} = L_s i_{qs} + L_m i_{qr} \end{cases} \tag{8.54}$$

from which the *dq*-axis stator currents are calculated to be

$$\begin{cases} i_{ds} = \dfrac{\lambda_{ds} - L_m i_{dr}}{L_s} \\ i_{qs} = \dfrac{\lambda_{qs} - L_m i_{qr}}{L_s} \end{cases} \tag{8.55}$$

Substituting (8.55) into (8.53) yields

$$T_e = \frac{3PL_m}{2L_s}(-i_{qr}\lambda_{ds} + i_{dr}\lambda_{qs}) \tag{8.56}$$

The above equation indicates that the electromagnetic torque is a function of rotor current and stator flux linkages. In the DFIG wind energy system, the stator voltage is constant since it is directly connected to the grid. The rotor current is controlled by the rotor-side converter. It is thus desirable to find the relationship between the torque, stator voltage, and rotor current. Referring to the induction generator model of Figure 3-6 (Chapter 3), the stator voltage vector for the steady-state operation of the generator is

$$\vec{v}_s = R_s\vec{i}_s + j\omega_s\vec{\lambda}_s \tag{8.57}$$

The representation in *dq*-axis is

$$(v_{ds} + jv_{qs}) = R_s(i_{ds} + ji_{qs}) + j\omega_s(\lambda_{ds} + j\lambda_{qs}) \tag{8.58}$$

from which the *dq*-axis stator flux linkages are

$$\begin{cases} \lambda_{ds} = \dfrac{v_{qs} - R_s i_{qs}}{\omega_s} \\[2ex] \lambda_{qs} = -\dfrac{v_{ds} - R_s i_{ds}}{\omega_s} \end{cases} \tag{8.59}$$

Substituting (8.59) into (8.56) gives

$$\begin{aligned} T_e &= \frac{3PL_m}{2\omega_s L_s}(-i_{qr}(v_{qs} - R_s i_{qs}) - i_{dr}(v_{ds} - R_s i_{ds})) \\[1ex] &= \frac{3PL_m}{2\omega_s L_s}(-i_{qr}v_{qs} + R_s i_{qs}i_{qr} + R_s i_{ds}i_{dr} - i_{dr}v_{ds}) \end{aligned} \tag{8.60}$$

With $v_{qs} = 0$ for the stator voltage oriented control, the torque equation can be simplified to

$$T_e = \frac{3PL_m}{2\omega_s L_s}(R_s i_{qs}i_{qr} + R_s i_{ds}i_{dr} - i_{dr}v_{ds}) \tag{8.61}$$

Ignoring the stator resistance R_s, which is normally very low for large DFIG, the torque equation can be further simplified:

$$T_e = -\frac{3PL_m}{2\omega_s L_s}i_{dr}v_{ds} \tag{8.62}$$

It can be observed from the above equation that the electromagnetic torque is a function of d-axis rotor current and stator voltage.

The stator active and reactive power can be calculated by [5]

$$\begin{cases} P_s = \dfrac{3}{2}(v_{ds}i_{ds} + v_{qs}i_{qs}) \\[2ex] Q_s = \dfrac{3}{2}(v_{qs}i_{ds} - v_{ds}i_{qs}) \end{cases} \tag{8.63}$$

Using the stator voltage oriented control ($v_{qs} = 0$), the above equation can be simplified to

$$\begin{cases} P_s = \dfrac{3}{2}v_{ds}i_{ds} \\[2ex] Q_s = -\dfrac{3}{2}v_{ds}i_{qs} \end{cases} \quad \text{for } v_{qs} = 0 \tag{8.64}$$

Substituting (8.55) into (8.64) yields

$$\begin{cases} P_s = \dfrac{3}{2} v_{ds} \left(\dfrac{\lambda_{ds} - L_m i_{dr}}{L_s} \right) \\[3ex] Q_s = -\dfrac{3}{2} v_{ds} \left(\dfrac{\lambda_{qs} - L_m i_{qr}}{L_s} \right) \end{cases} \tag{8.65}$$

from which

$$\begin{cases} i_{dr} = -\dfrac{2L_s}{3 v_{ds} L_m} P_s + \dfrac{1}{L_m} \lambda_{ds} \\[3ex] i_{qr} = \dfrac{2L_s}{3 v_{ds} L_m} Q_s + \dfrac{1}{L_m} \lambda_{qs} \end{cases} \tag{8.66}$$

Substituting the stator flux linkages from (8.59) into (8.66) gives

$$\begin{cases} i_{dr} = -\dfrac{2L_s}{3 v_{ds} L_m} P_s + \dfrac{v_{qs} - R_s i_{qs}}{\omega_s L_m} = -\dfrac{2L_s}{3 v_{ds} L_m} P_s - \dfrac{R_s}{\omega_s L_m} i_{qs} \\[3ex] i_{qr} = \dfrac{2L_s}{3 v_{ds} L_m} Q_s - \dfrac{v_{ds} - R_s i_{ds}}{\omega_s L_m} = \dfrac{2L_s}{3 v_{ds} L_m} Q_s + \dfrac{R_s}{\omega_s L_m} i_{ds} - \dfrac{v_{ds}}{\omega_s L_m} \end{cases} \quad \text{for } v_{qs} = 0 \tag{8.67}$$

Neglecting the stator resistance R_s, we have

$$\begin{cases} i_{dr} = -\dfrac{2L_s}{3 v_{ds} L_m} P_s & \text{(a)} \\[3ex] i_{qr} = \dfrac{2L_s}{3 v_{ds} L_m} Q_s - \dfrac{v_{ds}}{\omega_s L_m} & \text{(b)} \end{cases} \tag{8.68}$$

The above equations indicate that for a given stator voltage, the stator active power P_s and reactive power Q_s can be controlled by the dq-axis rotor currents.

8.5.2   System Block Diagram

Figure 8-14 shows the block diagram of DFIG wind energy system with stator voltage oriented control. The stator voltage vector angle θ_s is identified by the θ_s calculator, and the rotor position angle θ_r is measured by an encoder mounted on the shaft of the generator. The slip angle for the reference frame transformation is obtained by $\theta_{sl} = \theta_s - \theta_r$. The abc/dq and dq/abc transformation blocks transform the variables in the abc stationary reference frame to the dq synchronous reference frame, and vice versa. The angle of the stator voltage vector can be obtained by

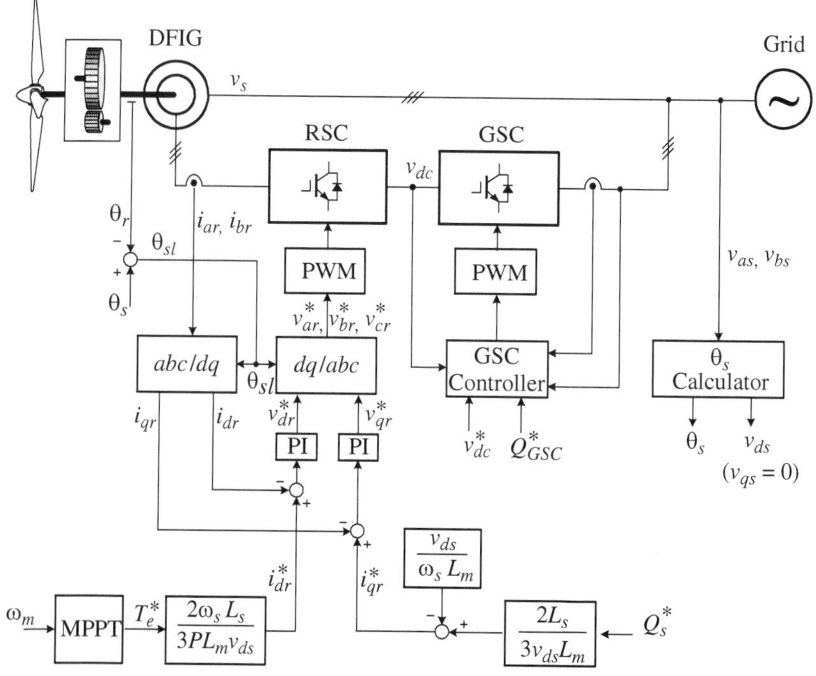

<u>Figure 8-14.</u> Bloc diagram of a DFIG wind energy system with stator voltage oriented control.

$$\theta_s = \tan^{-1} \frac{v_\beta}{v_\alpha} \tag{8.69}$$

in which the dq-axis stator voltages are given by

$$\begin{bmatrix} v_\alpha \\ v_\beta \end{bmatrix} = \frac{2}{3} \begin{bmatrix} 1 & -1/2 & -1/2 \\ 0 & \sqrt{3}/2 & -\sqrt{3}/2 \end{bmatrix} \cdot \begin{bmatrix} v_{as} \\ v_{bs} \\ v_{cs} \end{bmatrix} \tag{8.70}$$

where v_{as}, v_{bs}, and v_{cs} are the measured three-phase stator voltages.

The MPPT block generates the reference torque T_e^* based on the optimal torque method discussed in Chapter 2 for maximum power point tracking. The reference for the d-axis rotor current i_{dr}^*, which is the torque producing component of the rotor current, is calculated according to Equation (8.62). For a given stator reactive power reference Q_s^*, the q-axis rotor current reference i_{qr}^* is calculated by Equation (8.68b). The reference dq-axis currents, i_{dr}^* and i_{qr}^*, are then compared to the measured values, i_{dr}^* and i_{qr}^*, and the errors passed through PI controllers. The output of the PI controllers, v_{dr}^* and v_{qr}^*, are the dq-axis rotor voltage references in the synchronous frame, which are transformed into a three-phase reference for rotor voltages, v_{ar}^*, v_{br}^*, and v_{cr}^* in the stationary frame. The rotor reference voltages can serve as the three-phase modulating waveforms in carrier-based modulation schemes or be converted into a reference space vector for the space vector modulation (SVM). The PWM block generates gating signals for the rotor-side converter.

The grid-side converter performs two main functions: (1) it keeps the DC link voltage v_{dc} constant, and (2) provides reactive power to the grid when required. The reactive power reference, Q^*_{GSC}, can be set to zero for unity power factor operation of the converter. The overall power factor of the DFIG wind energy system is then controlled by the rotor-side converter through its reference Q^*_s. The operation principle of the grid-tied converter has been discussed in Chapter 4 and, therefore, is not repeated here.

8.5.3 Dynamic Performance of DFIG WECS

The dynamic performance of DFIG WECS with stator voltage oriented control is analyzed in the following case study.

Case Study 8-3—Transients from Supersynchronous to Subsynchronous Operation. The transients of a 1.5 MW/690 V DFIG wind energy system caused by a step change in wind speed are investigated. The torque and power versus rotor speed characteristics of the system at the wind speeds of 0.7 pu and 1.0 pu are illustrated in Figure 8-15. With the implementation of the MPPT scheme, the electromagnetic torque T_e of the generator is proportional to the square of the rotor speed $(\omega_m)^2$, whereas the mechanical power P_m is proportional to the cube of the rotor speed $(\omega_m)^3$. It is noted that the turbine speed is equal to the rotor mechanical speed in per-unit terms and, therefore, there is no need to consider the gear ratio of the gearbox.

The system initially operates in the supersynchronous mode with a rated rotor speed of 1.0 pu (1750 rpm) as indicated by Point A in Figure 8-15. The case study investigates the transition when the wind speed suddenly decreases to 0.7 pu and the operating point is moved from Point A to C. At this point, the generator operates at the subsynchronous speed of 1225 rpm. The generator and system parameters for the case study are given in Table 8.5.

With the rotational losses neglected, the mechanical torque of the generator T_m is equal to the mechanical torque generated by the turbine in per-unit terms. The simulated waveforms of the 1.5 MW DFIG WECS from the supersynchronous to subsynchronous mode are given in Figure 8-16.

During the period of $0 < t < 0.5$ sec, the wind energy system operates in steady state at 1.0 pu rotor speed, and the mechanical torque T_m from the turbine shaft is -1.0 pu. The electromagnetic torque of the generator T_e is equal to the mechanical torque. The system operates at Point A in Figures 8-15 and 8-16. It is noted that in the simulation, an ideal transformer with a turns ratio of 0.358 is added in the rotor circuit to match the rotor-side voltage with the grid voltage.

At $t = 0.5$ sec, the wind speed suddenly decreases to 0.7 pu. The turbine mechanical torque T_m is instantly decreased from -1.0 pu to about -0.22 pu. This value is set by the turbine torque-speed characteristics in Figure 8-15 (Point B). However, the mechanical speed of the turbine and generator cannot change instantaneously at $t = 0.5$ sec due to the moment of inertia. As a result, the torque reference T^*_e produced by the

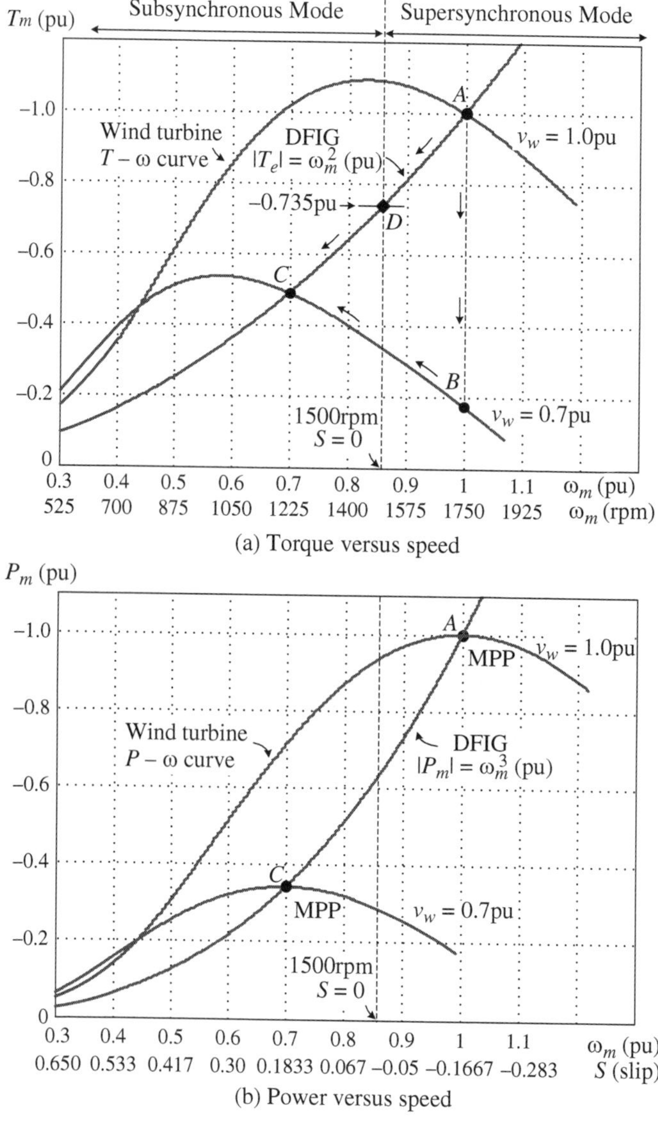

(a) Torque versus speed

(b) Power versus speed

Figure 8-15. Torque and power versus rotor speed of DIFG WECS.

MPPT block remains unchanged, and the generator torque T_e remains unchanged, as shown in Figure 8-15.

During the period of $0.5 < t < 3.75$ sec, the difference between the turbine mechanical torque T_m and the generator electromagnetic torque T_e decelerates the generator, and the rotor speed decreases accordingly. At $t = 1.0$ sec, the generator speed reaches 0.857 pu (1500 rpm), when the slip s becomes zero and the electromagnetic torque T_e is -0.735 pu. The generator is in transition from the supersynchronous mode to sub-

Table 8-5. Parameters and operating condition of the DFIG WECS

Induction generator	Generator ratings: 1.5 MW, 690 V, 50 Hz, 1750 rpm, 8.185 kN·m	Generator parameters: Table B-5 (Appendix B)
System input variables	Torque reference at $t = 0$ Reactive power reference for GSC DC link voltage reference	$T_e^* = -8.185$ kN·m (1.0 pu) $Q_{GSC}^* = 0$ $v_{dc}^* = 436.2$ V
Control scheme	Stator voltage oriented control (SVOC)	Figure 8-14
Rotor-side converter	Converter type Modulation scheme Switching sequence Switching frequency AC harmonic filter	Two-level VSC Space-vector modulation Seven segment; Table 4-4, Chapter 4 4 kHz No
DC link filter	Capacitor	130.73 mF
Grid-side converter	The same as RSC	
Electric grid	Voltage/frequency Line inductance (harmonic filter)	690 V/50 Hz 0.0775 mH
Moment of inertia	Reduced for short simulation time	98.26 kg·m^2 ($H = 1.1$ sec)
Reference frame transformation	abc/dq transformation dq/abc transformation	Eq. (3.1), Chapter 3 Eq. (3.3), Chapter 3

synchronous mode. This operating point is shown in Figures 8-15 and 8-16 (Point D). The system finally reaches its steady-state operating point C at $t = 3.75$ sec. The turbine mechanical torque T_m reaches Point C following the T_m-ω_m curve. The generator torque T_e reaches the same operating point via its own T_e-ω_m curve, as shown in Figures 8-15 and 8-16.

After $t > 3.75$ sec, the DFIG wind energy system operates in steady state. The generator electromagnetic torque T_e is equal to the turbine mechanical torque T_m of 0.49 pu, and the rotor speed is 0.7 pu.

The waveform of rotor phase-a voltage v_{ar}, which is generated by the rotor-side PWM converter, is also given in Figure 8-16 together with its fundamental-frequency component v_{ar1}. It can be observed that the frequency of the rotor voltage is not constant; it varies with the rotor speed. When the rotor operates at the synchronous speed, the rotor frequency is zero, and a DC current flows through the rotor and converter circuit. There is no voltage drop across the rotor leakage inductance L_{lr} or across the equivalent inductance L_{eq} of the rotor-side converter. Therefore the rotor voltage v_{ar1} at the synchronous speed is close to zero.

The rotor phase-a and phase-b currents, i_{ar} and i_{br}, are also shown in the figure. The phase-a current i_{ar} leads the phase-b current i_{br} in the supersynchronous mode, but the phase relation reverses in the subsynchronous mode. This is caused by the change in the slip, which is a negative value when the DFIG is in the supersynchronous mode and becomes positive in the subsynchronous mode. At the synchronous speed, both i_{ar} and i_{br} are DC currents. The DFIG operates just like a wound rotor synchronous generator, in which the DC rotor current is provided by its exciter.

The waveform of phase-a stator current i_{as} is also given in Figure 8-16. When the DFIG operates at rated speed and rated torque ($t < 0.5$ sec), the peak value of the stator current is 1.2 pu, which is also the rated stator current (1068.2 A, rms). Note that in

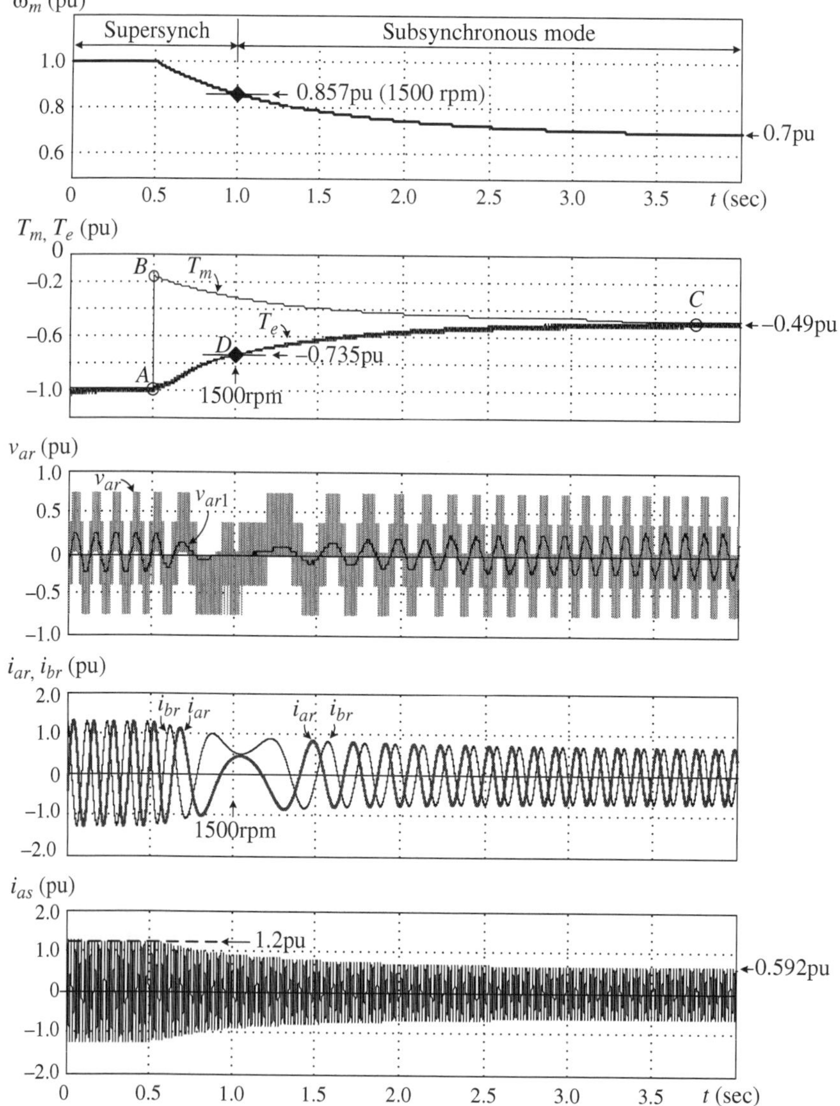

Figure 8-16. Transients of DFIG WECS from supersynchronous to subsynchronous mode.

DFIG WECS, the base current for the per-unit system is not the rated stator current (at rated condition both stator and rotor provide power to the grid; refer to Table B-5 in Appendix B for the definition of base current). The stator current decreases with the rotor speed and, finally, becomes 0.592 pu (peak). The detailed state-state analysis of the DFIG WECS will be given in Case Study 8-4.

8.5.4   Steady-State Performance of DFIG WECS

The steady-state performance of a DFIG WECS with stator voltage oriented control is analyzed in the following two case studies.

Case Study 8-4—Steady-State Analysis of the DFIG WECS Operating at 0.7 pu Rotor Speed. The operation of the 1.5 MW/690 V DFIG wind energy system is further investigated in this case study at rotor speed of 0.7 pu. The analysis starts with the steady-state calculations, followed by computer simulation. It is assumed that the stator power factor is unity. The generator parameters are given in Table B-5 in Appendix B.

With the generator operating at 0.7 pu speed, the mechanical torque of the generator is

$$T_m = -(\omega_{m,pu})^2 \times T_{m,R} = (0.7)^2 \times 8185 = -4010.7 \text{ N·m} \tag{8.71}$$

From Equation (8.7), the stator current can be calculated by

$$I_s = \frac{V_s \pm \sqrt{V_s^2 - \dfrac{4R_s T_m \omega_s}{3P}}}{2R_s} = -525.3 \text{ A} \quad (I_s = 1.509\times10^5 \text{A omitted}) \tag{8.72}$$

where $V_s = 690/\sqrt{3}$ V, $\omega_s = 2\pi \times 50$ rad/sec, $R_s = 6.25$ mΩ, and $P = 2$.

Using the equivalent circuit of Figure 8-4, the voltage across the magnetizing branch is

$$\begin{aligned}
\overline{V}_m &= \overline{V}_s - \overline{I}_s\left(R_s + j\omega_s L_{ls}\right) \\
&= 690/\sqrt{3}\angle 0° - 525.3\angle 180° \times (0.00625 + j100\pi \times 0.1687\times10^{-3}) \\
&= 399.8 + j27.8 = 400.7\angle 4.0° \text{ V}
\end{aligned} \tag{8.73}$$

The magnetizing current is calculated by

$$\overline{I}_m = \frac{\overline{V}_m}{j\omega_s L_m} = 16.19 - j232.42 = 233.0\angle -86.0° \text{ A} \tag{8.74}$$

The rotor current is

$$\overline{I}_r = \overline{I}_s - \overline{I}_m = -541.50 + j232.42 = 589.27\angle 156.8° \text{ A} \tag{8.75}$$

The rotor voltage is determined by

$$\overline{V}_r = s\overline{V}_m - \overline{I}_r\left(R_r + js\omega_s L_{lr}\right) = 76.99\angle 6.5° \text{ V} \tag{8.76}$$

where $s = (\omega_s - \omega_r)/\omega_s = 0.1833$. The slip frequency at the 0.7 pu rotor speed is

$$\omega_{sl} = s\omega_s = 0.1833 \times 2\pi \times 50 = 57.58 \text{ rad/sec} \quad (9.165 \text{ Hz}) \tag{8.77}$$

The peak values of the rotor voltage and rotor current are

$$\begin{cases} i_r = \sqrt{2}I_r = \sqrt{2} \times 589.27 = 833.25 \text{ A} \quad (0.664 \text{ pu}) \\ v_r = \sqrt{2}V_r = \sqrt{2} \times 76.99 = 108.9 \text{ V} \quad (0.273 \text{ pu}) \end{cases} \tag{8.78}$$

The rotor power factor angle is

$$\varphi_r = 6.5° - 156.8° = -150.3° \tag{8.79}$$

The equivalent impedance for the rotor-side converter is given by

$$\overline{Z}_{eq} = \overline{V}_r / \overline{I}_r = -0.11351\,\Omega + j - 0.06472\ \Omega \tag{8.80}$$

from which

$$\begin{cases} R_{eq} = -0.11351\ \Omega \\ X_{eq} = -0.06472\ \Omega \end{cases} \tag{8.81}$$

The mechanical power of the generator is

$$\begin{aligned} P_m &= 3I_m^2\left(R_{eq} + R_r\right)(1-s)/s \\ &= 3 \times 589.27^2\left(-0.11351 + 0.00263\right)(1 + 0.1833)/(-0.1833) \\ &= -514.50 \text{ kW} \end{aligned} \tag{8.82}$$

Alternatively, the input mechanical power can be calculated by

$$P_m = T_m\omega_m = -4010.7 \times 1750 \times 0.7 \times 2\pi /60 = -514.50 \text{ kW} \tag{8.83}$$

The rotor power is

$$P_r = 3(I_r)^2 R_{eq} = 3 \times 589.27^2 \times (-0.11351) = -118.24 \text{ kW} \tag{8.84}$$

The rotor and stator winding losses are

$$P_{cu,r} = 3(I_r)^2 R_r = 2.74 \text{ kW} \tag{8.85}$$

$$P_{cu,s} = 3(I_r)^2 R_s = 2.19 \text{ kW} \tag{8.86}$$

The stator active power is

$$P_s = 3V_s I_s \cos\varphi_s = 690/\sqrt{3}\times589.27\times\cos(180°) = -627.81 \ \text{kW} \qquad (8.87)$$

The stator power factor angle is 180° since the DFIG operates in the generating mode with a unity power factor.

The total power delivered to the grid is

$$\left|P_g\right| = \left|P_s\right| - \left|P_r\right| = 627.81 - 118.24 = 509.57 \,\text{kW} \qquad (8.88)$$

The difference between P_m and P_g is the sum of the losses of the stator and rotor windings, given by

$$\left|P_m\right| - \left|P_g\right| = P_{cu,r} + P_{cu,s} = 4.9 \ \text{kW} \qquad (8.89)$$

The efficiency of the DFIG is then

$$\eta = \left|P_g\right|/\left|P_m\right| = 509.57/514.50 = 99.0\% \qquad (8.90)$$

The grid current is calculated by

$$\left|I_g\right| = \frac{\left|P_g\right|}{3V_g} = \frac{509.6\times10^3}{3\times398.4} = 426.37 \ \text{A} \qquad (8.91)$$

and its peak value is

$$i_g = \sqrt{2}\left|I_g\right| = \sqrt{2}\times426.37 = 603.0 \ \text{A} \quad (0.48\,\text{pu}) \qquad (8.92)$$

The peak stator current is

$$i_s = \sqrt{2}\left|I_s\right| = \sqrt{2}\times525.3 = 742.9 \ \text{A} \quad (0.59\,\text{pu}) \qquad (8.93)$$

The grid current is lower than the stator current. This is due to the fact that the stator power is the sum of the grid power and rotor power ($|P_s| = |P_g| + |P_r|$) in the in the sub-synchronous mode of operation.

The above calculations can be verified by reviewing the steady-state waveforms of the DFIG WECS operating at 0.7 pu rotor speed, given in Figure 8-17. These are in fact, the expanded waveforms from 3.8 sec to 4.0 sec, previously shown in Figure 8-16. The peak values of the phase-a rotor current i_{ar} and rotor voltage v_{ar1} are around 0.664 pu and 0.273 pu, which are the same as those calculated by Equation (8.78). The rotor power factor angle φ_r is $-150.3°$, which is predicted by Equation (8.79).

Simulated waveforms for the grid-side variables are given in Figure 8.17b, including phase-a grid voltage v_{ag}, grid current i_{ag}, and stator current i_{as}. The peak values of the grid and stator currents are very close to those calculated by Equations (8.92) and

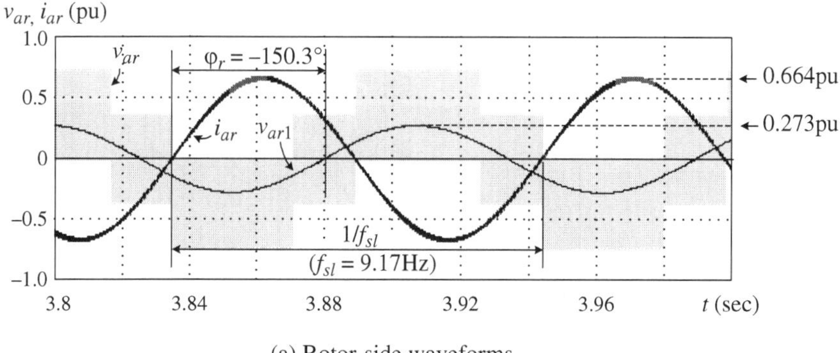

(a) Rotor-side waveforms

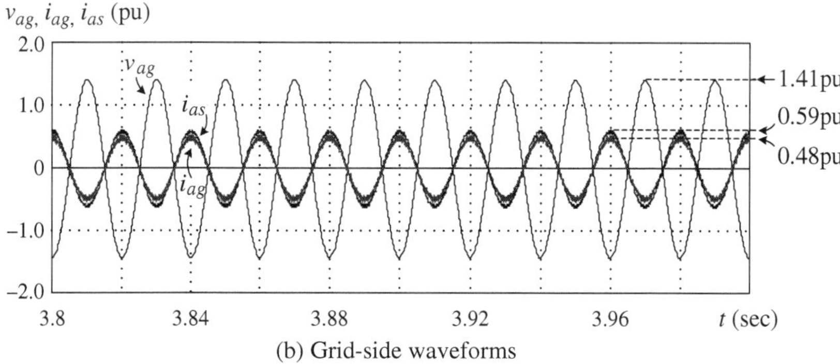

(b) Grid-side waveforms

Figure 8-17. Steady state waveforms of DFIG WECS operating at 0.7 pu rotor speed.

(8.93). The waveforms of the grid voltage and stator current are 180° out of phase. The DFIG operates in the generating mode with unity stator power factor.

Case Study 8-5—Steady-State Analysis of dq-axis Voltages and Currents Based on Phasor Diagram. In this case study, the relationship between the phasor diagram and dq-axis variables of the DFIG WECS is investigated. The dq-axis voltages and currents are calculated and results are then verified by time-domain simulations.

Figure 8-18 shows the phasor diagram of the 1.5 MW/690 V DFIG WECS operating at 0.7 pu rotor speed (subsynchronous mode) with unity stator power factor. The stator voltage $\overline{V}_s$ is used as the reference phasor. The stator current phasor $\overline{I}_s$ lags its voltage by 180°. Based on steady-state calculations in Case Study 8-4, the rotor voltage $\overline{V}_r$ leads $\overline{V}_s$ by 6.5°, whereas the rotor current $\overline{I}_r$ leads $\overline{V}_s$ by 156.8°.

In order to determine the dq-axis components of the rotor voltage and current, the d and q axes are added to the phasor diagram with the d-axis aligned with the stator voltage phasor $\overline{V}_s$. This alignment corresponds to the stator voltage oriented control. Note that the d and q axes in this diagram are stationary and, therefore, do not rotate in space at the synchronous speed. They are simply added to the steady-state phasor diagram to

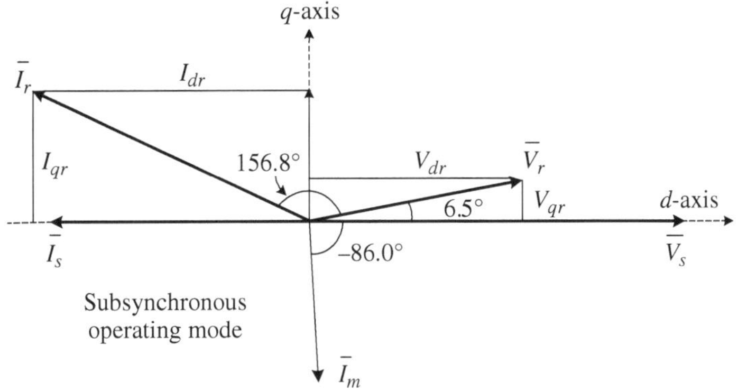

<u>Figure 8-18.</u> Phasor diagram of the DFIG superimposed with dq-axes.

determine the dq-axis components of the variables in the DFIG system. The peak values of the rotor voltage and current are calculated by

$$\begin{cases} v_r = \sqrt{2}V_r = \sqrt{2} \times 76.99 = 108.9\ \text{V} \\ i_r = \sqrt{2}I_r = \sqrt{2} \times 589.3 = 833.4\ \text{A} \end{cases} \tag{8.94}$$

where V_r and I_r are the rms values calculated in Case Study 8-4.

Based on Figure 8-18, the rotor dq-axis voltages are calculated by

$$\begin{cases} v_{dr} = 108.9\cos(6.5°) = 108.2\ \text{V} \quad (0.272\,\text{pu}) \\ v_{qr} = 108.9\sin(6.5°) = 12.33\ \text{V} \quad (0.031\,\text{pu}) \end{cases} \tag{8.95}$$

The rotor dq-axis currents are

$$\begin{cases} i_{dr} = 833.4\cos(156.8°) = -766.0\ \text{A} \quad (-0.610\ \text{pu}) \\ i_{qr} = 833.4\sin(156.8°) = 328.3\ \text{A} \quad (0.262\ \text{pu}) \end{cases} \tag{8.96}$$

The dq-axis rotor voltages and currents can also be found by simulation. With the same operating conditions as those in Case Study 8-3, the simulated results are given in Figure 8-19. It can be observed that when the DFIG WECS reaches steady state at 4.0 sec, the rotor speed is 0.7 pu and the dq-axis rotor voltages and currents are the same as those calculated by Equations (8.95) and (8.96).

8.6 DFIG WECS START-UP AND EXPERIMENTS

Figure 8-20 shows a simplified block diagram to help with the discussion on the start-up process of a DFIG WECS [6-7]. There are two switches in the system: SW1 in the stator circuit and SW2 in the rotor circuit. In practice, electromechanical switches are

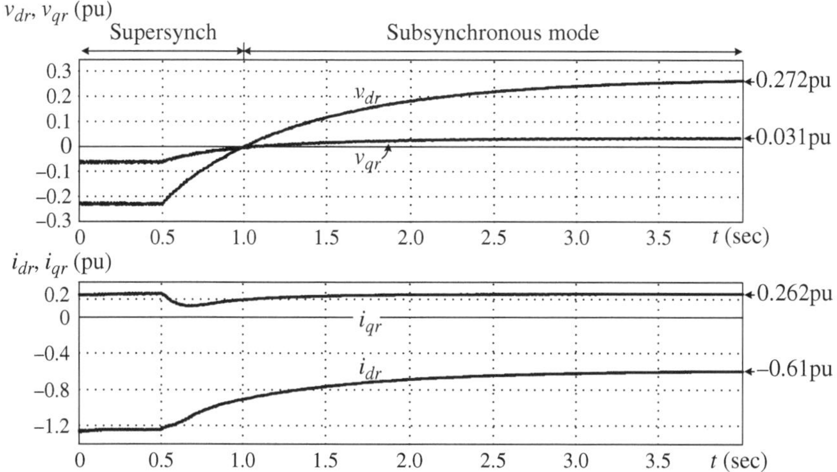

Figure 8-19. Simulated dq-axis rotor voltages and currents during DFIG transients from 1.0 pu to 0.7 pu speed.

normally used, which have little power losses in comparison to the solid-state switches. The pitch angle of the blade is adjusted during the system start-up, and is kept at an optimal value under the normal operating conditions. When the wind speed is higher than its rated value, the pitch angle is dynamically adjusted to keep the turbine's output power constant. When the wind speed exceeds its cut-out speed, the turbine blades are fully pitched, and no power is generated by the turbine. The system start-up process can be divided into the following steps.

Step 1—Initial parking state. In the initial stage with the wind speed below the cut-in speed, switches SW1 and SW2 are open, and both stator and rotor circuits are disconnected from the grid. The turbine blades are pitched out of the wind, and no torque is generated by the turbine. The system is in the parking state.

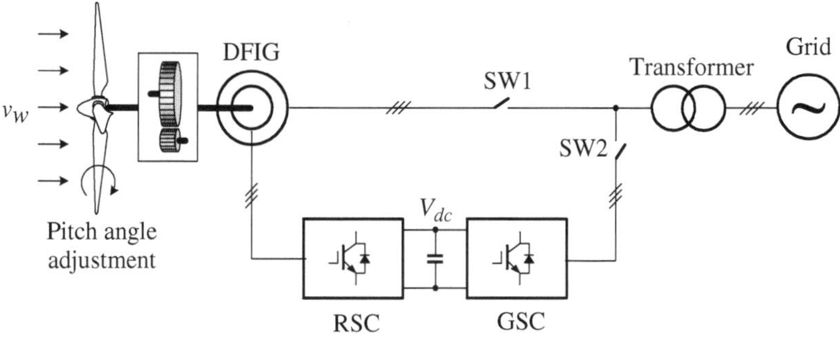

Figure 8-20. Start-up of a DFIG WECS.

Step 2—Turbine/generator acceleration and stator voltage generated. When the wind speed reaches the cut-in speed, the pitch angle of the blades is adjusted to provide starting torque, and the turbine starts to rotate. Switch SW2 in the rotor circuit is closed and the power converters are energized. The DC link voltage of the converters is controlled by the grid-side converter and kept at a fixed value. The rotor-side converter is controlled to provide excitation current to the DFIG. A three-phase balanced voltage is then induced in the stator, which is monitored for synchronization to the grid. The torque reference in the DFIG controller is set to zero. No power is generated or delivered to the grid.

Step 3—Synchronization of the voltage/frequency with the grid. During the rotor speed acceleration, both stator voltage and frequency are fully controlled by the rotor-side converter. When the generator accelerates to a speed that is set according to the measured wind speed, the stator voltage, frequency, and phase angle are adjusted to match those of the grid for synchronization. When the synchronization is achieved, SW1 is closed, and the DFIG WECS is connected to the grid.

Step 4—Power generation and optimal pitch angle. Once the DFIG is connected to the grid, the torque or power reference is increased from zero to a value generated from the MPPT algorithm according to the measured wind speed. The blade pitch angle is also adjusted to its optimal value, at which the maximum wind energy conversion efficiency is achieved. The start-up process is completed.

To investigate the DFIG wind energy system further, experimental waveforms obtained from a low-power 208 V/60 Hz laboratory DFIG system are provided. Figure 8-21

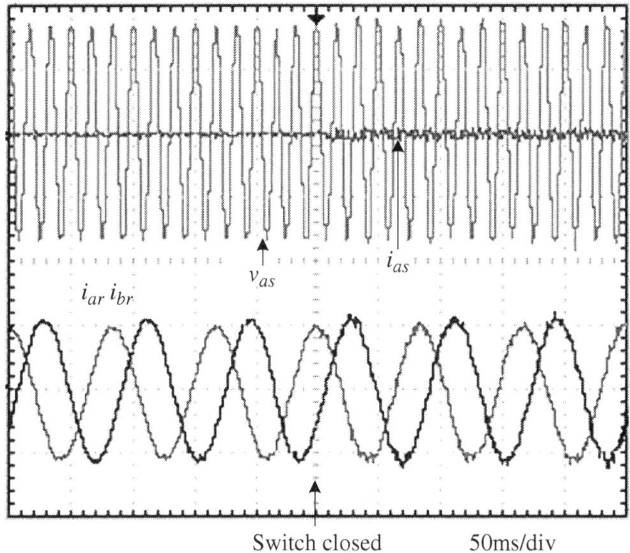

Figure 8-21. DFIG synchronization to the grid.

shows the stator phase voltage and current (v_{as} and i_{as}) and rotor phase currents (i_{ar} and i_{br}) during the system start-up. At the rotor speed of 0.8 pu, the rotor currents are adjusted such that the generated stator voltage, frequency, and phase angle are the same as those of the grid, and the DFIG is then connected to the grid. Due to the perfect matching of the generator voltage with the grid voltage, no in-rush stator current is observed during the transition. The only noticeable change is the small increase of the stator current due to switching harmonics after the DFIG is connected to the grid.

Figure 8-22 shows the measured rotor current waveforms when the DFIG transitions from the supersynchronous (10% above ω_s) to subsynchronous (10% below ω_s) mode. The transition is carried out when the DFIG does not deliver power to the grid, which is a little different from that in Case Study 8-3. The phase-a rotor current i_{ar} lags the phase-b current i_{br} in the supersynchronous mode but leads it in the subsynchronous mode, the same phenomenon as that discussed in Case Study 8-3.

The angles of the stator voltage vector θ_s, rotor position θ_r and the slip θ_{sl} are shown in Figure 8-23, from which the stator frequency f_s, rotor frequency f_r, and slip frequency f_{sl} can be measured. Note that the rotor frequency f_r relates to the rotor electrical speed by $\omega_r = 2\pi f_r$. It can be observed that the rotor frequency f_r is lower than the stator frequency f_s, which indicates that the DFIG operates in the subsynchronous mode.

8.7 SUMMARY

In this chapter, the operating principle of the doubly fed induction generator (DFIG) was discussed. The steady-state equivalent circuit for the DFIG was introduced and

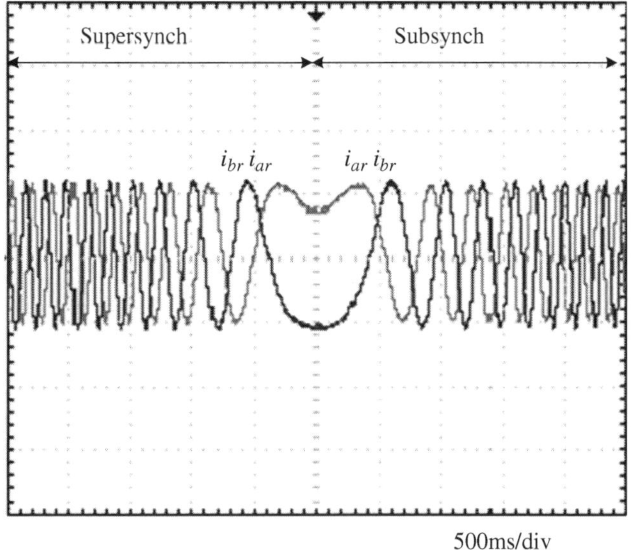

Figure 8-22. Rotor currents of the DFIG from supersynchronous to subsynchronous operation.

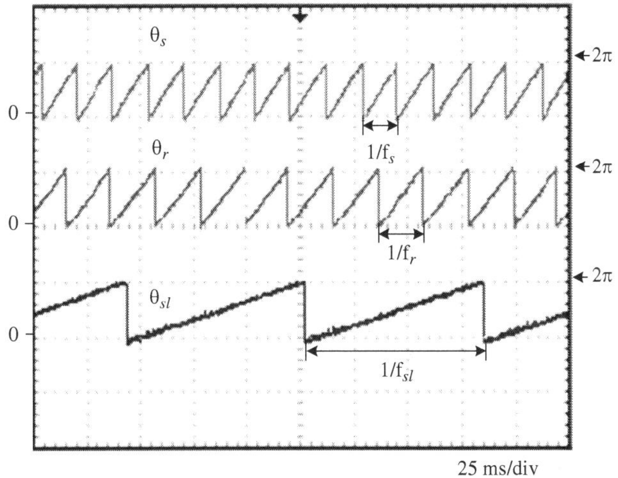

<u>Figure 8-23.</u> Angles of the stator voltage vector, rotor position and slip of the DFIG in the subsynchronous mode.

analyzed in detail, and the rotor-side converter was modeled by an equivalent impedance. The derived equivalent circuit was used to analyze the operation of the DFIG WECS under the super- and subsynchronous modes with leading, lagging, and unity power factor operation. A stator voltage oriented control (SVOC) scheme for the DFIG WECS was presented. The steady-state and dynamic performance of the system was investigated by numerical calculations and computer simulations. Important concepts for the operation of DFIG WECS were illustrated by case studies. The start-up process of the DFIG system was discussed and experimental results from a low-power laboratory DFIG system were provided.

REFERENCES

1. R. Datta and V. T. Ranganathan, Variable-Speed Wind Power Generation Using Doubly Fed Wound Rotor Induction Machine: A Comparison With Alternative Schemes, *IEEE Transactions on Energy Conversion,* Vol. 17. No. 3, 414–420, 2002.

2. V. Akhmatov, *Induction Generators for Wind Power,* Multi-Science Publishing Co. Ltd., 2005.

3. M. Godoy Simões and F. A. Farret, *Alternative Energy Systems: Design and Analysis with Induction Generators,* 2nd Edition, CRC Press, 2007.

4. S. Muller, M. Deicke, and R. W. De Doncker, Doubly Fed Induction Generator Systems for Wind Turbines, *IEEE Industry Applications Magazine,* Vol. 8, No. 3, 26–33, 2002.

5. M. Yamamoto and O. Motoyoshi, Active and Reactive Power Control for Doubly-Fed Wound Rotor Induction Generator, *IEEE Transactions on Power Electronics,* Vol. 6, No. 4, 624–629, 1991.

6. Y. Lang, X. Zhang, D. Xu, S.R. Hadianamrei, and Ma Hongfei, Stagewise Control of Connecting DFIG to the Grid, in *Proceeding of 2006 IEEE International Symposium on Industrial Electronics (ISIE)*, pp. 11129–11133, 2006.

7. W. Sadara and B. Neammanee, Implementation of a Three Phase Grid Synchronization for Doubly-fed Induction Generators in Wind Energy System, in *2010 International Conference on Electrical Engineering/Electronics Computer Telecommunications and Information Technology (ECTI-CON)*, pp. 1016–1020, 2010.

9

VARIABLE-SPEED WIND ENERGY SYSTEMS WITH SYNCHRONOUS GENERATORS

9.1 INTRODUCTION

Synchronous generators (SGs) have been widely used in variable-speed wind energy conversion systems (WECS). There are a number of alternative designs [1], including permanent magnet and wound rotor generators, salient and nonsalient pole generators, and generators with external and internal rotors. With power rating from a few kilowatts to a few megawatts, synchronous generators provide great flexibility to meet different technical requirements in practical wind energy systems.

The synchronous generator can be constructed with a large number of poles and operate at a speed that directly matches the turbine blade speed. Such a direct-drive system does not need a gearbox. This results a reduction in installation and maintenance costs and provides an advantage over induction generator (IG) based turbines where use of a gearbox is a must. The SG wind energy system is normally controlled by full-capacity power converters for variable-speed operation [2], ensuring maximum wind energy conversion efficiency throughout its operating range. With full-capacity converters, the system is able to meet various grid codes [3], including leading/lagging reactive power control and fault ride-through operation, without the need for additional equipment.

This chapter focuses on the control strategies for SG wind energy systems. It starts with an introduction and analysis of control schemes for the synchronous generator.

Power Conversion and Control of Wind Energy Systems. By B. Wu, Y. Lang, N. Zargari, and S. Kouro **275**
© 2011 the Institute of Electrical and Electronics Engineers, Inc. Published 2011 by John Wiley & Sons, Inc.

Operating principles and control schemes for voltage source converter (VSC) based SG WECS are discussed. The chapter also presents current source converter (CSC) technology for wind energy systems. Further analysis is carried out by computer simulations and case studies.

9.2 SYSTEM CONFIGURATION

The block diagram of a typical variable-speed synchronous generator WECS is shown in Figure 9-1. The system consists of a wind turbine, a gearbox, a synchronous generator, power converters, and a transformer for grid connection. As discussed in Chapter 3, two types of synchronous generators can be used: permanent magnet or wound rotor. The synchronous generator with the wound rotor needs a power supply for field excitation, whereas the permanent magnet synchronous generator (PMSG), as its name indicates, uses permanent magnets to produce the rotor field.

The rated speed of the wind turbine depends on its power rating and the number of blades. For three-blade horizontal-axis turbines, the rated speed of the turbine is approximately in the range of 20 to 300 rpm for small/medium size and 8 to 30 rpm for large megawatt turbines. On the other hand, the rated speed of the synchronous generator is dependent on its rated stator frequency and number of poles. A generator can be constructed with a few to around one hundred poles. For instance, with a rated stator frequency of 13.2 Hz, the rated speed of the generator is 22 rpm for a 72-pole generator.

As discussed in previous chapters, a gearbox with a high gear ratio is often required to match the low turbine speed to the high generator speed. The gearbox can be eliminated in direct-drive systems, in which a low-speed generator is used to match the turbine speed. The elimination of the gearbox, however, requires use of a generator with high number of poles. Such a generator is more expensive and heavier than one with a small number of poles for a given power rating [4]. A cost study is often required to evaluate the solution. The direct-drive wind turbine is competitive for offshore applications in which the system maintenance is costly and inconvenient.

The control of a wind energy system includes generator-side active power control with maximum power point tracking (MPPT), grid-side reactive power control, and

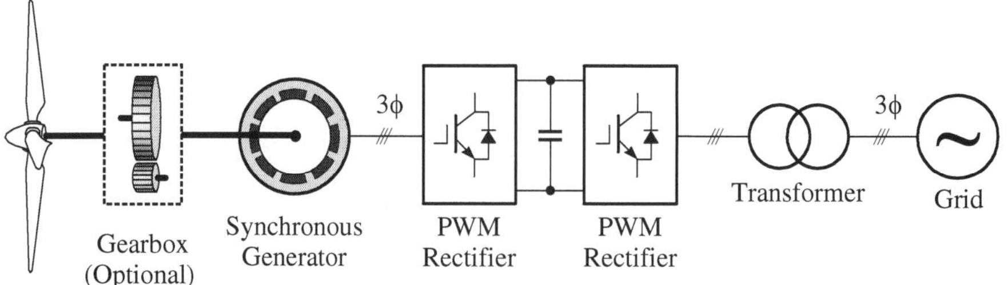

Figure 9-1. Block diagram of variable-speed SG wind energy system.

DC voltage control for voltage source converters or DC current control for current source converters. In contrast to the DFIG wind energy system, in which only around 30% of the total power flows through its converters, the converters in Figure 9-1 process full power from the generator to the grid. The use of full-capacity converters allows the control for the generator- and grid-side converters to be decoupled, which facilities the system design and increases the operating range of the generator.

The operating voltage for commercial wind generators is from a few hundred volts (typically 690 V) to a few thousands volts (e.g., 3000 V), whereas a high-voltage (e.g., 35 kV) system voltage is commonly used within a large-sized wind farm. Therefore, a transformer is normally required, as illustrated in Figure 9-1. The transformer also provides an electrical isolation between the individual wind turbines and the grid.

9.3 CONTROL OF SYNCHRONOUS GENERATORS

The synchronous generator can be controlled by a number of methods to achieve different objectives [5]. For instance, the d-axis stator current of the generator can be set to zero during the operation to achieve a linear relationship between the stator current and the electromagnetic torque. Alternatively, the generator can be controlled to produce maximum torque with a minimum stator current. Yet another approach is to operate the system with unity power factor. In this section, three control schemes to achieve the above objectives are presented and analyzed in detail.

9.3.1 Zero *d*-Axis Current (ZDC) Control

The zero d-axis current control can be realized by resolving the three-phase stator current in the stationary reference frame into d- and q-axis components in the synchronous reference frame. The d-axis component, i_{ds}, is then controlled to be zero [6]. With the d-axis stator current kept at zero, the stator current is equal to its q-axis component i_{qs}:

$$\begin{cases} \vec{i}_s = i_{ds} + j i_{qs} = j i_{qs} \\ i_s = \sqrt{i_{ds}^2 + i_{qs}^2} = i_{qs} \end{cases} \quad \text{for} \ \ i_{ds} = 0 \tag{9.1}$$

where $\vec{i}_s$ is the stator current space vector and i_s represents its magnitude, which is also the peak value of the three-phase stator current in the stationary reference frame.

The electromagnetic torque of the generator, derived in Chapter 3,

$$T_e = \frac{3}{2} P \left(\lambda_r i_{qs} - \left(L_d - L_q \right) i_{ds} i_{qs} \right) \tag{9.2}$$

can be simplified to

$$T_e = \frac{3}{2} P \lambda_r i_{qs} = \frac{3}{2} P \lambda_r i_s \tag{9.3}$$

where P is the pole pairs and λ_r is the rotor flux linkage produced by permanent magnets in the PMSG or by rotor winding in wound rotor synchronous generators (WRSGs). The above equation indicates that with $i_{ds} = 0$, the generator torque is proportional to the stator current i_s. With a constant rotor flux linkage λ_r, the torque exhibits a linear relationship with the stator current, which is similar to torque production in a DC machine with a constant field flux, where the electromagnetic torque is proportional to the armature current.

A space vector diagram for the generator operating with the zero d-axis stator current is illustrated in Figure 9-2. The vector diagram is derived under the assumption that the stator resistance R_s is negligible and the rotor flux linkage is aligned with the d-axis of the synchronous reference frame. All vectors in this diagram together with the dq-axis frame rotate in space at the synchronous speed, which is also the rotor speed of the generator ω_r. The stator current vector $\vec{i}_s$ is perpendicular to the rotor flux vector $\vec{\lambda}_r$. The magnitude of the stator voltage is given by

$$v_s = \sqrt{(v_{ds})^2 + (v_{qs})^2} = \sqrt{(\omega_r L_q i_{qs})^2 + (\omega_r \lambda_r)^2} \tag{9.4}$$

The stator power factor angle is defined by

$$\varphi_s = \theta_v - \theta_i \tag{9.5}$$

where θ_v and θ_i are the angles of the stator voltage and current vectors, given by

$$\begin{cases} \theta_v = \tan^{-1} \dfrac{v_{qs}}{v_{ds}} \\[2ex] \theta_i = \tan^{-1} \dfrac{i_{qs}}{i_{ds}} \end{cases} \tag{9.6}$$

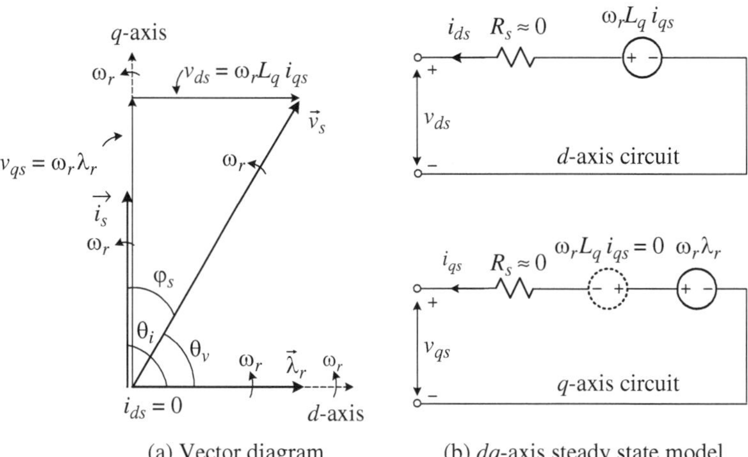

(a) Vector diagram   (b) dq-axis steady state model

Figure 9-2. Space vector diagram of synchronous generator with ZDC control.

For the ZDC scheme, $i_{ds} = 0$, the stator power factor angle is given by

$$\varphi_s = \theta_v - \theta_i = \left(\tan^{-1} \frac{v_{qs}}{v_{ds}} \right) - \frac{\pi}{2} \qquad \text{for } i_{ds} = 0 \tag{9.7}$$

9.3.2 Maximum Torque per Ampere (MTPA) Control

The maximum torque per ampere control generates a given torque with a minimum stator current [6]. For a given rotor flux linkage, the generator torque given in Equation (9.2) is a function of dq-axis stator currents i_{ds} and i_{qs}. This implies that the generator can produce a given torque with different values of i_{ds} and i_{qs}, which are the flux-producing and torque-producing components of the stator current, respectively. It is then possible to produce a torque with minimum stator current by adjusting the ratio of i_{ds} to i_{qs}. For a given stator current i_s, the magnitude of its d-axis current can be calculated by

$$i_{ds} = \sqrt{i_s^2 - i_{qs}^2} \tag{9.8}$$

Substituting the above equation into (9.2), the generator torque can be expressed as a function of i_{qs}:

$$T_e = \frac{3}{2} P \left(\lambda_r i_{qs} - \left(L_d - L_q \right) \left(\sqrt{i_s^2 - i_{qs}^2} \right) i_{qs} \right) \tag{9.9}$$

In the nonsalient generator, the d- and q-axis inductances are equal. The above equation can be simplified to Equation (9.3), where only the q-axis current contributes to the torque production. By setting the d-axis current to zero, the generator torque is produced by the minimum stator current ($i_s = i_{qs}$). Therefore, the ZDC control for the nonsalient generator is essentially the MTPA control.

For the salient-pole generator, the MTPA scheme can be derived through the following steps. Differentiating (9.9) with respect to i_{qs}, we have

$$\frac{dT_e}{di_{qs}} = \frac{3P}{2} \left(\lambda_r - \left(L_d - L_q \right) i_{ds} + \left(L_d - L_q \right) i_{qs}^2 \frac{1}{\sqrt{i_s^2 - i_{qs}^2}} \right) \tag{9.10}$$

To find the maximum torque per ampere, one can set the above derivative to zero:

$$\lambda_r - \left(L_d - L_q \right) i_{ds} + \left(L_d - L_q \right) \frac{i_{qs}^2}{i_{ds}} = 0 \tag{9.11}$$

from which

$$i_{ds} = \frac{\lambda_r}{2 \left(L_d - L_q \right)} \pm \sqrt{\frac{\lambda_r^2}{4 \left(L_q - L_d \right)^2} + i_{qs}^2} \qquad \text{for } L_d \neq L_q \tag{9.12}$$

The first term on the right side of the equation has a negative value since the d-axis inductance L_d of a PMSG is usually lower than the q-axis inductance L_q. To minimize the d-axis stator current $|i_{ds}|$ for the MTPA control, the positive sign for the second term of on the right side of (9.12) should be selected.

For a given electromagnetic torque T_e and rotor flux linkage λ_r, the dq-axis stator currents for the MTPA scheme can be obtained by solving

$$\begin{cases} T_e = \dfrac{3}{2} P \left(\lambda_r\, i_{qs} - \left(L_d - L_q \right) i_{ds}\, i_{qs} \right) \\[2mm] i_{ds} = \dfrac{\lambda_r}{2\left(L_d - L_q \right)} + \sqrt{\dfrac{\lambda_r^2}{4\left(L_q - L_d \right)^2} + i_{qs}^2} \end{cases} \quad \text{for } L_d \neq L_q \qquad (9.13)$$

The space vector diagram of the generator with MTPA control is given in Figure 9-3, where δ is angle of the stator current vector with respect to the q-axis, given by

$$\delta = \tan^{-1} \frac{i_{ds}}{i_{qs}} = \frac{\pi}{2} - \theta_i \qquad \text{for } 0 \leq \theta_i \leq \frac{\pi}{2} \qquad (9.14)$$

Figure 9-4a shows an example for trajectory of 0.8 pu torque with different combinations of the dq-axis rms stator currents I_{ds} and I_{qs} of a generator. The stator current I_s versus I_{ds} is also given in the figure. The stator current I_s reaches its lowest value (0.845 pu) at I_{ds} of 0.44 pu, at which point the MTPA operation is achieved. The trajectory of the generator torque produced by the MTPA scheme is illustrated in Figure 9-4b. It intersects with the 0.8 pu torque at a point where the torque is produced with a

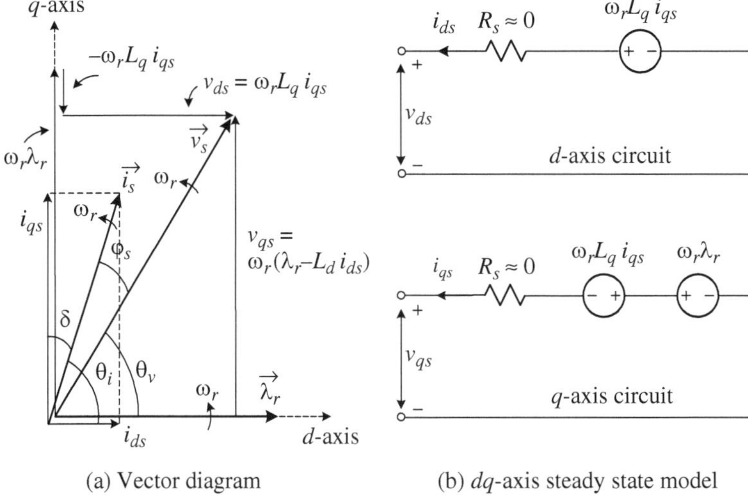

(a) Vector diagram   (b) dq-axis steady state model

Figure 9-3. Space vector diagram of synchronous generator with MTPA control.

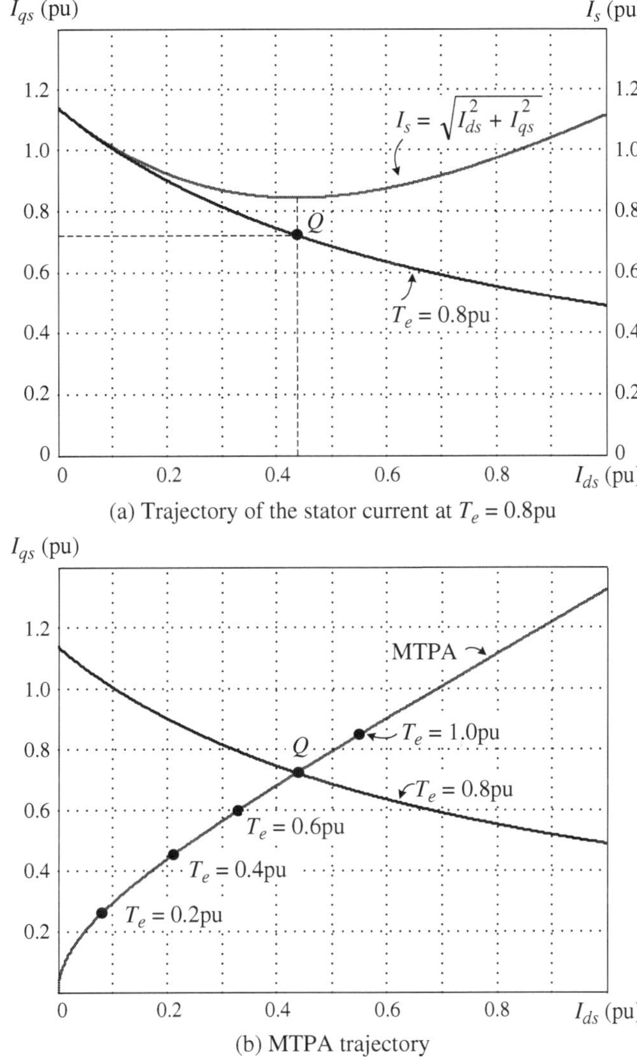

(a) Trajectory of the stator current at $T_e = 0.8$pu

(b) MTPA trajectory

Figure 9-4. Trajectory of maximum torque-per-ampere control.

minimum stator current. The main feature of the MTPA scheme is that it can produce a desired torque with a minimum stator current. This maximizes the utilization of the stator current and minimizes the losses dissipated in the stator winding.

9.3.3 Unity Power Factor (UPF) Control

To simplify the analysis, let us neglect the small voltage drop across the stator resistance R_s. The phase angles of the stator voltage and current can then be calculated by

$$\begin{cases} \theta_v = \tan^{-1}\left(\dfrac{v_{qs}}{v_{ds}}\right) = \tan^{-1}\dfrac{\omega_r\lambda_r - \omega_r L_d i_{ds}}{\omega_r L_q i_{qs}} \\[2em] \theta_i = \tan^{-1}\left(\dfrac{i_{qs}}{i_{ds}}\right) \end{cases} \tag{9.15}$$

Unity power factor operation can be realized when the stator power factor angle φ_s between the stator voltage and current is zero:

$$\varphi_s = \theta_v - \theta_i = 0 \tag{9.16}$$

The space vector diagram for the generator with UPF control is shown in Figure 9-5.
Substituting (9.15) into (9.16) yields

$$L_d i_{ds}^2 + L_q i_{qs}^2 - \lambda_r i_{ds} = 0 \tag{9.17}$$

Solving the above equation for i_{ds}, we have

$$i_{ds} = \begin{cases} \dfrac{\lambda_r + \sqrt{\lambda_r^2 - 4L_d L_q i_{qs}^2}}{2L_d} & \text{(a)} \quad \text{Not valid} \\[2em] \dfrac{\lambda_r - \sqrt{\lambda_r^2 - 4L_d L_q i_{qs}^2}}{2L_d} & \text{(b)} \end{cases} \tag{9.18}$$

The d-axis stator current calculated by Equation (9.18) has two possible values. It can be verified through numerical calculations that Equation (9.18a) is not valid since the d-axis stator current calculated by this equation normally exceeds its rated value. To have a valid solution to Equation (9.18b), $\lambda_r^2 - 4L_d L_q i_{qs}^2 > 0$, from which the q-axis stator current should satisfy the following constraint:

$$i_{qs} \leq \frac{\lambda_r}{2\sqrt{L_d L_q}} \tag{9.19}$$

9.3.4 Comparison of ZDC, MTPA, and UPF Controls

The performance of the ZDC, MTPA, and UPF schemes is further investigated and compared in this section for both salient and nonsalient synchronous generators. The investigation is carried out by three case studies. It is assumed that the generator operates with the MPPT control and, therefore, the generator's shaft input torque is proportional to the square of the rotor speed.

Case Study 9-1—Performance of Synchronous Generator with ZDC Control. In this case study, the steady-state performance of nonsalient and salient-

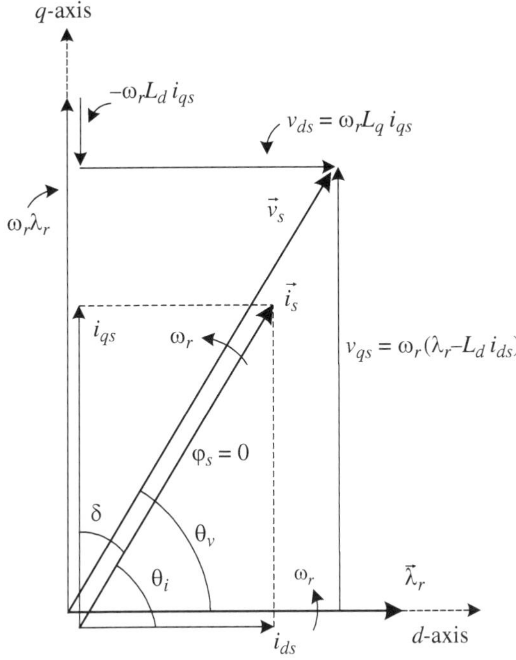

Figure 9-5. Space vector diagram of synchronous generator with UPF control.

pole synchronous generators with zero d-axis current control is investigated. The ZDC scheme is studied by a numerical example, followed by analysis of the generator operating in the full speed range.

Consider a 2.45 MW, 4000 V, 490 A nonsalient pole PMSG operating under the rated conditions. The parameters of the generator are given in Table B-10 in Appendix B. The electrical rotor speed ω_r and mechanical torque T_m of the generator are

$$\begin{cases} \omega_r = 2\pi f_s = 2\pi \times 53.3 = 334.89 \text{ rad/sec} \quad (1.0 \text{ pu}) \\ T_m = 58.459 \text{ kN·m} \quad (1.0 \text{ pu}) \end{cases} \tag{9.20}$$

The dq-axis stator currents of the generator with ZDC control are

$$\begin{cases} i_{ds} = 0 \quad \text{for ZDC} \\ i_{qs} = \dfrac{T_e}{1.5 \times P \times \lambda_r} = \dfrac{58.459 \times 10^3}{1.5 \times 8 \times 7.03} = 692.96 \text{ A} \end{cases} \tag{9.21}$$

where λ_r is the peak value of the rotor flux produced by the permanent magnets ($\lambda_r = 7.03$ Wb), and T_e is the electromagnetic torque, which is equal to the input mechanical torque T_m when the generator is in steady state,

$$T_e = T_m \tag{9.22}$$

The rms stator current is

$$I_s = i_s / \sqrt{2} = \sqrt{i_{ds}^2 + i_{qs}^2} / \sqrt{2} = 692.96 / \sqrt{2} = 490.0 \text{ A} \quad (1.0 \text{ pu}) \tag{9.23}$$

From the steady-state model of the generator in Figure 9-2b, the dq-axis stator voltages are

$$\begin{cases} v_{ds} = -I_d R_s + \omega_r L_q i_{qs} = 2279.38 \text{ V} \\ v_{qs} = -i_{qs} R_s - \omega_r L_d i_{ds} + \omega_r \lambda_r = 2338.97 \text{ V} \end{cases} \tag{9.24}$$

The rms stator voltage is

$$V_s = v_s / \sqrt{2} = \sqrt{v_{ds}^2 + v_{qs}^2} / \sqrt{2} = 3265.9 / \sqrt{2} = 2309.37 \text{ V} \quad (1.0 \text{ pu}) \tag{9.25}$$

The angles of the stator voltage and current vectors are

$$\begin{cases} \theta_v = \tan^{-1} \dfrac{v_{qs}}{v_{ds}} = 0.798 \text{ rad} \quad (45.74°) \\ \theta_i = \tan^{-1} \dfrac{i_{qs}}{i_{ds}} = \pi / 2 \text{ rad} \quad (90°) \end{cases} \tag{9.26}$$

The stator power factor angle is

$$\phi_s = \theta_v - \theta_i = -0.7725 \text{ rad} \quad (-44.26°) \tag{9.27}$$

From which the stator power factor of the generator is calculated as

$$PF_s = \cos\varphi_s = \cos(-44.26°) = 0.716 \tag{9.28}$$

The stator (output) active power is

$$P_s = 3V_s I_s \cos\varphi_s = 2431.2 \text{ kW} \tag{9.29}$$

Following the same steps, the stator voltage V_s, stator current I_s, stator power factor PF_s, mechanical torque T_m, and mechanical power P_m can be calculated for the entire speed range. Results are illustrated in Figure 9-6a. It is assumed in the above calculations that the SG based WECS operates with an MPPT scheme, whereby the mechanical power P_m of the generator is proportional to the cube of its mechanical speed ω_m.

The ZDC control scheme for a salient-pole PMSG of 2 MW, 690 V, 1868 A is also analyzed, and the results are given in Figure 9-6b. The parameters of the generator are given in Table B-11 of Appendix B. It is illustrated that at the rated rotor speed, the stator current and voltage exceed their rated values of 1.0 pu. This indicates that the

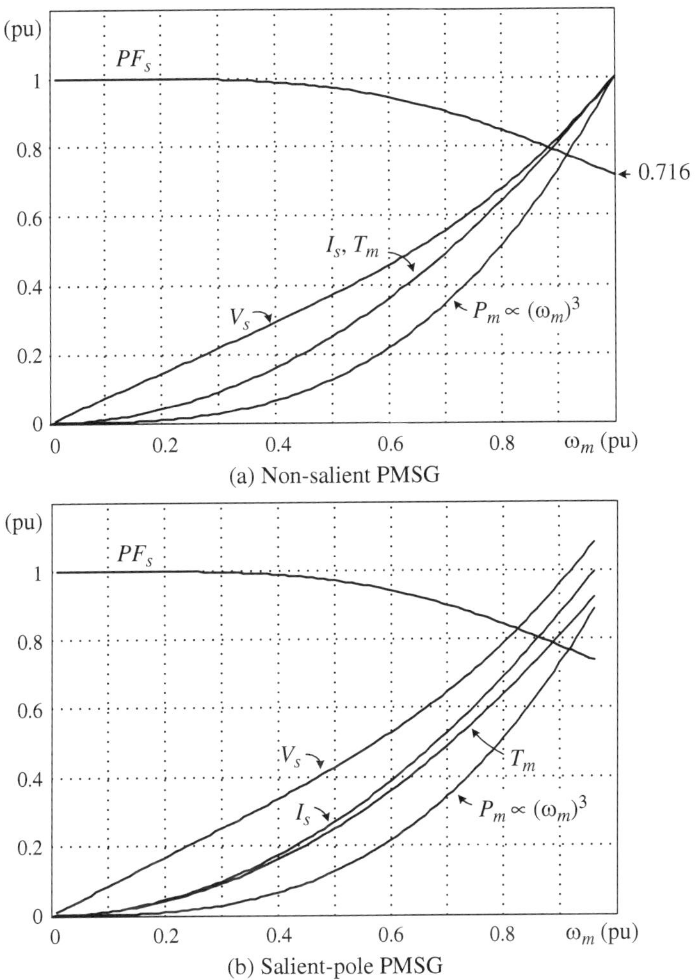

Figure 9-6. Performance of synchronous generator with ZDC scheme.

ZDC scheme may not be suitable for salient PMSG since the optimal operation of the generator cannot be achieved with $i_{ds} = 0$.

Case Study 9-2—Performance of Synchronous Generator with MTPA Scheme. As presented earlier, the MTPA scheme for nonsalient synchronous generators is equivalent to the ZDC control, which has been analyzed in Case Study 9-1. This case study, therefore, only investigates the performance of salient-pole synchronous generators with MTPA control.

Consider the 2 MW, 690 V, 1868 A salient-pole synchronous generator used in the previous case study. At the rated operating conditions, the electrical rotor speed ω_r and mechanical torque T_m of the generator are

$$\begin{cases} \omega_r = 2 \times \pi \times 11.25 \text{ Hz} = 70.69 \text{ rad/sec} \quad (1.0 \text{ pu}) \\ T_m = 852.78 \text{ kN·m} \quad (1.0 \text{ pu}) \end{cases} \tag{9.30}$$

For the MTPA control, Equation (9.13) must be satisfied. Solving the equation by numerical methods, the dq-axis stator currents are

$$\begin{cases} i_{ds} = 892.14 \text{ A} \quad (0.4777 \text{ pu}) \\ i_{qs} = 2486.1 \text{ A} \quad (1.3311 \text{ pu}) \end{cases} \tag{9.31}$$

The rms stator current is

$$I_s = i_s / \sqrt{2} = \sqrt{i_{ds}^2 + i_{qs}^2} / \sqrt{2} = 1867.8 \text{ A} \quad (1.0 \text{ pu}) \tag{9.32}$$

The dq-axis stator voltages are

$$\begin{cases} v_{ds} = -i_{ds} R_s + \omega_r L_q i_{qs} = 405.3 \text{ V} \\ v_{qs} = -i_{qs} R_s - \omega_r L_d i_{ds} + \omega_r \lambda_r = 391.3 \text{ V} \end{cases} \tag{9.33}$$

The rms stator voltage is

$$V_s = v_s / \sqrt{2} = \sqrt{v_{ds}^2 + v_{qs}^2} / \sqrt{2} = 398.4 \text{ V} \quad (1.0 \text{ pu}) \tag{9.34}$$

The angles of the stator voltage vector and current vectors are

$$\begin{cases} \theta_v = \tan^{-1} \dfrac{v_{qs}}{v_{ds}} = 0.7678 \text{ rad} \quad (43.99°) \\ \\ \theta_i = \tan^{-1} \dfrac{i_{qs}}{i_{ds}} = 1.2263 \text{ rad} \quad (70.26°) \end{cases} \tag{9.35}$$

The power factor angle is

$$\phi_s = \theta_v - \theta_i = 0.7678 - 1.2263 = -0.4585 \text{ rad} \quad (-26.27°) \tag{9.36}$$

from which the stator power factor of the generator is

$$PF_s = \cos \varphi_s = \cos(-26.27°) = 0.8967 \tag{9.37}$$

The angle of the stator current vector with respect to the q-axis is

$$\delta = \frac{\pi}{2} - \theta_i = \frac{\pi}{2} - 1.2263 = 0.3445 \text{ rad} \quad (19.74°) \tag{9.38}$$

The stator (output) active power is

$$P_s = 3 V_s I_s \cos \varphi_s = 2001.6 \text{ kW} \tag{9.39}$$

Figure 9-7 shows the characteristics of the 2 MW/690 V salient-pole synchronous generator with MTPA scheme. In order to achieve the MTPA control, the stator current angle δ (with respect to the q-axis) is adjusted according to the rotor speed of the generator.

Case Study 9-3—Performance of Synchronous Generator with UPF Control. The steady-state characteristics of synchronous generators with UPF scheme are investigated in this case study. To achieve the unity power factor operation at different rotor speeds, the constraints specified by Equations (9.18) and (9.19) must be satisfied. Following the same procedure as presented in previous case studies, the performance curves for the 2 MW/690 V salient-pole and 2.45 MW/4000 V nonsalient generator are depicted in Figure 9-8.

For the salient generator, the stator current I_s, stator voltage V_s, mechanical torque T_m, mechanical power P_m, and the stator current vector angle δ all increase nonlinearly with the rotor speed ω_m, as shown in Figure 9-8a. When the rotor speed ω_m increases to 0.92 pu, the stator current reaches its rated value and stator current vector angle δ approaches its maximum value of $\pi/4$. This implies that the generator under the UPF control cannot operate at a rotor speed higher than 0.92 pu.

Figure 9-8b shows the performance curves for the 2.45 MW nonsalient generator with the UPF control scheme. A similar phenomenon can be observed. The nonsalient generator reaches its limit specified by Equation (9.19), at which point the dq-axis currents are

$$
\begin{cases}
i_{qs} = \dfrac{\lambda_r}{2\sqrt{L_d L_q}} = \dfrac{\lambda_r}{2L_d} \\[3mm]
i_{ds} = \dfrac{\lambda_r - \sqrt{\lambda_r^2 - 4L_d L_q i_{qs}^2}}{2L_d} = \dfrac{\lambda_r}{2L_d}
\end{cases}
\qquad L_d = L_q \ \ (\text{nonsalient SG}) \qquad (9.40)
$$

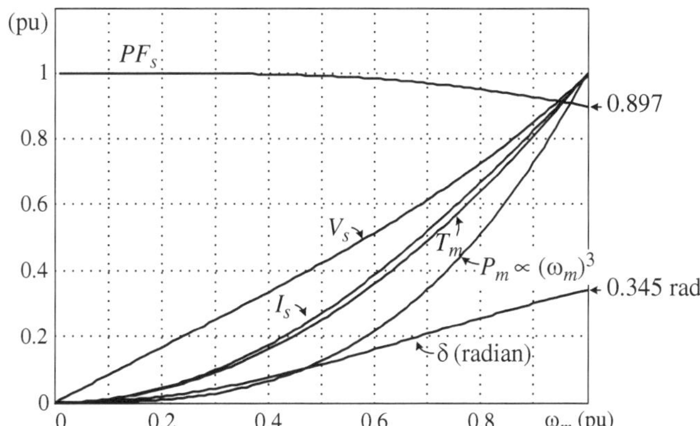

Figure 9-7. Characteristics of salient-pole synchronous generator with MTPA scheme.

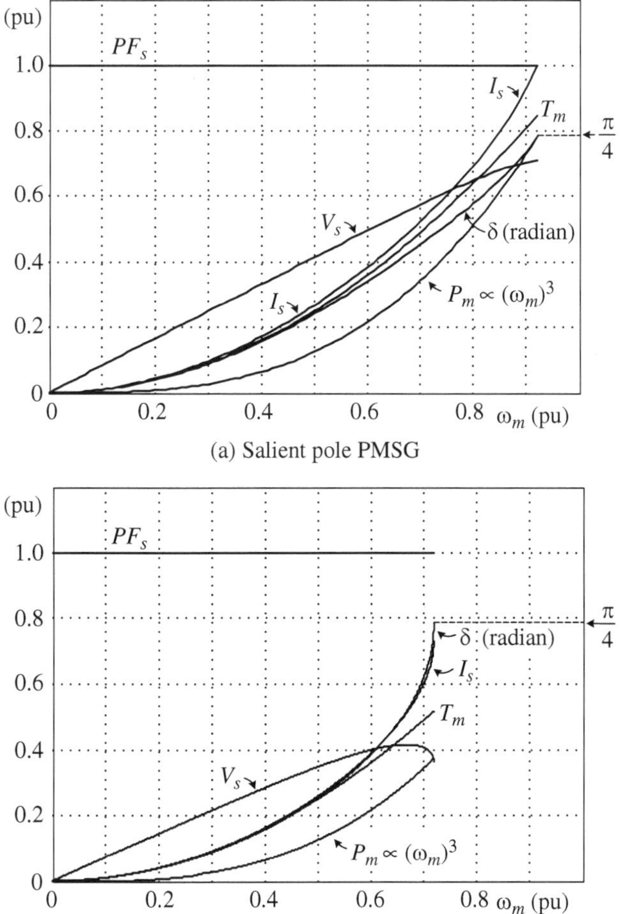

Figure 9-8. Performance of synchronous generator with UPF control.

and the stator current angles are

$$
\begin{cases}
\theta_i = \tan^{-1}\dfrac{i_{qs}}{i_{ds}} = \pi/4 \ \ \text{rad} \ \ (45°) \\[2mm]
\delta = \dfrac{\pi}{2} - \theta_i = \pi/4 \ \ \text{rad} \ \ (45°)
\end{cases}
\tag{9.41}
$$

This case study shows that the speed range of the salient and nonsalient generators with UPC control may be limited. It is noted that a PMSG can be specially designed to operate with unity power factor in the full speed range, if required.

Table 9-1 provides a summary for the operating range of synchronous generators with ZDC, MTPA, and UPF schemes. In summary, the ZDC scheme is valid for nonsalient synchronous generators, the MTPA scheme can be used for both salient-pole

Table 9-1. Operating range of synchronous generators with ZDC, MTPA, and UPF schemes

Generator Control Scheme	ZDC	MTPA	UPF
Nonsalient generator	Full	Full	Partial
Salient-pole generator	Partial	Full	Partial

and nonsalient generators, and the UPF scheme has limited operating range. The above conclusion is made based on two synchronous generators. Generators with different power and voltage ratings were examined as well and similar results were obtained.

9.4   SG WIND ENERGY SYSTEM WITH BACK-TO-BACK VSC

In the SG WECS, three system variables should be tightly controlled: (1) the maximum active power that can be produced by the wind turbine at a given wind speed, (2) the reactive power injected to the grid as dispatched by the supervisory controller or grid operator, and (3) the DC link voltage of the power converters. In most cases, the generator-side converter controls the active power of the generator with maximum power point tracking while the grid-side converter controls the DC voltage and reactive power to the grid.

In this section, the operating principle, dynamic performance, and steady-state operation of two typical SG wind energy systems with back-to-back PWM converters are investigated. These are (1) a nonsalient SG wind energy system with ZDC and optimal torque control (OTC), and (2) a salient-pole SG wind energy system with MTPA and rotor speed feedback control.

9.4.1   Nonsalient SG WECS with ZDC and Optimal Torque Control

The block diagram of a nonsalient PMSG wind energy system with ZDC and optimal torque control is shown in Figure 9-9, where the generator-side converter (rectifier) controls the active power of the system, and the grid-side converter (inverter) controls the DC voltage and the system reactive power. Depending on the power and voltage ratings of the system, the converters can be two-level voltage source converters, parallel converters, or three-level neutral point clamped (NPC) converters, as discussed in Chapters 4 and 5.

To facilitate the analysis of the control schemes, the following assumptions are made:

- There are no power losses in the converter system and the generator output power is equal to the active power delivered to the grid.
- The voltage and current harmonics produced by the converters do not affect the design of the control system, and, therefore, are neglected in the analysis.

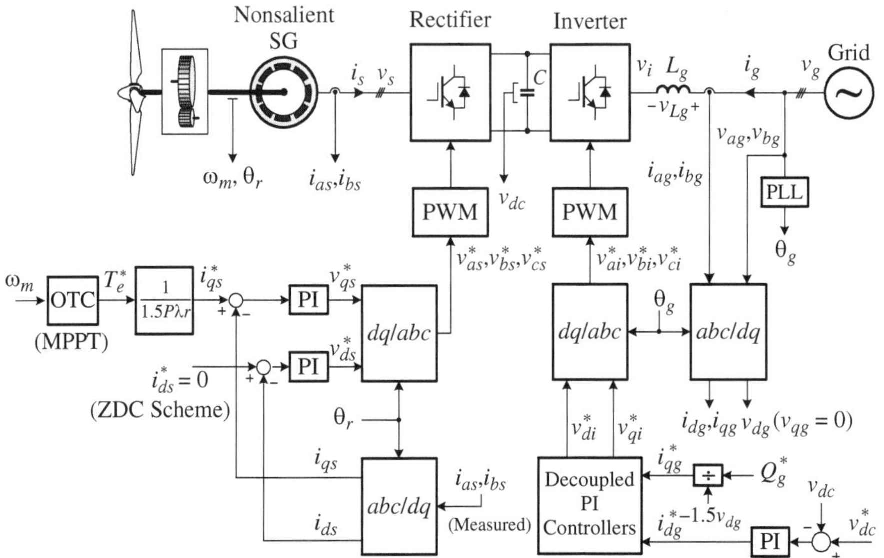

Figure 9-9. Control scheme of nonsalient SG wind energy system with ZDC control.

As shown in Figure 9-9, the MPPT operation is achieved by the optimal torque control introduced in Chapter 2. The active power control is realized by controlling the generator torque T_e through the torque-producing component i_{qs} of the stator current. The reference torque T_e^* is generated by the OTC block according to the measured rotor mechanical speed ω_m. The reference for the torque-producing stator current i_{qs}^* is calculated from T_e^* and according to Equation (9.3).

The d-axis stator current reference i_{ds}^* is set to zero to realize the ZDC control scheme. For rotor flux field orientation, the rotor flux position angle θ_r is detected by an encoder mounted on the shaft of the generator. The measured three-phase stator currents, i_{as}, i_{bs}, and i_{cs}, are transformed into the dq-axis currents, i_{ds} and i_{qs} (from stationary frame to rotor flux synchronous frame, using the abc/dq transformation). In practice, only two of the three stator currents need to be measured. The third current can be calculated by equating $i_{as} + i_{bs} + i_{cs} = 0$ for a three-phase balanced system.

The measured dq-axis stator currents are then compared with their reference currents, i_{ds}^* and i_{qs}^*, respectively. The errors are sent to two PI controllers, which generate the dq-axis reference voltages, v_{ds}^* and v_{qs}^* for the rectifier. These two synchronous-frame reference voltages are then transformed to three-phase reference voltages, v_{as}^*, v_{bs}^*, and v_{cs}^*, in the abc stationary frame via the dq/abc transformation. The transformation of the variables between the abc stationary frame and the dq synchronous frame can be performed by Equations (3.1) and (3.3) (Chapter 3).

The three-phase sinusoidal reference voltages are sent to the PWM generation block. Either carrier-based pulse width modulation or space vector modulation (SVM) techniques can be employed. Neglecting the harmonic components, the rectifier input voltages, v_{as}, v_{bs}, and v_{cs}, which are also the stator voltages of the generator, can then be adjusted according to their reference values such that the active power of the generator is controlled.

The main function of the inverter is to control the reactive power and the DC link voltage v_{dc}. The grid voltage (v_{ag} and v_{bg}), the grid current (i_{ag}, and i_{bg}), and the DC voltage (v_{dc}) are measured. A phase locked loop (PLL) is required to track the grid voltage vector and generate the grid voltage angle θ_g for voltage oriented control (VOC). The VOC scheme is discussed in detail in Chapter 4 and, thus, not repeated here.

9.4.2 Transient and Steady-State Analysis of Nonsalient SG WECS

The performance of the nonsalient SG wind energy system controlled by the ZDC scheme is investigated through the following two case studies.

Case Study 9-4—Start-up Transients Analysis of Nonsalient SG WECS with ZDC Control. The start-up transients of the SG wind energy system in Figure 9-9 are investigated in this case study. A 2.45 MW/4000 V nonsalient PM generator is employed. System parameters and operating conditions are summarized in Table 9-2.

It is assumed that the dynamic response of the turbine mechanical system is very slow in comparison to that of the control system of the power converter (this is due to

Table 9-2. Parameters and operating condition of a 2.45 MW/4000 V nonsalient PMSG WECS

Nonsalient PMSG	2.45 MW, 4000 V, 53.3 Hz, 400 rpm, 490 A	Table B-10, Appendix B
Control Scheme	For generator	Zero d-axis current (ZDC) scheme
	For wind energy system	Optimal torque control (OTC)
System input variables	d-axis stator current	$i_{ds}^* = 0$
	Torque reference	T_e^* from MPPT block
	DC link voltage reference	$v_{dc}^* = 6987$ V (3.06 pu)
Rectifier	Converter type	Three-level NPC inverter
	Modulation scheme/switching frequency	SVM/740 Hz
	Harmonic filter	No
Inverter	Converter type	Two-level VSC
	Modulation scheme/switching frequency	SVM/2040 Hz
	Harmonic filter (line inductance)	16.884 mH (0.2 pu)
DC link filter	Capacitor	1667 μF (4.0 pu)
Grid voltage	Three-phase balanced	4000 V/60 Hz

the large combined moment of inertia). This leads to a further assumption that the speeds of the turbine and generator are constant during the adjustment of the converter output power. Figure 9-10 shows the simulated waveforms during the start-up of the SG wind energy system. The system starts when the speed of the generator is brought to its rated value by the wind, at which point the power converter system and its controller are activated.

At $t = 0.25$ sec, the generator torque T_e starts to increase. It follows its reference T_e^* and ramps up to the rated value (1 pu) at $t = 1.25$ sec, and is then kept constant. With the ZDC scheme, the d-axis stator current i_{ds} is kept at zero. The q-axis stator current i_{qs} is proportional to the generator torque T_e. Note that these two currents are in the rotor flux synchronous frame and, therefore, are DC values in steady state. The phase-a stator current of the generator i_{as} is also illustrated in the figure. Its amplitude is proportional to that of the q-axis current i_{qs}. The DC voltage v_{dc} is kept at its reference value of 3.05 pu by the inverter during the transient and in steady state.

On the grid side, the reactive power Q_g is kept at zero by the controller for unity power factor operation. The active power to the grid P_g has a negative value, indicating that the power is fed from the inverter to the grid. This corresponds to the reference direction of the grid current i_g, which is from the grid to the inverter, as shown in Figure 9-9. The amplitude of the phase-a grid current i_{ag} varies proportionally with the active power. Due to use of decoupled PI controllers in the VOC scheme, the adjustment of active power P_g over time does not affect the reactive power Q_g of the system.

Case Study 9-5—Steady-State Analysis of Nonsalient SG WECS with ZDC Control.

The steady-state operation of the nonsalient 2.45 MW/4000 V SG wind energy system is further studied. Figure 9-11 shows the simulated steady-state waveforms of the rectifier and generator. The rectifier is a three-level NPC type, and is controlled using a conventional SVM scheme. The device switching frequency is 740 Hz, which leads to an equivalent rectifier switching frequency of 1480 Hz. The line-to-line voltage of the rectifier, v_{abs} (which is also the stator voltage), has five voltage levels, and an amplitude equal to the DC voltage v_{dc}. The fundamental component of the line-to-line voltage, v_{abs1}, and the phase voltage, v_{as1}, are also shown in Figure 9-11.

The stator current waveform i_{as} is near sinusoidal due to the large dq-axis inductance of the generator and use of the PWM technique. The phase-a rotor flux λ_{ar} has a sinusoidal waveform. The stator current angle θ_i (90°), stator voltage angle θ_v (45.74°), and stator power factor angle φ_s (−44.26°), are also shown in Figure 9-11. These values match very well with those calculated based on the steady-state equivalent circuit of the generator in Case Study 9-1. The corresponding space vector diagram for the generator with the ZDC control is shown in Figure 9-11b.

The generator operates at its rated speed of 400 rpm, which can be verified by measuring the fundamental frequency of the stator current waveform (53.3 Hz). The rotor flux angle θ_r varies with time periodically when the generator operates in steady state. When the rotor flux angle θ_r is zero, the phase-a rotor flux λ_{ar} reaches its peak value shown by Point A in Figure 9-11a and c.

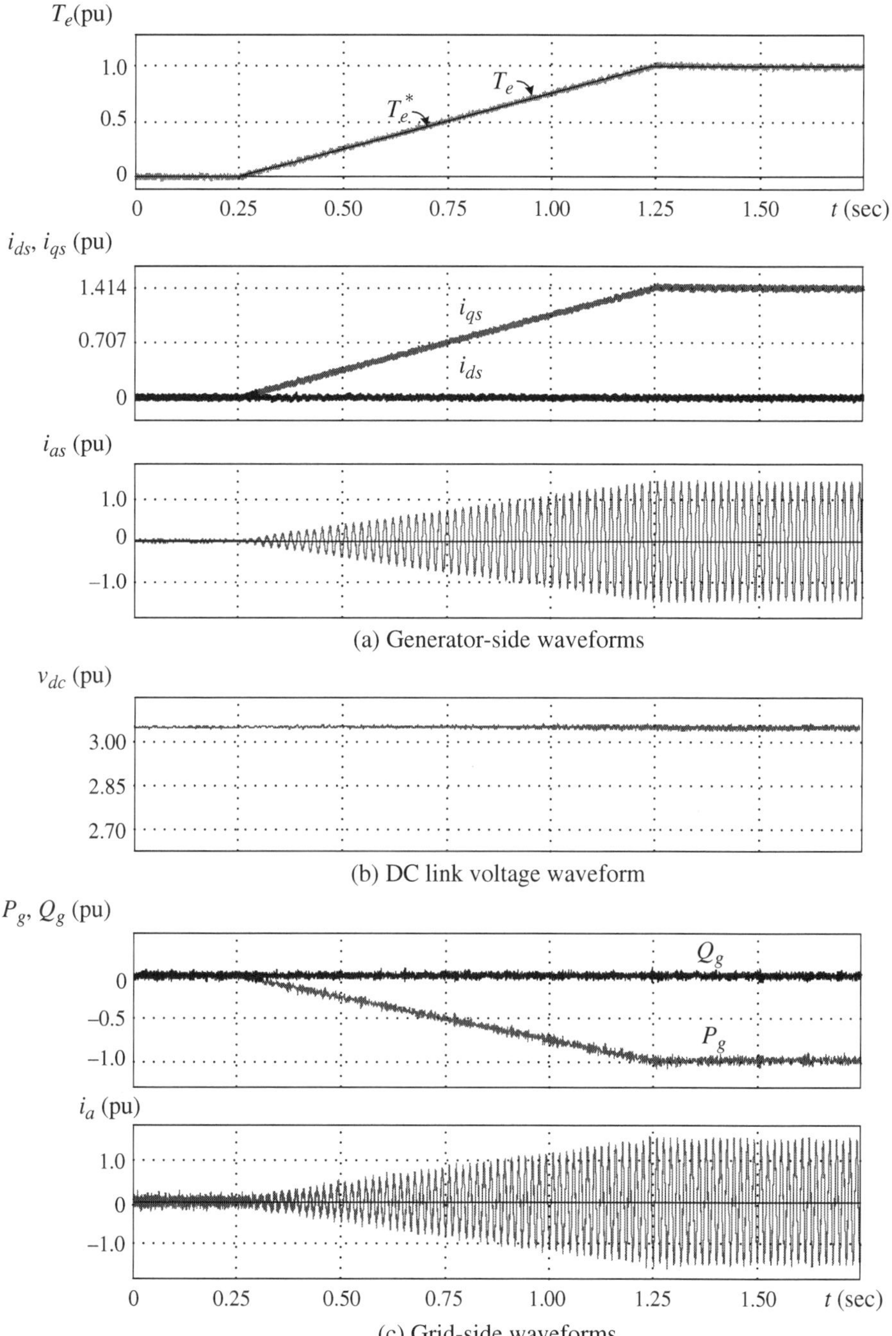

(a) Generator-side waveforms

(b) DC link voltage waveform

(c) Grid-side waveforms

Figure 9-10. Simulated waveforms of a nonsalient SG wind energy system during start-up.

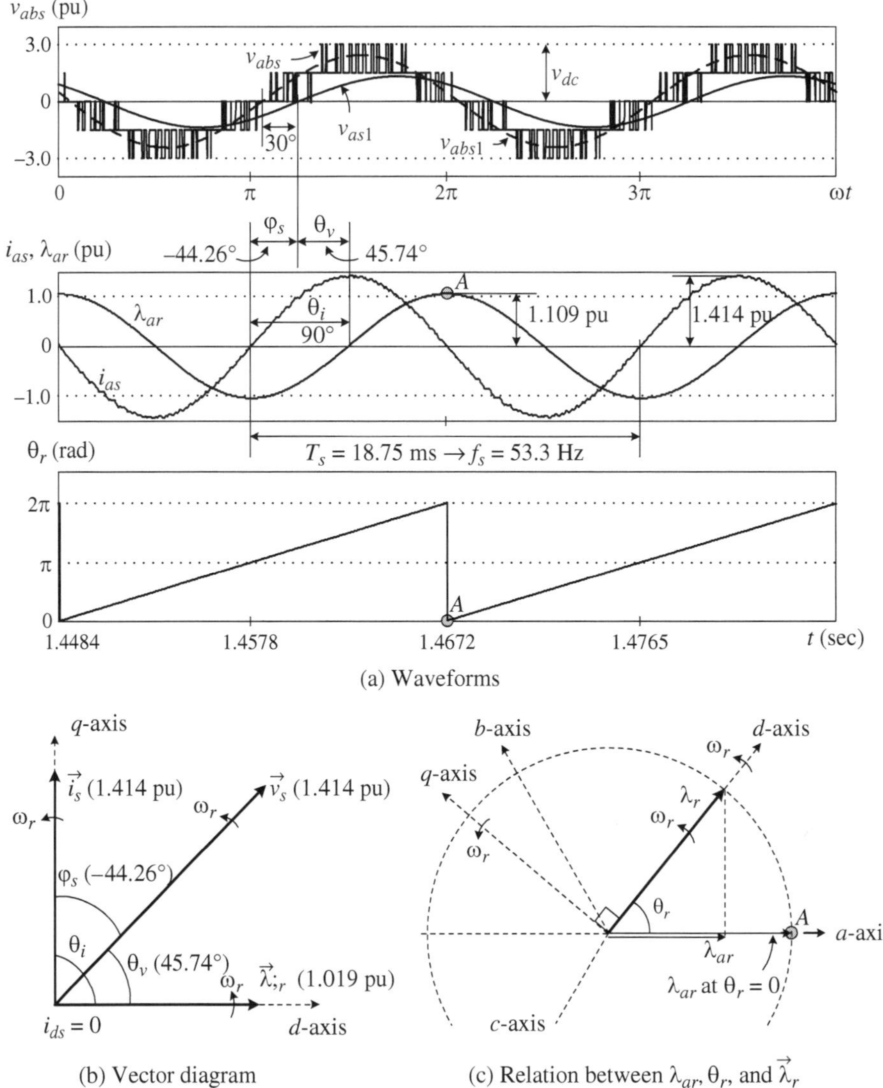

Figure 9-11. Steady-state analysis of a nonsalient SG with ZDC control.

9.4.3 Salient-Pole SG WECS with MTPA and Rotor Speed Feedback Controls

This section presents a rotor speed feedback control scheme for salient-pole SG wind energy systems [7–8]. Figure 9-12 shows the block diagram of such a system. There are three major differences in comparison with the control scheme presented in the previous section:

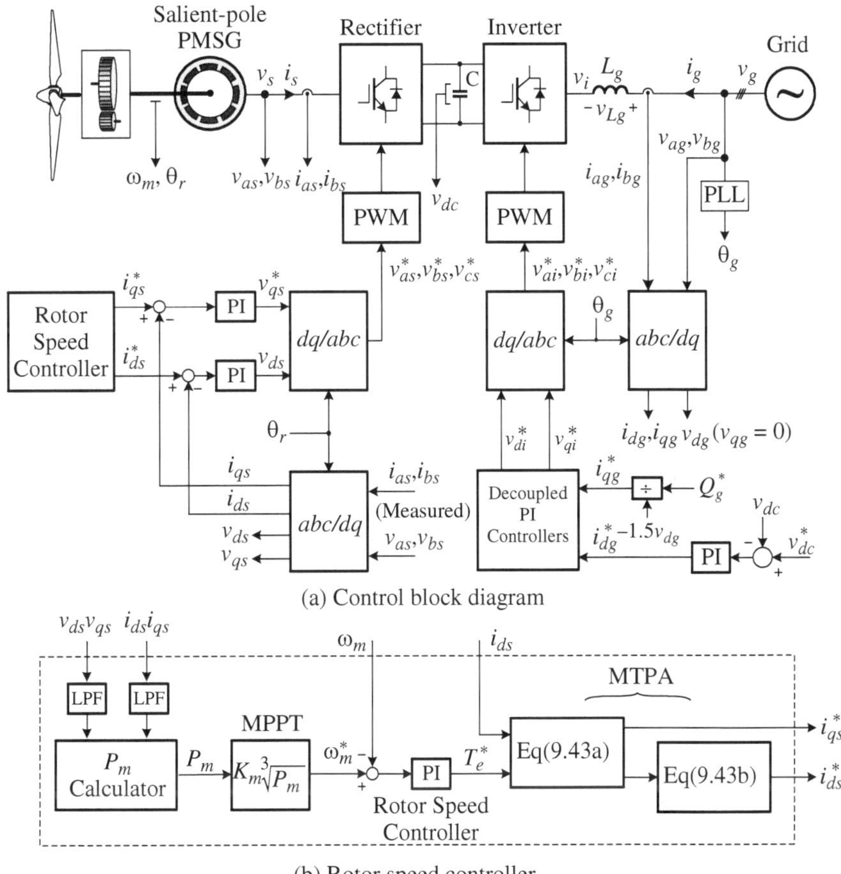

(a) Control block diagram

(b) Rotor speed controller

Figure 9-12. Block diagram of salient-pole SG WECS with MTPA and rotor speed feedback controls.

1. A salient-pole synchronous generator is employed instead of a nonsalient generator.

2. The MTPA control is used for the salient-pole generator versus the ZDC scheme for the nonsalient generator.

3. A rotor speed feedback control is employed with optimal power control for MPPT.

To implement the rotor speed feedback control, the rotor mechanical speed ω_m is detected and controlled by the rotor speed PI controller. The rotor speed reference ω_m^* is generated from the measured mechanical power P_m. To realize the MPPT operation, rotor speed reference is calculated by

$$\omega_m^* = K_m \sqrt[3]{P_m} \tag{9.42}$$

where coefficient K_m can be determined by the rated rotor speed and rated mechanical power of the generator.

To implement the MTPA for the salient-pole generator, the reference values for the dq-axis currents, i_{ds}^* and i_{qs}^*, are calculated according to Equation (9.13), given by

$$\begin{cases} i_{qs}^* = \dfrac{2T_e^*}{3P\left(\lambda_r - (L_d - L_q)\right)i_{ds}} & \text{(a)} \\[3mm] i_{ds}^* = \dfrac{\lambda_r}{2\left(L_d - L_q\right)} + \sqrt{\dfrac{\lambda_r^2}{4\left(L_q - L_d\right)^2} + \left(i_{qs}^*\right)^2} & \text{(b)} \end{cases} \qquad (9.43)$$

where i_{ds} is the measured d-axis stator current, T_e^* is generated by the rotor speed PI controller, and λ_r is rotor flux linkage, a constant value in PMSG.

The main function of the P_m calculator is to calculate the mechanical power of the generator based on measured dq-axis stator voltages and currents as shown in Figure 9-12:

$$P_m = P_s + P_{cu} = \frac{3}{2}\left(v_{ds}i_{ds} + v_{qs}i_{qs}\right) + \frac{3}{2}\left(i_s\right)^2 R_s \qquad (9.44)$$

where $i_s = \sqrt{(i_{ds})^2 + (i_{qs})^2}$ and P_{cu} is the power losses in the stator winding. The measured stator voltage contains a large amount of switching harmonics produced by the PWM rectifier. Low-pass filters (LPFs) are required to filter out the harmonics for accurate calculation of average mechanical power for the MPPT control.

9.4.4 Transient and Steady-State Analysis of Salient-pole SG WECS

The performance of a salient-pole SG wind energy system with MTPA and rotor speed feedback controls is examined through the following two case studies.

Case Study 9-6—Transient Analysis of Salient-Pole SG WECS with Rotor Speed Feedback Control. The transients of a 2 MW/690 V salient-pole PMSG wind energy system caused by a step change in wind speed are investigated. The power-versus-rotor speed characteristics of the system at wind speed of 0.7 pu and 1.0 pu are illustrated in Figure 9-13. The system initially operates at a rotor speed of 0.7 pu (15.75 rpm) as indicated by Point A. The case study investigates the transients when the wind speed suddenly increases to its rated speed of 1.0 pu, and the operating point is then moved from Point A to C. The generator and system parameters for the case study are given in Table 9-3.

The simulated waveforms of the 2 MW PMSG WECS operating from 0.7 to 1.0 pu rotor speed are given in Figure 9-14.

During $0 < t < 0.5$ sec, the system operates in steady state at 0.7 pu rotor speed, at which point the mechanical power P_m is 0.343 pu $(0.7)^3$ and the mechanical torque T_m is 0.49 pu. The electromagnetic torque of the generator T_e is equal to the mechanical torque. The system operates at Point A in Figures 9-13 and 9-14.

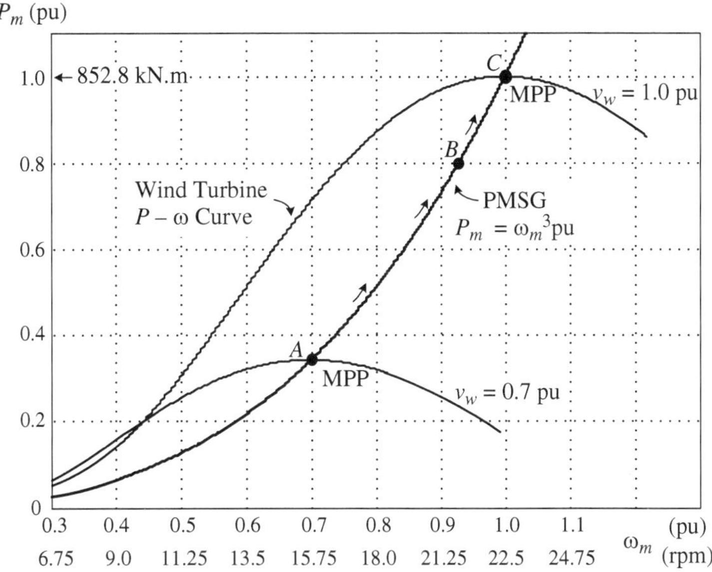

Figure 9-13. P_m versus ω_m characteristics of SG WECS at two different wind speeds.

Table 9-3. Parameters and operating condition of the salient-pole PMSG WECS

Synchronous generator	Generator ratings: 2 MW, 690 V, 50 Hz, 22.5 rpm 1867.8A, 852.8kN·m	Generator parameters: Table B-11 (Appendix B)
Control scheme	For generator For wind energy system	MTPA Rotor speed feedback with optimal power control
System input variables	Rotor speed reference produced by MPPT algorithm DC link voltage reference Grid-side reactive power reference	ω_m^* $v_{dc}^* = 1217$ V (3.06 pu) $Q_g^* = 0$
Rectifier	Converter type Modulation scheme / switching frequency AC harmonic filter	Two-level VSC SVM/2 kHz No
Inverter	Converter type Modulation scheme/switching frequency Harmonic filter (line inductance) Power factor	Two-level VSC SVM/2 kHz 0.1515 mH (0.2 pu) Unity ($PF = 1$)
Grid	Grid voltage	690 V, 50 Hz
DC link filter	Capacitor	53.49 mF (6.0 pu)
Moment of inertia	Reduced for short simulation time	321,660 kg·m² ($H = 0.4$ sec)

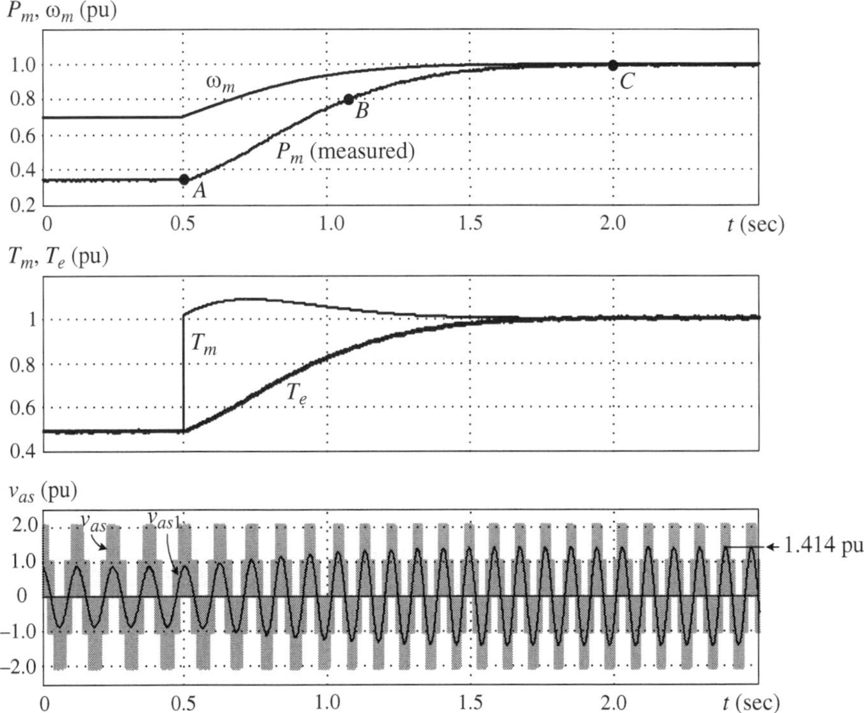

Figure 9-14. Transients of salient-pole SG wind energy system with rotor speed feedback control.

At $t = 0.5$ sec, the wind speed suddenly increases to the rated value of 1.0 pu. The turbine mechanical torque T_m is instantly increased from 0.49 pu to its rated value of 1.0 pu. However, the mechanical speed of the generator cannot change instantaneously due to the moment of inertia. As a result, the measured mechanical power P_m, and then the speed and torque references, ω_m^* and T_e^*, remain unchanged, which makes the generator torque T_e unchanged, as shown in Figure 9-14.

During $0.5 < t < 2.0$ sec, the difference between the mechanical torque T_m and the electromagnetic torque T_e causes the generator to accelerate (the rotor speed increases accordingly). The system finally reaches its steady-state operating point C at $t = 2.0$ sec, at which T_e is equal to T_m and the rotor speed reaches its rated value of 1 pu.

The rectifier phase-a PWM voltage v_{as} is also shown in Figure 9-14. The frequency and magnitude of its fundamental-frequency component v_{as1} vary with the rotor speed, accordingly.

The waveforms for the dq-axis stator currents i_{ds} and i_{qs}, stator power factor angle φ_s, and phase-a grid current i_{ag} during the system transients are illustrated in Figure 9-15. The dq-axis stator currents in the synchronous frame are of DC form. The stator power factor angle φ_s and the magnitude of the grid current i_{ag} vary with the rotor speed, accordingly.

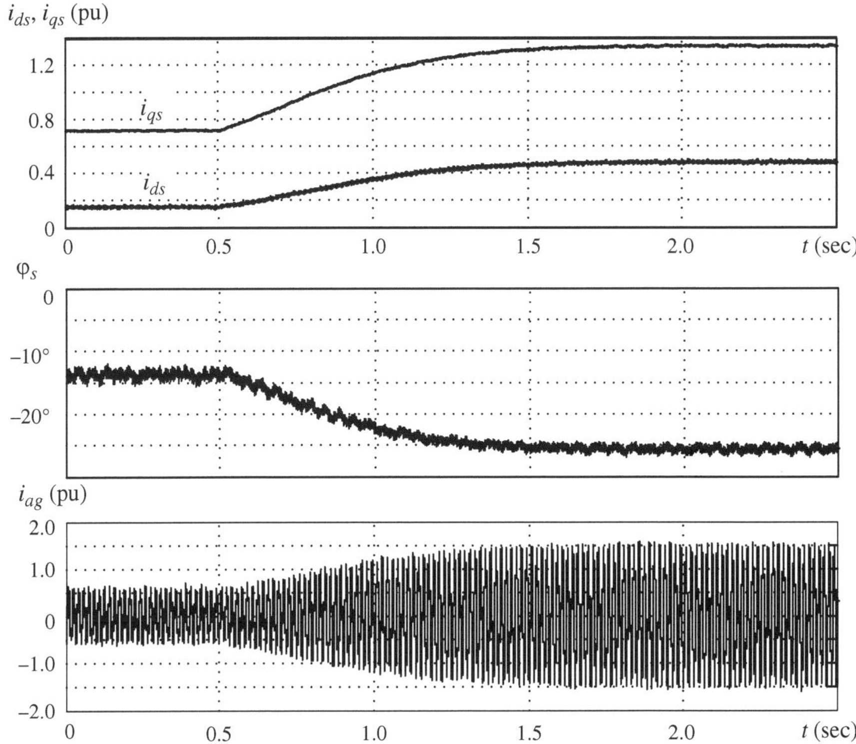

Figure 9-15. Transient waveforms for salient-pole SG WECS with rotor speed feedback control.

Case Study 9-7–Steady-State Analysis of 2 MW/690 V PMSG WECS with MTPA Scheme. The operation of the 2 MW/690 V salient-pole PMSG wind energy system is further analyzed at the rotor speed of 0.7 pu. The analysis starts with the steady-state calculations, followed by computer simulation. The generator parameters are given in Table B-11 of Appendix B. Note that switching harmonics produced by the power converters are not considered in the steady-state analysis.

At the rotor speed of 0.7 pu, the stator frequency f_s, the rotor speed ω_r in rad/sec, electromagnetic torque T_e, and mechanical power P_m of the generator are given by

$$\begin{cases} f_s = 11.25 \times 0.7 = 7.875 \text{ Hz} \\ \omega_r = 2\pi f_s = 49.48 \text{ rad/sec, or} \\ \omega_r = \dfrac{2\pi P}{60}(22.5 \text{ rpm}) \times 0.7 = 49.48 \text{ rad/sec} \\ T_e = (0.7)^2 \times T_{e,R} = 0.49 \times 852.78 = 417.86 \text{ kN·m} \\ P_m = (0.7)^3 \times P_{e,R} = 0.343 \times 2009.3 = 689.19 \text{ kW} \end{cases} \qquad (9.45)$$

Based on Equation (9.13), the dq-axis stator currents are obtained:

$$\begin{cases} i_{ds} = 282.34 \text{ A} \ \ (0.151\,\text{pu}) \\ i_{qs} = 1335.77 \text{ A} \ \ (0.715\,\text{pu}) \end{cases} \tag{9.46}$$

The peak and rms stator currents are

$$\begin{cases} i_s = \sqrt{i_{ds}^2 + i_{qs}^2} = 1365.3 \text{ A} \ \ (0.731\,\text{pu}) \\ I_s = i_s / \sqrt{2} = 965.4 \text{ A} \ \ (0.517\,\text{pu}) \end{cases} \tag{9.47}$$

Using the SG steady-state equivalent circuit, the dq-axis stator voltages are

$$\begin{cases} v_{ds} = -i_{ds}R_s + \omega_r L_q i_{qs} = 152.47 \text{ V} \\ v_{qs} = -i_{qs}R_s - \omega_r L_d i_{ds} + \omega_r \lambda_f = 310.72 \text{ V} \end{cases} \tag{9.\,48}$$

The peak and rms stator voltages are

$$\begin{cases} v_s = \sqrt{v_{ds}^2 + v_{qs}^2} = 346.1 \text{ V} \ \ (0.869\,\text{pu}) \\ V_s = v_s / \sqrt{2} = 244.74 \text{ V} \ \ (0.614\,\text{pu}) \end{cases} \tag{9.\,49}$$

The angles of the stator voltage and current vectors are

$$\begin{cases} \theta_v = \tan^{-1}\dfrac{v_{qs}}{v_{ds}} = 1.1146 \text{ rad} \ \ (63.9°) \\[3mm] \theta_i = \tan^{-1}\dfrac{i_{qs}}{i_{ds}} = 1.3624 \text{ rad} \ \ (78.1°) \end{cases} \tag{9.50}$$

The stator power factor angle and power factors are

$$\begin{cases} \phi_s = \theta_v - \theta_i = -0.248 \text{ rad} \ \ (-14.2°) \\ PF_s = \cos\phi_s = \cos(-14.2°) = 0.9694 \end{cases} \tag{9.51}$$

The stator (output) active power is

$$P_s = 3V_s I_s \cos\varphi_s = 687.1 \text{ kW} \tag{9.52}$$

The steady-state waveforms of the PMSG WECS operating at 0.7 pu rotor speed are shown in Figure 9-16a. The peak values of the phase-a stator voltage v_{as1} and stator current i_{as}, and angles of the stator voltage θ_v, stator current θ_i, and stator power factor φ_s, match the steady-state calculations and can be easily verified. The corresponding space vector diagram is given in Figure 9-16b.

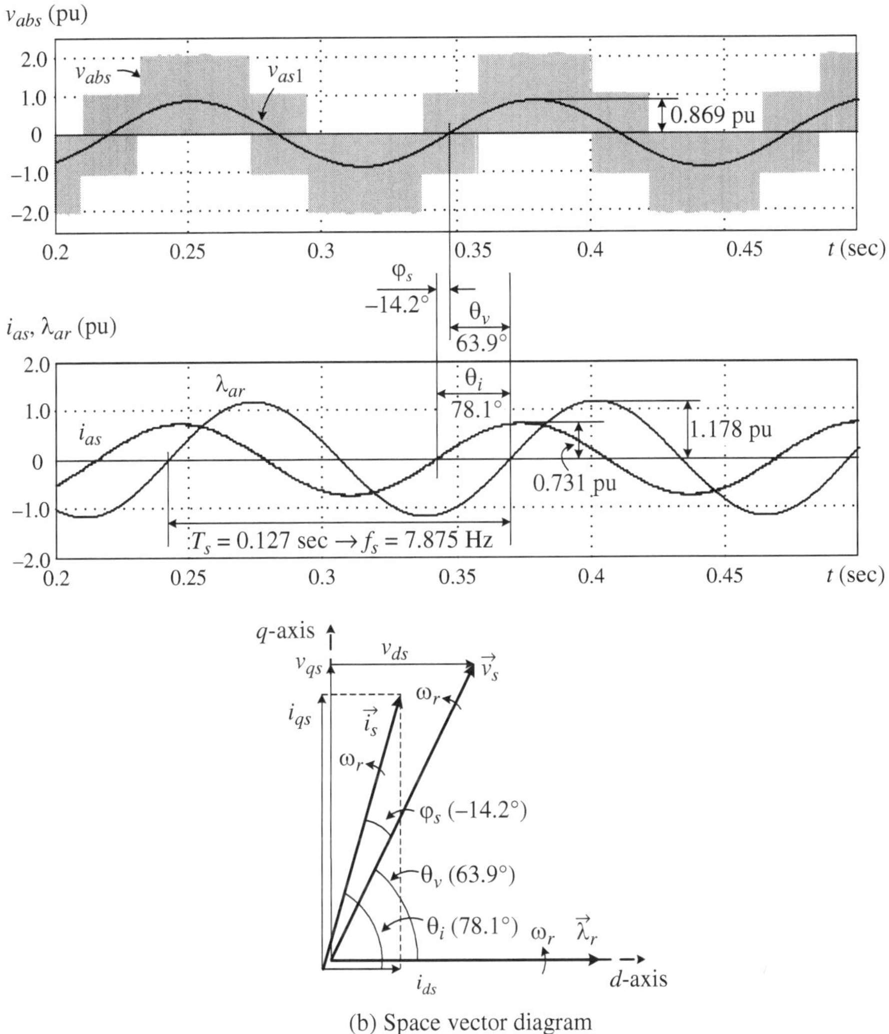

Figure 9-16. Steady-state waveforms and space vector diagram.

9.4.5   Grid-Side MPPT Control Scheme

In the control schemes presented above, the active power of the wind energy system is controlled by the rectifier to achieve MPPT control, whereas the DC link voltage is regulated by the inverter. An alternative approach would be to adjust the active power by the inverter and control the DC voltage by the rectifier, as shown in Figure 9-17.

The control of the DC voltage is achieved by a PI controller that compares the measured DC voltage v_{dc} with its reference v_{dc}^*. The output of the PI controller is the refer-

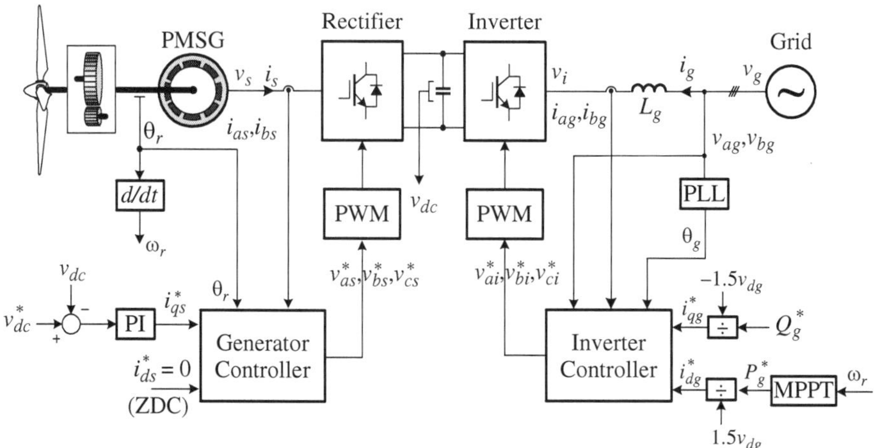

Figure 9-17. Control scheme of SG wind energy system with grid-side MPPT control.

ence for the q-axis stator current i_{qs}^*. The reference for the d-axis stator current i_{ds}^* is zero to satisfy the ZDC scheme. The dq-axis current reference signals, i_{ds}^* and i_{qs}^*, are then sent to the generator controller block, which closely controls the dq-axis stator currents (refer to Figure 9-9 for detailed control scheme).

For the inverter control, the reference for the active power to the grid, P_g^*, is calculated by the MPPT block according to the measured wind velocity or rotor speed ω_r. The grid current reference signals, i_{dg}^* and i_{qg}^*, are generated according to the active and reactive power requirements. The current references are then sent to the inverter controller block, through which the active and reactive power of the system is regulated. The DC link voltage control by the rectifier (Figure 9-17) may not perform as well as the DC link voltage control by the inverter (Figure 9-9) due to the fact that the rectifier input voltage varies with the rotor speed, whereas the grid voltage is constant in most cases.

9.5 DC/DC BOOST CONVERTER INTERFACED SG WIND ENERGY SYSTEM

A DC/DC boost converter with a three-phase diode rectifier can be used to replace the PWM rectifier discussed in the previous sections. This could simplify the control and potentially reduce the cost of the system. There are two distinctive characteristics of this type of wind energy system:

1. The rotor flux of the synchronous generator is produced by permanent magnets or rotor field winding, and, therefore, the generator does not require the rectifier to provide magnetization as it does for the induction generators.
2. The output DC voltage of the boost converter can be regulated (increased) to a

level required by the inverter. This is especially important at low wind speeds, when the output voltage of the diode rectifier is too low for the inverter to operate properly.

Figure 9-18 shows the block diagram of a typical wind energy system interfaced using a DC/DC boost converter. Similar to the other wind energy systems, three variables need to be tightly controlled: the DC voltage, the generator active power, and the grid reactive power. The DC voltage v_{dc1} produced by the diode rectifier varies with the generator speed.The DC voltage v_{dc2}, however, is maintained at a constant value. Due to the boost nature of the converter, the DC voltage v_{dc2} is normally higher than v_{dc1}. The active power of the generator is controlled by adjusting the duty cycle D of the boost converter through a PI controller, whereas the reactive power is adjusted by the inverter in the same manner as that presented in Chapter 4.

Figure 9-19 shows the simulated steady-state waveforms. The waveform is distorted, caused by the nonlinear nature of the diode rectifier. The current waveform contains fifth and seventh harmonics with a magnitude of about 15% and 7%, respectively, as shown in the harmonic spectrum.

Due to the harmonics in the stator current, the generator torque waveform contains ripples. The dominant harmonic torque is at the sixth harmonic (10%), which is mainly produced by the fifth- and seventh-order harmonic currents. This is one of the main drawbacks of the diode-rectifier-based wind energy system. The torque ripples may also cause additional mechanical vibrations and torsional resonances in large wind energy systems.

It is noted that due to the use of a diode rectifier, the PMSG operates with a power factor close to unity. However, a PMSG may not be able to produce the rated power with unity power factor operation as illustrated in Figure 9-8. To solve the problem, the PMSG should be redesigned. For simplicity, the problem was mitigated in the simulation by reducing dq-axis inductance of the PMSG to 20% of its original value given in Table B-11.

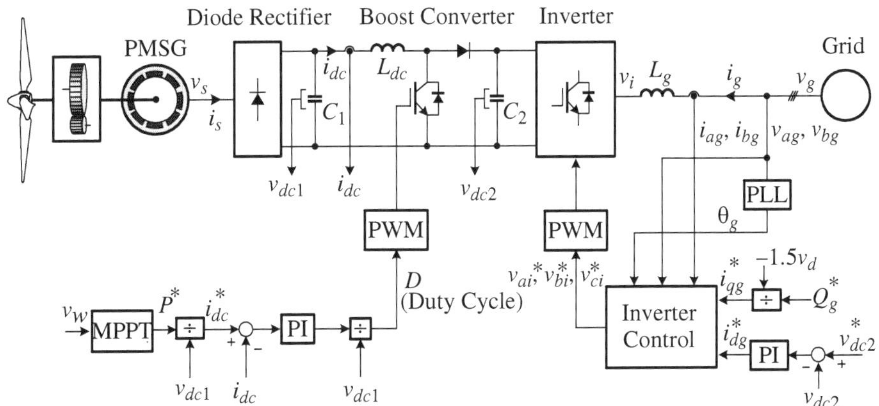

Figure 9-18. DC/DC boost converter interfaced SG wind energy system.

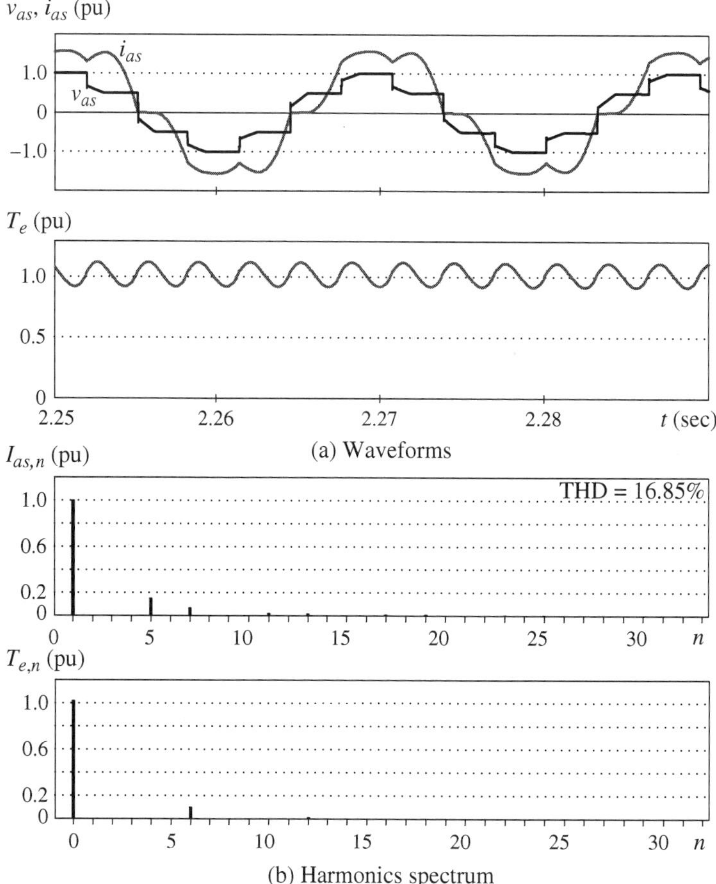

Figure 9-19. Generator current and voltage waveforms at the rated operating conditions.

9.6 REACTIVE POWER CONTROL OF SG WECS

This section focuses on a few issues related to the reactive power control of the SG wind energy system, including (1) the decoupled control of active and reactive power, and (2) the effect of the grid voltage and converter rating on maximum reactive power to the grid.

Case Study 9-8—Reactive Power Analysis of SG Wind Energy Systems. In this case study, the control of the grid-side reactive power of an SG wind energy system is studied, and the effects of grid voltage variation and the inverter power rating are examined. To illustrate the reactive power control, the SG wind energy system shown in Figure 9-9 is used as an example. The system parameters are listed in Table 9-3, except that the reference for the DC voltage is set to 2.94 pu. In the following analysis, it is assumed that:

- The inverter is ideal with no power losses.
- The power and voltage rating for the inverter is 2 MVA/690 V (1.0 pu).
- The effect of current/voltage harmonics on active and reactive power is neglected.

It is assumed that for a given wind speed, the generator operates at a rotor speed of 0.8 pu. The grid-side active power P_g and maximum reactive power $Q_{g,\max}$ that the inverter can provide are calculated by

$$\begin{cases} P_g = -\left(\omega_{r,pu}\right)^3 = -0.8^3 = -0.512 \ \text{pu} & (-1.024\,\text{MW}) \\ Q_{g,\max} = \pm\sqrt{1 - P_g^2} = \pm 0.859 \ \text{pu} & (\pm 1.718\,\text{MVAR}) \end{cases} \tag{9.53}$$

Figure 9-20 shows the simulated waveforms. The active power of the system is kept at –0.512 pu (the minus sign indicates that the power is delivered from the inverter to the grid). The reactive power to the grid Q_g follows its reference Q_g^* (not shown) and ramps up at $t = 0.5$ sec to its maximum positive value of +0.859 pu (lagging). It ramps down at $t = 1.5$ sec to its maximum negative value of –0.859 pu (leading). The active power to the grid P_g is kept constant during the adjustment of the reactive power.

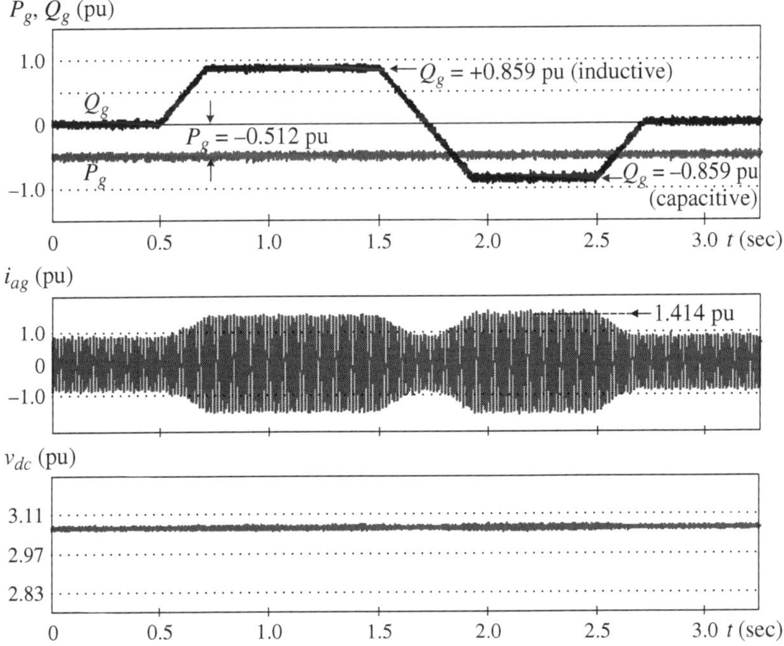

Figure 9-20. Decoupled active and reactive power control at $\omega_r = 0.8$ pu.

The waveform of the phase-*a* grid current i_{ag} varies with the reactive power. Its amplitude is $\sqrt{2}$ pu (rated) when $P_g = -0.512$ pu and $Q_g = \pm0.859$ pu, $[S_g = 1.0$ pu (rated)]. The DC voltage v_{dc} is closely controlled by the inverter, and does not vary with the change in the reactive power.

Figure 9-21 shows the expanded steady-state waveforms of the grid current and voltage around $t = 1.0$ sec ($Q_g = +0.859$ pu) and around $t = 2.0$ sec ($Q_g = -0.859$ pu). Lagging or leading power factor operations are shown in Figure 9-21, where the grid current lags the voltage by 120.8° or leads the voltage by the same angle (Figure 9-21b). The inverter operates under the rated conditions when both current and voltage

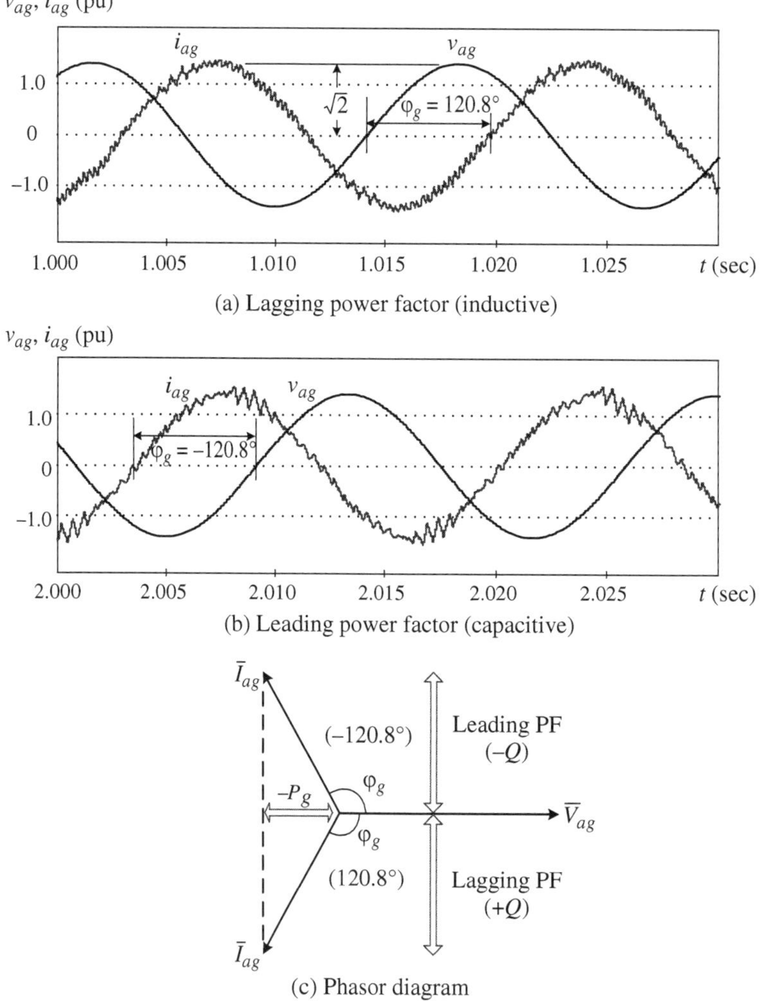

Figure 9-21. Grid-side waveforms and phasor diagram at $\omega_r = 0.8$ pu.

are at 1 pu, rms. The grid-side active, reactive and apparent power can be calculated by

$$\left\{\begin{array}{l} P_g = V_{ag}I_{ga}\cos\varphi_g = -0.512\text{ pu} \quad (-1.024\text{ MW}) \\ Q_g = V_{ag}I_{ag}\sin\varphi_g = \pm 0.859\text{ pu} \quad (\pm 1.718\text{ MVAR}) \\ S_g = \sqrt{\left(Q_g\right)^2 + \left(P_g\right)^2} = 1.0\text{ pu} \quad (2\text{ MVA}) \end{array}\right. \qquad (9.54)$$

The corresponding phasor diagram of the system is shown in Figure 9-21c.

The maximum reactive power $Q_{g,\max}$ that can be injected to the grid is limited by the power rating of the inverter. Figure 9-22 illustrates the relationship between $Q_{g,\max}$ and the speed of the generator ω_r under the MPPT scheme. In general, $Q_{g,\max}$ decreases with the increase of ω_r. When the generator operates at 0.5 pu speed with the rated grid voltage at Point A of Figure 9-22a, the active power delivered to the grid is 0.125 pu and the maximum capacity for the reactive power is 0.992 pu. At $\omega_r = 0.8$ pu, the maximum reactive power to the grid is 0.859 pu. At Point C, the inverter delivers its rated active power to the grid and, therefore, does not have extra capacity to produce reactive power.

The maximum capacity of the reactive power decreases with the reduction of the grid voltage. Figure 9-22a shows such a case, where the grid voltage V_g is reduced to 0.8 pu. Since the active power of the system is kept constant by the MPPT scheme, the reduction of the grid voltage causes an increase in the active component of the grid current, and reduces the reactive power capacity.

To increase the reactive power capacity of the wind energy system, the power converters can be oversized. Figure 9-22b shows the relationship between $Q_{g,\max}$ and ω_r when the inverter current rating is increased by 20% to satisfy higher reactive power requirements for the system.

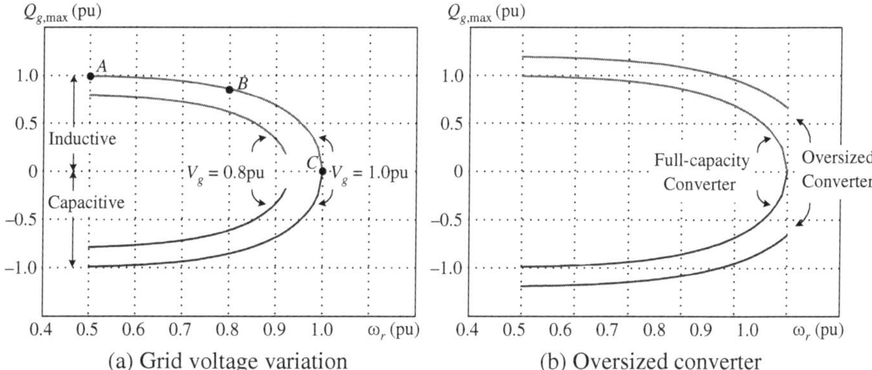

Figure 9-22. Effect of grid voltage variation and converter rating on $Q_{g,\max}$.

9.7   CURRENT SOURCE CONVERTER BASED SG WIND ENERGY SYSTEMS

In this section, two control schemes for SG wind energy systems using current source converter technology are presented and analyzed.

9.7.1   CSC Wind Energy Systems with Delay Angle Control

As discussed in Section 7.5 of Chapter 7, a current source converter can be controlled by delay angle, modulation index, or both. Figure 9-23 shows a control scheme of the CSC SG wind turbine system with delay angle control only. This system is different from that presented in Figure 7-21 (Chapter 7) in the following aspects:

- A PMSG is employed instead of an SCIG.
- The ZDC scheme is implemented for the PMSG (Figure 7-21 included a FOC scheme for the SCIG).
- The rotor flux angle θ_r is measured directly by a digital encoder instead of flux angle θ_f being obtained by measured stator voltages and currents in the SCIG WECS.

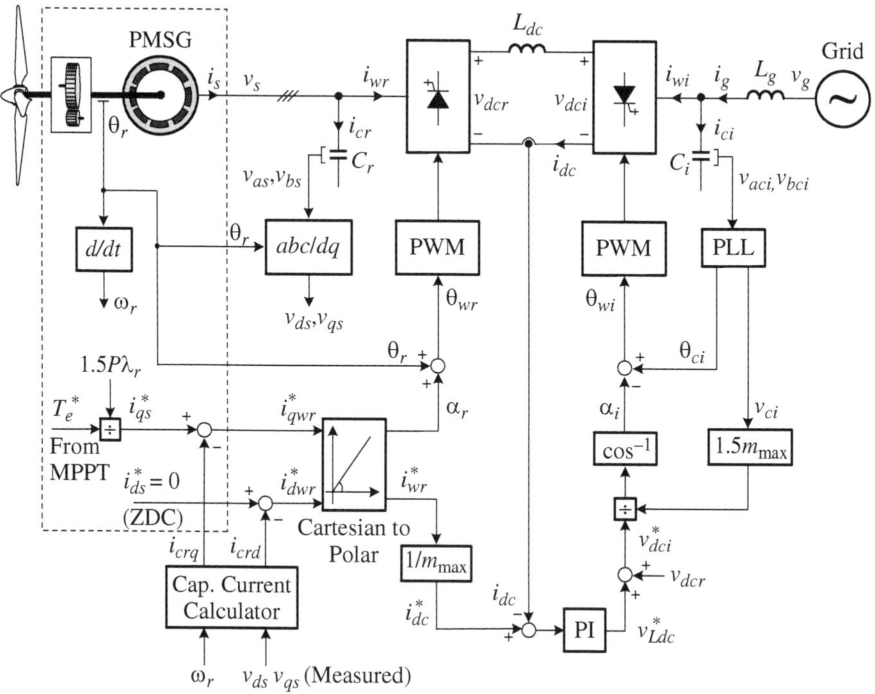

Figure 9-23. CSC SG wind energy system with delay angle control.

- The reference direction for the stator current flows out of the synchronous generator (generator convention) compared with the SCIG stator current with its reference direction flowing into the machine (motor convention).
- There is no need to detect or control the rotor flux λ_r for the PMSG.

The major differences are indicated by a dashed box in Figure 9-23.

Case Study 9-9—Steady-State Analysis of a CSC SG Wind Energy System. The purpose of the case study is to analyze the steady-state performance of a CSC PMSG wind energy system controlled by the ZDC scheme. Consider the wind energy system of Figure 9-23 with a 2.45 MW, 4000 V, 400 rpm nonsalient pole PMSG operating at the rotor speed of 0.8 pu. The parameters of the generator are given in Table B-10 in Appendix B. Figure 9-24 shows the simulated steady-state wave-

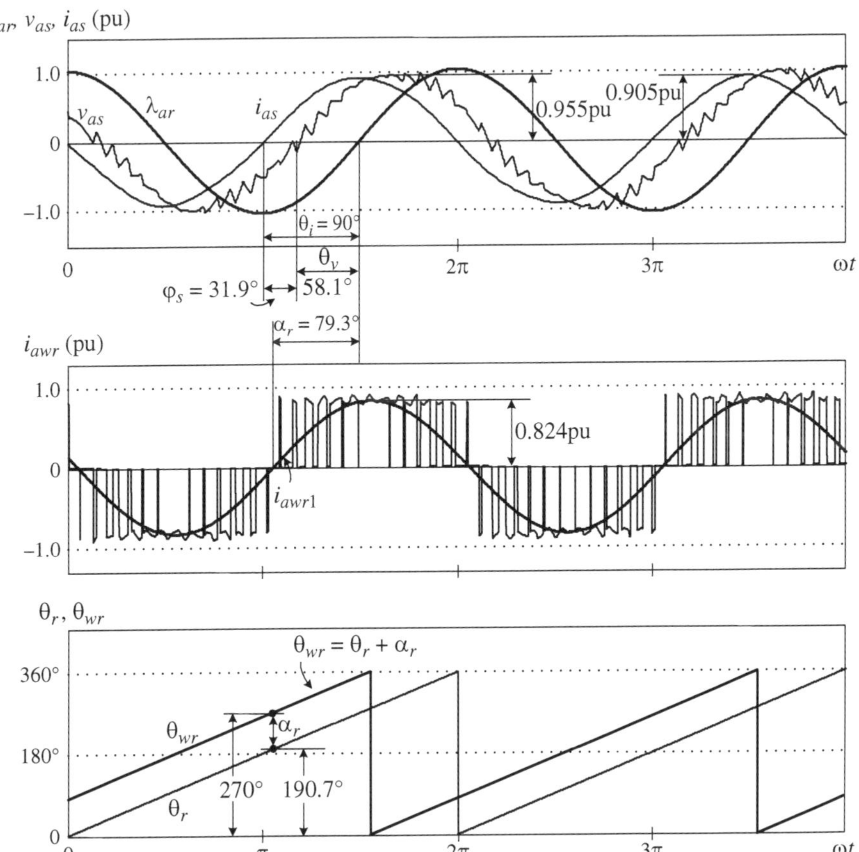

Figure 9-24. Waveforms of the CSC PMSG wind energy system with delay angle control.

forms for phase-*a* stator current i_{as}, stator voltage v_{as}, rotor flux λ_{as}, and rectifier PWM current i_{awr}.

The waveforms of the rotor flux angle θ_r, and the delay angle for the PWM current θ_{wr} are also illustrated. The definitions for the rotor flux angle θ_r and the angle of the rectifier PWM current vector θ_{wr} are given in the space vector diagram of Figure 9-25 together with the stator power factor angle φ_s and the stator voltage angle θ_v. Note that the rotor flux vector $\vec{\lambda}_r$ and rectifier PWM current vector $\vec{i}_{wr}$ rotate in space at the synchronous speed of ω_r and their relevant angles are referenced to the *a*-axis of the stator frame. Therefore, θ_r and θ_{wr}, vary periodically from zero to 2π with time. The stator power factor angle φ_s and the stator voltage angle θ_v are constant since the vectors $\vec{v}_s$ and $\vec{i}_s$ and the *d*-axis of the synchronous frame all rotate at the same speed in space when the system operates in steady state.

At the rotor speed of 0.8 pu, the electrical rotor speed ω_r and mechanical torque T_m of the generator are

$$\begin{cases} \omega_r = 2\pi f_{s,R} \times 0.8 = 2\pi \times 53.3 \times 0.8 = 267.92 \text{ rad/sec} \\ T_m = T_{m,R} \times 0.8 = 37.413 \text{ kN·m} \end{cases} \tag{9.55}$$

The stator currents of the generator with ZDC control are

$$\begin{cases} i_{ds} = 0 \quad \text{for ZDC} \\ i_{qs} = \dfrac{T_e}{1.5 \times P \times \lambda_r} = \dfrac{37.413 \times 10^3}{1.5 \times 8 \times 7.03} = 443.5 \text{ A} \\ i_s = \sqrt{i_{ds}^2 + i_{qs}^2} = 443.5 \text{ A} \quad (0.905 \text{ pu}) \end{cases} \tag{9.56}$$

where $\lambda_r = 7.03$ Wb, and $T_e = T_m = 37.413$ kN·m.

The *dq*-axis stator voltages are

$$\begin{cases} v_{ds} = -i_{ds} R_s + \omega_r L_q i_{qs} = 1166.3 \text{ V} \\ v_{qs} = -i_{qs} R_s - \omega_r L_d i_{ds} + \omega_r \lambda_r = 1872.7 \text{ V} \\ v_s = \sqrt{v_{ds}^2 + v_{qs}^2} = 2206.2 \text{ V} \quad (0.955 \text{ pu}) \end{cases} \tag{9.57}$$

The angles of the stator voltage, current, and power factor are

$$\begin{cases} \theta_v = \tan^{-1} \dfrac{v_{qs}}{v_{ds}} = 1.014 \text{ rad} \quad (58.1°) \\ \theta_i = \tan^{-1} \dfrac{i_{qs}}{i_{ds}} = \pi/2 \text{ rad} \quad (90.0°) \\ \phi_s = \theta_v - \theta_i = -0.557 \text{ rad} \quad (-31.9°) \end{cases} \tag{9.58}$$

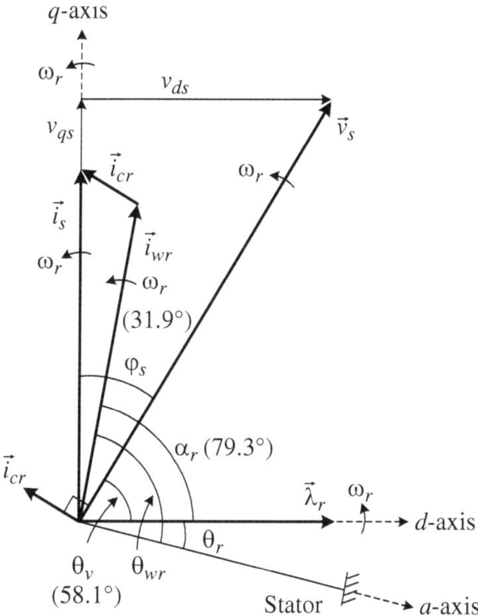

Figure 9-25. Space vector diagram for CSC PMSG wind energy system.

The stator power factor is

$$PF_s = \cos\varphi_s = \cos(-31.9°) = 0.849 \quad \text{(leading)} \tag{9.59}$$

From which the stator (output) active power can be calculated:

$$P_s = 3V_s I_s \cos\varphi_s = 1246.0 \text{ kW} \tag{9.60}$$

To calculate the capacitor current and rectifier PWM currents, the steady-state equivalent circuit of Figure 9-26 can be used. The *dq*-axis currents in the filter capacitor C_r are

$$\begin{cases} i_{crd} = -\omega_r v_{qs} C_r = -267.92 \times 1872.71 \times 149.0 \times 10^{-6} = -74.76 \text{ A} \\ i_{crq} = \omega_r v_{ds} C_r = 267.92 \times 1166.33 \times 149.0 \times 10^{-6} = 46.56 \text{ A} \end{cases} \tag{9.61}$$

where C_r is equal to 149 μF (0.233 pu). The capacitor is selected such that the LC resonant frequency is around 0.25 pu, which is sufficiently low to avoid possible LC resonances that may be excited by the switching harmonics in the system. It should be noted that the subscripts *crd* and *crq* in Equation (9.61) are used instead of *dcr* and *qcr* to avoid confusion with the subscript *dc* for the DC link variables.

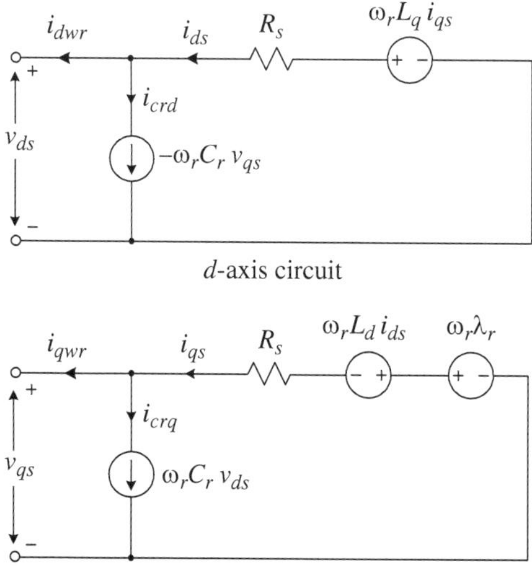

Figure 9-26. Steady-state equivalent circuit for CSC SG wind energy system.

The rectifier input currents can be calculated by

$$\begin{cases} i_{dwr} = i_{ds} - i_{crd} = 0 - (-74.76) = 74.76 \text{ A} \\ i_{qwr} = i_{qs} - i_{crq} = 443.5 - 46.56 = 396.9 \text{ A} \\ i_{wr} = \sqrt{i_{dwr}^2 + i_{qwr}^2} = 403.9 \text{ A} \ (0.824 \text{ pu}) \end{cases} \tag{9.62}$$

from which the rectifier delay angle is

$$\alpha_r = \tan^{-1} \frac{i_{qwr}}{i_{dwr}} = \tan^{-1} \frac{396.9}{74.76} = 79.3° \tag{9.63}$$

The calculated peak values for the stator current is (0.905 pu), stator voltage v_s (0.955 pu), and fundamental rectifier PWM current i_{wr} (0.824 pu), as well as angles, including θ_v (58.1°), φ_s (−31.9°), and α_r (79.3°), correlate well with simulated results in Figure 9-24. The operation and control of the current source inverter for the PMSG WECS remain the same as that discussed in Chapter 7 and, therefore, are not repeated here.

9.7.2   CSC Wind Energy System with Reactive Power Control

To control the reactive power of the CSC wind energy system in addition to the active power and DC current control, both delay angle and modulation index control can be used for the current source rectifier and inverter. Figure 9-27 shows a block diagram of such a system, in which the active power control is achieved by the rectifier through

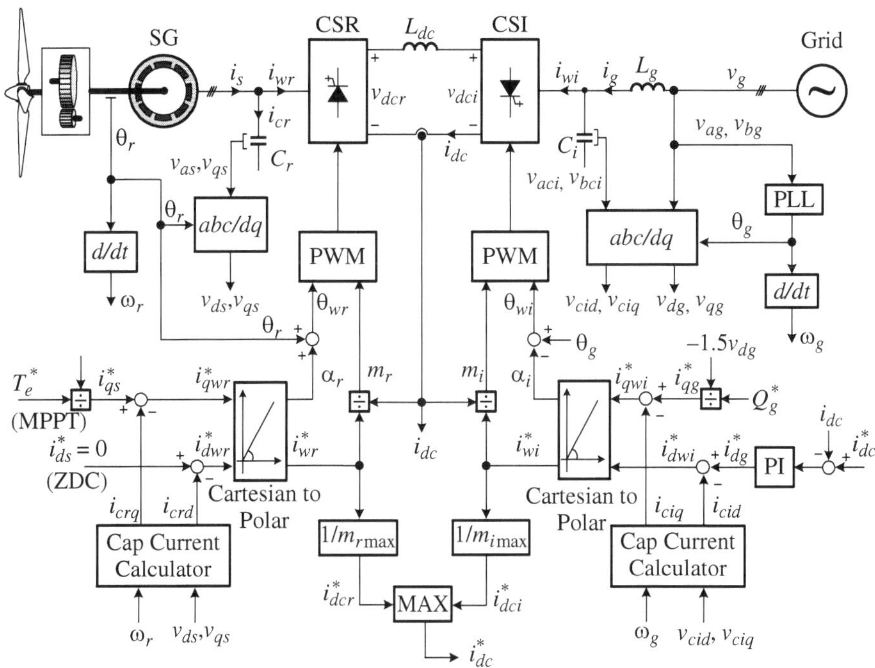

Figure 9-27. CSC SG wind energy system with active and reactive power control.

generator torque regulation, whereas the DC current and the reactive power control are realized by the inverter [9].

The control of the rectifier and generator is the same as that given in Figure 9-23 except that a modulation index control is added. The rectifier modulation index m_r can be calculated by dividing the rectifier current reference i_{wr}^* by the measured DC current i_{dc}.

For the inverter control, the three-phase grid and capacitor voltages are measured. The measured grid voltages in the stationary frame are transformed into dq-axis variables in the grid voltage reference frame using the grid voltage angle θ_g and the *abc/dq* transformation block. The reference for the q-axis (reactive) grid current i_{qg}^* is obtained according to the reactive power reference Q_g^*. The d-axis (active) grid current reference i_{dg}^* is generated by the DC current PI controller.

The references for the dq-axis inverter PWM currents, i_{dwi}^* and i_{qwi}^*, can be found from

$$\begin{cases} i_{dwi}^* = i_{dg}^* - i_{dpi}^* = i_{dg}^* + \omega_g C_i v_{ciq} \\ i_{qwi}^* = i_{qg}^* - i_{qpi}^* = i_{qg}^* - \omega_g C_i v_{cid} \end{cases} \tag{9.64}$$

where ω_g is the grid frequency, and i_{cid} and i_{ciq} are the dq-axis capacitor currents obtained from its steady-state model. Based on i_{dwi}^* and i_{qwi}^*, the amplitude of the inverter PWM current i_{wi}^* and the inverter delay angle α_i can be calculated by the Cartesian-to-polar transformation:

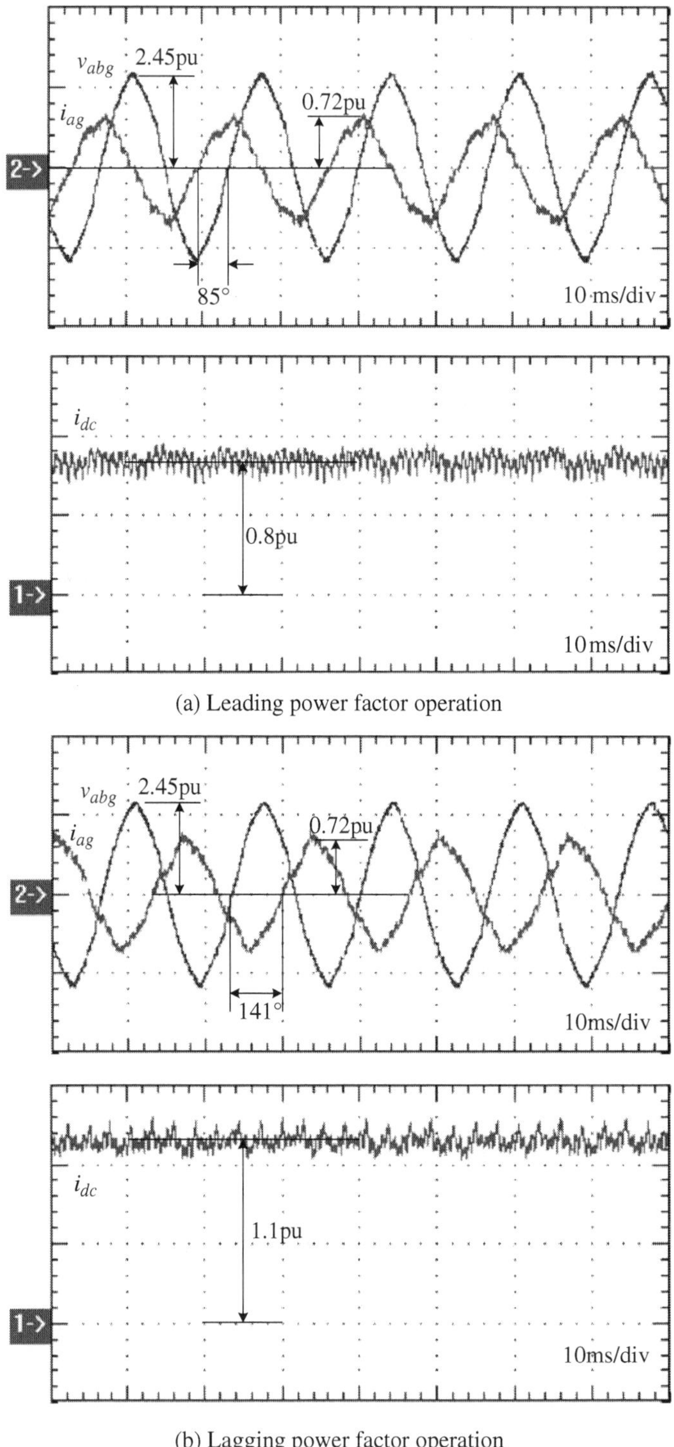

(a) Leading power factor operation

(b) Lagging power factor operation

<u>Figure 9-28.</u> Experimental results obtained from a laboratory CSC PMSG wind energy system.

$$\begin{cases} i_{wi}^* = \sqrt{\left(i_{dwi}^*\right)^2 + \left(i_{qwi}^*\right)^2} \\ \alpha_i = \tan^{-1}\left(i_{qwi}^* \big/ i_{dwi}^*\right) \end{cases} \qquad (9.65)$$

The inverter modulation index m_i is calculated by dividing the PWM current reference i_{wi}^* by the measured DC current i_{dc}. It is fed to the inverter PWM generation block together with θ_{wi}, where θ_{wi} is the angle of the inverter PWM current. The CSC SVM scheme presented in Chapter 4 can be used for the inverter control.

To determine the DC current reference i_{dc}^*, the rectifier- and inverter-side DC current references, i_{dcr}^* and i_{dci}^*, should be considered. The rectifier-side reference i_{dcr}^* is derived by dividing the rectifier PWM current reference i_{wr}^* by its maximum modulation index $m_{r\max}$. The inverter-side reference i_{dci}^* is calculated by dividing the inverter current reference i_{wi}^* by its maximum modulation index $m_{i\max}$. The use of maximum modulation indices leads to minimum DC current and can reduce the overall power converter loss. The two calculated DC current references (i_{dcr}^* and i_{dci}^*) are usually different. The larger value is selected as the DC current reference i_{dc}^* by the MAX block to ensure the proper operation of the system.

Figure 9-28 shows experimental results obtained from a small laboratory CSC PMSG wind energy system, where v_{abg} and i_{ag} are the line-to-line grid voltage and phase-a grid current, respectively. The experimental system was developed according to the control scheme in Figure 9-27. The system operated at the rotor speed of 0.7 pu with a leading and lagging power factor.

9.8 SUMMARY

This chapter provided a comprehensive analysis of various configurations and control schemes for synchronous generator based wind energy systems. Three control schemes for the salient and nonsalient synchronous generators (SG) were analyzed, including zero d-axis current (ZDC) control, maximum torque per ampere (MTPA) control, and unity power factor (UPF) control. The performance of these SG control schemes were analyzed and compared.

Advanced control schemes for SG wind energy systems using voltage- and current source converters were discussed. Three main control tasks, generator-side active power control, grid-side reactive power control, and DC voltage/current control, were analyzed in detail. The operating principle and important concepts were illustrated by case studies.

REFERENCES

1. H. Polinder, F. F. A. Van der Pijl, G.-J. de Vilder, and P. J. Tavner, Comparison of Direct-Drive and Geared Generator Concepts for Wind Turbines, *IEEE Transactions on Energy Conversion, V*ol. 21, No. 3, 725–733, 2006.

2. M. Chinchilla, S. Arnaltes, and J. Burgos, Control of Permanent-Magnet Generators Applied to Variable-Speed Wind-Energy Systems Connected to the Grid, *IEEE Transactions on Energy Conversion,* Vol. 21, No. 1, 130–135, 2006.

3. A.D. Hansen and G. Michalke, Multi-pole Permanent Magnet Synchronous Generator Wind Turbines' Grid Support Capability in Uninterrupted Operation During Grid Faults, *IET Renewable Power Generation,* Vol. 3, No. 3, 333–348, 2009.

4. H. Li, Z. Chen, and H. Polinder, Optimization of Multibrid Permanent-Magnet Wind Generator Systems, *IEEE Transactions on Energy Conversion,* Vol. 24, No. 1, 82–92, 2009.

5. B. K. Bose, *Power Electronics and Motor Drives: Advances and Trends,* Academic Press, 2006.

6. M. P. Kazmierkowski, R. Krishnan, and F. Blaabjerg, *Control in Power Electronics: Selected Problems,* Academic Press, 2002.

7. Y. Abdel-Rady, I. Mohamed, and Tsing K. Lee, Adaptive Self-Tuning MTPA Vector Controller for IPMSM Drive System, *IEEE Transactions on Energy Conversion,* Vol. 21, No. 3, 636–644, 2006.

8. K. Clark, N. M. Miller, and J. J. Sanchez-Gasca, *Modeling of GE Wind Turbine-Generators for Grid Studies,* General Electric International Inc., 2009.

9. Y. Lang, B. Wu, and N. Zargari, A Novel Reactive Power Control Scheme for CSC Based PMSG Wind Energy System, in *Proceedings of 2008 IEEE Industry Applications Society Annual Meeting (IAS),* pp. 1–6, 2008.

APPENDIX A

PER-UNIT SYSTEM

Table A-1 provides definitions of base values for the per-unit system, in which the rated values of the wind generator or power converter are selected as the base values. The per-unit system in this book is mainly used to present simulation and calculation results. In other words, all computer simulations and numerical calculations are conducted with SI units, but presented or displayed in per-unit terms for the convenience of analysis and discussion.

Table A-1. Definitions of base values for the per-unit system

Base Apparent Power	$S_B = S_R$ (VA)	S_R—Rated apparent power of the generator or power converter
Base Active Power	$P_B = P_{m,R}$ (W)	$P_{m,R}$—Rated mechanical power of the generator. Refer to Figure A-1 for the definition of mechanical power P_m for induction generators.
Base Voltage	$V_B = V_R$ (V)	V_R—Rated phase voltage of the generator or power converter
Base Current	$I_B = I_R$ (A)	I_R—Rated phase current of the generator or power converter

(*continued*)

Table A-1. Continued

Base Frequency	$\omega_B = 2\pi f_R$ (rad/s)	f_R—Rated stator frequency of the generator or nominal frequency of the electric grid
Base Rotor Speed	$\omega_{m,B} = \omega_{m,R}$ (rad/s) $\omega_{r,B} = \omega_{r,R}$ (rad/s)	$\omega_{m,R}$—Rated rotor mechanical speed; $\omega_{m,R}$ = (rated rpm) $\times\, 2\pi/60$ $\omega_{r,R}$—Rated rotor electrical speed; $\omega_{r,R} = \omega_{m,R} \times P$
Base Torque	$T_B = T_{m,R}$ (N–m)	$T_{m,R}$—Rated mechanical torque; $T_{m,R} = P_{m,R}/\omega_{m,R}$
Base Flux Linkage	$\Lambda_B = \dfrac{V_R}{\omega_B}$ (Wb)	$\dfrac{V_R}{\omega_B}$—Nominal flux linkage
Base Wind Speed	$v_{w,B} = v_{w,R}$ (m/s)	$v_{w,R}$—Rated wind speed
Base Impedance	$Z_B = \dfrac{V_B}{I_B}$ (Ω)	
Base Inductance	$L_B = \dfrac{Z_B}{\omega_B}$ (H)	
Base Capacitance	$C_B = \dfrac{1}{\omega_B Z_B}$ (F)	

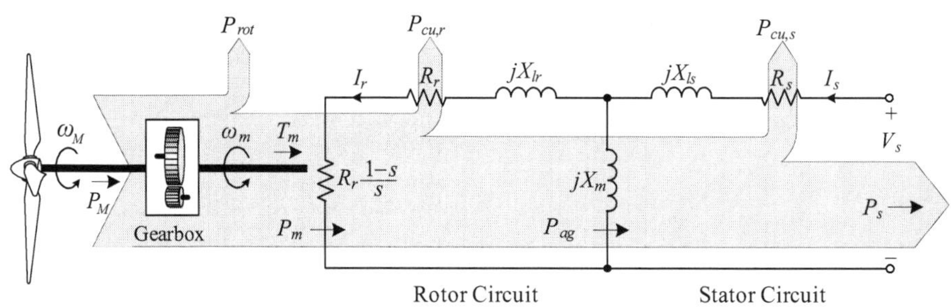

<u>Figure A-1.</u> Power flow of an induction generator in a wind energy system.

APPENDIX B

GENERATOR PARAMETERS

This appendix provides detailed parameters for a number of generators used in case studies in various chapters and Appendix C. It is noted that all the paratmeters listed in this appendix are referred to the stator side.

B.1   SQUIRREL CAGE INDUCTION GENERATORS

Table B-1. 2.3 MW, 690 V, 50 Hz squirrel cage induction generator (SCIG) parameters

Generator Type	SCIG, 2.3 MW, 690 V, 50 Hz	
Rated Output Power	2.30 MW	
Rated Mechanical Power	2.3339 MW	1.0 pu
Rated Apparent Power	2.59 MVA	1.0 pu
Rated Line-to-line Voltage	690 V (rms)	
Rated Phase Voltage	398.4 V (rms)	1.0 pu
Rated Stator Current	2168 A (rms)	1.0 pu
Rated Stator Frequency	50 Hz	1.0 pu
Rated Power Factor	0.888	
Rated Rotor Speed	1512 rpm	1.0 pu
Rated Slip	−0.008	
Number of Pole Pairs	2	
Rated Mechanical Torque	14.74 kN·m	1.0 pu

(*continued*)

Power Conversion and Control of Wind Energy Systems. By B. Wu, Y. Lang, N. Zargari, and S. Kouro   **319**
© 2011 the Institute of Electrical and Electronics Engineers, Inc. Published 2011 by John Wiley & Sons, Inc.

Table B-1. *Continued*

Generator Type	SCIG, 2.3 MW, 690 V, 50 Hz	
Rated Stator Flux Linkage	1.2748 Wb (rms)	1.0053 pu
Rated Rotor Flux Linkage	1.2096 Wb (rms)	0.9539 pu
Stator Winding Resistance, R_s	1.102 mΩ	0.006 pu
Rotor Winding Resistance, R_r	1.497 mΩ	0.008 pu
Stator Leakage Inductance, L_{ls}	0.06492 mH	0.111 pu
Rotor Leakage Inductance, L_{lr}	0.06492 mH	0.111 pu
Magnetizing Inductance, L_m	2.13461 mH	3.6481 pu
Moment of Inertia, J	1200 kg·m^2	
Inertia Time Constant, H	5.8078 sec	
Base Flux Linkage, Λ_B	1.2681 Wb (rms)	1.0 pu
Base Impedance, Z_B	0.1838 Ω	1.0 pu
Base Inductance, L_B	0.58513 mH	1.0 pu
Base Capacitance, C_B	17316.17 µF	1.0 pu

Note: $H = J(\omega_m)^2/(2S_B)$

Table B-2. 1.45 MW, 575 V, 50 Hz SCIG parameters

Generator Type	SCIG, 1.45 MW, 575 V, 50 Hz	
Rated Output Power	1.45 MW	
Rated Mechanical Power	1.4707 MW	1.0 pu
Rated Apparent Power	1.72 MVA	1.0 pu
Rated Line-to-line Voltage	575 V (rms)	
Rated Phase Voltage	331.98 V (rms)	1.0 pu
Rated Stator Current	1723 A (rms)	1.0 pu
Rated Stator Frequency	50 Hz	1.0 pu
Rated Power Factor	0.8439	
Rated Rotor Speed	1007.2 rpm	1.0 pu
Rated Slip	−0.0072	
Number of Pole Pairs	3	
Rated Mechanical Torque	13.944 kN·m	1.0 pu
Rated Stator Flux Linkage	1.063 Wb (rms)	1.006 pu
Rated Rotor Flux Linkage	0.9757 Wb (rms)	0.9233 pu
Stator Winding Resistance, R_s	1.354 mΩ	0.007 pu
Rotor Winding Resistance, R_r	1.39 mΩ	0.0072 pu
Stator Leakage Inductance, L_{ls}	0.1044 mH	0.1706 pu
Rotor Leakage Inductance, L_{lr}	0.0498 mH	0.0814 pu
Magnetizing Inductance, L_m	1.77016 mH	2.8931 pu
Base Flux Linkage, Λ_B	1.0567 Wb (rms)	1.0 pu
Base Impedance, Z_B	0.1922 Ω	1.0 pu
Base Inductance, L_B	0.61187 mH	1.0 pu
Base Capacitance, C_B	16599.34 µF	1.0 pu

Table B-3. 3 MW, 3000 V, 60 Hz SCIG parameters

Generator Type	SCIG, 3.0 MW, 3000 V, 60 Hz	
Rated Output Power	3.0 MW	
Rated Mechanical Power	3.0402 MW	1.0 pu
Rated Apparent Power	3.36 MVA	1.0 pu
Rated Line-to-line Voltage	3000 V (rms)	
Rated Phase Voltage	1732.1 V (rms)	1.0 pu
Rated Stator Current	646.1 A (rms)	1.0 pu
Rated Stator Frequency	60 Hz	1.0 pu
Rated Power Factor	0.8934	
Rated Rotor Speed	1812 rpm	1.0 pu
Rated Slip	−0.0067	
Number of Pole Pairs	2	
Rated Mechanical Torque	16.022 kN·m	1.0 pu
Rated Stator Flux Linkage	4.6199 Wb (rms)	1.0056 pu
Rated Rotor Flux Linkage	4.3916 Wb (rms)	0.9559 pu
Stator Winding Resistance, R_s	16.623 mΩ	0.0062 pu
Rotor Winding Resistance, R_r	18.152 mΩ	0.0068 pu
Stator Leakage Inductance, L_{ls}	0.782 mH	0.1101 pu
Rotor Leakage Inductance, L_{lr}	0.782 mH	0.1101 pu
Magnetizing Inductance, L_m	27.168 mH	3.8237 pu
Base Flux Linkage, Λ_B	4.5944 Wb (rms)	1.0 pu
Base Impedance, Z_B	2.6786 Ω	1.0 pu
Base Inductance, L_B	7.105 mH	1.0 pu
Base Capacitance, C_B	990.3 μF	1.0 pu

Table B-4. 4 MW, 4000 V, 50 Hz SCIG parameters

Generator Type	SCIG, 4.0 MW, 4000 V, 50 Hz	
Rated Output Power	4.0 MW	
Rated Mechanical Power	4.0606 MW	1.0 pu
Rated Apparent Power	4.842 MVA	1.0 pu
Rated Line-to-line Voltage	4000 V (rms)	
Rated Phase Voltage	2309.4 V (rms)	1.0 pu
Rated Stator Current	698.88 A (rms)	1.0 pu
Rated Stator Frequency	50 Hz	1.0 pu
Rated Power Factor	0.8261	
Rated Rotor Speed	1510.5 rpm	1.0 pu
Rated Slip	−0.007	
Number of Pole Pairs	2	
Rated Mechanical Torque	25.671 kN·m	1.0 pu
Rated Stator Flux Linkage	7.3917 Wb (rms)	1.0055 pu
Rated Rotor Flux Linkage	6.7114 Wb (rms)	0.913 pu
Stator Winding Resistance, R_s	22.104 mΩ	0.0067 pu
Rotor Winding Resistance, R_r	23.1515 mΩ	0.007 pu

(continued)

Table B-4. *Continued*

Generator Type	SCIG, 4.0 MW, 4000 V, 50 Hz	
Stator Leakage Inductance, L_{ls}	1.698 mH	0.1614 pu
Rotor Leakage Inductance, L_{lr}	1.698 mH	0.1614 pu
Magnetizing Inductance, L_m	33.597 mH	3.1942 pu
Base Flux Linkage, Λ_B	7.3511 Wb (rms)	1.0 pu
Base Impedance, Z_B	3.3044 Ω	1.0 pu
Base Inductance, L_B	10.518 mH	1.0 pu
Base Capacitance, C_B	963.29 μF	1.0 pu

B.2 DOUBLY FED INDUCTION GENERATORS

Table B-5. 1.5 MW, 690 V, 50 Hz doubly fed induction generator (DFIG) parameters

Generator Type	DFIG, 1.5 MW, 690 V, 50 Hz	
Rated Mechanical Power	1.5 MW	1.0 pu
Rated Stator Line-to-line Voltage	690 V (rms)	
Rated Stator Phase Voltage	398.4 V (rms)	1.0 pu
Rated Rotor Phase Voltage	67.97 V (rms)	0.1706 pu
Rated Stator Current	1068.2 A (rms)	0.8511 pu
Rated Rotor Current	1125.6 A (rms)	0.8968 pu
Rated Stator Frequency	50 Hz	1.0 pu
Rated Rotor Speed	1750 rpm	1.0 pu
Nominal Rotor Speed Range	1200–1750 rpm	0.686–1.0 pu
Rated Slip	−0.1667	
Number of Pole Pairs	2	
Rated Mechanical Torque	8.185 kN·m	1.0 pu
Stator Winding Resistance, R_s	2.65 mΩ	0.0084 pu
Rotor Winding Resistance, R_r	2.63 mΩ	0.0083 pu
Stator Leakage Inductance, L_{ls}	0.1687 mH	0.167 pu
Rotor Leakage Inductance, L_{lr}	0.1337 mH	0.1323 pu
Magnetizing Inductance, L_m	5.4749 mH	5.419 pu
Base Current, $I_B = 1.5$ MW/$(\sqrt{3} \times 690$ V$)$	1255.1 A (rms)	1.0 pu
Base Flux Linkage, Λ_B	1.2681 Wb (rms)	1.0 pu
Base Impedance, Z_B	0.3174 Ω	1.0 pu
Base Inductance, L_B	1.0103 mH	1.0 pu
Base Capacitance, C_B	10028.7 μF	1.0 pu

Table B-6. 1 MW, 575 V, 60 Hz DFIG parameters

Generator Type	DFIG, 1.0 MW, 575 V, 60 Hz	
Rated Mechanical Power	1.0 MW	1.0 pu
Rated Stator Line-to-line Voltage	575 V (rms)	
Rated Stator Phase Voltage	331.98 V (rms)	1.0 pu
Rated Rotor Phase Voltage	67.97 V (rms)	0.2047 pu

Table B-6. *Continued*

Generator Type	DFIG, 1.0 MW, 575 V, 60 Hz	
Rated Stator Current	829.2 A (rms)	0.8258 pu
Rated Rotor Current	882.2 A (rms)	0.8786 pu
Rated Stator Frequency	60 Hz	1.0 pu
Rated Rotor Speed	2160 rpm	1.0 pu
Nominal Rotor Speed Range	1350–2160 rpm	0.625–1.0 pu
Rated Slip	−0.2	
Number of Pole Pairs	2	
Rated Mechanical Torque	4.421 kN·m	1.0 pu
Stator Winding Resistance, R_s	3.654 mΩ	0.0111 pu
Rotor Winding Resistance, R_r	3.569 mΩ	0.0108 pu
Stator Leakage Inductance, L_{ls}	0.1304 mH	0.1487 pu
Rotor Leakage Inductance, L_{lr}	0.1198 mH	0.1366 pu
Magnetizing Inductance, L_m	4.12 mH	4.6978 pu
Base Current, $I_B = 1\ \text{MW}/(\sqrt{3} \times 575\ \text{V})$	1004.1 A (rms)	1.0 pu
Base Flux Linkage, Λ_B	0.8806 Wb (rms)	1.0 pu
Base Impedance, Z_B	0.3306 Ω	1.0 pu
Base Inductance, L_B	0.877 mH	1.0 pu
Base Capacitance, C_B	8022.93 μF	1.0 pu

Table B-7. 5 MW, 950 V, 50 Hz DFIG parameters

Generator Type	DFIG, 5.0 MW, 950 V, 50 Hz	
Rated Mechanical Power	5.0 MW	1.0 pu
Rated Stator Line-to-line Voltage	950 V (rms)	
Rated Stator Phase Voltage	548.48 V (rms)	1.0 pu
Rated Rotor Phase Voltage	381.05 V (rms)	0.6947 pu
Rated Stator Current	2578.4 A (rms)	0.8485 pu
Rated Rotor Current	3188.7 A (rms)	1.0494 pu
Rated Stator Frequency	50 Hz	1.0 pu
Rated Rotor Speed	1170 rpm	1.0 pu
Nominal Rotor Speed Range	670–1170 rpm	0.573–1.0 pu
Rated Slip	−0.17	
Number of Pole Pairs	3	
Rated Mechanical Torque	40.809 kN·m	1.0 pu
Stator Winding Resistance, R_s	1.552 mΩ	0.0086 pu
Rotor Winding Resistance, R_r	1.446 mΩ	0.008 pu
Stator Leakage Inductance, L_{ls}	1.2721 mH	2.2141 pu
Rotor Leakage Inductance, L_{lr}	1.1194 mH	1.9483 pu
Magnetizing Inductance, L_m	5.5182 mH	9.6044 pu
Base Current, $I_B = 5\ \text{MW}/(\sqrt{3} \times 950\ \text{V})$	3038.7 A (rms)	1.0 pu
Base Flux Linkage, Λ_B	1.7459 Wb (rms)	1.0 pu
Base Impedance, Z_B	0.1805 Ω	1.0 pu
Base Inductance, L_B	0.5746 mH	1.0 pu
Base Capacitance, C_B	17634.9 μF	1.0 pu

Table B-8. 6 MW, 4000 V, 50 Hz DFIG parameters

Generator Type	DFIG, 6.0 MW, 4000 V, 50 Hz	
Rated Mechanical Power	6.0 MW	1.0 pu
Rated Stator Line-to-line Voltage	4000 V (rms)	
Rated Stator Phase Voltage	2309.4 V (rms)	1.0 pu
Rated Rotor Phase Voltage	381.05 V (rms)	0.165 pu
Rated Stator Current	733.9 A (rms)	0.8474pu
Rated Rotor Current	793.9 A (rms)	0.9167 pu
Rated Stator Frequency	50 Hz	1.0 pu
Rated Rotor Speed	1170rpm	1.0 pu
Nominal Rotor Speed Range	800–1170 rpm	0.684–1.0 pu
Rated Slip	−0.17	
Number of Pole Pairs	3	
Rated Mechanical Torque	48.971 kN·m	1.0 pu
Stator Winding Resistance, R_s	26.86 mΩ	0.0101 pu
Rotor Winding Resistance, R_r	25.74 mΩ	0.0097 pu
Stator Leakage Inductance, L_{ls}	0.23142 mH	0.0273 pu
Rotor Leakage Inductance, L_{lr}	0.2183 mH	0.0257 pu
Magnetizing Inductance, L_m	25.908 mH	3.0522 pu
Base Current, $I_B = 6$ MW/$(\sqrt{3} \times 4000$ V$)$	866.03 A (rms)	1.0 pu
Base Flux Linkage, Λ_B	7.3511 Wb (rms)	1.0 pu
Base Impedance, Z_B	2.667 Ω	1.0 pu
Base Inductance, L_B	8.488 mH	1.0 pu
Base Capacitance, C_B	1193.66 μF	1.0 pu

B.3 SYNCHRONOUS GENERATORS

Table B-9. 2 MW, 690 V, 9.75 Hz nonsalient pole permanent magnet synchronous generator (PMSG) parameters

Generator Type	PMSG, 2.0 MW, 690 V, 9.75 Hz, nonsalient pole	
Rated Mechanical Power	2.0 MW	1.0 pu
Rated Apparent Power	2.2419 MVA	1.0 pu
Rated Line-to-line Voltage	690 V (rms)	
Rated Phase Voltage	398.4 V (rms)	1.0 pu
Rated Stator Current	1867.76 A (rms)	1.0 pu
Rated Stator Frequency	9.75 Hz	1.0 pu
Rated Power Factor	0.8921	
Rated Rotor Speed	22.5 rpm	1.0 pu
Number of Pole Pairs	26	
Rated Mechanical Torque	848.826 kN·m	1.0 pu
Rated Rotor Flux Linkage	5.8264 Wb (rms)	0.896 pu
Stator Winding Resistance, R_s	0.821 mΩ	0.00387 pu
d-axis Synchronous Inductance, L_d	1.5731 mH	0.4538 pu
q-axis Synchronous Inductance, L_q	1.5731 mH	0.4538 pu

Table B-9. *Continued*

Generator Type	PMSG, 2.0 MW, 690 V, 9.75 Hz, nonsalient pole	
Base Flux Linkage, Λ_B	6.5029 Wb	1.0 pu
Base Impedance, Z_B	0.2124 Ω	1.0 pu
Base Inductance, L_B	3.4666 mH	1.0 pu
Base Capacitance, C_B	76865.87 μF	1.0 pu

Table B-10. 2.45 MW, 4000 V, 53.33 Hz nonsalient pole PMSG parameters

Generator Type	PMSG, 2.45 MW, 4000 V, 53.33 Hz, nonsalient pole	
Rated Mechanical Power	2.4487 MW	1.0 pu
Rated Apparent Power	3.419 MVA	1.0 pu
Rated Line-to-line Voltage	4000 V (rms)	
Rated Phase Voltage	2309.4 V (rms)	1.0 pu
Rated Stator Current	490 A (rms)	1.0 pu
Rated Stator Frequency	53.33 Hz	1.0 pu
Rated Power Factor	0.7162	
Rated Rotor Speed	400 rpm	1.0 pu
Number of Pole Pairs	8	
Rated Mechanical Torque	58.4585 kN·m	1.0 pu
Rated Rotor Flux Linkage	4.971 Wb (rms)	0.7213 pu
Stator Winding Resistance, R_s	24.21 mΩ	0.00517 pu
d-axis Synchronous Inductance, L_d	9.816 mH	0.7029 pu
q-axis Synchronous Inductance, L_q	9.816 mH	0.7029 pu
Base Flux Linkage, Λ_B	6.892 Wb (rms)	1.0 pu
Base Impedance, Z_B	4.6797 Ω	1.0 pu
Base Inductance, L_B	13.966 mH	1.0 pu
Base Capacitance, C_B	637.72 μF	1.0 pu

Table B-11. 2 MW, 690 V, 11.25 Hz salient pole PMSG parameters

Generator Type	PMSG, 2.0 MW, 690 V, 11.25 Hz, salient pole	
Rated Mechanical Power	2.0093 MW	1.0 pu
Rated Apparent Power	2.2408 MVA	1.0 pu
Rated Line-to-line Voltage	690 V (rms)	
Rated Phase Voltage	398.4 V (rms)	1.0 pu
Rated Stator Current	1867.76 A (rms)	1.0 pu
Rated Stator Frequency	11.25 Hz	1.0 pu
Rated Power Factor	0.8967	
Rated Rotor Speed	22.5 rpm	1.0 pu
Number of Pole Pairs	30	

(continued)

Table B-11. *Continued*

Generator Type	PMSG, 2.0 MW, 690 V, 11.25 Hz, salient pole	
Rated Mechanical Torque	852.77 kN·m	1.0 pu
Rated Rotor Flux Linkage	4.696 Wb (rms)	0.8332 pu
Stator Winding Resistance, R_s	0.73051 mΩ	0.00344 pu
d-axis Synchronous Inductance, L_d	1.21 mH	0.4026 pu
q-axis Synchronous Inductance, L_q	2.31 mH	0.7685 pu
Optimal Angle of Stator Current (with respect to q-axis)	19.738°	
Base Flux Linkage, Λ_B	5.6358 Wb (rms)	1.0 pu
Base Impedance, Z_B	0.2125 Ω	1.0 pu
Base Inductance, L_B	3.006 mH	1.0 pu
Base Capacitance, C_B	66584.41 μF	1.0 pu

Table B-12. 2.5 MW, 4000 V, 40 Hz salient pole PMSG parameters

Generator Type	PMSG, 2.5 MW, 4000 V, 40 Hz, salient pole	
Rated Mechanical Power	2.5 MW	1.0 pu
Rated Apparent Power	3.383 MVA	1.0 pu
Rated Line-to-line Voltage	4000 V (rms)	
Rated Phase Voltage	2309.4 V (rms)	1.0 pu
Rated Stator Current	485 A (rms)	1.0 pu
Rated Stator Frequency	40 Hz	1.0 pu
Rated Power Factor	0.739	
Rated Rotor Speed	400 rpm	1.0 pu
Number of Pole Pairs	6	
Rated Mechanical Torque	59.6831 kN·m	1.0 pu
Rated Rotor Flux Linkage	4.759 Wb (rms)	0.5179 pu
Stator Winding Resistance, R_s	24.25 mΩ	0.00513 pu
d-axis Synchronous Inductance, L_d	8.9995 mH	0.4782 pu
q-axis Synchronous Inductance, L_q	21.8463 mH	1.161 pu
Optimal Angle of Stator Current (with respect to q-axis)	32.784°	
Base Flux Linkage, Λ_B	9.1888 Wb (rms)	1.0 pu
Base Impedance, Z_B	4.7295 Ω	1.0 pu
Base Inductance, L_B	18.818 mH	1.0 pu
Base Capacitance, C_B	841.283 μF	1.0 pu

PROBLEMS AND SOLUTIONS

PROBLEMS AND ANSWERS FOR CHAPTER 3—WIND GENERATORS AND MODELING

Topic: Steady-state and Power Flow Analysis of Induction Generators Using Conventional Equivalent Circuit

3-1 (Solved Problem). A 2.3 MW, 690 V, 50 Hz squirrel cage induction generator (SCIG) is used in a fixed-speed wind energy conversion system (WECS). The SCIG is directly connected to the grid of 690 V/50 Hz. The generator parameters are given in Table B-1 of Appendix B. At a given wind speed, the SCIG operates at the rated speed of 1512 rpm. Using the induction generator equivalent circuit of Figure 3-13b, determine the following:

a) The slip, and rotor mechanical and electrical speeds

b) The stator and rotor currents

c) The mechanical power and torque

d) The stator and rotor winding losses

e) The generator efficiency and power factor

f) The stator and rotor flux linkages

Power Conversion and Control of Wind Energy Systems. By B. Wu, Y. Lang, N. Zargari, and S. Kouro
© 2011 the Institute of Electrical and Electronics Engineers, Inc. Published 2011 by John Wiley & Sons, Inc.

Solution:

a) The slip:

$$s = \frac{n_s - n_r}{n_s} = \frac{1500 - 1512}{1500} = -0.008 \ \ \text{(negative slip)}$$

The rotor mechanical speed:

$$\omega_m = 1512 \times (2\pi/60) = 158.336 \ \text{rad/sec}$$

The rotor electrical speed:

$$\omega_r = \omega_m \times P = 158.336 \times 2 = 316.67 \ \text{rad/sec}$$

The stator frequency:

$$\omega_s = 2\pi \times 50 = 314.16 \ \text{rad/sec}$$

b) The stator voltage:

$$\overline{V}_s = 690/\sqrt{3}\angle 0° = 398.37\angle 0° \ \text{V (rms)}$$

The equivalent impedance of SCIG:

$$\overline{Z}_s = R_s + jX_{ls} + jX_m \ // \left(\frac{R_r}{s} + jX_{lr} \right) = 0.1838\angle 152.6° \ \Omega$$

where $X_{ls} = \omega_s L_{ls} = 0.0204 \ \Omega$, $X_{lr} = \omega_s L_{lr} = 0.0204 \ \Omega$, and $X_m = \omega_s L_m = 0.6706 \ \Omega$
The stator current:

$$\overline{I}_s = \frac{\overline{V}_s}{\overline{Z}_s} = \frac{690/\sqrt{3}\angle 0°}{0.1838\angle 152.6°} = 2168\angle -152.6° \ \text{A (rms)}$$

The rotor current can be calculated by

$$\overline{I}_r = \frac{jX_m \overline{I}_s}{jX_m + \left(\dfrac{R_r}{s} + jX_{lr} \right)} = 2030.8\angle -167.7° \ \text{A (rms)}$$

c) The mechanical power:

$$P_m = \left| 3I_r^2 \frac{R_r}{s}(1-s) \right| = \left| 3 \times 2030.8^2 \times \frac{1.497 \times 10^{-3}}{-0.008}(1+0.008) \right| = 2.3339 \times 10^6 \ \text{W}$$

The mechanical torque:

$$T_m = P_m \times P / \omega_r = 14740 \text{ N·m}$$

d) The stator winding loss:

$$P_{cu,s} = 3 \times I_s^2 R_s = 15.538 \times 10^3 \text{ W}$$

The rotor winding loss:

$$P_{cu,r} = 3 \times I_r^2 R_r = 18.521 \times 10^3 \text{ W}$$

e) The stator output power of SCIG:

$$P_s = \left| P_m \right| - \left| P_{cu,s} \right| - \left| P_{cu,r} \right| = 2.3339 \times 10^6 - 15.538 \times 10^3 - 18.521 \times 10^3 = 2300 \times 10^3 \text{ W}$$

The SCIG efficiency:

$$\eta = \frac{P_s}{P_m} = 98.54\%$$

The stator power factor angle:

$$\varphi_s = \angle V_s - \angle I_s = 0° - (-152.6°) = 152.6°$$

The stator power factor:

$$PF_s = \cos \phi_s = -0.888 \text{ (Lagging)}$$

f) The magnetizing flux linkage can be expressed as

$$\overline{\Lambda}_m = (\overline{I}_s - \overline{I}_r)L_m = 1.2168 \angle -83.9° \text{ Wb (rms)}$$

Alternatively,

$$\overline{\Lambda}_m = \frac{\overline{V}_s - \overline{I}_s(R_s + jX_{ls})}{j\omega_s} = 1.2168 \angle -83.9° \text{ Wb (rms)}$$

The stator flux linkage can be obtained by

$$\overline{\Lambda}_s = \overline{\Lambda}_m + L_{ls}\overline{I}_s = 1.2748 \angle -89.8° \text{ Wb (rms)}$$

The peak value of stator flux:

$$\lambda_s = \sqrt{2}\Lambda_s = 1.8028 \text{ Wb}$$

The rotor flux linkage can then be obtained by

$$\overline{\Lambda}_r = \overline{\Lambda}_m - L_{lr}\overline{I}_r = 1.2096\angle-77.7°\ \text{Wb (rms)}$$

The peak value of rotor flux:

$$\lambda_r = \sqrt{2}\Lambda_r = 1.7106\ \text{Wb}$$

3-2. Repeat Problem 3-1 when the SCIG operates at a speed of 1508 rpm:

Answers:

a) $s = -0.00533$, $\omega_m = 157.92\ \text{rad/sec}$, $\omega_r = 315.83\ \text{rad/sec}$

b) $\overline{Z}_s = 0.2617\angle149.62°\ \Omega$, $\overline{I}_s = 1521.9\angle-149.62°\ \text{A}$, $\overline{I}_r = 1368.4\angle-171.73°\ \text{A}$

c) $P_m = 1.585\times10^6\ \text{W}$, $T_m = 10038\ \text{N·m}$

d) $P_{cu,s} = 7.6577\times10^3\ \text{W}$, $P_{cu,r} = 8.41\times10^3\ \text{W}$

e) $P_s = 1.5692\times10^6\ \text{W}$, $\eta = 98.97\%$, $PF_s = -0.8627$

f) $\overline{\Lambda}_m = 1.2259\angle-85.89°\ \text{Wb}$, $\overline{\Lambda}_s = 1.2727\angle-89.88°\ \text{Wb}$,
$\overline{\Lambda}_r = 1.2226\angle-81.73°\ \text{Wb}$

Topic: Steady-State and Power Flow Analysis of Induction Generators using Approximate Equivalent Circuit

3-3 (Solved Problem). Consider a 2.3 MW, 690 V, 50 Hz squirrel cage induction generator in a wind energy conversion system. The generator is directly connected to the grid of 690 V/50 Hz. The generator parameters are given in Table B-1 of Appendix B. At a given wind speed, the SCIG operates at a speed of 1510 rpm. Using the approximate induction generator equivalent circuit of Figure C-1, calculate the following:

a) The slip, and rotor mechanical and electrical speeds
b) The rotor and stator currents
c) The mechanical power and torque

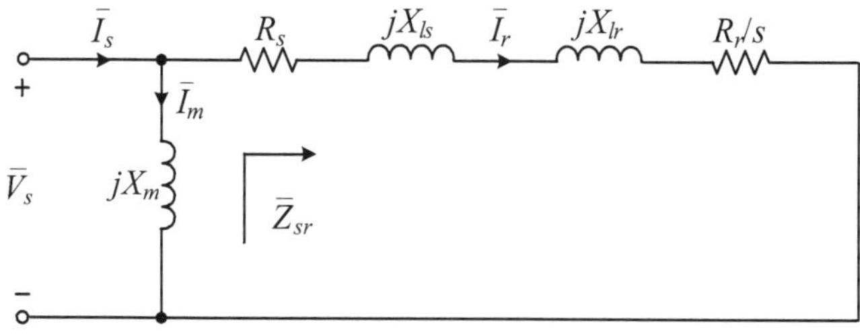

Figure C-1. Approximate induction generator equivalent circuit.

d) The stator and rotor winding losses

e) The generator efficiency and power factor

Solution:

a) The slip:

$$s = \frac{n_s - n_r}{n_s} = \frac{1500 - 1510}{1500} = -0.00667 \quad \text{(negative slip)}$$

The rotor mechanical speed:

$$\omega_m = 1510 \times (2\pi / 60) = 158.127 \text{ rad/sec}$$

The rotor electrical speed:

$$\omega_r = \omega_m \times P = 158.127 \times 2 = 316.254 \text{ rad/sec}$$

The stator frequency:

$$\omega_s = 2\pi \times 50 = 314.16 \text{ rad/sec}$$

b) The stator voltage:

$$\overline{V}_s = 690 / \sqrt{3} \angle 0° = 398.37 \angle 0° \text{ V (rms)}$$

The equivalent impedance of SCIG:

$$\overline{Z}_{sr} = R_s + jX_{ls} + \frac{R_r}{s} + jX_{lr} = 0.22714 \angle 169.65° \ \Omega$$

where $X_{ls} = \omega_s L_{ls} = 0.0204\ \Omega$, $X_{lr} = \omega_s L_{lr} = 0.0204\ \Omega$, and $X_m = \omega_s L_m = 0.6706\ \Omega$
The rotor current:

$$\overline{I}_r = \frac{\overline{V}_s}{\overline{Z}_{sr}} = 1753.855 \angle -169.65° \text{ A (rms)}$$

The magnetizing current:

$$\overline{I}_m = \frac{\overline{V}_s}{jX_m} = 594.05 \angle -90° \text{ A (rms)}$$

The stator current:

$$\overline{I}_s = \overline{I}_r + \overline{I}_m = 1950.1 \angle -152.22° \text{ A (rms)}$$

c) The mechanical power:

$$P_m = \left| 3I_r^2 \frac{R_r}{s}(1-s) \right| = 2.086 \times 10^6 \text{ W}$$

The generator mechanical torque:

$$T_m = P_m \times P / \omega_r = 13191.7 \text{ N·m}$$

d) The stator winding loss:

$$P_{cu,s} = 3 \times I_s^2 R_s = 12.573 \times 10^3 \text{ W}$$

The rotor winding loss:

$$P_{cu,r} = 3 \times I_r^2 R_r = 13.814 \times 10^3 \text{ W}$$

e) The stator output power of SCIG:

$$P_s = \left| P_m \right| - \left| P_{cu,s} \right| - \left| P_{cu,r} \right| = 2.0596 \times 10^6 \text{ W}$$

The SCIG efficiency:

$$\eta = \frac{P_s}{P_m} = 98.74\%$$

The stator power factor angle:

$$\varphi_s = \angle V_s - \angle I_s = 0° - (-152.22°) = 152.22°$$

The stator power factor:

$$PF_s = \cos \varphi_s = -0.8847$$

3-4. Repeat Problem 3-3 when the SCIG operates at a speed of 1504 rpm.

Answers:

a) $s = -0.00267$, $\omega_m = 157.5$ rad/sec, $\omega_r = 314.997$ rad/sec
b) $\overline{Z}_{sr} = 0.5618\angle175.84° \ \Omega$, $\overline{I}_r = 709.15\angle-175.84°$ A,
    $\overline{I}_s = 957.59\angle-137.61°$ A
c) $P_m = 849.205 \times 10^3$ W, $T_m = 5391.83$ N·m
d) $P_{cu,s} = 3.0315 \times 10^3$ W, $P_{cu,r} = 2.2585 \times 10^3$ W
e) $P_s = 843.915 \times 10^3$ W, $\eta = 99.38\%$, $PF_s = -0.7386$

Topic: Steady-State Analysis of Permanent Magnet Synchronous Generators

3-5 (Solved Problem). A 2.0 MW, 690 V, 11.25 Hz salient-pole, permanent-magnet synchronous generator (PMSG) is used in a stand-alone wind energy system. The parameters of the generator are given in Table B-11 of Appendix B. The generator is connected to an *RL* load as shown in Figure 3-25. The load parameters are given by R_L = 0.19125 Ω and L_L = 1.3104 mH. At a wind speed of 12.0 m/s, the generator operates at 1.0 pu rotor speed. Neglecting the rotational and stator core losses, determine the following:

a) The rotor mechanical and electrical speeds, and load impedance value
b) The *dq*-axis and rms stator currents
c) The *dq*-axis and rms stator voltages
d) The mechanical power and torque
e) The stator winding loss and power factor
f) The stator active and reactive powers, and generator efficiency

Solution:

a) The rated rotor mechanical speed:

$$\omega_{m,R} = 22.5 \times (2\pi / 60) = 2.356 \text{ rad/sec}$$

The rotor mechanical speed:

$$\omega_m = \omega_{m,R} \times \omega_{m,\text{pu}} = 2.356 \times 1.0 = 2.356 \text{ rad/sec} \quad (22.5 \text{ rpm})$$

The rotor electrical speed:

$$\omega_r = \omega_m \times P = 2.356 \times 30 = 70.69 \text{ rad/sec}$$

The load impedance value can be calculated by

$$\bar{Z}_L = R_L + jX_L = 0.19125 + j92.626 \times 10^{-3} = 0.2125 \,\angle 25.84° \; \Omega$$

b) The *q*-axis stator current:

$$i_{qs} = \frac{\omega_r \lambda_r (R_L + R_s)}{(R_L + R_s)^2 + \omega_r^2 (L_L + L_d)(L_L + L_q)} = 1093.1 \text{ A}$$

The *d*-axis stator current:

$$i_{ds} = \frac{\omega_r (L_L + L_q)}{R_L + R_s} i_{qs} = 1457.1 \text{ A}$$

The rms stator current:

$$I_s = \sqrt{i_{ds}^2 + i_{qs}^2}\,/\sqrt{2} = 1288 \text{ A}$$

c) The q-axis stator voltage:

$$v_{qs} = -R_s i_{qs} - \omega_r L_d i_{ds} + \omega_r \lambda_r = 344 \text{ V}$$

The d-axis stator voltage:

$$v_{ds} = -R_s i_{ds} + \omega_r L_q i_{qs} = 177.42 \text{ V}$$

The rms stator voltage:

$$V_s = \sqrt{v_{ds}^2 + v_{qs}^2}\,/\sqrt{2} = 273.7 \text{ V}$$

d) The electromagnetic torque developed by the PMSG:

$$T_e = T_m = \frac{3}{2}P\left[\lambda_r\, i_{qs} - \left(L_d - L_q\right)i_{ds}i_{qs}\right] = 405.50 \times 10^3 \text{ N·m}$$

The mechanical power:

$$P_m = T_m \omega_r / P = 955.44 \times 10^3 \text{ W}$$

e) The stator winding loss:

$$P_{cu,s} = 3(I_s)^2 R_s = 3.635 \times 10^3 \text{ W}$$

The rotor winding losses are zero for PMSG.
The angles of the stator voltage and current vectors:

$$\theta_v = \tan^{-1}\frac{v_{qs}}{v_{ds}} = 62.72°$$

$$\theta_i = \tan^{-1}\frac{i_{qs}}{i_{ds}} = 36.88°$$

The stator power factor angle:

$$\varphi_s = \theta_v - \theta_i = 25.84° \qquad \Rightarrow PF_s = 0.9 \text{ (lagging)}$$

Alternatively, the stator power factor can be calculated by

$$\varphi_s = \tan^{-1}\left(\frac{\omega_r L_L}{R_L}\right) = 25.84° \qquad \Rightarrow PF_s = 0.9 \text{ (lagging)}$$

f) The stator active and reactive powers:

$$P_s = 3V_s I_s \cos\varphi_s = 951.81 \times 10^3 \text{ W}$$

$$Q_s = 3V_s I_s \sin\varphi_s = 460.98 \times 10^3 \text{ VAR}$$

Alternatively,

$$P_s = 1.5(v_{ds} i_{ds} + v_{qs} i_{qs}) = 951.81 \times 10^3 \text{ W}$$

$$Q_s = 1.5(v_{qs} i_{ds} - v_{ds} i_{qs}) = 460.98 \times 10^3 \text{ VAR}$$

Alternatively, stator active power can be calculated by

$$P_s = P_m - P_{cu,s} = (955.44 - 3.635) \times 10^3 = 951.81 \times 10^3 \text{ W}$$

The efficiency of the PMSG is then

$$\eta = P_s / P_m = 951.81 / 955.44 = 99.62\%$$

Note: In the calculation of efficiency, only the stator winding losses are considered. The rotational and core losses of the generator are ignored.

3-6. Repeat Problem 3-5 when the PMSG is connected to 0.8 pu pure R load. At a wind speed of 10.8 m/s, the generator operates at 0.9 pu rotor speed.

Answers:

a) $\omega_m = 2.121 \text{ rad/sec}$ (20.25 rpm), $\quad \omega_r = 63.62 \text{ rad/sec}$, $\quad \overline{Z}_L = 0.17 + j0 \ \Omega$

b) $i_{qs} = 1782.7 \text{ A}$, $\quad i_{ds} = 1534.5 \text{ A}$, $\quad I_s = 1663.2 \text{ A}$

c) $v_{qs} = 303.07 \text{ V}$, $\quad v_{ds} = 260.87 \text{ V}$, $\quad V_s = 282.76 \text{ V}$

d) $T_e = 668.19 \times 10^3 \text{ N·m}$, $\quad P_m = 1416.9 \times 10^3 \text{ W}$

e) $P_{cu,s} = 6.063 \times 10^3 \text{ W}$, $\quad PF_s = 1.0$

f) $P_s = 1.411 \times 10^3 \text{ W}$, $\quad Q_s = 0 \text{ VAR}$, $\quad \eta = 99.57\%$

3-7. Repeat Problem 3-5 when the PMSG is connected to 0.7 pu pure L load. At a wind speed of 9.6 m/s, the generator operates at 0.8 pu rotor speed.

Answers:

a) $\omega_m = 1.885 \text{ rad/sec}$ (18 rpm), $\quad \omega_r = 56.55 \text{ rad/sec}$, $\quad \overline{Z}_L = 0 + j0.1488 \ \Omega$

b) $i_{qs} = 5.864 \text{ A}$, $\quad i_{ds} = 2003.7 \text{ A}$, $\quad I_s = 1416.92 \text{ A}$

c) $v_{qs} = 238.4 \text{ V}$, $\quad v_{ds} = -0.69 \text{ V}$, $\quad V_s = 168.63 \text{ V}$

d) $T_e = 233.4 \times 10^3 \text{ N·m}$, $\quad P_m = 439.96 \times 10^3 \text{ W}$

e) $P_{cu,s} = 4.399 \times 10^3$ W, $\qquad PF_s = 0$ (lagging)

f) $P_s = 0$ W, $\qquad Q_s = 716.66 \times 10^3$ VAR, $\qquad \eta = $ not applicable

3-8 (Solved Problem). Repeat Problem 3-5 when the PMSG is connected to *RC* load. The equivalent load impedance is: $\overline{Z}_L = 0.19125 - j92.6266 \times 10^{-3}$ Ω.

Solution:

a) The rated rotor mechanical speed:

$$\omega_{m,R} = 22.5 \times (2\pi / 60) = 2.356 \text{ rad/sec}$$

The rotor mechanical speed:

$$\omega_m = \omega_{m,R} \times \omega_{m,\text{pu}} = 2.356 \times 1.0 = 2.356 \text{ rad/sec} \quad (22.5 \text{ rpm})$$

The rotor electrical speed:

$$\omega_r = \omega_m \times P = 2.356 \times 30 = 70.69 \text{ rad/sec}$$

The load impedance:

$$\overline{Z}_L = R_L \,//\, \frac{1}{j\omega_r C_L} = R_{eq} - jX_{eq} = 0.19125 - j92.62 \times 10^{-3} = 0.2125 \angle -25.84° \text{ Ω}$$

b) The *q*-axis stator current:

$$i_{qs} = \frac{\omega_r \lambda_r (R_{eq} + R_s)}{(R_{eq} + R_s)^2 + (X_{eq} + \omega_r L_d)(X_{eq} + \omega_r L_q)} = 2478.9 \text{ A}$$

The *d*-axis stator current:

$$i_{ds} = \frac{X_{eq} + \omega_r L_q}{R_{eq} + R_s} i_{qs} = 912.37 \text{ A}$$

The rms stator current:

$$I_s = \sqrt{i_{ds}^2 + i_{qs}^2} / \sqrt{2} = 1867.8 \text{ A}$$

c) The *q*-axis stator voltage:

$$v_{qs} = -R_s i_{qs} - \omega_r L_d i_{ds} + \omega_r \lambda_r = 389.59 \text{ V}$$

The *d*-axis stator voltage:

$$v_{ds} = -R_s i_{ds} + \omega_r L_q i_{qs} = 404.11 \text{ V}$$

The rms stator voltage:

$$V_s = \sqrt{v_{ds}^2 + v_{qs}^2} \big/ \sqrt{2} = 396.2 \text{ V}$$

d) The electromagnetic torque developed by the PMSG:

$$T_e = T_m = \frac{3}{2} P \left[\lambda_r i_{qs} - \left(L_d - L_q \right) i_{ds} i_{qs} \right] = 852.79 \times 10^3 \text{ N·m}$$

The mechanical power:

$$P_m = T_m \omega_r \big/ P = 2009.3 \times 10^3 \text{ W}$$

e) The stator winding loss:

$$P_{cu,s} = 3(I_s)^2 R_s = 7.646 \times 10^3 \text{ W}$$

The rotor winding losses are zero for PMSG.
The angles of the stator voltage and current vectors:

$$\theta_v = \tan^{-1} \frac{v_{qs}}{v_{ds}} = 43.95°$$

$$\theta_i = \tan^{-1} \frac{i_{qs}}{i_{ds}} = 69.794°$$

The stator power factor angle:

$$\varphi_s = \theta_v - \theta_i = -25.84° \qquad \Rightarrow PF_s = 0.9 \text{ (leading)}$$

Alternatively, the stator power factor can be calculated by

$$\varphi_s = \tan^{-1} \left(\frac{X_{eq}}{R_{eq}} \right) = -25.84° \qquad \Rightarrow PF_s = 0.9 \text{ (leading)}$$

f) The stator active and reactive powers:

$$P_s = 3 V_s I_s \cos \varphi_s = 2001.7 \times 10^3 \text{ W}$$

$$Q_s = 3 V_s I_s \sin \varphi_s = -969.47 \times 10^3 \text{ VAR}$$

Alternatively,

$$P_s = 1.5(v_{ds} i_{ds} + v_{qs} i_{qs}) = 2001.7 \times 10^3 \text{ W}$$

$$Q_s = 1.5(v_{qs} i_{ds} - v_{ds} i_{qs}) = -969.47 \times 10^3 \text{ VAR}$$

Alternatively, stator active power can be calculated by

$$P_s = P_m - P_{cu,s} = (2009.3 - 7.646) \times 10^3 = 2001.7 \times 10^3 \ \text{W}$$

The efficiency of the PMSG neglecting rotational and core losses is then

$$\eta = P_s / P_m = 2001.7 / 2009.3 = 99.62\%$$

3-9 (Solved Problem). A 2.5 MW, 4000 V, 40 Hz salient-pole PMSG is employed in a stand-alone wind power system. The parameters of the generator are given in Table B-12 of Appendix B. The generator is connected to an *RL* load. The load parameters are given by R_L = 3.5944 Ω and L_L = 10.726 mH. At given wind and generator speeds, the stator current is found to be $\bar{i}_s$ = 226.83 + j125.33 A. Neglecting rotational and stator core losses, find the following:

a) The rotor electrical and mechanical speeds, and load impedance value
b) The *dq*-axis and rms stator voltages
c) The mechanical power and torque
d) The stator winding loss and power factor
e) The stator active and reactive powers, and generator efficiency

Solution:

a) The *d*-axis stator current can be related to rotor electrical speed as

$$i_{ds} = \frac{\omega_r (L_L + L_q)}{R_L + R_s} i_{qs}$$

from which, the rotor electrical speed can be calculated by

$$\omega_r = \frac{i_{ds}}{i_{qs}} \frac{(R_L + R_s)}{(L_L + L_q)} = 201.06 \ \text{rad/sec}$$

The rotor mechanical speed:

$$\omega_m = \omega_r / P = 201.06 / 6 = 33.51 \ \text{rad/sec} \quad (320 \ \text{rpm})$$

The output impedance value can be calculated by

$$\bar{Z}_L = R_L + jX_L = 3.5944 + j2.6958 = 4.493\angle 36.87° \ \Omega$$

b) The *q*-axis stator voltage:

$$v_{qs} = -R_s i_{qs} - \omega_r L_d i_{ds} + \omega_r \lambda_r = 939.71 \ \text{V}$$

The d-axis stator voltage:

$$v_{ds} = -R_s i_{ds} + \omega_r L_q i_{qs} = 545.03 \text{ V}$$

The rms stator voltage:

$$V_s = \sqrt{v_{ds}^2 + v_{qs}^2} / \sqrt{2} = 768.15 \text{ V}$$

c) The electromagnetic torque developed by the PMSG:

$$T_e = T_m = \frac{3}{2} P \left[\lambda_r i_{qs} - \left(L_d - L_q \right) i_{ds} i_{qs} \right] = 10.88 \times 10^3 \text{ N·m}$$

The mechanical power:

$$P_m = T_e \omega_r / P = 364.56 \times 10^3 \text{ W}$$

d) The stator winding loss:

$$P_{cu,s} = 3(I_s)^2 R_s = 2.443 \times 10^3 \text{ W}$$

where $I_s = \sqrt{i_{ds}^2 + i_{qs}^2}/\sqrt{2} = 183.25$ A. The rotor winding losses are zero for PMSG. The angles of the stator voltage and current vectors:

$$\theta_v = \tan^{-1} \frac{v_{qs}}{v_{ds}} = 59.87°$$

$$\theta_i = \tan^{-1} \frac{i_{qs}}{i_{ds}} = 28.92°$$

The stator power factor angle:

$$\varphi_s = \theta_v - \theta_i = 30.964° \qquad \Rightarrow PF_s = 0.8575 \text{ (lagging)}$$

Alternatively, the stator power factor can be calculated by

$$\varphi_s = \tan^{-1}\left(\frac{\omega_r L_L}{R_L} \right) = 30.964° \qquad \Rightarrow PF_s = 0.8575 \text{ (lagging)}$$

e) The stator active and reactive powers:

$$P_s = 3V_s I_s \cos\varphi_s = 362.12 \times 10^3 \text{ W}$$

$$Q_s = 3V_s I_s \sin\varphi_s = 217.27 \times 10^3 \text{ VAR}$$

Alternatively,

$$P_s = 1.5(v_{ds}i_{ds} + v_{qs}i_{qs}) = 362.12 \times 10^3 \text{ W}$$

$$Q_s = 1.5(v_{qs}i_{ds} - v_{ds}i_{qs}) = 217.27 \times 10^3 \text{ VAR}$$

Alternatively, stator active power can be calculated by

$$P_s = P_m - P_{cu,s} = (364.56 - 2.443) \times 10^3 = 362.12 \times 10^3 \text{ W}$$

The efficiency of the PMSG neglecting rotational and core losses is then

$$\eta = P_s / P_m = 362.12/364.56 = 99.62\%$$

3-10. Repeat Problem 3-9 when the PMSG is connected to an RC load. The equivalent load impedance is $\overline{Z}_L = 3.2161 - j1.9931$ Ω. At given wind and generator speeds, the stator current is found to be $\bar{i}_s = 422.43 - j464.26$ A.

Answers:

a) $\omega_r = 226.195$ rad/sec, $\omega_m = 37.7$ rad/sec (360 rpm),
   $\overline{Z}_L = 3.2161 - j1.9931$ Ω
b) $v_{qs} = 651.16$ V, $v_{ds} = 2283.92$ V, $V_s = 1679.3$ V
c) $T_e = 50.797 \times 10^3$ N·m, $P_m = 1915 \times 10^3$ W
d) $P_{cu,s} = 14.332 \times 10^3$ W, $PF_s = 0.85$ (leading)
e) $P_s = 1900.7 \times 10^3$ W, $Q_s = -1177.9 \times 10^3$ VAR, $\eta = 99.25\%$

3-11 (Solved Problem). A stand-alone wind energy system employs a 2.5 MW, 4000 V, 40 Hz salient PMSG. The parameters of the generator are given in Table B-12 of Appendix B. The generator is connected to an *RC* load. At given wind and generator speeds, the peak stator voltage and currents are found to be $1773.965 + j1031.742$ V and $288.213 + j324.366$ A, respectively. Neglecting the stator core and rotational losses, calculate the following:

a) The rms stator voltage and current
b) The stator active and reactive powers
c) The stator winding loss and power factor
d) The rotor electrical and mechanical speeds
e) The mechanical power and torque
f) The load parameters

Solution:

a) The rms stator voltage:

$$V_s = \sqrt{v_{ds}^2 + v_{qs}^2}\,/\sqrt{2} = 1451.1 \text{ V}$$

The rms stator current:

$$I_s = \sqrt{i_{ds}^2 + i_{qs}^2}\,/\sqrt{2} = 306.8 \text{ A}$$

b) The stator active and reactive powers:

$$P_s = 1.5(v_{ds}i_{ds} + v_{qs}i_{qs}) = 1268.9 \times 10^3 \text{ W}$$

$$Q_s = 1.5(v_{qs}i_{ds} - v_{ds}i_{qs}) = -417.08 \times 10^3 \text{ VAR}$$

c) The stator winding loss:

$$P_{cu,s} = 3(I_s)^2 R_s = 6.849 \times 10^3 \text{ W}$$

The rotor winding losses are zero for PMSG.
    The angles of the stator voltage and current vectors:

$$\theta_v = \tan^{-1}\frac{v_{qs}}{v_{ds}} = 30.18°$$

$$\theta_i = \tan^{-1}\frac{i_{qs}}{i_{ds}} = 48.3775°$$

The stator power factor:

$$\varphi_s = \theta_v - \theta_i = -18.195° \qquad \Rightarrow PF_s = 0.95 \text{ (leading)}$$

Alternatively, the stator power factor can be calculated by

$$\cos\varphi_s = \frac{P_s}{3V_s I_s} = 0.95 \text{ (leading)}$$

d) The d-axis stator voltage can be related to rotor electrical speed as

$$v_{ds} = -R_s i_{ds} + \omega_r L_q i_{qs}$$

from which, the rotor electrical speed can be calculated by

$$\omega_r = \frac{V_{ds} + R_s I_{ds}}{L_q I_{qs}} = 251.327 \text{ rad/sec}$$

The rotor mechanical speed:

$$\omega_m = \omega_r / P = 251.327/6 = 41.888 \text{ rad/sec} \quad (400 \text{ rpm})$$

e) The electromagnetic torque developed by the PMSG:

$$T_e = T_m = \frac{3}{2} P \left[\lambda_r i_{qs} - \left(L_d - L_q \right) i_{ds} i_{qs} \right] = 30.457 \times 10^3 \text{ N·m}$$

The mechanical power:

$$P_m = T_e \omega_r / P = 1275.76 \times 10^3 \text{ W}$$

f) The stator power factor angle can be related to load as

$$\varphi_s = \tan^{-1} \left(\frac{X_{eq}}{R_{eq}} \right)$$

from which

$$X_{eq} = \tan \varphi_s \times R_{eq} = 0.3287 \times R_{eq} \quad (1)$$

The *d*-axis stator current can be related to load as

$$i_{ds} = \frac{X_{eq} + \omega_r L_q}{R_{eq} + R_s} i_{qs} \quad (2)$$

By substituting (1) in (2), the load values can be found from

$$R_{eq} = 4.493 \ \Omega$$

$$X_{eq} = 0.3287 \times R_{eq} = 1.4768 \ \Omega$$

The output impedance value can be represented as

$$\overline{Z}_L = R_L \ // \left(\frac{1}{j\omega_r C_L} \right) = R_{eq} - jX_{eq} = 4.493 - j1.4768 = 4.7295 \angle -18.195° \ \Omega$$

3-12. Repeat Problem 3-14 if the peak stator voltage and current are $744.91 + j1085.10$ V and $241.22 + j143.93$ A, respectively.

Answers:

a) $V_s = 930.68$ V, $\qquad I_s = 198.63$ A

b) $P_s = 503.81 \times 10^3$ W, $\qquad Q_s = 231.80 \times 10^3$ VAR

c) $P_{cu,s} = 2.87 \times 10^3$ W, $\qquad PF_s = 0.9085$ (lagging)

d) $\omega_r = 238.76$ rad/sec, $\qquad \omega_m = 39.794$ rad/sec (380 rpm)

e) $T_e = 12.733 \times 10^3$ N·m, $\qquad P_m = 506.68 \times 10^3$ W

f) $\overline{Z}_L = 4.2566 + j2.0615$ Ω

PROBLEMS AND ANSWERS FOR CHAPTER 4—POWER CONVERTERS IN WIND ENERGY CONVERSION SYSTEMS

Topic: AC Voltage Controllers (Soft Starter)

4-1 (Solved Problem). A single-phase, full-wave AC voltage controller in Figure 4-2 has an input voltage of 220 V (rms), 50 Hz, and a load resistance of 10 Ω. The converter operates at a firing angle of 45°. Assuming that the converter is ideal, calculate/answer the following:

a) The rms output voltage and current
b) The load apparent, active, and reactive powers
c) The rms input and thyristor currents
d) The apparent, active, reactive powers, and input power factor
e) Draw the waveforms for v_s, i_{g1}, i_{g2}, v_o, i_o, i_{T1}, and i_{T2}

Solution:

a) The output voltage:

$$V_o = V_s \left(1 - \frac{\alpha}{\pi} + \frac{\sin 2\alpha}{2\pi} \right)^{1/2} = 220 \left(1 - \frac{\pi/4}{\pi} + \frac{\sin 2(\pi/4)}{2\pi} \right)^{1/2} = 209.77 \text{ V (rms)}$$

(α is in radians)

The output current:

$$I_o = \frac{V_o}{R} = \frac{209.77}{10} = 20.977 \text{ A (rms)}$$

b) The load apparent power:

$$S_o = V_o \times I_o = 209.77 \times 20.977 = 4400.3 \text{ VA}$$

The load active power:

$$P_o = I_o^2 \times R = 20.977^2 \times 10 = 4400.3 \text{ W}$$

The load active power:

$$Q_o = \sqrt{S_o^2 - P_o^2} = 0$$

c) The input current:

$$I_s = I_o = 20.977 \text{ A (rms)} \quad \text{(ideal converter, no losses)}$$

The thyristor currents:

$$I_{T1} = I_{T2} = \frac{I_o}{\sqrt{2}} = \frac{20.977}{\sqrt{2}} = 14.833 \text{ A (rms)}$$

(each thyristor carries half the load current)

d) The input apparent power:

$$S_s = V_s \times I_s = 220 \times 20.977 = 4614.9 \text{ VA}$$

The input active power:

$$P_s = P_o = 4400.31 \text{ W} \quad \text{(ideal converter, no losses)}$$

The input reactive power:

$$Q_s = \sqrt{S_s^2 - P_s^2} = \sqrt{4614.92^2 - 4400.31^2} = 1390.9 \text{ VAR}$$

The input power factor:

$$PF_s = \frac{P_s}{S_s} = \frac{4400.3}{4614.9} = 0.9535$$

e) Waveforms not provided.

4-2. A single-phase full-wave AC voltage controller with an input voltage of 120 V (rms) and 60 Hz feeds a 1.0 kW/120 V resistive load. Repeat Problem P4-1 for a firing angle of 60°.

Answers:

a) $R = 14.4\ \Omega$, $V_o = 107.63$ V, $\qquad I_o = 7.48$ A
b) $S_o = 804.5$ VA, $\qquad P_o = 804.5$ W, $\qquad Q_o = 0$ VAR
c) $I_s = 7.48$ A, $\qquad I_{T1} = I_{T2} = 5.285$ A
d) $S_s = 896.94$ VA, $\qquad P_s = 804.5$ W, $\qquad Q_s = 396.59$ VAR, $\qquad PF_s = 0.8969$ (lag)

4-3 (Solved Problem). A three-phase 690 V/2.3 MVA AC voltage controller (Figure 4-6) is loaded with a star-connected R load of 1.0 pu per phase. The controller is supplied by a three-phase utility voltage of 690 V and 50 Hz and operates at a firing angle of 120°. Assuming that the power converter is ideal, calculate/answer the following:

a) The load resistance value
b) The rms load phase voltage and line current
c) The three-phase apparent, active, and reactive powers of the load
d) The rms input line current
e) The three-phase input apparent, active, and reactive powers, and power factor. State the reason why the input power factor is not unity even though the controller is loaded with a pure resistive load.
f) Using the template given in Figure C-2, draw the waveform for the phase-*a* load voltage v_{an}.

Solution:

a) The load resistance value:

$$S_B = \frac{2300 \times 10^3}{3} = 766.67 \times 10^3 \text{ VA (1.0 pu)}, \qquad V_s = V_B = \frac{690}{\sqrt{3}} = 398.38 \text{ V (1.0 pu)}$$

$$I_B = \frac{S_B}{V_B} = \frac{766.67 \times 10^3}{398.38} = 1924.5 \text{ A (1.0 pu)}, \quad Z_B = \frac{V_B}{I_B} = \frac{398.38}{1924.5} = 0.207 \text{ }\Omega \text{ (1.0 pu)}$$

$$R = Z_B \times R_{pu} = 0.207 \times 1.0 = 0.207 \text{ }\Omega$$

b) The load phase voltage:

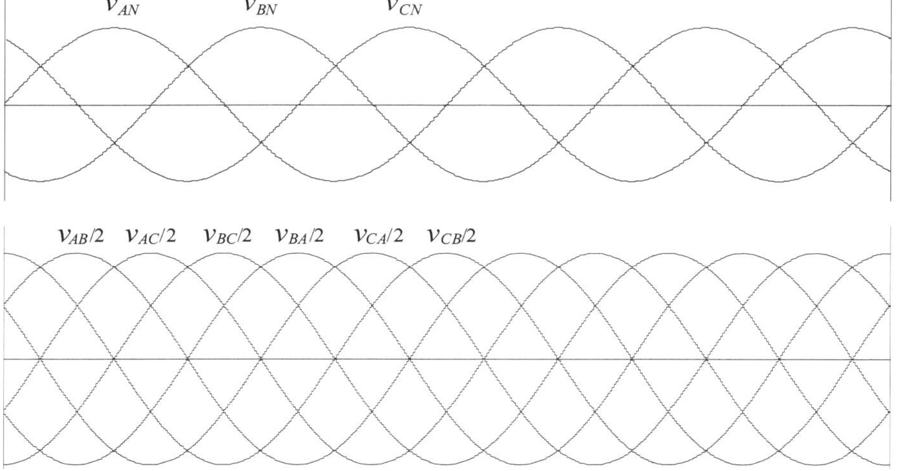

<u>Figure C-2.</u> Template for three-phase AC voltage controller waveforms.

$$V_{an} = V_s \left(\frac{5}{4} - \frac{3\alpha}{2\pi} + \frac{3\sin(2\alpha + \pi/3)}{4\pi} \right)^{1/2}$$

$$= 398.37 \times \left(\frac{5}{4} - \frac{3 \times (2\pi/3)}{2\pi} + \frac{3\sin(2 \times (2\pi/3) + \pi/3)}{4\pi} \right)^{1/2} = 82.85 \text{ V (rms)}$$

(α is in radians)
The load line current:

$$I_a = \frac{V_{an}}{R} = \frac{82.85}{0.207} = 400.24 \text{ A (rms)}$$

c) The three-phase apparent power of the load:

$$S_o = 3 \times V_{an} \times I_a = 3 \times 82.85 \times 400.24 = 99.48 \times 10^3 \text{ VA}$$

The three-phase active power consumed by the load:

$$P_o = 3 \times I_a^2 \times R = 3 \times 400.24^2 \times 0.207 = 99.48 \times 10^3 \text{ W}$$

The three-phase reactive power consumed by the load:

$$Q_o = \sqrt{S_o^2 - P_o^2} = \sqrt{99478^2 - 99478^2} = 0 \text{ VAR}$$

d) The input line current:

$$I_A = I_a = 400.24 \text{ A (rms)}$$

e) Three-phase input apparent power:

$$S_s = 3 \times V_s \times I_A = 3 \times 398.38 \times 400.24 = 478.33 \times 10^3 \text{ VA}$$

The input active power:

$$P_s = S_s \times \cos\varphi_s = 3 \times V_s \times I_A \times \cos\varphi_s = 3 \times 398.38 \times 400.24 \times 0.208 = 99.48 \times 10^3 \text{ W}$$

The input reactive power:

$$Q_s = \sqrt{S_s^2 - P_s^2} = \sqrt{478330^2 - 99478^2} = 467.87 \times 10^3 \text{ VAR}$$

The input power factor:

$$PF_s = \frac{P_s}{S_s} = \frac{99.48 \times 10^3}{478.33 \times 10^3} = 0.208$$

f) For the waveforms, see Figure 4-9a.

4-4. Repeat the Problem 4-3 with a firing angle of 75°. Compare the input power factor and explain why the input power factor is improved.

Answers:

a) $R = 0.207\ \Omega$

b) $V_{an} = 281.69$ V, $\qquad I_a = 1360.8$ A

c) $S_o = 1150 \times 10^3$ VA, $\qquad P_o = 1150 \times 10^3$ W, $\qquad Q_o = 0$ VAR

d) $I_A = 1360.8$ A

e) $S_s = 1626.3 \times 10^3$ VA, $\qquad P_s = 1150 \times 10^3$ W, $\qquad Q_s = 1150 \times 10^3$ VAR, $PF_s = 0.7071$

f) For the waveforms, see Figure 4-9b.

4-5. Repeat the Problem 4-3 with a firing angle of 30° and compare the input power factor and explain why the input power factor is improved.

Answers:

a) $R = 0.207\ \Omega$

b) $V_{an} = 389.7$ V, $\qquad I_a = 1882.4$ A

c) $S_o = 2200.5 \times 10^3$ VA, $\qquad P_o = 2200.5 \times 10^3$ W, $\qquad Q_o = 0$ VAR

d) $I_A = 1882.4$ A

e) $S_s = 2249.7 \times 10^3$ VA, $\qquad P_s = 2200.5 \times 10^3$ W, $\qquad Q_s = 467.8 \times 10^3$ VAR, $PF_s = 0.978$

f) Waveforms: See Figure 4-9c.

Topic: Interleaved Boost Converters

4-6 (Solved Problem). A single-channel boost converter is used in a 600 kW, 690 V, 50 Hz PMSG wind energy conversion system as shown in Figure C-3. The boost

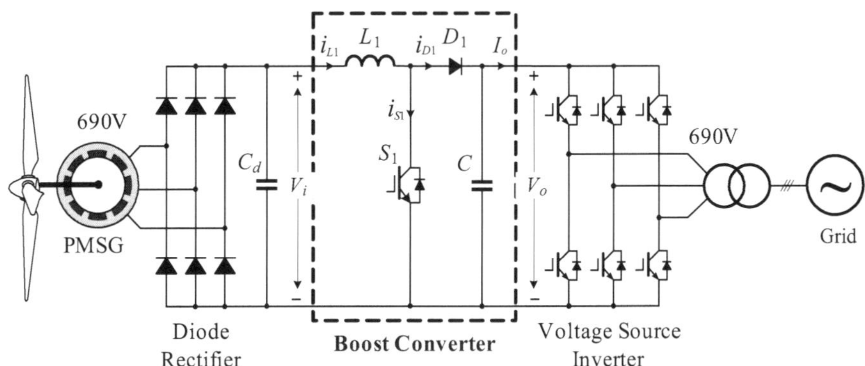

Figure C-3. PMSG wind energy conversion system using a single-channel boost converter.

converter transfers the power from the generator to the grid of 690 V/50 Hz via a diode rectifier and a two-level voltage source inverter. The line-to-line stator voltages of the generator operating at the minimum and rated wind speeds are 80 V and 690 V, respectively. The output voltage of the boost converter, v_o, is set by the inverter with a modulation index m_a of 0.8, which leaves 20% margin for adjustments. The switching harmonics generated by the inverter are neglected, and thus the inverter AC-side fundamental-frequency voltage is equal to the grid voltage of 690 V. The switching frequency of the boost converter is 2000 Hz. Calculate/answer the following:

a) The minimum and maximum input voltages and output voltage of boost converter
b) The minimum and maximum duty ratios
c) The average output current and maximum inductor current when the generator delivers the rated power to the grid
d) Derive expression for the calculation of maximum boundary inductor current $I_{LB,\max}$, boundary output current I_{oB}, and the maximum boundary output current $I_{oB,\max}$
e) The minimum value of inductance to operate the boost converter in the CCM mode
f) The minimum output capacitor value assuming that the maximum allowable output voltage ripple is 8% of the average output voltage

Solution:

a) The minimum and maximum input voltages for the boost converter can be calculated by

$$V_{i(\min)} = \frac{3\sqrt{2}}{\pi} \times V_{LL,\min} \approx 1.35 \times 80 = 108 \ \text{V}$$

$$V_{i(\max)} = \frac{3\sqrt{2}}{\pi} \times V_{LL,\max} \approx 1.35 \times 690 = 931.5 \ \text{V}$$

where $3\sqrt{2}/\pi$ is the voltage gain of the diode rectifier.

The output voltage of boost converter is set by the inverter:

$$V_o = \frac{\sqrt{2}V_{grid}}{m_a} = \frac{\sqrt{2} \times 690}{0.8} \approx 1220 \ \text{V}$$

b) The minimum and maximum duty ratios:

$$D_{\min} = 1 - \frac{V_{i(\max)}}{V_o} = 1 - \frac{931.5}{1220} = 0.2365$$

$$D_{\max} = 1 - \frac{V_{i(\min)}}{V_o} = 1 - \frac{108}{1220} = 0.9115$$

c) The average output and maximum inductor current:

$$I_o = \frac{P_o}{V_o} = \frac{P_i}{V_o} = \frac{600 \times 10^3}{1220} = 491.8 \text{ A}$$

$$I_{L1(\max)} = \frac{I_o}{1 - D_{\max}} = \frac{491.8}{1 - 0.9115} = 5556 \text{ A}$$

d) The boundary inductor current: $I_{LB} = D(1 - D)\dfrac{V_o T_s}{2L_1}$

$$\frac{dI_{LB}}{dD} = 0 \rightarrow D = 0.5; \qquad I_{LB,\max} = \frac{V_o T_s}{8L_1} \quad \text{for } D = 0.5$$

$$I_o = (1 - D)I_{L1}, \qquad I_{oB} = (1 - D)I_{LB} = D(1 - D)^2 \times \frac{V_o T_s}{2L_1}$$

$$\frac{dI_{oB}}{dD} = 0 \rightarrow D = 0.333; \qquad I_{oB,\max} = 0.074 \frac{V_o T_s}{L_1} \quad \text{for } D = 0.333$$

e) The CCM mode of operation can be achieved when $I_o > I_{oB,\max}$.
The maximum boundary output current $I_{oB,\max}$ occurs at $D = 0.333$.
The minimum value of the inductance to operate boost converter at the boundary of the CCM can be obtained by equating I_o to $I_{oB,\max}$, from which

$$L_{1,\min} = 0.074 \frac{V_o T_s}{I_o} = 0.074 \frac{1220 \times 0.5 \times 10^{-3}}{491.8} = 91.785 \text{ μH}$$

f) The minimum output capacitor value:

$$C_{\min} = \frac{I_o D_{\max} T_s}{\Delta V_o} = \frac{491.8 \times 0.9115 \times 0.5 \times 10^{-3}}{0.08 \times 1220} = 2296.45 \text{ μF}$$

4-7. Repeat Problem 4-6 with the following changes: a 500 kW, 575 V, 60 Hz PMSG is connected to the grid of 575 V/60 Hz via power converters, its line-to-line stator voltages at the minimum and rated wind speeds are 70 V and 575 V, and the switching frequency of the boost converter is 2200 Hz.

Answers:

a) $V_{i(\min)} = 94.5$ V, $\qquad V_{i(\max)} = 776.2$ V, $\qquad V_o \approx 1020$ V
b) $D_{\min} = 0.239$, $\qquad D_{\max} = 0.9074$
c) $I_o = 490.2$ A, $\qquad I_{L1(\max)} = 5291$ A
d) $L_{1,\min} = 70$ μH
e) $C_{\min} = 2477.6$ μF

4-8 (Solved Problem). A single-channel boost converter is used in a 600 kW, 690 V, 50 Hz PMSG wind energy conversion system as shown in Figure C-3. The boost converter transfers the power from the generator to a grid of 690V/50Hz via a diode rectifier and a two-level voltage source inverter. The inductance L_1 and capacitance C of the boost converter are 270 μH and 2300 μF, respectively. The switching frequency of the boost converter is 2000 Hz. The output voltage v_0 of the boost converter is set by the inverter to 1220 V. The generator operates with an MPPT scheme and its stator active power is proportional to the cube of the rotor speed. At a given wind speed, the PMSG is operating at 0.9 pu rotor speed and its line-to-line stator voltage is 640 V. Calculate/answer the following:

a) The input voltage and power to the boost converter

b) The boundary inductor and output currents, operating mode, and duty cycle

c) The peak-to-peak and average inductor currents

d) The percentage inductor current ripple

e) The percentage output voltage ripple

f) Draw the waveforms for gate signals, i_{L1}, v_{L1}, i_{S1}, v_{S1}, i_{D1}, and v_{D1}, assuming the output filter C is very large and the output voltage is ripple free. Indicate values on the waveforms.

Solution:

a) The input voltage to the boost converter:

$$V_i = \frac{3\sqrt{2}}{\pi} \times V_{LL} \approx 1.35 \times 640 = 864 \ \text{V}$$

The input power to the boost converter at 0.9 pu rotor speed:

$$P_i = P_{m,R} \times (\omega_{m,\text{pu}})^3 = 600 \times 10^3 \times (0.9)^3 = 437.4 \times 10^3 \ \text{W}$$

b) Assume that the converter is operating in the continuous conduction mode (CCM). Based on this assumption, the duty cycle can be calculated by

$$D = 1 - \frac{V_i}{V_o} = 1 - \frac{864}{1220} = 0.2918$$

The boundary inductor current:

$$I_{LB} = D(1-D)\frac{V_o T_s}{2L_1} = 0.2918 \times 0.7082 \times \frac{1220 \times 0.5 \times 10^{-3}}{2 \times 270 \times 10^{-6}} = 233.44 \ \text{A}$$

The boundary and average output currents:

$$I_{oB} = (1-D)I_{LB} = (0.7082) \times 233.44 = 165.32 \ \text{A}$$

$$I_o = \frac{P_o}{V_o} = \frac{P_i}{V_o} = \frac{437.4 \times 10^3}{1220} = 358.52 \ \text{A}$$

Since $I_o > I_{oB}$, the converter operates in the CCM. Therefore, the assumption is valid.

The duty cycle thus is $D = 0.2918$.

c) The peak-to-peak inductor current:

$$\Delta I_{L1} = D(1-D)\frac{V_o T_s}{L_1} = 0.2918 \times 0.7082 \times \frac{1220 \times 0.5 \times 10^{-3}}{270 \times 10^{-6}} = 466.89 \text{ A}$$

Alternatively,

$$\Delta I_{L1} = D\frac{V_i T_s}{L_1} = 0.2918 \times \frac{864 \times 0.5 \times 10^{-3}}{270 \times 10^{-6}} = 466.89 \text{ A}$$

or

$$\Delta I_{L1} = (1-D)\frac{(V_o - V_i)T_s}{L_1} = 0.7082 \times \frac{(1220 - 864) \times 0.5 \times 10^{-3}}{270 \times 10^{-6}} = 466.89 \text{ A}$$

The average inductor current:

$$I_{L1} = \frac{I_o}{1-D} = \frac{358.52}{1-0.2918} = 506.25 \text{ A}$$

d) The percentage inductor current ripple:

$$\frac{\Delta I_{L1}}{I_{L1}} = \frac{466.89}{506.25} \times 100\% = 92.22\%$$

e) The output voltage ripple:

$$\Delta V_o = D\frac{I_o T_s}{C} = 0.2918 \times \frac{358.52 \times 0.5 \times 10^{-3}}{2300 \times 10^{-6}} = 22.74 \text{ V}$$

The percentage output ripple voltage:

$$\frac{\Delta V_o}{V_o} = \frac{22.74}{1220} \times 100\% = 1.86\ \%$$

f) Waveforms not provided.

4-9. Repeat Problem 4-8 when the PMSG operates at 0.5 pu rotor speed with its line-to-line voltage of 150 V.

Answers:

a) $V_i = 202.5$ V, $\qquad P_i = 75 \times 10^3$ W

b) $D = 0.834$, $\qquad I_{LB} = 156.38$A, $\qquad I_{oB} = 25.95$A, $\qquad I_o = 61.48$ A, CCM

c) $\Delta I_{L1} = 312.76$ A

d) $I_{L1} = 370.37$ A, $\dfrac{\Delta I_{L1}}{I_{L1}} = 84.44\%$

e) $\dfrac{\Delta V_o}{V_o} = 0.914\,\%$

4-10 (Solved Problem). A two-channel boost converter is used in a 1.2 MW, 690 V, 50 Hz PMSG wind energy conversion system as shown in Figure 4-22. The boost converter transfers the power from the generator to the grid of 690 V/50 Hz via a diode rectifier and a two-level voltage source inverter. The inductance L_1 ($L_2 = L_1$) and capacitance C_o of the boost converter are 270 μH and 2300 μF, respectively. The switching frequency of the boost converter is 2000 Hz. The output voltage v_o of the boost converter is set by the inverter to 1220 V. At a given wind speed, the generator is delivering the rated power to the grid and its line-to-line stator voltage is found to be 690 V. Calculate/answer the following:

a) The input voltage and power to the boost converter
b) The boundary inductor and output currents for each channel, operating mode, and duty cycle
c) The peak-to-peak and average inductor currents in each channel
d) The peak-to-peak and average input currents
e) The percentage inductor and input current ripples
f) The percentage output voltage ripple if the boost converters are interleaved and not interleaved

Solution:

a) The input voltage to the two-channel boost converter:

$$V_i = \frac{3\sqrt{2}}{\pi} \times V_{LL} \approx 1.35 \times 690 = 931.5 \text{ V}$$

The input power to the two-channel boost converter at 1.0 pu rotor speed:

$$P_i = P_{m,R} \times (\omega_{m,pu})^3 = 1200 \times 10^3 \times (1.0)^3 = 1200 \times 10^3 \text{ W}$$

b) Assume that the converter is operating in the continuous conduction mode (CCM). Therefore, based on this assumption, the duty cycle can be calculated by

$$D = 1 - \frac{V_i}{V_o} = 1 - \frac{931.5}{1220} = 0.2365$$

The boundary inductor currents for each channel:

$$I_{LB} = I_{LB1} = I_{LB2} = D(1-D)\frac{V_o T_s}{2L} = 0.2365 \times 0.7635 \times \frac{1220 \times 0.5 \times 10^{-3}}{2 \times 270 \times 10^{-6}} = 203.96 \text{ A}$$

The boundary output currents for each channel:

$$I_{oB} = I_{oB1} = I_{oB2} = (1-D)I_{LB} = 0.7635 \times 203.96 = 155.73 \text{ A}$$

The total average output current of the interleaved converter is

$$I_o = \frac{P_o}{V_o} = \frac{P_i}{V_o} = \frac{1200 \times 10^3}{1220} = 983.6 \text{ A}$$

from which the average output current for each channel can be obtained by

$$I_{o1} = I_{o2} = I_o / 2 = 983.6 / 2 = 491.8 \text{ A}$$

Since $I_{o1} = I_{o2} > I_{oB}$, the converter operates in the CCM. Therefore, the assumption is valid. Therefore, the duty cycle $D = 0.2365$.

c) The peak-to-peak inductor currents in each channel:

$$\Delta I_L = \Delta I_{L1} = \Delta I_{L2} = D(1-D)\frac{V_o T_s}{L} = 0.2365 \times 0.7635 \times \frac{1220 \times 0.5 \times 10^{-3}}{270 \times 10^{-6}} = 407.92 \text{ A}$$

The average inductor currents in each channel:

$$I_{L1} = I_{L2} = \frac{I_o}{2 \times (1-D)} = \frac{983.6}{2 \times 0.7635} = 644.12 \text{ A}$$

d) The total input current ripple:

$$\Delta I_i = D(1-2D)\frac{V_o T_s}{L} = 0.2365 \times (1 - 2 \times 0.2365) \times \frac{1220 \times 0.5 \times 10^{-3}}{270 \times 10^{-6}} = 281.58 \text{ A}$$

The total input current of the boost converter is given by

$$I_i = \begin{cases} P_i / V_i = 1200 \times 10^3 / 931.5 = 1288.2 \text{ A} \\ I_o / (1-D) = 983.6 / (1 - 0.2365) = 1288.2 \text{ A} \end{cases}$$

e) The percentage inductor current ripple in each channel:

$$\frac{\Delta I_{L1}}{I_{L1}} = \frac{\Delta I_{L2}}{I_{L2}} = \frac{407.92}{644.12} \times 100\% = 63.33\%$$

The total input current ripple can be found from

$$\frac{\Delta I_i}{I_i} = \frac{281.58}{1288.2} \times 100\% = 21.86\%$$

which is much lower than the inductor ripple current of 63.33% in each of the channels.

The ratio of the total input ripple current ΔI_i to the inductor ripple current ΔI_L of each channel is

$$\frac{\Delta I_i}{\Delta I_L} = \frac{281.58}{407.92} \times 100\% = 69.03\%$$

f) The load of the boost converter, which is the inverter, can be modeled by an equivalent resistor as follows:

$$R_{eq} = \frac{V_o}{I_o} = \frac{1220}{983.6} = 1.24\ \Omega$$

Making use of Figure 4-20, the percentage output ripple voltage can be obtained by

$$\frac{\Delta V_o}{V_o} = 0.11 \left(\frac{T_s}{R_{eq}C_o} \right) = 0.11 \times \frac{0.5 \times 10^{-3}}{1.24 \times 2300 \times 10^{-6}} = 0.18\%$$

If the operation of the two-channel converters were not interleaved, switches S_1 and S_2 would be turned on and off simultaneously. The output ripple voltage would then be

$$\frac{\Delta V_o}{V_o} = D \left(\frac{T_s}{R_{eq}C_o} \right) = 0.2365 \times \frac{0.5 \times 10^{-3}}{1.24 \times 2300 \times 10^{-6}} = 4.15\%$$

which is around 23.7 times higher than that for the interleaved boost converter.

4-11. Repeat Problem 4-10 when the PMSG operates at 0.6 pu rotor speed with line-line stator voltage of 350 V, assuming that the stator active power is proportional to the cube of the rotor speed due to the use of MPPT control schemes.

Answers:

a) $V_i = 472.5$ V, $\quad P_i = 259.2 \times 10^3$ W

b) $D = 0.6127,\quad I_{LB} = I_{LB1} = I_{LB2} = 268.06$ A, $\quad I_{oB} = I_{oB1} = I_{oB2} = 103.82$ A,

$\quad I_o = 212.5$ A, $\quad I_{o1} = I_{o2} = 106.23$ A, $\quad$ CCM

c) $\Delta I_L = \Delta I_{L1} = \Delta I_{L2} = 536.12$ A, $\quad I_{L1} = I_{L2} = 274.29$ A

d) $\Delta I_i = 197.23$ A, $\quad I_i = 548.57$ A

e) $\dfrac{\Delta I_{L1}}{I_{L1}} = \dfrac{\Delta I_{L2}}{I_{L2}} = 195.5\%,\qquad \dfrac{\Delta I_i}{I_i} = 35.95\%,\qquad \dfrac{\Delta I_i}{\Delta I_L} = 36.8\%$

f) $R_{eq} = 5.742\ \Omega,\qquad \dfrac{\Delta V_o}{V_o} = 0.04\%,\qquad \dfrac{\Delta V_o}{V_o} = 2.32\%$ (noninterleaved operation)

Topic: Two-level Voltage Source Inverter with Space Vector Modulation

4-12 (Solved Problem). Consider a two-level voltage source inverter shown in Figure 4-23a. The inverter is connected to a three-phase balanced RL load, and its DC voltage is 1220V. The inverter is modulated by space vector modulation and its simulation algorithm is given in Figure 4-30. The switching frequencey of the converter is 720 Hz. At a given instance of time, the three-phase reference voltages, v_a^*, v_b^*, and v_c^* are found to be 398.37 V, 145.81 V, and −544.18 V, respectively. Determine the following:

a) The α-β components of the three-phase reference voltages
b) The angle of the reference voltage vector and sector number
c) The reference voltage vector and modulation index
d) The dwell times
e) Three stationary space vectors that are used to synthesize the reference voltage vector $\vec{v}_{ref}$
f) The waveforms of v_{aN}, v_{bN}, v_{cN}, and v_{ab} with a seven-segment switching pattern that can satisfy the switching sequence design requirements given in Section 4.4.2.

Solution:

a) The α-β components of the three-phase reference voltages based on the $abc/\alpha\beta$ transformation:

$$\begin{bmatrix} v_\alpha \\ v_\beta \end{bmatrix} = \frac{2}{3}\begin{bmatrix} 1 & -1/2 & -1/2 \\ 0 & \sqrt{3}/2 & -\sqrt{3}/2 \end{bmatrix}\begin{bmatrix} v_a^* \\ v_b^* \\ v_c^* \end{bmatrix}$$

from which

$$v_\alpha = v_a^* = 398.37 \text{ V}$$

$$v_\beta = \frac{\sqrt{3}}{3}\left(v_a^* + 2v_b^*\right) = \frac{\sqrt{3}}{3}(398.37 + 2\times145.81) = 398.37 \text{ V}$$

b) The angle of the reference voltage vector:

$$\theta = \tan^{-1}\frac{v_\beta}{v_\alpha} = \tan^{-1}\frac{398.37}{398.37} = 45° \quad (\pi/4 \text{ rad})$$

The sector number: $k = \text{I}$ (refer to Figure 4-27).

c) The reference voltage vector:

$$v_{ref} = \sqrt{v_\alpha^2 + v_\beta^2} = \sqrt{398.37^2 + 398.37^2} = 563.38 \text{ V}$$

$$\vec{v}_{ref} = v_{ref}\, e^{j\theta} = 563.38\angle 45° \text{ V}$$

The modulation index:

$$m_a = \frac{\sqrt{3}\, v_{ref}}{V_{dc}} = \frac{\sqrt{3} \times 563.38}{1220} = 0.8$$

d) The modified reference voltage angle for dwell times calculation:

$$\theta' = \theta - (k-1)\pi/ = 45^\circ \qquad \text{for } k = 1$$

The dwell times for voltage vectors $\vec{V}_1$, $\vec{V}_2$, and $\vec{V}_0$:

$$\begin{cases} T_a = \dfrac{\sqrt{3}\,T_s\, v_{ref}}{V_{dc}} \sin\left(\dfrac{\pi}{3} - \theta'\right) = \dfrac{\sqrt{3} \times (1/720) \times 563.38}{1220} \sin\left(\dfrac{\pi}{3} - \dfrac{\pi}{4}\right) = 287.52 \;\; \mu s \\[4mm] T_b = \dfrac{\sqrt{3}\,T_s\, v_{ref}}{V_{dc}} \sin\theta' = \dfrac{\sqrt{3} \times (1/720) \times 563.38}{1220} \sin\left(\dfrac{\pi}{4}\right) = 785.52 \;\; \mu s \\[4mm] T_0 = T_s - T_a - T_b = 315.85 \;\; \mu s \end{cases}$$

Note: Since $\dfrac{\pi}{6} < \theta' < \dfrac{\pi}{3}$, the dwell time T_a is less than T_b (refer to Table 4-3).

e) The voltage space vectors. Since the reference voltage vector lies in sector I, it can be synthesized by vectors $\vec{V}_1$, $\vec{V}_2$, and $\vec{V}_0$, respectively:

$$\vec{V}_1 = \frac{2}{3}V_{dc}e^{j0} = \frac{2}{3}V_{dc} = 813.33 \;\; \text{V}$$

$$\vec{V}_2 = \frac{2}{3}V_{dc}\, e^{j\frac{\pi}{3}} = \frac{2}{3}V_{dc}\,(\cos\pi/3 + j\sin\pi/3) = 406.67 + j704.37 = 813.33\angle 60^\circ \;\; \text{V}$$

$$\vec{V}_0 = 0$$

f) For the waveforms of v_{aN}, v_{bN}, v_{cN}, and v_{ab}, see Figure 4-29.

4-13. Repeat Problem 4-12 if the three-phase reference voltages are found to be 0V, 487.90 V, and –487.90 V, respectively.

Answers:

a) $v_\alpha = 0$ V, $\qquad v_\beta = 563.38$ V

b) $\theta = 90^\circ$, $\qquad k = \text{II}$

c) $v_{ref} = 563.38$ V, $\qquad m_a = 0.8$

d) $\theta' = 30^\circ$, $\qquad T_a = 555.45\mu s$, $\qquad T_b = 555.45\mu s$, $\qquad T_0 = 278\mu s$

e) $\vec{V}_2 = 406.67 + j704.37 = 813.33\angle 60^\circ$ V, $\qquad \vec{V}_3 = 813.33\angle 120^\circ$ V, $\qquad \vec{V}_0 = 0$

f) See Figure C-4.

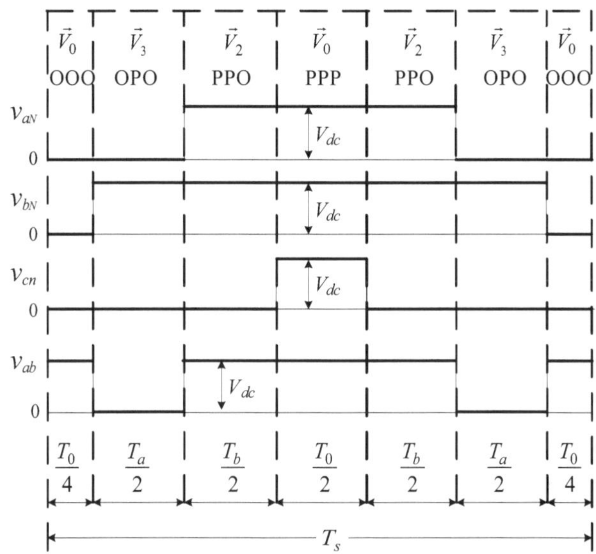

<u>Figure C-4.</u> Waveforms of v_{aN}, v_{bN}, v_{cN}, and v_{ab} with $\vec{v}_{ref}$ in Sector II.

4-14. Repeat Problem 4-12 if the three-phase reference voltages are found to be -398.37 V, -145.81 V, and 544.18 V, respectively.

Answers:

a) $v_\alpha = -398.37$ V, $\qquad v_\beta = -398.37$ V
b) $\theta = 225°$, $\qquad k = IV$
c) $v_{ref} = 563.38$ V, $\qquad m_a = 0.8$
d) $\theta' = 45°$, $\qquad T_a = 287.52\,\mu s$, $\qquad T_b = 785.52\,\mu s$, $\qquad T_0 = 315.85\,\mu s$
e) $\vec{V}_4 = 813.33\angle180°$ V, $\qquad \vec{V}_5 = -406.67 - j704.37 = 813.33\angle240°$ V, $\qquad \vec{V}_0 = 0$
f) See Figure C-5.

Topic: PWM Current Source Inverter with Space-Vector Modulation

4-15 (Solved Problem). A PWM three-phase current source inverter shown in Figure 4-38 is connected to a three-phase, balanced RL load, and its DC link current is 827.6 A. The inverter is modulated by space vector modulation whose space vector diagram is given in Figure 4-40. The switching frequency of the inverter is 720 Hz. At a given instate of time, the three-phase references currents are fund to be -413.80 A, 565.26 A, and -151.460 A, respectively. Find the following:

a) The α-β components of the three-phase reference currents
b) The angle of the reference current vector and sector number

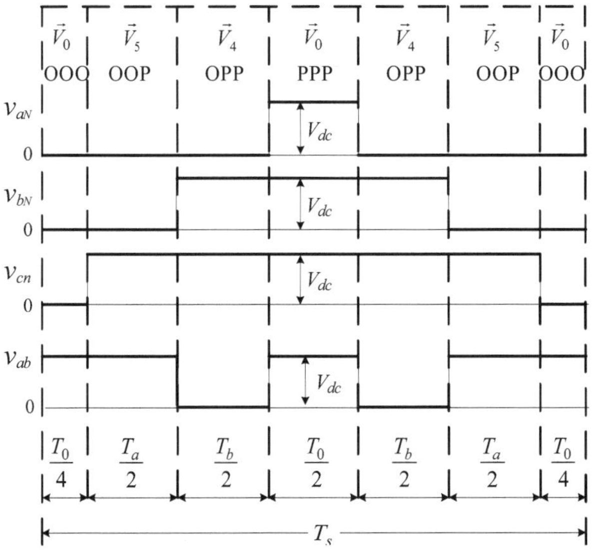

Figure C-5. Waveforms of v_{aN}, v_{bN}, v_{cN}, and v_{ab} with $\vec{v}_{ref}$ in Sector IV.

c) The reference current vector and the modulation index
d) The dwell times
e) Three stationary space vectors that are used to synthesize the reference current vector $\vec{i}_{ref}$

Solution:

a) The α-β components of the three-phase reference currents:

$$\begin{bmatrix} i_\alpha \\ i_\beta \end{bmatrix} = \frac{2}{3}\begin{bmatrix} 1 & -1/2 & -1/2 \\ 0 & \sqrt{3}/2 & -\sqrt{3}/2 \end{bmatrix}\begin{bmatrix} i_a^* \\ i_b^* \\ i_c^* \end{bmatrix}$$

from which

$$i_\alpha = i_a^* = -413.80 \ \text{A}$$

$$i_\beta = \frac{\sqrt{3}}{3}\left(i_a^* + 2i_b^*\right) = \frac{\sqrt{3}}{3}\left(-413.8 + 2\times 565.26\right) = 413.8 \ \text{A}$$

b) The angle of the reference current vector:

$$\theta = \tan^{-1}\frac{i_\beta}{i_\alpha} = \tan^{-1}\frac{413.8}{-413.8} = 135° \quad (3\pi/4 \ \text{rad})$$

The sector number, k = III (refer to Figure 4-41)

c) The reference current vector:

$$i_{ref} = \sqrt{i_\alpha^2 + i_\beta^2} = \sqrt{413.8^2 + 413.8^2} = 585.2 \text{ A}$$

$$\vec{i}_{ref} = i_{ref}\, e^{j\theta} = 585.2\angle 135° \text{ A}$$

The modulation index:

$$m_a = \frac{\hat{I}_{w1}}{I_{dc}} = \frac{\sqrt{2}\times 585.2}{827.6} = 1.0$$

d) The modified reference vector angle for dwell times calculation:

$$\theta' = \theta - (k-1)\pi/3 = 15° \text{ for } k = 3$$

The dwell times for current vectors $\vec{I}_1$, $\vec{I}_2$, and $\vec{I}_0$:

$$\begin{cases} T_1 = m_a T_s \sin(\pi/6 - \theta') = 1.0\times(1/720)\times\sin(\pi/6 - \pi/12) = 254.18 \ \mu s \\ T_2 = m_a T_s \sin(\pi/6 + \theta') = 1.0\times(1/720)\times\sin(\pi/6 + \pi/12) = 694.44 \ \mu s \\ T_0 = T_s - T_1 - T_2 = 440.26 \ \mu s \end{cases}$$

e) The current space vectors:
Since the reference current vector lies in Sector III, it can be synthesized by current vectors $\vec{I}_3$, $\vec{I}_4$, and $\vec{I}_0$, respectively.

$$\vec{I}_3 = \frac{2}{\sqrt{3}} I_{dc} e^{j\frac{2\pi}{3}-\frac{\pi}{6}} = \frac{2}{\sqrt{3}} I_{dc}(\cos\pi/2 + j\sin\pi/2) = j955.63 = 955.63\angle 90° \text{ A}$$

$$\vec{I}_4 = \frac{2}{\sqrt{3}} I_{dc} e^{j\frac{3\pi}{3}-\frac{\pi}{6}} = \frac{2}{\sqrt{3}} I_{dc}(\cos 5\pi/6 + j\sin 5\pi/6) = j955.63 = 955.63\angle 150° \text{ A}$$

$$\vec{I}_0 = 0$$

4-16. Repeat Problem 4-15 if the three-phase reference currents are found to be 540.65A, −76.38A, and −464.27A respectively.

Answers:

a) $i_\alpha = 540.65$ A, $i_\beta = 223.94$ A
b) $\theta = 2.5°$, $k = I$
c) $i_{ref} = 585.2$ A, $m_a = 1.0$
d) $\theta' = 22.5°$, $T_1 = 128.19 \ \mu s$, $T_2 = 779.14 \ \mu s$, $T_0 = 481.56 \ \mu s$
e) $\vec{I}_1 = 955.63\angle -30°$ A, $\vec{I}_2 = 955.63\angle 30°$ A, $\vec{I}_0 = 0$

Topic: Control of Grid-Connected Converters

4-17 (Solved Problem). Consider a grid-connected two-level voltage source inverter with voltage-oriented control (VOC) shown in Figure 4-48. The inverter is connected to the grid of 690 V/50 Hz and delivers 2.3 MW to the grid with unity power-factor operation. The inverter is modulated by an SVM scheme with modulation index of 0.8 and operates under steady-state conditions. The line inductance L_g is 0.1098 mH. To simplify the analysis, all the harmonics produced by the inverter are neglected. When the grid voltage vector angle θ_g is –45°, determine the following:

a) The instantaneous three-phase grid voltages and currents
b) The grid voltage angle
c) The dq-axis grid voltages and currents
d) The active and reactive powers delivered to the grid (using dq-axis grid voltages and currents)
e) The dc-link voltage and current
f) The dq-axis and three-phase reference voltages for the PWM modulator

Solution:

a) The grid voltage, current and frequency:

$$v_g = \frac{690}{\sqrt{3}} \times \sqrt{2} = 563.38 \ \text{V (peak)},$$

$$i_g = \sqrt{2} \frac{P_g}{3(v_g/\sqrt{2})} = \frac{2P_g}{3v_g} = \frac{2 \times 2.3 \times 10^6}{3 \times 563.38} = 2721.6 \ \text{A (peak)}$$

$$\varphi_g = 180^\circ \quad \text{(generating mode; refer to Figure 4-46)}$$

$\omega_g = 2\pi \times 50 = 314.16$ rad/sec. At $\theta_g = -45^\circ$, $\omega_g t = \theta_g = -\pi/4$
The instantaneous three-phase grid voltages:

$$\begin{cases} v_{ag} = v_g \cos \omega_g t = 563.38 \times \cos(-\pi/4) = 398.37 \ \text{V (peak)} \\ v_{bg} = v_g \cos\left(\omega_g t - 2\pi/3\right) = 563.38 \times \cos(-\pi/4 - 2\pi/3) = -544.19 \ \text{V (peak)} \\ v_{cg} = v_g \cos\left(\omega_g t + 2\pi/3\right) = 563.38 \times \cos(-\pi/4 + 2\pi/3) = 145.81 \ \text{V (peak)} \end{cases}$$

The instantaneous three-phase grid currents:

$$\begin{cases} i_{ag} = i_g \cos(\omega_g t - \varphi_g) = 2721.6 \times \cos(-\pi/4 - \pi) = -1924.5 \ \text{A (peak)} \\ i_{bg} = i_g \cos\left(\omega_g t - \varphi_g - 2\pi/3\right) = 2721.6 \times \cos(-\pi/4 - \pi - 2\pi/3) = 2628.9 \ \text{A (peak)} \\ i_{cg} = i_g \cos\left(\omega_g t - \varphi_g + 2\pi/3\right) = 2721.6 \times \cos(-\pi/4 - \pi + 2\pi/3) = -704.42 \ \text{A (peak)} \end{cases}$$

b) The α–β components of the three-phase grid voltages:

$$\begin{bmatrix} v_\alpha \\ v_\beta \end{bmatrix} = \frac{2}{3}\begin{bmatrix} 1 & -1/2 & -1/2 \\ 0 & \sqrt{3}/2 & -\sqrt{3}/2 \end{bmatrix}\begin{bmatrix} v_{ag} \\ v_{bg} \\ v_{cg} \end{bmatrix}$$

from which

$$v_\alpha = v_{ag} = 398.37 \text{ V (peak)}$$

$$v_\beta = \frac{\sqrt{3}}{3}\left(v_{ag} + 2v_{bg}\right) = \frac{\sqrt{3}}{3}(398.37 + 2\times -544.19) = -398.37 \text{ V (peak)}$$

Verify the grid voltage vector angle:

$$\theta_g = \tan^{-1}\frac{v_\beta}{v_\alpha} = \tan^{-1}\frac{-398.37}{398.37} = -\pi/4 \text{ rad}$$

c) The dq/abc transformation with voltage oriented control is given by

$$\begin{bmatrix} x_{dg} \\ x_{qg} \end{bmatrix} = \frac{2}{3}\begin{bmatrix} \cos\theta_g & \cos(\theta_g - 2\pi/3) & \cos(\theta_g - 4\pi/3) \\ -\sin\theta_g & -\sin(\theta_g - 2\pi/3) & -\sin(\theta_g - 4\pi/3) \end{bmatrix} \cdot \begin{bmatrix} x_{ag} \\ x_{bg} \\ x_{cg} \end{bmatrix}$$

from which, the dq-axis grid voltages can be calculated by

$$v_{dg} = \frac{2}{3}\left(v_{ag}\cos\theta_g + v_{bg}\cos(\theta_g - 2\pi/3) + v_{cg}\cos(\theta_g - 4\pi/3)\right)$$

$$= \frac{2}{3}\left(398.37\times\cos(-\pi/4) - 544.19\times\cos(-\pi/4 - 2\pi/3) + 145.81\times\cos(-\pi/4 - 4\pi/3)\right)$$

$$= 563.38 \text{ V (peak)}$$

$$v_{qg} = -\frac{2}{3}\left(v_{ag}\sin\theta_g + v_{bg}\sin(\theta_g - 2\pi/3) + v_{cg}\sin(\theta_g - 4\pi/3)\right)$$

$$= -\frac{2}{3}\left(398.37\times\sin(-\pi/4) - 544.19\times\sin(-\pi/4 - 2\pi/3) + 145.81\times\sin(-\pi/4 - 4\pi/3)\right)$$

$$= 0 \text{ V}$$

The dq-axis grid currents can be found in a similar way:

$$i_{dg} = i_{dg}^* = \frac{2}{3}\left(i_{ag}\cos\theta_g + i_{bg}\cos(\theta_g - 2\pi/3) + i_{cg}\cos(\theta_g - 4\pi/3)\right)$$

$$= \frac{2}{3}\left(-1924.5\times\cos(-\pi/4) - 2628.9\times\cos(-\pi/4 - 2\pi/3) - 704.42\times\cos(-\pi/4 - 4\pi/3)\right)$$

$$= -2721.65 \text{ A (peak)}$$

$$i_{qg} = i_{qg}^* = -\frac{2}{3}\left(i_{ag}\sin\theta_g + i_{bg}\sin(\theta_g - 2\pi/3) + i_{cg}\sin(\theta_g - 4\pi/3)\right)$$

$$= -\frac{2}{3}\left(-1924.5 \times \sin(-\pi/4) - 2628.9 \times \sin(-\pi/4 - 2\pi/3) - 704.42 \times \sin(-\pi/4 - 4\pi/3)\right)$$

$$= 0 \ \text{A (peak)}$$

d) The active and reactive powers delivered to the grid:

$$
\begin{cases}
P_g = \dfrac{3}{2}(v_{dg}i_{dg} + v_{qg}i_{qg}) = \dfrac{3}{2}(563.38 \times -2721.6 + 0) = -2300 \times 10^3 \ \text{W} \quad (-1.0\,\text{pu}) \\[3mm]
Q_g = \dfrac{3}{2}(v_{qg}i_{dg} - v_{dg}i_{qg}) = -\dfrac{3}{2}(0 - 0) = 0 \ \text{VAR}
\end{cases}
$$

e) The DC-link voltage and current:

$$V_{dc} = \frac{\sqrt{2}V_{abil}}{m_a} = \frac{\sqrt{2}\times690}{0.8} = 1220 \ \text{V}$$

The modulation index m_a is 0.8, leaving 20% margin for adjustments.

$$I_{dc} = \frac{P_g}{V_{dc}} = \frac{-2300\times10^3}{1220} = -1885.6$$

f) The dq-axis reference voltages for the PWM modulator:

$$
\begin{cases}
v_{di} = -(k_1 + k_2/S)(i_{dg}^* - i_{dg}) + \omega_g L_g i_{qg} + v_{dg} \\[2mm]
v_{qi} = -(k_1 + k_2/S)(i_{qg}^* - i_{qg}) - \omega_g L_g i_{dg} + v_{qg}
\end{cases}
$$

In steady-state $i_{dg}^* = i_{dg}$ and $i_{qg}^* = i_{qg}$, thus

$$
\begin{cases}
v_{di} = \omega_g L_g i_{qg} + v_{dg} = 0 + 563.38 = 563.38 \ \text{V (peak)} \\[2mm]
v_{qi} = -\omega_g L_g i_{dg} + v_{qg} = -314.16\times0.1098\times10^{-3}\times-(2721.65) + 0 = 93.88 \ \text{V (peak)}
\end{cases}
$$

Note: Though the q-axis grid voltage v_{qg} is zero, its reference v_{qi} is not zero because of the $-\omega_g L_g i_{dg}$ term used in the decoupled controller.

The three-phase reference voltages for the PWM modulator can be obtained by dq/abc transformation:

$$
\begin{bmatrix} x_{ai} \\ x_{bi} \\ x_{ci} \end{bmatrix} =
\begin{bmatrix}
\cos\theta_g & -\sin\theta_g \\
\cos(\theta_g - 2\pi/3) & -\sin(\theta_g - 2\pi/3) \\
\cos(\theta_g - 4\pi/3) & -\sin(\theta_g - 4\pi/3)
\end{bmatrix} \cdot
\begin{bmatrix} x_{di} \\ x_{qi} \end{bmatrix}
$$

from which

$$v_{ai}^* = v_{di} \cos\theta_g - v_{qi} \sin\theta_g$$
$$= 563.38 \times \cos(-\pi/4) - 93.88 \times \sin(-\pi/4) = 464.76 \ \text{V (peak)}$$

$$v_{bi}^* = v_{di} \cos(\theta_g - 2\pi/3) - v_{qi} \sin(\theta_g - 2\pi/3)$$
$$= 563.38 \times \cos(-\pi/4 - 2\pi/3) - 93.88 \times \sin(-\pi/4 - 2\pi/3) = -519.89 \ \text{V (peak)}$$

$$v_{ci}^* = v_{di} \cos(\theta_g - 4\pi/3) - v_{qi} \sin(\theta_g - 4\pi/3)$$
$$= 563.38 \times \cos(-\pi/4 - 4\pi/3) - 93.88 \times \sin(-\pi/4 - 4\pi/3) = 55.13 \ \text{V (peak)}$$

Cross Check:

$$P_g = 3V_g I_g \cos\varphi_g = 3 \times \frac{690}{\sqrt{3}} \times 1924.5 \times \cos(180°) = -2300 \times 10^3 \ \text{W}$$

$$Q_g = 3V_g I_g \sin\varphi_g = 3 \times \frac{690}{\sqrt{3}} \times 1924.5 \times \sin(180°) = 0 \ \text{VAR}$$

4-18. Repeat Problem 4-17 if the grid-side power factor is 0.8 leading. Perform the calculations when the grid voltage vector angle θ_g is 90°.

Answers:

a) $\varphi_g = 216.87°$ (3.785 rad), $\quad v_{ag} = 0$ V, $\quad v_{bg} = 487.9$ V, $\quad v_{cg} = -487.9$ V
$\quad i_{ag} = -1633$ A, $\quad i_{bg} = -1069.1$ A, $\quad i_{cg} = 2702.1$ A
b) $v_\alpha = 0$ V, $\quad v_\beta = 563.38$ V, $\quad \theta_g = 90°$ $\quad (\pi/2 \text{ rad})$
c) $v_{dg} = 563.38$ V, $\quad v_{qg} = 0$ V, $\quad i_{dg} = -2177.3$ A, $\quad i_{qg} = 1633$ A
d) $P_g = -1840 \times 10^3$ W $(-0.8\,\text{pu})$, $\quad Q_g = -1380 \times 10^3$ VAR $(-0.6\,\text{pu})$
e) $V_{dc} = 1220$ V, $\quad I_{dc} = -1508.5$ A
f) $v_{di} = 619.71$ V, $\quad v_{qi} = 75.11$ V, $\quad v_{ai}^* = -75.11$ V, $\quad v_{bi}^* = 574.24$ V,
$\quad v_{ci}^* = -499.13$ V

4-19. Repeat Problem 4-17 if the grid-side power factor is 0.9 lagging and the inverter delivers 50% of the rated power to the grid. Perform the calculations when the grid voltage vector angle θ_g is 180°.

Answers:

a) $i_g = 1360.8$ A, $\quad \varphi_g = 154.16°$ (2.691 rad), $\quad v_{ag} = -563.38$ V,
$\quad v_{bg} = 281.69$ V, $\quad v_{cg} = 281.69$ V, $\quad i_{ag} = 1224.7$ A, $\quad i_{bg} = -98.67$ A,
$\quad i_{cg} = -1126.1$ A
b) $v_\alpha = -563.38$ V, $\quad v_\beta = 0$ V, $\quad \theta_g = -180°$ $\quad (-\pi \text{ rad})$
c) $v_{dg} = 563.38$ V, $\quad v_{qg} = 0$ V, $\quad i_{dg} = -1224.7$ A, $\quad i_{qg} = -593.171$ A

d) $P_g = -1035 \times 10^3$ W $(-0.45 \, \text{pu})$, $Q_g = 501.27 \times 10^3$ VAR $(0.2179 \, \text{pu})$

e) $V_{dc} = 1220$ V, $I_{dc} = -848.53$ A

f) $v_{di} = 542.92$ V, $v_{qi} = 42.25$ V, $v_{ai}^* = -542.92$ V, $v_{bi}^* = 234.87$ V,

   $v_{ci}^* = 308.05$ V

PROBLEMS AND ANSWERS FOR CHAPTER 6—FIXED-SPEED INDUCTION GENERATOR WECS

Topic: Steady-State and Power Flow Analysis of Induction Generators Using Conventional Equivalent Circuit

6-1 (Solved Problem). This problem together with Problems 6-2 and 6-3 are designed (1) to investigate the transient response of a squirrel cage induction generator directly connected to the grid when the speed of the generator is brought by the wind to below, at, and above its synchronous speed, respectively, and (2) to verify that a large SCIG cannot be connected to the grid directly due to the excessive inrush current and torque oscillations.

Consider a 1.45 MW, 575 V, 50 Hz squirrel cage induction generator WECS shown in Figure C-6. The nameplate and parameters of the generator are listed in Table B-2 of Appendix B. The rated stator current, electromagnetic torque, and rotor speed of the generator are 1723 A, 13.94 kN·m, and 1007.2 rpm, respectively. When the wind speed is above its cut-in speed, the pitch angle of the blade is slightly adjusted into the wind from the position in which the turbine was parked such that a small amount of torque is produced to accelerate the turbine and the generator.

When the rotor speed reaches 977 rpm (0.97 pu), the generator is connected to the grid by switch S. Assuming that (1) the mechanical torque applied to the generator shaft is negligibly small ($T_m = 0$) during the startup process, and (2) the total moment of inertia of the system is 1000 kg·m². Develop a Simulink program to investigate the start-up transients, and answer the following questions.

a) Plot the simulated waveforms for the stator currents i_{as}, i_{bs}, and i_{cs} (kA), electromagnetic torque T_e (kN·m), and rotor mechanical speed n_r (rpm).

b) Determine the maximum positive and negative peak values of the stator current (kA) and electromagnetic torque (kN·m).

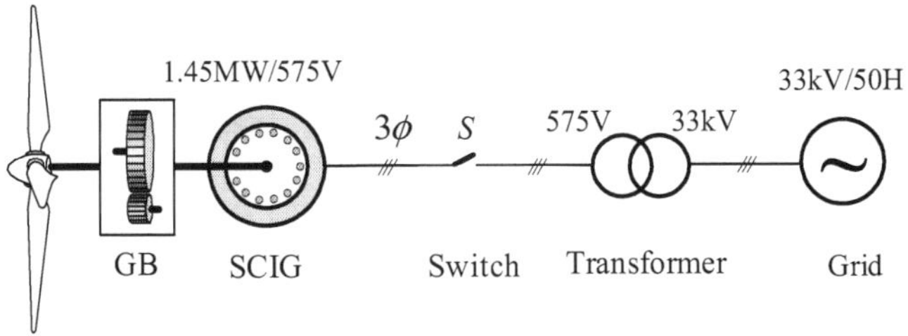

Figure C-6. Single-line diagram of a three-phase SCIG WECS with direct grid connection.

c) Determine rotor speed overshoot (%) and the time at which the rotor speed reaches its maximum value.

Solution:

a) The simulated waveforms for the stator currents, electromagnetic torque, and rotor mechanical speed are given in Figure C-7.

b) The maximum positive and negative peak values of the stator current are 17.57 kA and −13.58 kA, respectively. The maximum positive and negative peak values of electromagnetic torque are 20.77 kN·m and −10.96 kN·m, respectively. The torque is oscillatory during the transients, which may cause excessive stress to the mechanical system of the wind turbine.

c) The rotor speed overshoot is given by $(n_r - n_{syn})/n_{syn} = (1004.9 - 1000)/1000 = 0.49\%$, and the time at which the rotor speed reaches its maximum value is 0.36 sec.

6-2. Repeat Problem 6-1 with switch S closed when the generator speed reaches its synchronous speed of 1000 rpm.

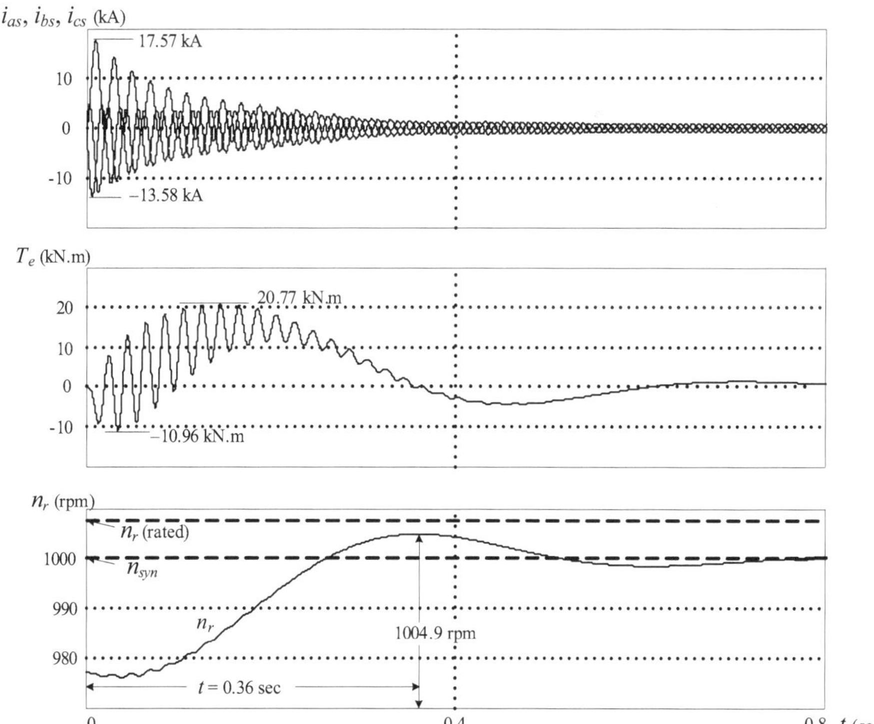

Figure C-7. Dynamic response of 1.45 MW/575 V fixed-speed SCIG WECS with direct grid connection.

Solution:

a) Not provided.

b) The maximum positive and negative peak values of the stator current are 17.59 kA and −13.60 kA, respectively. The maximum positive and negative peak values of electromagnetic torque are 10.82 kN·m and −13.60 kN·m, respectively.

c) The rotor speed overshoot is given by $(n_r - n_{syn})/n_{syn} = 0.037\%$, and the time at which the rotor speed reaches its maximum value is 0.41 sec.

6-3. Repeat Problem 6-1 with the following modifications: Switch S is closed when the generator speed reaches 1037 rpm (1.03pu). Compare your results with those in Problems 6-1 and 6-2, and draw conclusions.

Solution:

a) Not provided.

b) The maximum positive and negative peak values of the stator current are 17.62 kA and −13.61 kA, respectively. The maximum positive and negative peak values of electromagnetic torque are 5.39 kN·m and −27.09 kN·m, respectively.

c) The rotor speed overshoot is given by $(n_r - n_{syn})/n_{syn} = 0.21\%$, and the time at which the rotor speed reaches its maximum value is 0.615 sec.

6-4 (Solved Problem). Consider a fixed-speed SCIG wind energy conversion system in Figure C-8, where a soft starter is employed. A 1.45 MW, 575 V, 50 Hz SCIG is used, and its parameters are given in Table B-2 of Appendix B. The rated stator current, electromagnetic torque, and rotor speed of the generator are 1723 A, 13.94 kN·m, and 1007.2 rpm, respectively. When the rotor speed is brought by the wind to 977 rpm (0.97 pu), the generator is connected to the grid by switch S. In the meantime the soft starter is activated with a large firing angle of 120°. The firing angle is then linearly decreased from 120° to 0° in 0.5 sec. Assuming that (1) the mechanical torque applied to the generator shaft is negligibly small ($T_m = 0$) during the startup process, and (2) the total moment of inertia of the mechanical system is 1000 kg·m², develop a Simulink program to investigate the start-up transients, and answer the following questions.

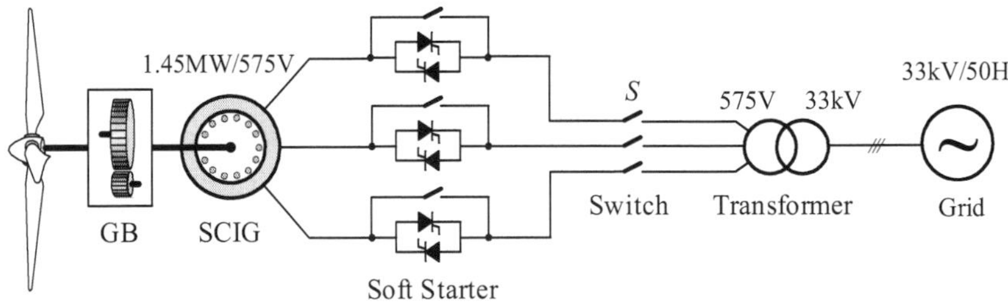

Figure C-8. SCIG fixed-speed wind energy system with a soft starter.

a) Plot the simulated waveforms for the stator currents i_{as}, i_{bs}, and i_{cs} (kA), electromagnetic torque T_e (kN·m), and rotor mechanical speed n_r (rpm).

b) Determine the maximum positive and negative peak values of the stator current (kA), and maximum positive peak value electromagnetic torque (kN·m).

c) Determine rotor speed overshoot (%) and the time at which the rotor speed reaches its maximum value.

d) Compare your results with those given in Problem 6-1 and draw conclusions.

Solution:

a) The simulated waveforms for the stator currents, electromagnetic torque, and rotor mechanical speed are given in Figure C-9.

b) The maximum positive and negative peak values of the stator current are 4.52 kA and –4.52 kA, respectively. The maximum positive peak value of electromagnetic torque is 14.45 kN·m.

c) The rotor speed overshoot is $(n_r - n_{syn})/n_{syn} = 0.445$ %, and the time at which the rotor speed reaches its maximum value is 0.435 sec.

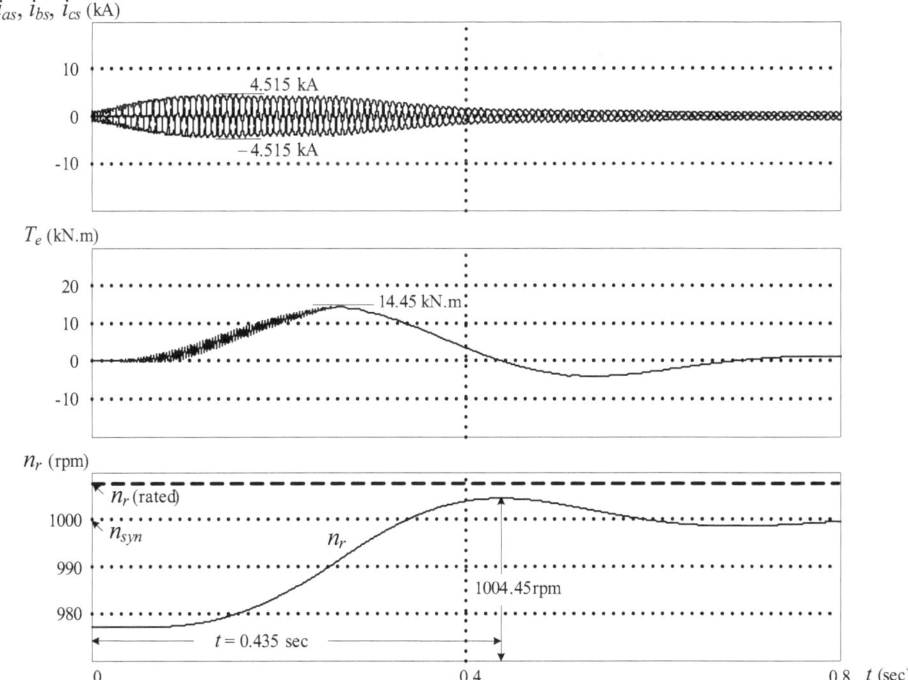

Figure C-9. Dynamic response of 1.45 MW/575 V fixed-speed SCIG WECS with soft starter.

6-5. Repeat Problem 6-4 with the following modifications: the firing angle is linearly decreased from $120°$ to $0°$ in 0.25 sec.

Solution:

a) Not provided.

b) The maximum positive and negative peak values of the stator current are 7.075 kA and -7.075 kA, respectively. The maximum positive peak value of electromagnetic torque is 14.15 kN·m.

c) The rotor speed overshoot is $(n_r - n_{syn})/n_{syn} = 0.444\%$, and the time at which the rotor speed reaches its maximum value is 0.4 sec.

6-6. Repeat Problem 6-4 with the firing angle linearly decreased from $120°$ to $0°$ in 1.0 sec. Compare your results with those in Problems 6-4 and 6-5, and make conclusions.

Solution:

a) Not provided.

b) The maximum positive and negative peak values of the stator current are 3.51 kA and -3.51 kA, respectively. The maximum positive peak value of electromagnetic torque is 12.65 kN·m.

c) The rotor speed overshoot is $(n_r - n_{syn})/n_{syn} = 0.6\%$, and the time at which the rotor speed reaches its maximum value is 0.57 sec.

6-7. Repeat Problem 6-4 using a 3.0 MW, 3000 V, 60 Hz generator with its parameters given in Table B-3 of Appendix B.

Solution:

a) Not provided.

b) The maximum positive and negative peak values of the stator current are 2.13 kA and -2.13 kA, respectively. The maximum positive peak value of electromagnetic torque is 25.8 kN·m.

c) The rotor speed overshoot is $(n_r - n_{syn})/n_{syn} = (1805.1 - 1800)/1800 = 0.51\%$, and the time at which the rotor speed reaches its maximum value is 0.51 sec.

Topic: Reactive Power Compensation

6-8 (Solved Problem). A medium voltage fixed-speed SCIG WECS using a 3.0 MW, 3000 V, 60 Hz, 1812 rpm squirrel cage induction generator is shown in Figure C-10, where a capacitor C is installed as a power factor compensator. The nameplate and parameters of the generator are listed in Table B-3 of Appendix B. The system is connected to the transformer station of 3 kV/60 Hz. Assuming that the generator operates under its rated conditions with the rated speed of 1812 rpm, answer the following questions.

a) Determine the total impedance of the generator $\overline{Z}_s$.

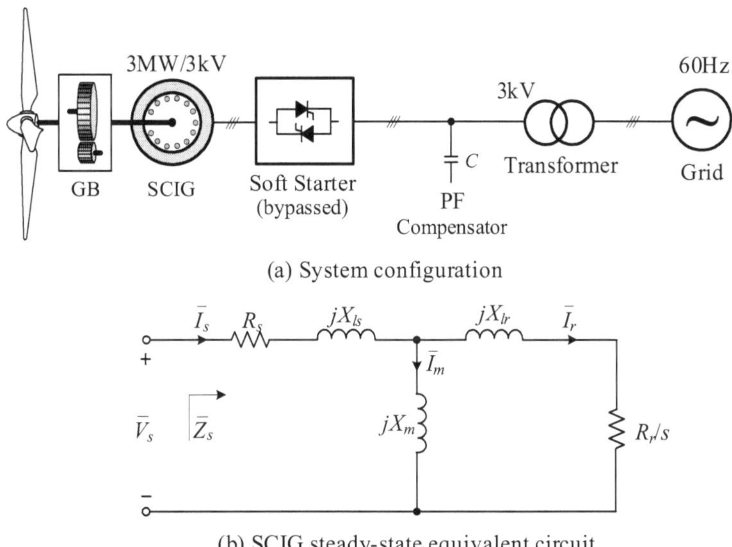

Figure C-10. A fixed-speed medium voltage WECS with PF compensation.

b) Calculate the stator current $\bar{I}_s$ and stator power factor PF_s.

c) Determine the stator apparent, active and reactive power S_s, P_s, and Q_s.

d) Find the value of the capacitor C (per phase, wye-connection) for the system to operate at unity power factor.

Solution:

a) The total impedance of the generator $\overline{Z}_s$ is calculated by

$$\overline{Z}_s = R_s + jX_{ls} + jX_m \,//\left(\frac{R_r}{s} + jX_{lr}\right) = 2.681\angle153.30°\ \Omega$$

where $s = \dfrac{1800 - 1812}{1800} = -0.0067$

b) The stator current is found from

$$\bar{I}_s = \frac{\bar{V}_s}{\overline{Z}_s} = \frac{3000/\sqrt{3}\angle0°}{2.681\angle153.30°} = 646.1\angle-153.30°\ A$$

and the stator power factor is given by

$$PF_s = \cos\varphi_s = -0.8934$$

where $\varphi_s = \angle\bar{V}_s - \angle\bar{I}_s = 153.30°$

c) The stator apparent, active and reactive power can be calculated by

$$S_s = 3V_s I_s = 3 \times 1732.1 \times 646.1 = 3.357 \text{ MVA}$$

$$P_s = S_s \cos(\varphi_s) = -2.999 \text{MW}$$

$$Q_s = S_s \sin(\varphi_s) = 1.508 \text{ MVAR}$$

d) The value of the capacitor C for unity power factor operation is obtained by

$$C = \frac{Q_c}{3(V_s)^2 \omega_s} = \frac{1.508 \times 10^6}{3 \times (1732.1)^2 \times (2\pi \times 60)} = 445 \ \mu\text{F} \ \ (0.45 \text{ pu per phase, wye-connection})$$

6-9. Repeat Problem 6-8 with the rotor speed of 1806 rpm.

Solution:

a) $\overline{Z}_s = 4.851\angle146.47° \ \Omega,$ $s = -0.0033$
b) $\overline{I}_s = 357.0\angle-146.47° \ \text{A},$ $PF_s = -0.8336$
c) $S_s = 1.855 \text{ MVA},$ $P_s = -1.546 \text{ MW},$ $Q_s = 1.025 \text{ MVAR}$
d) $C = 302 \ \mu\text{F}$

6-10. Repeat Problem 6-8 with the rotor speed of 1800 rpm (synchronous speed).

Solution:

a) $\overline{Z}_s = 10.537\angle89.91° \ \Omega,$ $s = 0$
b) $\overline{I}_s = 164.4\angle-89.91° \ \text{A},$ $PF_s = 0.0016$
c) $S_s = 0.854 \text{ MVA},$ $P_s = -1.347 \text{ kW},$ $Q_s = 0.854 \text{ MVAR}$
d) $C = 252 \ \mu\text{F}$

6-11. Repeat Problem 6-8 with the following modifications: (1) replace the 3.0 MW, 3000 V, 60 Hz generator with a 1.45 MW, 575 V, 50 Hz generator operating at the rotor speed of 1007.2 rpm; and (2) the generator is connected to the transformer station of 575 V/50 Hz. The nameplate and parameters of the generator are given in Table B-2 of Appendix B.

Solution:

a) The total impedance of the generator is calculated by

$$\overline{Z}_s = R_s + jX_{ls} + jX_m \ // \left(\frac{R_r}{s} + jX_{lr} \right) = 0.1927\angle147.55° \ \Omega$$

$$\text{where} \ \ s = \frac{1000 - 1007.2}{1000} = -0.0072$$

b) The stator current is found from

$$\bar{I}_s = \frac{\bar{V}_s}{\bar{Z}_s} = \frac{575/\sqrt{3}\angle 0^\circ}{0.1927\angle 147.55^\circ} = 1723.0\angle -147.55^\circ \text{ A}$$

and the stator power factor is given by

$$PF_s = \cos\varphi_s = -0.8439$$

where $\varphi_s = \angle\bar{V}_s - \angle\bar{I}_s = 147.55^\circ$

c) The stator apparent, active, and reactive powers are calculated by

$$S_s = 3V_s I_s = 3\times 331.98 \times 1723.0 = 1.716 \text{ MVA}$$

$$P_s = S_s \cos(\varphi_s) = -1.448\text{MW}$$

$$Q_s = S_s \sin(\varphi_s) = 0.921 \text{ MVAR}$$

d) The value of the capacitor C for unity power factor operation is obtained by

$$C = \frac{Q_c}{3(V_s)^2 \omega_s} = \frac{0.921\times 10^6}{3\times(331.98)^2 \times(2\pi\times 50)} = 8864 \text{ }\mu\text{F}$$

6-12. Repeat Problem 6-11 with the rotor speed of 1003.6 rpm (synchronous speed).

Solution:

a) $\bar{Z}_s = 0.3310\angle 138.93^\circ \text{ }\Omega,$ $s = -0.0036$
b) $\bar{I}_s = 1003.1\angle -138.93^\circ \text{ A},$ $PF_s = -0.7539$
c) $S_s = 0.999 \text{ MVA},$ $P_s = -0.753 \text{ MW},$ $Q_s = 0.656 \text{ MVAR}$
d) $C = 6391 \text{ }\mu\text{F}$

6-13. In Problems 6-8 to 6-12, the speed of the generator was given, based on which the capacitor value for reactive power compensation was calculated. In this problem, the calculation of reactive power compensation is based on given stator active power, but the generator speed is not known. Repeat Problem 6-8 when the stator active power is $P_s = -1$ pu. Use the simplified steady-state equivalent circuit of Figure C-11 to assist the calculation. Compare the results with those in Problem 6-8 and draw conclusions.

Solution:

a) From the simplified equivalent circuit, the following two equations can be obtained by

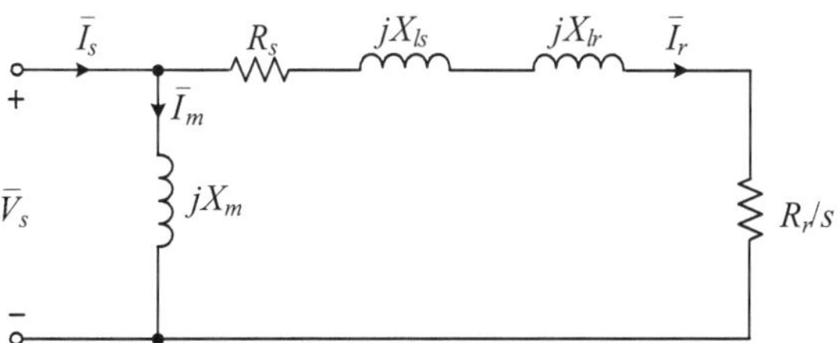

Figure C-11. Simplified equivalent circuit of a SCIG.

$$\begin{cases} 3I_s^2 R = P_s \\ I_s \times \sqrt{R^2 + X^2} = V_s \end{cases}$$

where $R = R_s + R_r/s$ and $X = X_{ls} + X_{lr}$

The sum of the resistance can be determined by solving

$$R^2 - \frac{3V_s^2}{P_s}R + X^2 = 0$$

$$R = \begin{cases} -2.880\,\Omega \\ -0.1207\,\Omega \end{cases}$$

The slip can be found from

$$s = \frac{R_r}{R - R_s} = \begin{cases} -0.0063 \\ -0.1322 \ \ (\text{omitted}) \end{cases}$$

The total impedance of the generator $\overline{Z}_s$ is calculated by

$$\overline{Z}_s = jX_m \,//\, \left(\frac{R_r}{s} + R_s + jX_{lr} + jX_{ls} \right) = 2.686\angle 153.54°\ \Omega$$

b) The stator current $\overline{I}_s$ is found from

$$\overline{I}_s = \frac{\overline{V}_s}{\overline{Z}_s} = \frac{3000/\sqrt{3}\angle 0°}{2.686\angle 153.54°} = 644.7\angle -153.54°\ \text{A}$$

and the stator power factor is given by

$$PF_s = \cos \varphi_s = -0.8952$$

where $\varphi_s = \angle \overline{V}_s - \angle \overline{I}_s = 153.54°$

c) The stator apparent, active and reactive power can be calculated by

$$S_s = 3V_s I_s = 3 \times 1732.1 \times 644.7 = 3.350 \text{ MVA}$$

$$P_s = S_s \cos(\varphi_s) = -2.999 \text{MW}$$

$$Q_s = S_s \sin(\varphi_s) = 1.493 \text{ MVAR}$$

d) The value of the capacitor C for unity power factor operation is obtained by

$$C = \frac{Q_c}{3(V_s)^2 \omega_s} = \frac{1.493 \times 10^6}{3 \times (1732.1)^2 \times (2\pi \times 60)} = 440 \ \mu\text{F} \quad (0.44 \text{ pu})$$

6-14. Repeat Problem 6-13 with the stator active power $P_s = -0.5$ pu.

Solution:

a) $s = -0.003, \qquad \overline{Z}_s = 4.951 \angle 145.58° \ \Omega$
b) $\overline{I}_s = 349.9 \angle -145.58° \text{ A}, \qquad PF_s = -0.8239$
c) $S_s = 1.818 \text{ MVA}, \qquad P_s = -1.50 \text{ MW}, \qquad Q_s = 1.028 \text{ MVAR}$
d) $\quad C = 303 \ \mu\text{F}$

PROBLEMS AND ANSWERS FOR CHAPTER 7–VARIABLE-SPEED INDUCTION GENERATOR BASED WECS

Assumptions for all the problems in this chapter:

1. Switching harmonics produced by power converters are neglected.
2. Power converters are ideal without power losses.
3. Core and rotational losses of the generator are neglected.

Topic: Steady-State Analysis of SCIG WECS with Direct FOC

7-1 (Solved Problem). Consider a 2.3 MW, 690 V, 50 Hz SCIG wind energy system operating with direct field-oriented control (FOC). At a wind speed of 12 m/s, the generator operates at 1.0 pu rotor speed. With the rated stator voltage of 690 V and rated slip of –0.008, the generator parameters are given in Table B-1 of Appendix B. The rotor flux λ_r is kept at its rated value of 1.7106 Wb (peak) by the direct field-oriented controller. Calculate the following:

a) The generator mechanical torque and power
b) The stator and rotor currents
c) The magnetizing, stator and rotor flux linkages
d) The *dq*-axis stator currents
e) The *dq*-axis stator voltages

Solution:

a) The generator mechanical torque at 1.0 pu rotor speed:

$$T_m = T_{m,R} \times (\omega_{m,pu})^2 = -14740 \times 1^2 = -14740 \text{ N·m}$$

The rated mechanical power:

$$P_{m,R} = \omega_{m,R} \times T_{m,R} = 1512(2\pi)/60 \times (-14740) = -2.3339 \times 10^6 \text{ W}$$

The generator mechanical power at 1.0 pu rotor speed:

$$P_m = P_{m,R} \times (\omega_{m,pu})^3 = -2.3339 \times 10^6 \times 1^3 = -2.3339 \times 10^6 \text{ W}$$

b) Selecting the stator voltage as a reference phasor, the stator voltage is

$$\overline{V}_s = 690/\sqrt{3}\angle 0° = 398.37\angle 0° \text{ V} \quad \text{(rms)}$$

The equivalent impedance of SCIG:

$$\overline{Z}_s = R_s + jX_{ls} + jX_m \,//\, \left(\frac{R_r}{s} + jX_{lr} \right) = 0.1838\angle 152.6° \ \Omega \ \text{ for } s = -0.008 \text{ (rated slip)}$$

where $X_{ls} = \omega_s L_{ls} = 2 \times 3.1416 \times 50 \times 0.06492 \times 10^{-3} = 0.0204$ Ω, $X_{lr} = \omega_s L_{lr} = 0.0204$ Ω, and $X_m = \omega_s L_m = 0.6706$ Ω
   The stator current:

$$\overline{I}_s = \frac{\overline{V}_s}{\overline{Z}_s} = \frac{690/\sqrt{3}\angle 0°}{0.1838\angle 152.6°} = I_s\angle\varphi_s = 2168\angle -152.6° \text{ A} \quad \text{(rms)}$$

The rotor current can be calculated by

$$\overline{I}_r = \frac{jX_m\overline{I}_s}{jX_m + \left(\dfrac{R_r}{s} + jX_{lr} \right)} = 2030.8\angle -167.7° \text{ A} \quad \text{(rms)}$$

c) The magnetizing flux linkage can be expressed as

$$\overline{\Lambda}_m = (\overline{I}_s - \overline{I}_r)L_m = 1.2168\angle -83.9° \text{ Wb} \quad \text{(rms)}$$

Alternatively, $\overline{\Lambda}_m = \dfrac{\overline{V}_s - \overline{I}_s(R_s + jX_{ls})}{j\omega_s} = 1.2168\angle -83.9° \text{ Wb (rms)}$

The stator flux linkage can be obtained by

$$\overline{\Lambda}_s = \overline{\Lambda}_m + L_{ls}\,\overline{I}_s = 1.2748\angle -89.8° \text{ Wb } (\text{rms})$$

The peak value of stator flux: $\lambda_s = \sqrt{2}\Lambda_s = 1.8028$ Wb
The rotor flux linkage is then can be obtained by

$$\overline{\Lambda}_r = \overline{\Lambda}_m - L_{lr}\,\overline{I}_r = \Lambda_r\angle\theta_2 = 1.2096\angle -77.7° \text{ Wb } (\text{rms})$$

The peak value of rotor flux, $\lambda_r = \sqrt{2}\Lambda_r = 1.7106$ Wb

d) To find *dq*-axis stator currents and voltages, the *dq* axes are added to the phasor diagram with the *d*-axis aligned with the rotor flux phasor $\overline{\Lambda}_r$, as shown in Figure C-12. The following angles are then defined:

φ_s—the stator power factor angle
θ_2—the angle between $\overline{V}_s$ and $\overline{\Lambda}_r$
θ_1—the angle between $\overline{I}_s$ and $\overline{\Lambda}_r$

The angle between the stator current $\overline{I}_s$ and rotor flux $\overline{\Lambda}_r$ is

$$\theta_1 = \varphi_s - \theta_2 = -74.85°$$

The *dq*-axis stator currents can be calculated by

$$\begin{cases} I_{ds} = I_s \cos(\angle\overline{I}_s - \angle\overline{\Lambda}_r) = 2168\cos(-74.85°) \text{ A} = 566.66 \text{ A} \\ I_{qs} = I_s \sin(\angle\overline{I}_s - \angle\overline{\Lambda}_r) = 2168\sin(-74.85°) \text{ A} = -2092.5 \text{ A} \end{cases} (\text{rms})$$

e) The *dq*-axis stator voltages can be calculated by

$$\begin{cases} V_{ds} = V_s \sin(90 - |\theta_2|) = 398.37\sin(12.3°) \text{ V} = 84.72 \text{ V} \\ V_{qs} = V_s \cos(90 - |\theta_2|) = 398.37\cos(12.3°) \text{ V} = 389.3 \text{ V} \end{cases} (\text{rms})$$

Cross Check:

$$T_m = \frac{3PL_m}{2L_r}\left(i_{qs}\lambda_{dr}\right) = \frac{3PL_m}{2L_r}\left(i_{qs}\lambda_r\right) ; \; i_{qs} \text{ and } \lambda_r \text{ are the peak values.}$$

$$= \frac{3\times 2\times 2.1346\times 10^{-3}}{2\times 2.2\times 10^{-3}}\left(-2959.3\times 1.7106\right) = -14740 \text{ N·m, verified.}$$

$$P_m = 3I_r^2\frac{R_r}{s}(1-s) = 3\times 2030.8^2\times\frac{1.497\times 10^{-3}}{-0.008}(1+0.008) = -2.3336\times 10^6 \text{ W, verified.}$$

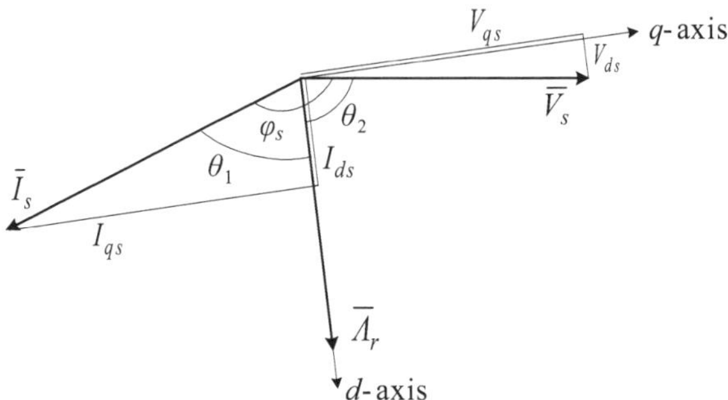

<u>Figure C-12.</u> Phasor diagram for the analysis.

7-2. Repeat Problem 7-1 for a wind speed of 7.2 m/s for which the generator operates at 0.6 pu rotor speed with the stator voltage of 235.61 V and slip of −0.0048.

Answers:

a) $T_m = -5306.4$ N·m, $\quad P_{m,R} = -2.3339 \times 10^6$ W, $\quad P_m = -504.12 \times 10^3$ W

b) $\overline{V}_s = 235.61\angle 0°$ V (rms), $\quad \overline{Z}_s = 0.2494\angle 138.5°\ \Omega$ for $s = -0.0048$

$X_{ls} = 0.0123\ \Omega$, $\quad X_{lr} = 0.0123\ \Omega$ and $X_m = 0.4037\ \Omega$

$\overline{I}_s = 944.55\angle -138.5°$ A (rms), $\quad \overline{I}_r = 733.4\angle -175.4°$ A (rms)

c) $\overline{\Lambda}_m = 1.2105\angle -87.7°$ Wb (rms), $\quad \overline{\Lambda}_s = 1.2501\angle -89.8°$ Wb (rms),

$\lambda_s = 1.7679$ Wb, $\quad \overline{\Lambda}_r = 1.210\angle -85.4°$ Wb (rms), $\quad \lambda_r = 1.7106$ Wb

d) $I_{ds} = 566.7$ A (rms), $\quad I_{qs} = -755.7$ A (rms)

e) $V_{ds} = 18.91$ V (rms), $\quad V_{qs} = 234.9$ V (rms), $\quad T_m = -5322.5$ N·m,

$P_m = -505.66 \times 10^3$ W, verified.

Topic: Steady-State Analysis of SCIG WECS with Indirect FOC

7-3 (Solved Problem). A 2.3 MW, 690 V, 50 Hz SCIG is used in a wind energy system controlled by an indirect FOC scheme. At a wind speed of 10.8 m/s, the generator operates at 0.9 pu rotor speed. The generator parameters are given in Table B-1 of Appendix B. The rotor flux λ_r is kept at its rated value of 1.7106 Wb (peak) by the indirect FOC controller. Neglecting rotational losses, calculate the following:

a) The generator mechanical torque and power

b) The dq-axis and rms stator current

c) The slip and stator frequency

d) The rms stator voltage

e) The magnetizing, stator and rotor flux linkage values

f) The dq-axis rms stator voltages

Solution:

Refer to Table 7-2 in Chapter 7.

a) The generator mechanical torque at 0.9 pu rotor speed:

$$T_m = T_{m,R} \times (\omega_{m,\text{pu}})^2 = -14740 \times 0.9^2 = -11939 \text{ N·m}$$

The rated mechanical power:

$$P_{m,R} = \omega_{m,R} \times T_{m,R} = 1512(2\pi)/60 \times (-14740) = -2.3339 \times 10^6 \text{ W}$$

The mechanical power at 0.9 pu rotor speed:

$$P_m = P_{m,R} \times (\omega_{m,\text{pu}})^3 = -2.3339 \times 10^6 \times 0.9^3 = -1.7014 \times 10^6 \text{ W}$$

b) The d-axis stator current:

$$i_{ds} = \frac{\lambda_r}{L_m} = \frac{1.7106}{2.1346 \times 10^{-3}} = 801.36 \text{ A (peak)}$$

$$I_{ds} = i_{ds}/\sqrt{2} = 566.65 \text{ A (rms)}$$

The q-axis stator current:

$$i_{qs} = \frac{T_e}{K_T \lambda_r} = \frac{2L_r}{3PL_m} \frac{T_e}{\lambda_r} = \frac{2 \times 2.2 \times 10^{-3}}{3 \times 2 \times 2.1346 \times 10^{-3}} \cdot \frac{-11939}{1.7106} = -2397.3 \text{ A (peak)}$$

$$I_{qs} = i_{qs}/\sqrt{2} = -1695.2 \text{ A (rms)}$$

The stator current:

$$i_s = \sqrt{i_{ds}^2 + i_{qs}^2} = \sqrt{801.36^2 + 2397.3^2} = 2527.7 \text{ A (peak)}$$

Set the stator current as a reference phasor:

$$\bar{I}_s = i_s/\sqrt{2}\angle 0° = 1787.4\angle 0° \text{ A (rms)}$$

c) The slip frequency:

$$\omega_{sl} = \frac{R_r L_m}{L_r \lambda_r} i_{qs} = \frac{1.497 \times 10^{-3} \times 2.1346 \times 10^{-3}}{2.2 \times 10^{-3} \times 1.7106} \times (-2397.3) = -2.036 \text{ rad/s } (-0.324 \text{ Hz})$$

The stator frequency can be found from

$$\omega_r = (1512 \times 0.9) \times (2\pi/60)P = 285 \text{ rad/s} \quad (1360.8 \text{ rpm})$$

$$\omega_s = \omega_r + \omega_{sl} = 285 - 2.036 = 282.97 \text{ rad/s} \quad (45.036 \text{Hz})$$

The slip:

$$s = \frac{\omega_{sl}}{\omega_s} = -0.0072$$

d) The equivalent impedance of SCIG:

$$\bar{Z}_s = R_s + jX_{ls} + jX_m \; // \left(\frac{R_r}{s} + jX_{lr} \right) = 0.1993 \angle 151.5° \; \Omega$$

where $X_{ls} = \omega_s L_{ls} = 2 \times 3.1416 \times 45.036 \times 0.06492 \times 10^{-3} = 0.0814 \; \Omega$, $X_{lr} = \omega_s L_{lr} = 0.0184 \; \Omega$ and $X_m = \omega_s L_m = 0.604 \; \Omega$

The stator voltage:

$$\bar{V}_s = \bar{I}_s \times \bar{Z}_s = 1787.4 \angle 0° \times 0.1993 \angle 151.5° = 356.25 \angle 151.5° \text{ V (rms)}$$

e) The rotor current can be calculated by

$$\bar{I}_r = \frac{jX_m \bar{I}_s}{jX_m + \left(\dfrac{R_r}{s} + jX_{lr} \right)} = 1645.1 \angle -18.5° \text{ A (rms)}$$

The magnetizing flux linkage can be expressed as

$$\bar{\Lambda}_m = (\bar{I}_s - \bar{I}_r)L_m = 1.2143 \angle 66.5° \text{ Wb (rms)}$$

Alternatively, $\bar{\Lambda}_m = \dfrac{\bar{V}_s - \bar{I}_s(R_s + jX_{ls})}{j\omega_s} = 1.2143 \angle 66.5° \text{ Wb (rms)}$

The stator flux linkage can be obtained by

$$\bar{\Lambda}_s = \bar{\Lambda}_m + L_{ls}\bar{I}_s = 1.2651 \angle 61.65° \text{ Wb (rms)}$$

The peak value of stator flux: $\lambda_s = \sqrt{2}\Lambda_s = 1.7891 \text{ Wb}$
The rotor flux linkage is then can be obtained by

$$\bar{\Lambda}_r = \bar{\Lambda}_m - L_{lr}\bar{I}_r = \Lambda_r \angle \theta_1 = 1.2096 \angle 71.5° \text{ Wb (rms)}$$

The peak value of rotor flux, $\lambda_r = \sqrt{2}\Lambda_r = 1.7106 \text{ Wb}$

f) Referring to Figure 7-12 in Chapter 7, the angle between the stator voltage $\overline{V}_s$ and rotor flux $\overline{\Lambda}_r$ is

$$\theta_2 = \varphi_s - \theta_1 = 79.98°$$

where φ_s is the stator power factor, given by

$$\varphi_s = \angle \overline{V}_s - \angle \overline{I}_s = 151.5°$$

The rms dq-axis stator voltages can be calculated by

$$\begin{cases} V_{ds} = V_s \cos(\angle \overline{V}_s - \angle \overline{\Lambda}_r) = 356.25\cos(79.98°)\ V = 61.99\ V \\ V_{qs} = V_s \sin(\angle \overline{V}_s - \angle \overline{\Lambda}_r) = 356.25\sin(79.98°)\ V = 350.8\ V \end{cases} \quad \text{(rms)}$$

Cross Check:

$$T_m = \frac{3PL_m}{2L_r}(i_{qs}\lambda_{dr}) = \frac{3PL_m}{2L_r}(i_{qs}\lambda_r);\ i_{qs}\ \text{and}\ \lambda_r\ \text{are the peak values.}$$

$$= \frac{3\times2\times2.1346\times10^{-3}}{2\times2.2\times10^{-3}}(-2397.3\times1.7106) = -11939\ \text{N·m, verified.}$$

$$P_m = 3I_r^2\frac{R_r}{s}(1-s) = 3\times1645.1^2\times\frac{1.497\times10^{-3}}{-0.0072}(1+0.0072) = -1.7014\times10^6\ \text{W, verified.}$$

7-4. Repeat Problem 7-3 for a wind speed of 6 m/s for which the generator operates at 0.5 pu rotor speed.

Answers:

a) $T_m = -3685$ N·m, $\quad P_{m,R} = -2.3339\times10^6$ W, $\quad P_m = -291.73\times10^3$ W

b) $I_{ds} = 566.65$ A (rms), $\quad I_{qs} = -523.2$ A (rms), $\quad \overline{I}_s = 771.25\angle0°$ A (rms)

c) $\omega_{sl} = -0.6284$ rad/s (−0.1 Hz), $\quad \omega_s = 157.71$ rad/s (25.1 Hz), $\quad s = -0.004$

d) $\overline{Z}_s = 0.2545\angle129.45°\ \Omega$, $\quad X_{ls} = 0.0102\ \Omega$, $\quad X_{lr} = 0.0102\ \Omega$,
$\quad X_m = 0.3366\ \Omega$, $\quad \overline{V}_s = 196.3\angle129.45°$ V (rms)

e) $\overline{I}_r = 507.75\angle-47.3°$ A (rms), $\quad \overline{\Lambda}_m = 1.21\angle41.16°$ Wb (rms),
$\quad \overline{\Lambda}_s = 1.2482\angle39.64°$ Wb (rms), $\quad \lambda_s = 1.7652$ Wb,
$\quad \overline{\Lambda}_r = 1.210\angle42.72°$ Wb (rms), $\quad \lambda_r = 1.7106$ Wb

f) $V_{ds} = 11.18$ V (rms), $\quad V_{qs} = 195.98$ V (rms), $\quad T_m = -3685$ N·m,
$\quad P_m = -291.73\times10^3$ W, verified.

7-5. Consider a 3.0 MW, 3000 V, 60 Hz SCIG wind energy system. At a wind speed of 10.8 m/s, the generator operates at 0.9 pu rotor speed. The generator parameters are given in Table B-3 of Appendix B. The rotor flux λ_r is kept at its rated value of 6.2107 Wb (peak) by the indirect field-oriented controller. Repeat Problem 7-3.

Answers:

a) $T_m = -12978 \text{ N·m},$ $\quad P_{m,R} = 3.0402 \times 10^6 \text{ W},$ $\quad P_m = -2.2163 \times 10^6 \text{ W}$

b) $I_{ds} = 161.6 \text{ A (rms)},$ $\quad I_{qs} = -506.7 \text{ A (rms)},$ $\quad \overline{I}_s = 531.86\angle 0° \text{ A (rms)}$

c) $\omega_{sl} = -2.0358 \text{ rad/s } (-0.324 \text{ Hz}),$ $\quad \omega_s = 339.52 \text{ rad/s } (54.036 \text{ Hz}),$
$\quad s = -0.006$

d) $\overline{Z}_s = 2.912\angle 152.3° \text{ } \Omega,$ $\quad X_{ls} = 0.2655 \text{ } \Omega,$ $\quad X_{lr} = 0.2655 \text{ } \Omega,$
$\quad X_m = 9.224 \text{ } \Omega,$ $\quad \overline{V}_s = 1548.9\angle 152.3° \text{ V (rms)}$

e) $\overline{I}_r = 492.5\angle -17.7° \text{ A (rms)},$ $\quad \overline{\Lambda}_m = 4.4085\angle 67.3° \text{ Wb (rms)},$
$\quad \overline{\Lambda}_s = 4.5851\angle 62.5° \text{ Wb (rms)},$ $\quad \lambda_s = 6.4843 \text{ Wb},$
$\quad \overline{\Lambda}_r = 4.3916\angle 72.3° \text{ Wb (rms)},$ $\quad \lambda_r = 6.2107 \text{ Wb}$

f) $V_{ds} = 267.99 \text{ V (rms)},$ $\quad V_{qs} = 1525.5 \text{ V (rms)},$ $\quad T_m = -12978 \text{ N·m},$
$\quad P_m = -2.2163 \times 10^6 \text{ W},$ verified.

7-6. Repeat Problem 7-5 for a wind speed of 6 m/s, at which the generator operates at 0.5 pu rotor speed.

Answers:

a) $T_m = -4005.5 \text{ N·m},$ $\quad P_{m,R} = -3.0402 \times 10^6 \text{ W},$ $\quad P_m = -380.03 \times 10^3 \text{ W}$

b) $I_{ds} = 161.65 \text{ A (rms)},$ $\quad I_{qs} = -156.4 \text{ A (rms)},$ $\quad \overline{I}_s = 224.92\angle 0° \text{ A (rms)}$

c) $\omega_{sl} = -0.6283 \text{ rad/s } (-0.1 \text{ Hz}),$ $\quad \omega_s = 189.12 \text{ rad/s } (30.1 \text{ Hz}),$ $\quad s = -0.0033$

d) $\overline{Z}_s = 3.7936\angle 130.8° \text{ } \Omega,$ $\quad X_{ls} = 0.1479 \text{ } \Omega,$ $\quad X_{lr} = 0.1479 \text{ } \Omega,$
$\quad X_m = 5.1381 \text{ } \Omega,$ $\quad \overline{V}_s = 853.23\angle 130.8° \text{ V (rms)}$

e) $\overline{I}_r = 152\angle -45.95° \text{ A (rms)},$ $\quad \overline{\Lambda}_m = 4.3932\angle 42.5° \text{ Wb (rms)},$
$\quad \overline{\Lambda}_s = 4.5244\angle 41° \text{ Wb (rms)},$ $\quad \lambda_s = 6.3985 \text{ Wb},$
$\quad \overline{\Lambda}_r = 4.3916\angle 44.05° \text{ Wb (rms)},$ $\quad \lambda_r = 6.2107 \text{ Wb}$

f) $V_{ds} = 48.3 \text{ V (rms)},$ $\quad V_{qs} = 851.86 \text{ V (rms)},$ $\quad T_m = -4005.5 \text{ N·m},$
$\quad P_m = -380.03 \times 10^3 \text{ W},$ verified.

Topic: Steady-State Analysis of SCIG WECS with DTC

7-7 (Solved Problem). A 1.45 MW, 575 V, 50 Hz SCIG is used in a wind energy system controlled by the DTC scheme. At a wind speed of 10.8 m/s, the generator operates at 0.9 pu rotor speed. The generator parameters are given in Table B-2 of Appendix B. The stator flux λ_s is kept at its rated value of 1.5033 Wb (peak) by the DTC controller. Calculate the following:

a) The generator mechanical torque and power

b) The dq-axis and rms stator current

c) The stator frequency and slip
d) The rms stator voltage
e) The magnetizing, stator, and rotor flux linkage values
f) The dq-axis rms stator voltages

Solution:

Refer to Table 7-6 in Chapter 7.

a) The generator mechanical torque at 0.9 pu rotor speed:

$$T_m = T_{m,R} \times \left(\omega_{m,\mathrm{pu}}\right)^2 = -13944 \times 0.9^2 = -11295 \text{ N·m}$$

The rated mechanical power:

$$P_{m,R} = \omega_{m,R} \times T_{m,R} = 1007.2 \times (2\pi)/60 \times (-13944) = -1.4707 \times 10^6 \text{ W}$$

The generator mechanical power at 0.9 pu rotor speed:

$$P_m = P_{m,R} \times \left(\omega_{m,\mathrm{pu}}\right)^3 = -1.4707 \times 10^6 \times 0.9^3 = -1.0722 \times 10^6 \text{ W}$$

b) The q-axis stator current:

$$i_{qs} = \frac{2T_e}{3P\lambda_s} -- = \frac{2 \times (-11295)}{3 \times 3 \times 1.5033} = -1669.6 \text{ A} \text{ (peak)}$$

$$I_{qs} = i_{qs}/\sqrt{2} = -1180.6 \text{ A} \text{ (rms)}$$

The slip frequency ω_{sl} is determined by

$$(\tau_r\sigma)^2 \omega_{sl}^2 - \frac{(1-\sigma)\tau_r\lambda_s}{L_s i_{qs}}\omega_{sl} + 1 = 0 \quad \rightarrow \quad 0.0114\omega_{sl}^2 + 0.5776\omega_{sl} + 1 = 0$$

where τ_r is the rotor time constant, defined by

$$\tau_r = L_r/R_r = 1.8199 \times 10^{-3}/1.39 \times 10^{-3} = 1.3093$$

and σ is the total leakage factor, defined by

$$\sigma = 1 - \frac{L_m^2}{L_s L_r} = 1 - \frac{(1.77016 \times 10^{-3})^2}{1.8745 \times 10^{-3} \times 1.8199 \times 10^{-3}} = 0.0815$$

from which

$$\omega_{sl} = \begin{cases} -1.7948 \text{ rad/sec} & (-0.2857 \text{ Hz}) \\ -48.891 \text{ rad/sec} & (-7.7813 \text{ Hz}) \quad \text{(omitted)} \end{cases}$$

The *d*-axis stator current:

$$i_{ds} = \frac{\lambda_s + (\sigma L_s \tau_r \omega_{sl} i_{qs})}{L_s}$$

$$= \frac{1.5033 + (0.0815 \times 1.8745 \times 10^{-3} \times 1.3093 \times (-1.7948) \times (-1669.6))}{1.8745 \times 10^{-3}} = 1121.8 \text{ A}$$

$$= 1121.8 \text{ A (peak)}$$

$$I_{ds} = i_{ds} / \sqrt{2} = 793.26 \text{ A (rms)}$$

The stator current:

$$i_s = \sqrt{i_{ds}^2 + i_{qs}^2} = \sqrt{1121.8^2 + 1669.6^2} = 2011.5 \text{ A (peak)}$$

$$\overline{I}_s = i_s / \sqrt{2} \angle 0° = 1422.3 \angle 0° \text{ A (rms)}$$

c) The stator frequency can be found from

$$\omega_r = (1007.2 \times 0.9) \times (2\pi/60) P = 284.78 \text{ rad/s} \quad (906.48 \text{ rpm})$$

$$\omega_s = \omega_r + \omega_{sl} = 284.78 - 1.7948 = 282.98 \text{ rad/s} \quad (45.0383 \text{ Hz})$$

The slip:

$$s = \frac{\omega_{sl}}{\omega_s} = -0.0063$$

d) The equivalent impedance of SCIG:

$$\overline{Z}_s = R_s + jX_{ls} + jX_m \; // \left(\frac{R_r}{s} + jX_{lr} \right) = 0.2104 \angle 145.9° \; \Omega$$

where $X_{ls} = \omega_s L_{ls} = 2 \times 3.1416 \times 45.038 \times 0.1044 \times 10^{-3} = 0.0295 \; \Omega$, $X_{lr} = \omega_s L_{lr} = 0.0141 \; \Omega$ and $X_m = \omega_s L_m = 0.5009 \; \Omega$

The stator voltage:

$$\overline{V}_s = \overline{I}_s \times \overline{Z}_s = 1422.3 \angle 0° \times 0.2104 \angle 145.9° = V_s \angle \varphi_s = 299.2 \angle 145.9° \text{ V (rms)}$$

e) The rotor current can be calculated by

$$\overline{I}_r = \frac{jX_m \overline{I}_s}{jX_m + \left(\dfrac{R_r}{s} + jX_{lr} \right)} = 1273 \angle -23.1° \text{ A (rms)}$$

The magnetizing flux linkage can be expressed as

$$\overline{\Lambda}_m = (\overline{I}_s - \overline{I}_r)L_m = 0.9879\angle 63.3° \text{ Wb (rms)}$$

Alternatively, $\overline{\Lambda}_m = \dfrac{\overline{V}_s - \overline{I}_s(R_s + jX_{ls})}{j\omega_s} = 0.9789\angle 63.3° \text{ Wb (rms)}$

The stator flux linkage can be obtained by

$$\overline{\Lambda}_s = \overline{\Lambda}_m + L_{ls}\overline{I}_s = 1.063\angle 56.1° \text{ Wb (rms)}$$

The peak value of stator flux: $\lambda_s = \sqrt{2}\Lambda_s = 1.5033$ Wb
The rotor flux linkage is then can be obtained by

$$\overline{\Lambda}_r = \overline{\Lambda}_m - L_{lr}\overline{I}_r = \Lambda_r\angle\theta_1 = 0.9859\angle 66.95° \text{ Wb (rms)}$$

The peak value of rotor flux, $\lambda_r = \sqrt{2}\Lambda_r = 1.3942$ Wb
f) The torque angle:

$$\theta_T = \angle\overline{\Lambda}_s - \angle\overline{\Lambda}_r = \angle\overline{\Lambda}_s - \theta_1 = 56.1° - 66.95° = -10.85°$$

Cross Check:

$$T_e = \frac{3P}{2}\frac{L_m}{\sigma L_s L_r}\lambda_s\lambda_r \sin\theta_T = \frac{3\times 3}{2}\frac{1.7702\times 10^{-3}}{0.0815\times 1.8745\times 10^{-3}\times 1.8199\times 10^{-3}}$$

$$\times 1.5033\times 1.3942\times \sin(-10.85°) = -1129.5 \text{ N·m, verified.}$$

$$P_m = 3I_r^2\frac{R_r}{s}(1-s) = 3\times 1273^2\times \frac{1.39\times 10^{-3}}{-0.0063}(1+0.0063) = -1.0722\times 10^6 \text{ W, verified.}$$

7-8. Repeat Problem 7-7 for a wind speed of 6 m/s, at which the generator operates at 0.5 pu rotor speed.

Answers:

a) $T_m = -3486$ N·m., $\qquad P_{m,R} = 1.4707\times 10^6$ W, $\qquad P_m = -183.84\times 10^3$ W

b) $I_{qs} = -364.38$ A (rms), $\qquad \omega_{sl} = -0.5361$ rad/s, $\qquad I_{ds} = 587.92$ A (rms),
$\quad \overline{I}_s = 691.68\angle 0°$ A (rms)

c) $\omega_s = 157.67$ rad/s (501.89 rpm) (25.09 Hz), $\qquad s = -0.0034$

d) $\overline{Z}_s = 0.2416\angle 121.52°$ Ω, $\qquad X_{ls} = 0.0165$ Ω, $\qquad X_{lr} = 0.0079$ Ω,
$\quad X_m = 0.2791$ Ω, $\qquad \overline{V}_s = 167.12\angle 121.52°$ V (rms)

e) $\overline{I}_r = 386.5\angle -54.9°$ A (rms), $\overline{\Lambda}_m = 1.0023\angle 33.96°$ Wb (rms),
   $\overline{\Lambda}_s = 1.063\angle 31.79°$ Wb (rms), $\lambda_s = 1.5033$ Wb
   $\overline{\Lambda}_r = 1.0022\angle 35.1°$ Wb (rms), $\lambda_r = 1.4173$ Wb
f) $\theta_T = -3.28°$, $T_m = -3486$ N·m, $P_m = -183.84\times10^3$ W, verified.

7-9. Consider a 4.0 MW, 4000 V, 50 Hz SCIG wind energy system. At a wind speed of 10.8 m/s, the generator operates at 0.9 pu rotor speed. The generator parameters are given in Table B-4 of Appendix B. The stator flux λ_s is kept at its rated value of 10.4535 Wb (peak) by the DTC controller. Repeat Problem 7-7:

Answers:

a) $T_m = -20794$ N·m, $P_{m,R} = 4.0606\times10^6$ W, $P_m = -2.9602\times10^6$ W
b) $I_{qs} = -468.85$ A (rms), $\omega_{sl} = -1.7188$ rad/s, $I_{ds} = 324.8$ A (rms),
   $\overline{I}_s = 570.35\angle 0°$ A (rms)
c) $\omega_s = 283$ rad/s (45.042 Hz), $s = -0.0061$
d) $\overline{Z}_s = 3.6496\angle 145.1°$ Ω, $X_{ls} = 0.4805$ Ω, $X_{lr} = 0.4805$ Ω,
   $X_m = 9.5081$ Ω, $\overline{V}_s = 2081.5\angle 145.1°$ V (rms)
e) $\overline{I}_r = 507.23\angle -20.89°$ A (rms), $\overline{\Lambda}_m = 6.8864\angle 61.9°$ Wb (rms),
   $\overline{\Lambda}_s = 7.3917\angle 55.3°$ Wb (rms), $\lambda_s = 10.4535$ Wb,
   $\overline{\Lambda}_r = 6.8323\angle 69.11°$ Wb (rms), $\lambda_r = 9.6624$ Wb
f) $\theta_T = -13.82°$, $T_m = -20794$ N·m, $P_m = -2.9602\times10^6$ W, verified.

7-10. Repeat Problem 7-9 for a wind speed of 6 m/s, at which the generator operates at 0.5 pu rotor speed.

Answers:

a) $T_m = -6417.8$ N·m, $P_{m,R} = -4.0606\times10^6$ W, $P_m = -507.58\times10^3$ W
b) $I_{qs} = -144.71$ A (rms), $\omega_{sl} = -0.5028$ rad/s, $I_{ds} = 219.84$ A (rms),
   $\overline{I}_s = 263.2\angle 0°$ A (rms)
c) $\omega_s = 157.68$ rad/s (25.095 Hz), $s = -0.0032$
d) $\overline{Z}_s = 4.4162\angle 123.1°$ Ω, $X_{ls} = 0.2677$ Ω, $X_{lr} = 0.2677$ Ω,
   $X_m = 5.2975$ Ω, $\overline{V}_s = 1162.3\angle 123.1°$ V (rms)
e) $\overline{I}_r = 152.4\angle -52.53°$ A (rms), $\overline{\Lambda}_m = 7.0227\angle 35.36°$ Wb (rms),
   $\overline{\Lambda}_s = 7.3917\angle 33.35°$ Wb (rms), $\lambda_s = 10.454$ Wb,
   $\overline{\Lambda}_r = 7.018\angle 37.47°$ Wb (rms), $\lambda_r = 9.925$ Wb
f) $\theta_T = -4.12°$, $T_m = -6417.8$ N·m, $P_m = -507.58\times10^3$ W, verified.

Topic: Steady-State Analysis of PWM CSC Interfaced SCIG WECS with Direct FOC

7-11 (Solved Problem). A 3.0 MW, 3000 V, 60 Hz current source converter is used in SCIG wind energy system operating with direct FOC. At a wind speed of 9.6 m/s, the generator operates at 0.8 pu rotor speed with slip of –0.0053 and stator current of 429.93 A (rms). The generator parameters are given in Table B-3 of Appendix B. The rotor flux λ_r is kept at its rated value of 6.2107 Wb (peak) by the direct FOC controller. The rectifier filter capacitor value is 296.8 μF. Neglecting rotational losses, determine the following:

a) The generator mechanical torque and power
b) The stator voltage and rotor current
c) The magnetizing, stator, and rotor flux linkages
d) The *dq*-axis stator currents and voltages
e) The rms capacitor and rectifier PWM currents
f) The rectifier and inverter firing angles

Solution:

a) The generator mechanical torque at 0.8 pu rotor speed:

$$T_m = T_{m,R} \times (\omega_{m,\text{pu}})^2 = -16022 \times 0.8^2 = -10254 \text{ N·m}$$

The rated mechanical power:

$$P_{m,R} = \omega_{m,R} \times T_{m,R} = 1812(2\pi)/60 \times (-16022) = -3.0402 \times 10^6 \text{ W}$$

Generator mechanical power at 0.8 pu rotor speed:

$$P_m = P_{m,R} \times (\omega_{m,\text{pu}})^3 = -3.0402 \times 10^6 \times 0.8^3 = -1.5566 \times 10^6 \text{ W}$$

b) Selecting the stator current as a reference phasor, the stator current is: $\bar{I}_s = 429.93 \angle 0°$ A (rms)
   The equivalent impedance of SCIG:

$$\bar{Z}_s = R_s + jX_{ls} + jX_m // \left(\frac{R_r}{s} + jX_{lr} \right) = 3.1884 \angle 150° \text{ ; for } s = -0.0053$$

where $X_{ls} = \omega_s L_{ls} = 2 \times 3.1416 \times 48.064 \times 0.782 \times 10^{-3} = 0.2362$ Ω, $X_{lr} = \omega_s L_{lr} = 0.2362$ Ω and $X_m = \omega_s L_m = 8.2046$ Ω
The stator voltage:

$$\bar{V}_s = \bar{I}_s \times \bar{Z}_s = 429.93 \angle 0° \times 3.1884 \angle 150° = V_s \angle \varphi_s = 1370.8 \angle 150° \text{ V (rms)}$$

The rotor current can be calculated by

$$\bar{I}_r = \frac{jX_m \bar{I}_s}{jX_m + \left(\frac{R_r}{s} + jX_{lr} \right)} = 387.24 \angle -22.1° \text{ A (rms)}$$

c) The magnetizing flux linkage can be expressed as

$$\overline{\Lambda}_m = (\overline{I}_s - \overline{I}_r)L_m = 4.4021\angle 63.97°\ \text{Wb}\ (\text{rms})$$

Alternatively, $\overline{\Lambda}_m = \dfrac{\overline{V}_s - \overline{I}_s(R_s + jX_{ls})}{j\omega_s} = 4.4021\angle 63.97°\ \text{Wb}\ (\text{rms})$

The stator flux linkage can be obtained by

$$\overline{\Lambda}_s = \overline{\Lambda}_m + L_{ls}\overline{I}_s = 4.5596\angle 60.17°\ \text{Wb}\ (\text{rms})$$

The peak value of stator flux: $\lambda_s = \sqrt{2}\Lambda_s = 6.4483$ Wb
The rotor flux linkage is then can be obtained by

$$\overline{\Lambda}_r = \overline{\Lambda}_m - L_{lr}\overline{I}_r = \Lambda_r\angle\theta_1 = 4.3916\angle 67.9°\ \text{Wb}\ (\text{rms})$$

The peak value of rotor flux, $\lambda_r = \sqrt{2}\Lambda_r = 6.2107$ Wb

d) To find dq-axis stator currents and voltages, the dq axes are added to the phasor diagram with the d-axis aligned with the rotor flux phasor $\overline{\Lambda}_r$ as shown in Figure C-13. The following angles are then defined:

φ_s—the stator power factor angle
θ_1—the angle between $\overline{I}_s$ and $\overline{\Lambda}_r$
θ_2—the angle between $\overline{V}_s$ and $\overline{\Lambda}_r$
θ_3—the angle between $\overline{I}_s$ and $\overline{I}_{wr}$
α_r—the rectifier firing angle

The dq-axis stator current can then be determined by

$$\begin{cases} I_{ds} = I_s\cos\theta_1 = 429.93\cos(67.92°) = 161.65\ \text{A} \\ I_{qs} = I_s\sin\theta_1 = -429.93\sin(67.92°) = -398.38\ \text{A} \end{cases} \quad (\text{rms})$$

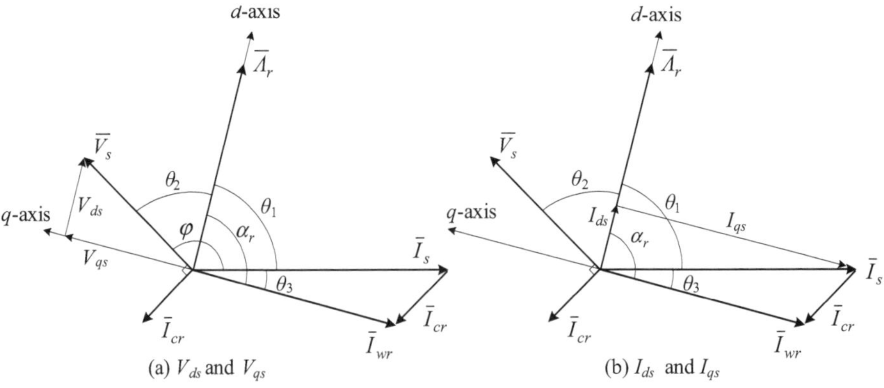

Figure C-13. Phasor diagram for the analysis.

The *dq*-axis stator voltages can be calculated by

$$\begin{cases} V_{ds} = V_s \cos\theta_2 = V_s \cos(\angle \overline{V}_s - \angle \overline{\Lambda}_r) = 1370.8\cos(82.11°) = 188.22 \text{ V} \\ V_{qs} = V_s \sin\theta_2 = V_s \sin(\angle \overline{V}_s - \angle \overline{\Lambda}_r) = 1370.8\sin(82.11°) = 1357.8 \text{ V} \end{cases} \text{ (rms)}$$

e) Method #1:
The rms capacitor current can be calculated by

$$\overline{I}_{cr} = (j\omega_s C_r)\overline{V}_s = (j302\times296.8\times10^{-6})(1370.8\angle150°) = 122.87\angle-119.98° \text{ A} \text{ (rms)}$$

The rms rectifier PWM current can be found from

$$\overline{I}_{wr} = \overline{I}_s + \overline{I}_{cr} = 429.93\angle0° + 122.87\angle-119.98° = 383.6\angle-16.11° \text{ A} \text{ (rms)}$$

Method #2:
The *dq*-axis currents of the rectifier-side capacitor C_r can be found from

$$\begin{cases} I_{crd} = -2\pi f_s V_{qs} C_r = -2\pi\times40.064\times1357.8\times296.8\times10^{-6} = -121.7 \text{ A} \\ I_{crq} = 2\pi f_s V_{ds} C_r = 2\pi\times40.064\times188.22\times296.8\times10^{-6} = 16.87 \text{ A} \end{cases} \text{ (rms)}$$

The rms capacitor current:

$$I_{cr} = \sqrt{I_{crd}^2 + I_{crq}^2} = 122.87 \text{ A} \text{ (rms)}$$

The *dq*-axis PWM currents of the rectifier can be calculated by

$$\begin{cases} I_{dwr} = I_{ds} + I_{crd} = 161.65 - 121.7 = 39.95 \text{ A} \\ I_{qwr} = I_{qs} + I_{crq} = -398.38 + 16.87 = -381.51 \text{ A} \end{cases} \text{ (rms)}$$

The rms rectifier PWM current:

$$I_{wr} = \sqrt{I_{dwr}^2 + I_{qwr}^2} = 383.6 \text{ A} \text{ (rms)}$$

f) The firing angle for the rectifier:

$$\alpha_r = \angle\overline{\Lambda}_r - \angle\overline{I}_{wr} = 67.92° - (-16.11°) = 84.02°$$

The rectifier firing angle can also be found from

$$\alpha_r = \tan^{-1}\frac{I_{qwr}}{I_{dwr}} = \tan^{-1}\frac{-381.51}{39.95} = 84.02°$$

The active power delivered from the generator to the DC link circuit is

$$P_{dc} = 3V_s I_s \cos\varphi_s = 3\times1370.8\times429.93\cos(150°) = -1.5315\times10^6 \text{ W}$$

The DC current I_{dc} is

$$I_{dc} = \sqrt{2}\, I_{wr1}\, m_r = \sqrt{2} \times 383.6 \times 1.0 = 542.5 \text{ A}$$

where rectifier modulation index m_r is set to its maximum value of 1.0.
The rectifier-side DC voltage:

$$V_{dcr} = \frac{P_{dc}}{I_{dc}} = \frac{-1.5315 \times 10^6}{542.5} = -2823.1 \text{ V}$$

The rectifier-side DC voltage can also be calculated by

$$V_{dcr} = \sqrt{3/2}\, V_{LL} m_r \cos\left(\left|\varphi_s\right| + \left|\theta_3\right|\right) = \sqrt{3/2} \times 3000 \times 1.0 \times \cos\left(150° + 16.1°\right) = -2823.1 \text{ V}$$

where $\theta_3 = \alpha_r - \theta_1 = 84.02° - 67.92° = 16.1°$
Neglecting the winding resistance of the DC choke L_{dc}, the inverter-side DC voltage is

$$V_{dci} = V_{dcr} = -2823.1 \text{ V}$$

The inverter firing angle:

$$\alpha_i = \cos^{-1}\left(\frac{V_{dci}}{\sqrt{3/2}\, m_i V_{LL}}\right) = \cos^{-1}\left(\frac{-2823.1}{\sqrt{3/2} \times 1.08 \times 3000}\right) = 135.35°$$

where m_i is the modulation index of the inverter with SHE scheme.

Cross Check:

$$T_m = \frac{3PL_m}{2L_r}(i_{qs}\lambda_{dr}) = \frac{3PL_m}{2L_r}(i_{qs}\lambda_r); \; i_{qs} \text{ and } \lambda_r \text{ are the peak values.}$$

$$= \frac{3 \times 2 \times 27.168 \times 10^{-3}}{2 \times 27.95 \times 10^{-3}}(-563.39 \times 6.2107) = -10204 \text{ N·m, verified.}$$

$$P_m = 3I_r^2 \frac{R_r}{s}(1-s) = 3 \times 387.24^2 \times \frac{18.152 \times 10^{-3}}{-0.0053}(1+0.0053)$$

$$= -1.5489 \times 10^6 \text{ W, verified.}$$

7-12. A 4.0 MW, 4000 V, 50 Hz SCIG is used in a variable-speed CSC wind energy system with indirect FOC. At a wind speed of 9.6 m/s, the generator operates at 0.8 pu rotor speed with slip of −0.0056 and stator current of 473.43 A (rms). The rotor flux λ_r is kept at its rated value of 9.4914 Wb (peak) by the indirect field-oriented controller. The rectifier filter capacitor value is 289 μF. Repeat Problem 7-11.

Answers:

a) $T_m = -16429$ N·m, $\quad P_{m,R} = -4.061 \times 10^6$ W, $\quad P_m = -2.079 \times 10^6$ W

b) $\bar{I}_s = 473.43\angle 0°$ A (rms), $\quad \bar{Z}_s = 3.8059\angle 143.4°$ Ω for $s = -0.0056$

$\quad X_{ls} = 0.4274$ Ω, $\quad X_{lr} = 0.4274$ Ω, $\quad X_m = 8.4557$ Ω

$\quad \bar{V}_s = 1801.8\angle 143.4°$ V (rms), $\quad \bar{I}_r = 408.57\angle -24.96°$ A (rms)

c) $\bar{\Lambda}_m = 6.7472\angle 59.14°$ Wb (rms), $\quad \bar{\Lambda}_s = 7.1927\angle 53.64°$ Wb (rms),

$\quad \lambda_s = 10.172$ Wb, $\quad \bar{\Lambda}_r = 6.7114\angle 65°$ Wb (rms), $\quad \lambda_r = 9.4914$ Wb

d) $I_{ds} = 199.7617$ A (rms), $\quad I_{qs} = -429.22$ A (rms), $\quad V_{ds} = 362.45$ V (rms),

$\quad V_{qs} = 1765$ V (rms)

e) $\bar{I}_{cr} = 131.1\angle -126.6°$ A (rms), $\quad \bar{I}_{wr} = 409.13\angle -14.91°$ A (rms)

f) $\alpha_r = 79.95°$, $\quad V_{dci} = V_{dcr} = -3552.6$ V, $\quad \alpha_i = 132.18°$,

$\quad T_m = -16453$ N·m, $\quad P_m = -2.082 \times 10^6$ W, verified.

Topic: Steady-State Analysis of PWM CSC WECS with Indirect FOC

7-13 (Solved Problem). Consider a 3.0 MW, 3000 V, 60 Hz CSC interfaced SCIG wind energy system. At a wind speed of 12 m/s, the generator operates at 1.0 pu rotor speed. The generator parameters are given in Table B-3 of Appendix B. The rotor flux λ_r is kept at its rated value of 6.2107 Wb (peak) by the indirect FOC controller. The rectifier filter capacitor value is 296.8 μF. Find the following:

a) The generator mechanical torque and power

b) The dq-axis and rms stator current

c) The slip and stator frequency

d) The rms stator voltage

e) The magnetizing, stator, and rotor flux linkage values,

f) The dq-axis rms stator voltages,

g) The rms capacitor and rectifier PWM currents

h) The rectifier and inverter firing angles

Solution:

a) The generator mechanical torque at 1.0 pu rotor speed:

$$T_m = T_{m,R} \times (\omega_{m,pu})^2 = -16022 \times 1^2 = -16022 \text{ N·m}$$

The rated mechanical power:

$$P_{m,R} = \omega_{m,R} \times T_{m,R} = 1812 \times (2\pi)/60 \times (-16022) = -3.0402 \times 10^6 \text{ W}$$

The generator mechanical power at 1.0 pu rotor speed:

$$P_m = P_{m,R} \times (\omega_{m,pu})^3 = -3.0402 \times 10^6 \times 1^3 = -3.0402 \times 10^6 \text{ W}$$

b) The d-axis stator current:

$$i_{ds} = \frac{\lambda_r}{L_m} = \frac{6.2107}{27.168 \times 10^{-3}} = 228.6 \text{ A (peak)}$$

$$I_{ds} = i_{ds}/\sqrt{2} = 161.65 \text{ A (rms)}$$

The q-axis stator current:

$$i_{qs} = \frac{T_e}{K_T \lambda_r} = \frac{2L_r}{3PL_m} \frac{T_e}{\lambda_r} = \frac{2 \times 27.95 \times 10^{-3}}{3 \times 2 \times 27.168 \times 10^{-3}} \cdot \frac{-16022}{6.2107} = -884.67 \text{ A (peak)}$$

$$I_{qs} = i_{qs}/\sqrt{2} = -625.56 \text{ A (rms)}$$

The stator current:

$$i_s = \sqrt{i_{ds}^2 + i_{qs}^2} = \sqrt{228.6^2 + 884.67^2} = 913.73 \text{ A (peak)}$$

$$\overline{I}_s = i_s/\sqrt{2}\angle 0° = 646.1\angle 0° \text{ A (rms)}$$

c) The slip frequency:

$$\omega_{sl} = \frac{R_r L_m}{L_r \lambda_r} i_{qs} = \frac{16.623 \times 10^{-3} \times 27.168 \times 10^{-3}}{27.95 \times 10^{-3} \times 6.2107} \times (-884.67)$$

$$= -2.5133 \text{ rad/s } (-12 \text{ rpm}) (-0.4 \text{ Hz})$$

The stator frequency can be found from

$$\omega_r = (1812 \times 1.0) \times (2\pi/60)P = 379.5 \text{ rad/s } (1812 \text{ rpm}) (60.4 \text{ Hz})$$

$$\omega_s = \omega_r + \omega_{sl} = 379.5 - 2.5133 = 376.99 \text{ rad/s } (1800 \text{ rpm}) (60 \text{ Hz})$$

The slip:

$$s = \frac{\omega_{sl}}{\omega_s} = -0.0067 \text{ (negative slip)}$$

d) The equivalent impedance of SCIG:

$$\overline{Z}_s = R_s + jX_{ls} + jX_m \ // \left(\frac{R_r}{s} + jX_{lr} \right) = 2.6807\angle 153.3° \ \Omega$$

where $X_{ls} = \omega_s L_{ls} = 2 \times 3.1416 \times 60 \times 0.782 \times 10^{-3} = 0.2948\ \Omega$, $X_{lr} = \omega_s L_{lr} = 0.2948$ Ω, and $X_m = \omega_s L_m = 10.2421\ \Omega$

The stator voltage:

$$\overline{V}_s = \overline{I}_s \times \overline{Z}_s = 646.1\angle 0° \times 2.6807\angle 153.3° = V_s \angle \varphi_s = 1732\angle 153.3°\ \text{V (rms)}$$

e) The rotor current can be calculated by

$$\overline{I}_r = \frac{jX_m \overline{I}_s}{jX_m + \left(\dfrac{R_r}{s} + jX_{lr}\right)} = 608.05\angle -14.5°\ \text{A (rms)}$$

The magnetizing flux linkage can be expressed as

$$\overline{\Lambda}_m = (\overline{I}_s - \overline{I}_r)L_m = 4.4173\angle 69.3°\ \text{Wb (rms)}$$

Alternatively, $\overline{\Lambda}_m = \dfrac{\overline{V}_s - \overline{I}_s(R_s + jX_{ls})}{j\omega_s} = 4.4173\angle 69.3°\ \text{Wb (rms)}$

The stator flux linkage can be obtained by

$$\overline{\Lambda}_s = \overline{\Lambda}_m + L_{ls}\overline{I}_s = 4.6198\angle 63.5°\ \text{Wb (rms)}$$

The peak value of stator flux: $\lambda_s = \sqrt{2}\Lambda_s = 6.5335$ Wb
The rotor flux linkage is then can be obtained by

$$\overline{\Lambda}_r = \overline{\Lambda}_m - L_{lr}\overline{I}_r = \Lambda_r \angle \theta_1 = 4.3916\angle 75.5°\ \text{Wb (rms)}$$

The peak value of rotor flux, $\lambda_r = \sqrt{2}\Lambda_r = 6.2107$ Wb

f) Referring to Figure C-13, the rms dq-axis stator voltages can be calculated by

$$\begin{cases} V_{ds} = V_s \cos(\angle \overline{V}_s - \angle \overline{\Lambda}_r) = 1732\cos(77.78°)\ \text{V} = 366.36\ \text{V} \\ V_{qs} = V_s \sin(\angle \overline{V}_s - \angle \overline{\Lambda}_r) = 1732\sin(77.78°)\ \text{V} = 1692.8\ \text{V} \end{cases} \text{(rms)}$$

g) Method #1:
The rms capacitor current can be calculated by

$$\overline{I}_{cr} = (j\omega_s C_r)\overline{V}_s = (j376.99 \times 296.8 \times 10^{-6})(1732\angle 153.3°) = 193.8\angle -116.7°\ \text{A (rms)}$$

The rms rectifier PWM current can be found from

$$\overline{I}_{wr} = \overline{I}_s + \overline{I}_{cr} = 646.1\angle 0° + 193.8\angle -116.7° = 585.2\angle -17.2°\ \text{A (rms)}$$

Method #2:

The dq-axis currents of the rectifier-side capacitor C_r can be found from

$$\begin{cases} I_{crd} = -2\pi f_s V_{qs} C_r = -2\pi \times 60 \times 1692.2 \times 296.8 \times 10^{-6} = -189.42 \text{ A} \\ I_{crq} = 2\pi f_s V_{ds} C_r = 2\pi \times 60 \times 366.36 \times 296.8 \times 10^{-6} = 40.99 \text{ A} \end{cases} \quad \text{(rms)}$$

The rms capacitor current:

$$I_{cr} = \sqrt{I_{crd}^2 + I_{crq}^2} = 193.8 \text{ A (rms)}$$

The dq-axis PWM currents of the rectifier can be calculated by

$$\begin{cases} I_{dwr} = I_{ds} + I_{crd} = 161.65 - 189.42 = -27.77 \text{ A} \\ I_{qwr} = I_{qs} + I_{crq} = -625.55 + 40.99 = -584.55 \text{ A} \end{cases} \quad \text{(rms)}$$

The rms rectifier PWM current:

$$I_{wr} = \sqrt{I_{dwr}^2 + I_{qwr}^2} = 585.2 \text{ A (rms)}$$

h) The firing angle for the rectifier:

$$\alpha_r = \angle \overline{\Lambda}_r - \angle \overline{I}_{wr} = 75.5° - (-17.2°) = 92.7°$$

The rectifier firing angle can also be found from

$$\alpha_r = \tan^{-1} \frac{I_{qwr}}{I_{dwr}} = \tan^{-1} \frac{-584.55}{-27.77} = 92.7°$$

The active power delivered from the generator to the DC link circuit is

$$P_{dc} = 3V_s I_s \cos\varphi_s = 3 \times 1732 \times 646.1 \cos(153.3°) = -3 \times 10^6 \text{ W}$$

The DC current I_{dc} is

$$I_{dc} = \sqrt{2} \, I_{wr1} \, m_r = \sqrt{2} \times 585.2 \times 1.0 = 827.6 \text{ A}$$

where rectifier modulation index m_r is set to its maximum value of 1.0. The rectifier-side DC voltage:

$$V_{dcr} = \frac{P_{dc}}{I_{dc}} = \frac{-3 \times 10^6}{827.6} = -3623.9 \text{ V}$$

The rectifier-side dc voltage can also be calculated by

$$V_{dcr} = \sqrt{3/2}\, V_{LL} m_r \cos\left(\left|\varphi_s\right| + \left|\theta_3\right|\right) = \sqrt{3/2}\times 3000 \times 1.0 \times \cos\left(153.3^\circ + 17.2^\circ\right) = -3623.9 \text{ V}$$

where, $\theta_3 = \alpha_r - \theta_1 = 92.7^\circ - 75.5^\circ = 17.2^\circ$

Neglecting the winding resistance of the dc choke L_{dc}, the inverter-side DC voltage is

$$V_{dci} = V_{dcr} = -3623.9 \text{ V}$$

The inverter firing angle:

$$\alpha_i = \cos^{-1}\left(\frac{V_{dci}}{\sqrt{3/2}\, m_i V_{LL}}\right) = \cos^{-1}\left(\frac{-3623.9}{\sqrt{3/2}\times 1.08 \times 3000}\right) = 156.0^\circ$$

where m_i is the modulation index of the inverter with SHE scheme.

Cross Check:

$$T_m = \frac{3PL_m}{2L_r}(i_{qs}\lambda_{dr}) = \frac{3PL_m}{2L_r}(i_{qs}\lambda_r);\ i_{qs} \text{ and } \lambda_r \text{ are the peak values.}$$

$$= \frac{3\times 2\times 27.168\times 10^{-3}}{2\times 27.95\times 10^{-3}}(-884.67\times 6.2107) = -16022 \text{ N·m}, \text{ verified.}$$

$$P_m = 3I_r^2\frac{R_r}{s}(1-s) = 3\times 608.05^2 \times \frac{18.152\times 10^{-3}}{-0.0067}(1+0.0067)$$

$$= -3.0402\times 10^6 \text{ W}, \text{ verified.}$$

7-14. Repeat Problem 7-13 for a wind speed of 7.2 m/s, at which the generator operates at 0.6 pu rotor speed.

Answers:

a) $T_m = -5767.9$ N·m, $\quad P_{m,R} = -3.0402\times 10^6$ W, $\quad P_m = -656.69\times 10^3$ W

b) $I_{ds} = 161.65$ A (rms), $\quad I_{qs} = -225.2$ A (rms), $\quad \bar{I}_s = 277.2\angle 0^\circ$ A (rms)

c) $\omega_{sl} = -0.9048$ rad/s (-0.144 Hz), $\quad \omega_s = 226.79$ rad/s (36.096 Hz), $\quad s = -0.004$

d) $\bar{Z}_s = 3.6946\angle 139.77^\circ\ \Omega$, $\quad X_{ls} = 0.1774\ \Omega$, $\quad X_{lr} = 0.1774\ \Omega$,

$\quad X_m = 6.1616\ \Omega$, $\quad \bar{V}_s = 1024.2\angle 139.77^\circ$ V (rms)

e) $\bar{I}_r = 218.9\angle -35.67^\circ$ A (rms), $\quad \bar{\Lambda}_m = 4.3949\angle 52.1^\circ$ Wb (rms),

$\quad \bar{\Lambda}_s = 4.5313\angle 49.93^\circ$ Wb (rms), $\quad \lambda_s = 6.4083$ Wb,

$\overline{\Lambda}_r = 4.3916\angle 54.33°$ Wb (rms), $\lambda_r = 6.2107$ Wb

f) $V_{ds} = 81.45$ V (rms), $V_{qs} = 1020.9$ V (rms)

g) $\overline{I}_{cr} = 114.6\angle -130.23°$ A (rms), $\overline{I}_{wr} = 221.23\angle -23.3°$ A (rms)

h) $\alpha_r = 77.63°$, $V_{dci} = V_{dcr} = -2078.4$ V, $\alpha_i = 121.58°$

7-15. Consider a 4.0 MW, 4000 V, 50 Hz SCIG wind energy system. At a wind speed of 12 m/s, the generator operates at 1.0 pu rotor speed. The generator parameters are given in Table B-4 of Appendix B. The rotor flux λ_r is kept at its rated value of 9.4914 Wb (peak) by the indirect field-oriented controller. The rectifier filter capacitor value is 289 μF. Repeat Problem 7-13:

Answers:

a) $T_m = -25671$ N·m, $P_{m,R} = -4.0606 \times 10^6$ W, $P_m = -4.0606 \times 10^6$ W

b) $I_{ds} = 199.76$ A (rms), $I_{qs} = -669.72$ A (rms), $\overline{I}_s = 698.88\angle 0°$ A (rms)

c) $\omega_{sl} = -2.1991$ rad/s (−0.35 Hz), $\omega_s = 314.16$ rad/s (50 Hz), $s = -0.007$

d) $\overline{Z}_s = 3.3045\angle 145.7°$ Ω, $X_{ls} = 0.5334$ Ω, $X_{lr} = 0.5334$ Ω,

   $X_m = 10.5548$ Ω, $\overline{V}_s = 2309.4\angle 145.7°$ V (rms)

e) $\overline{I}_r = 637.5\angle -16.61°$ A (rms), $\overline{\Lambda}_m = 6.7981\angle 64.23°$ Wb (rms),

   $\overline{\Lambda}_s = 7.3917\angle 55.92°$ Wb (rms), $\lambda_s = 10.4535$ Wb,

   $\overline{\Lambda}_r = 6.7114\angle 73.39°$ Wb (rms), $\lambda_r = 9.4914$ Wb

f) $V_{ds} = 701.74$ V (rms), $V_{qs} = 2200.2$ V (rms)

g) $\overline{I}_{cr} = 209.68\angle -124.3°$ A (rms), $\overline{I}_{wr} = 606\angle -16.61°$ A (rms)

h) $\alpha_r = 89.99°$, $V_{dci} = V_{dcr} = -4667.3$ V, $\alpha_i = 151.9°$

7-16. Repeat Problem 7-15 for a wind speed of 7.2 m/s, at which the generator operates at 0.6 pu rotor speed.

Answers:

a) $T_m = -9241.6$ N·m, $P_{m,R} = -4.0606 \times 10^6$ W, $P_m = -877.09 \times 10^3$ W

b) $I_{ds} = 199.76$ A (rms), $I_{qs} = -241.1$ A (rms), $\overline{I}_s = 313.1\angle 0°$ A (rms)

c) $\omega_{sl} = -0.7917$ rad/s (−0.126 Hz), $\omega_s = 189.02$ rad/s (30.084 Hz),

   $s = -0.0042$

d) $\overline{Z}_s = 4.2686\angle 133.68°$ Ω, $X_{ls} = 0.321$ Ω, $X_{lr} = 0.321$ Ω,

   $X_m = 6.3506$ Ω, $\overline{V}_s = 1336.5\angle 133.68°$ V (rms)

e) $\overline{I}_r = 229.5\angle -39.64°$ A (rms), $\overline{\Lambda}_m = 6.7227\angle 47.03°$ Wb (rms),

   $\overline{\Lambda}_s = 7.0957\angle 43.89°$ Wb (rms), $\lambda_s = 10.035$ Wb ,

   $\overline{\Lambda}_r = 6.7114\angle 50.36°$ Wb (rms), $\lambda_r = 9.4914$ Wb

f) $V_{ds} = 155.5$ V (rms), $V_{qs} = 1327.4$ V (rms)

g) $\overline{I}_{cr} = 121.3\angle -136.3°$ A (rms), $\overline{I}_{wr} = 240.42\angle -20.4°$ A (rms)

h) $\alpha_r = 70.75°$, $V_{dci} = V_{dcr} = -2549.8$ V, $\alpha_i = 118.81°$

PROBLEMS AND ANSWERS FOR CHAPTER 8—DOUBLY FED INDUCTION GENERATOR BASED WECS

Assumptions for all the problems in this chapter:

1. Switching harmonics produced by power converters are neglected.
2. Power converters are ideal without power losses.
3. Core and rotational losses of the generator are neglected.

Topic: Steady-State Equivalent Circuit of DFIG with Rotor-Side Converter

8-1 (Solved Problem). A 1.0 MW, 575 V, 60 Hz, 2160 rpm DFIG is used in a variable-speed wind energy conversion system. The parameters of the generator are given in Table B-6 of Appendix B. The generator operates with an MPPT scheme and its stator power factor is unity. Assuming that the DFIG operates at a supersynchronous speed of 2160 rpm, determine the following:

a) The generator mechanical torque and power
b) The rms stator current
c) The rms magnetizing voltage and current
d) The rms rotor current and voltage
e) The equivalent resistance and reactance for the rotor side converter
f) The maximum torque and the corresponding slip

Solution:

a) The rotor mechanical speed:

$$\omega_m = 2160 \times (2\pi / 60) = 226.19 \text{ rad/sec}$$

The rotor electrical speed:

$$\omega_r = \omega_m \times P = 226.19 \times 2 = 452.39 \text{ rad/sec}$$

The rated rotor mechanical speed:

$$\omega_{m,R} = 2160 \times (2\pi / 60) = 226.19 \text{ rad/sec}$$

The stator frequency:

$$\omega_s = 2\pi \times 60 = 376.99 \text{ rad/sec}$$

The pu rotor speed $\omega_{m,\text{pu}} = \omega_m / \omega_{m,R} = 226.19/226.19 = 1.0$ pu

The generator mechanical torque at 1.0 pu rotor speed:

$$T_m = T_{m,R} \times (\omega_{m,\mathrm{pu}})^2 = -4421 \times (1.0)^2 = -4421 \text{ N·m}$$

The rated mechanical power:

$$P_{m,R} = \omega_{m,R} \times T_{m,R} = 226.19 \times (-4421) = -1000 \times 10^3 \text{ W}$$

The generator mechanical power at 1.0 pu rotor speed:

$$P_m = P_{m,R} \times (\omega_{m,\mathrm{pu}})^3 = -1000 \times 10^3 \times (1.0)^3 = -1000 \times 10^3 \text{ W}$$

b) The stator current from Equation (8.7):

$$I_s = \frac{V_s \pm \sqrt{V_s^2 - \dfrac{4R_s T_m \omega_s}{3P}}}{2R_s} = -829.18 \text{ A (rms)} \quad (I_s = 91.682 \times 10^3 \text{A omitted})$$

where $V_s = 575/\sqrt{3}$ V, $T_m = -4421$ N·m, $\omega_s = 376.99$ rad/sec, $R_s = 3.654$ mΩ, and $P = 2$

c) The magnetizing branch voltage:

$$\begin{aligned}
\overline{V}_m &= \overline{V}_s - \overline{I}_s \left(R_s + j\omega_s L_{ls} \right) \\
&= 575/\sqrt{3}\angle 0° - 829.18\angle 180° \times (3.654 \times 10^{-3} + j120\pi \times 0.1304 \times 10^{-3}) \\
&= 337.48\angle 6.94° \text{ V (rms)}
\end{aligned}$$

The magnetizing current can be calculated by

$$\overline{I}_m = \frac{\overline{V}_m}{j\omega_s L_m} = 217.28\angle -83.1° \text{ A (rms)}$$

d) The rotor current:

$$\overline{I}_r = \overline{I}_s - \overline{I}_m = 829.18\angle 180° - 217.28\angle -83.1° = 882.19\angle 165.85° \text{ A (rms)}$$

The rotor voltage:

$$\overline{V}_r = s\overline{V}_m - \overline{I}_r \left(R_r + js\omega_s L_{lr} \right) = 67.97\angle -165.83° \text{ V (rms)}$$

where $s = (\omega_s - \omega_r)/\omega_s = (376.99 - 452.39)/376.99 = -0.2$

e) The equivalent impedance for the rotor side converter is given by

$$\overline{Z}_{eq} = \overline{V}_r / \overline{I}_r = 0.06782 + j0.03656 \ \Omega$$

from which $R_{eq} = 0.06782 \ \Omega$ and $X_{eq} = 0.03656 \ \Omega$

f) The slip at which the maximum torque occurs can be obtained from Equation (2.34):

$$s_{T\max} = \pm\sqrt{\frac{(R_r + R_{eq})^2 + X_{eq}^2}{R_s^2 + (X_{ls} + X_{lr})^2}} = -0.8497$$

($s = +0.8497$ is omitted because of the supersynchronous mode of operation.)

The maximum torque:

$$T_{\max} = \frac{1}{2\omega_s / P}\times\frac{3V_s^2}{R_s + \dfrac{(X_{ls} + X_{lr})X_{eq}}{R_r + R_{eq}} - \sqrt{\left((X_{ls} + X_{lr})^2 + R_s^2\right)\times\left(1 + \dfrac{X_{eq}^2}{(R_r + R_{eq})^2}\right)}}$$

$$= -16214 \text{ N·m} \quad (3.6675 \text{ pu})$$

Cross Check:

$$T_m = \frac{1}{\omega_s / P}\times 3I_r^2 (R_{eq} + R_r) / s = \frac{1}{2\pi \times 60 / 2}$$

$$\times 3\times 882.19^2(0.06782 + 3.569\times 10^{-3}) / (-0.2) = -4421 \text{ N· m, verified.}$$

$$P_m = 3I_r^2(R_{eq} + R_r)(1 - s) / s = 3\times 882.19^2(0.06782 + 3.569\times 10^{-3})(1 + 0.2) / (-0.2)$$

$$= -1000\times 10^3 \text{ W, verified.}$$

8-2. Repeat Problem 8-1 when the DFIG operates at synchronous speed of 1800 rpm.

Answers:

a) $T_m = -3070.1$ N·m, $\qquad P_m = -578.71\times10^3$ W

b) $I_s = -577.41$ A (rms)

c) $\bar{V}_m = 335.29\angle 4.86°$ V (rms), $\qquad \bar{I}_m = 215.87\angle -85.14°$ A (rms)

d) $\bar{I}_r = 633.32\angle 160.15°$ A (rms), $\qquad \bar{V}_r = 2.26\angle -19.85°$ V (rms)

e) $R_{eq} = -0.00357$ Ω, $\qquad X_{eq} = 0$ Ω

f) $s_{T\max} = 0$, $\qquad T_{\max} = -6752$ N·m (1.5272 pu)

8-3. Repeat Problem 8-1 when the DFIG operates at a subsynchronous speed of 1350 rpm.

Answers:

a) $T_m = -1726.9$ N·m, $\qquad P_m = -244.14\times10^3$ W

b) $I_s = -325.69$ A (rms)

c) $\overline{V}_m = 333.55\angle 2.75°$ V (rms), $\overline{I}_m = 214.75\angle -87.25°$ A (rms)

d) $\overline{I}_r = 398.63\angle 147.45°$ A (rms), $\overline{V}_r = 87.2\angle 4.63°$ V (rms)

e) $R_{eq} = -0.17428$ Ω, $X_{eq} = -0.13219$ Ω

f) $s_{T\max} = 2.2873$, $T_{\max} = -20542$ N·m (4.6465 pu)

8-4 (Solved Problem). Consider a 1.0 MW, 575 V, 60 Hz, 2160 rpm DFIG WECS. The parameters of the generator are given in Table B-6 of Appendix B. The generator operates with an MPPT scheme and its stator power factor is unity. At a given wind and generator speed, the equivalent resistance R_{eq} and reactance X_{eq} for the rotor side converter and the maximum torque to mechanical torque ratio $T_{\max}/T_m$ are found to be 0.03778 Ω, 0.02165 Ω, and 4.4123 respectively. Calculate the following:

a) The maximum torque and the corresponding slip

b) The generator mechanical torque and power

c) The rotor mechanical and electrical speeds and slip

d) The rms stator current

e) The rms rotor current and voltage

Solution:

a) The positive values for R_{eq} and X_{eq} indicate supersynchronous mode of operation. The negative values indicate subsynchronous operation. Zero value for X_{eq} indicate synchronous operation.

   The slip at which the maximum torque occurs can be obtained from Equation (8.34):

$$s_{T\max} = \pm\sqrt{\frac{(R_r + R_{eq})^2 + X_{eq}^2}{R_s^2 + (X_{ls} + X_{lr})^2}} = -0.4944$$

($s = +0.4944$ is omitted because of the supersynchronous mode of operation.)

$$T_{\max} = \frac{1}{2\omega_s / P} \times \frac{3V_s^2}{R_s + \dfrac{(X_{ls} + X_{lr})X_{eq}}{R_r + R_{eq}} - \sqrt{\left((X_{ls} + X_{lr})^2 + R_s^2\right) \times \left(1 + \dfrac{X_{eq}^2}{(R_r + R_{eq})^2}\right)}}$$

$$= -16391 \text{ N·m}   (3.708 \text{ pu})$$

b) The generator mechanical torque:

$$T_m = \frac{T_{\max}}{T_{\max}/T_m} = \frac{-16391}{4.4124} = -3714.8 \text{ N·m}$$

The generator mechanical torque can be related to pu rotor speed as

$$T_m = T_{m,R} \times (\omega_{m,\text{pu}})^2 \text{ N·m}$$

from which pu rotor speed can be calculated by

$$\omega_{m,\text{pu}} = \sqrt{\frac{T_m}{T_{m,R}}} = \sqrt{\frac{-3714.8}{-4421}} = 0.9167$$

The rated mechanical power:

$$P_{m,R} = \omega_{m,R} \times T_{m,R} = 2160(2\pi)/60 \times (-4421) = -1000 \times 10^3 \text{ W}$$

The generator mechanical power at 0.9167 pu rotor speed:

$$P_m = P_{m,R} \times (\omega_{m,\text{pu}})^3 = -1000 \times 10^3 \times (0.9167)^3 = -770.26 \times 10^3 \text{ W}$$

c) The rotor mechanical and electrical speeds:

$$\omega_m = \omega_{m,R} \times \omega_{m,\text{pu}} = 2160(2\pi)/60 \times 0.9167 = 207.35 \text{ rad/sec } (1980 \text{ rpm})$$

$$\omega_r = \omega_m \times P = 207.35 \times 2 = 414.7 \text{ rad/sec}$$

The slip can be obtained as:

$$s = (\omega_s - \omega_r)/\omega_s = (376.99 - 414.7)/376.99 = -0.1$$

d) The stator current:

$$I_s = \frac{V_s \pm \sqrt{V_s^2 - \dfrac{4R_s T_m \omega_s}{3P}}}{2R_s} = -697.74 \text{ A (rms)} \quad (I_s = 91.55 \times 10^3 \text{ A omitted})$$

where $V_s = 575/\sqrt{3}$ V, $T_m = -3741.8$ N·m, $\omega_s = 376.99$ rad/sec, $R_s = 3.654$ mΩ, and $P = 2$

e) From the steady-state equivalent circuit of DFIG with the rotor side converter as shown in Figure 8-4, the rotor current can be calculated by

$$\bar{I}_r = \frac{jX_m \bar{I}_s}{jX_m + \left(\dfrac{R_r}{s} + jX_{lr}\right) + \left(\dfrac{R_{eq}}{s} + j\dfrac{X_{eq}}{s}\right)} = 751.35\angle163.34° \text{ A (rms)}$$

Alternatively, the rms rotor current can be found from the mechanical torque equation:

$$T_m = \frac{1}{\omega_s / P} \times 3 I_r'^2 \left(R_{eq} + R_r \right) / s \quad \text{N·m}$$

from the above

$$I_r = \sqrt{\frac{T_m \times (\omega_s / P)}{3(R_{eq} + R_r)/s}} = 751.35 \quad \text{A (rms)}$$

The equivalent impedance for the rotor side converter:

$$\overline{Z}_{eq} = \overline{V}_r / \overline{I}_r = 0.0378 + j0.02165 = 0.04354 \angle 29.82° \ \Omega \ \text{(given)}$$

The rotor voltage:

$$\overline{V}_r = \overline{Z}_{eq} \times \overline{I}_r = 32.72 \angle -166.84° \ \text{V (rms)}$$

Cross Check:

$$T_m = \frac{1}{\omega_s / P} \times 3 I_r^2 (R_{eq} + R_r) / s = \frac{1}{2\pi \times 60 / 2}$$

$$\times 3 \times 751.35^2 (0.0378 + 3.569 \times 10^{-3}) / (-0.1) = -3714.87 \quad \text{N·m, verified.}$$

$$P_m = 3 I_r^2 (R_{eq} + R_r)(1 - s) / s = 3 \times 751.35^2 (0.0378 + 3.569 \times 10^{-3})(1 + 0.1)/(-0.\ 1) =$$

$$- 770.26 \times 10^3 \ \text{W, verified.}$$

8-5. Repeat Problem 8-4 if the R_{eq}, X_{eq}, and T_{max}/T_m are −0.0295 Ω, −0.0150 Ω, and 6.227, respectively.

Answers:

a) $s_{T\max} = 0.3175$, $T_{\max} = -17254$ N·m (3.903 pu)
b) $T_m = -2770.8$ N·m, $P_m = -496.18 \times 10^3$ W
c) $\omega_m = 179.07$ rad/sec (1710 rpm), $\omega_r = 358.14$ rad/sec, $s = 0.05$
d) $I_s = -521.43$ A (rms)
e) $\overline{I}_r = 579.9 \angle 158.22°$ A (rms), $\overline{V}_r = 19.18 \angle 5.17°$ V (rms)

8-6. Repeat Problem 8-4 if the R_{eq}, X_{eq}, and T_{max}/T_m are −0.0937 Ω, −0.05887 Ω and 8.322, respectively.

Answers:

a) $s_{T\max} = 1.14$, $T_{\max} = -18459$ N·m (4.1754 pu)
b) $T_m = -2218$ N·m, $P_m = -355.4 \times 10^3$ W

c) $\omega_m = 160.22$ rad/sec (1530 rpm), $\omega_r = 320.44$ rad/sec, $s = 0.15$

d) $I_s = -417.9$ A (rms)

e) $\bar{I}_r = 481.64\angle153.5°$ A (rms), $\bar{V}_r = 53.28\angle5.64°$ V (rms)

Topic: Steady-State Analysis of DFIG WECS with $PF_s = 1$

8-7 (Solved Problem). A 5.0 MW, 950 V, 50 Hz, 1170 rpm DFIG is employed in a variable-speed WECS. The parameters of the generator are given in Table B-7 of Appendix B. The DFIG WECS is connected to a grid (line–line voltage $V_{AB} = 950$ V, 50 Hz). The generator operates with an MPPT scheme and its stator power factor is unity. The corresponding equivalent resistance and reactance for the rotor-side converter when the DFIG operates at subsynchronous speeds of 670 rpm are -0.13059 Ω and -0.2624 Ω, respectively. Calculate the following:

a) The generator mechanical torque and power
b) The rms stator and rotor currents
c) The stator and rotor winding losses
d) The stator and rotor active powers
e) The net power delivered to the grid and efficiency of the DFIG
f) The fundamental grid current

Solution:

a) The rotor mechanical speed:

$$\omega_m = 670\times(2\pi/60) = 70.162 \text{ rad/sec}$$

The rotor electrical speed:

$$\omega_r = \omega_m \times P = 70.16\times3 = 210.48 \text{ rad/sec}$$

The rated rotor mechanical speed:

$$\omega_{m,R} = 1170\times(2\pi/60) = 122.52 \text{ rad/sec}$$

The stator frequency:

$$\omega_s = 2\pi\times50 = 314.16 \text{ rad/sec}$$

The pu rotor speed $\omega_{m,\text{pu}} = \omega_m/\omega_{m,R} = 70.162/122.52 = 0.5727\,\text{pu}$
The generator mechanical torque at pu rotor speed:

$$T_m = T_{m,R}\times(\omega_{m,\text{pu}})^2 = -40809\times(0.5727)^2 = -13382.4 \text{ N·m}$$

The rated mechanical power:

$$P_{m,R} = \omega_{m,R} \times T_{m,R} = -122.52 \times 40809 = -5000 \times 10^3 \text{ W}$$

The generator mechanical power at pu rotor speed:

$$P_m = P_{m,R} \times (\omega_{m,\text{pu}})^3 = -5000 \times 10^3 \times (0.5727)^3 = -938.94 \times 10^3 \text{ W}$$

b) The rms stator current using conventional expression:

$$I_s = \frac{V_s \pm \sqrt{V_s^2 - \dfrac{4R_s T_m \omega_s}{3P}}}{2R_s} = -849.64 \text{ A (rms)} \quad (\ I_s = 35.43 \ \times 10^4 \text{ A omitted})$$

where $V_s = 950/\sqrt{3}$ V, $T_m = -13382.4$ N·m, $\omega_s = 314.16$ rad/sec, $R_s = 1.552$ mΩ, and $P = 3$

From the steady-state equivalent circuit of DFIG with the rotor-side converter as shown in Figure 8-4, the rotor current can be calculated by

$$\bar{I}_r = \frac{jX_m \bar{I}_s}{jX_m + \left(\dfrac{R_r}{s} + jX_{lr}\right) + \left(\dfrac{R_{eq}}{s} + j\dfrac{X_{eq}}{s}\right)} = 1092.55\angle 163.13° \text{ A (rms)}$$

where $s = (\omega_s - \omega_r)/\omega_s = (314.16 - 210.48)/314.16 = 0.33$

Alternatively, the rms rotor current can be found from the mechanical torque equation:

$$T_m = \frac{1}{\omega_s / P} \times 3I_r^2 \left(R_{eq} + R_r\right)/s \text{ N·m}$$

from which

$$I_r = \sqrt{\frac{T_m \times (\omega_s / P)}{3(R_{eq} + R_r)/s}} = 1092.55 \text{ A (rms)}$$

c) The stator and rotor winding losses:

$$P_{cu,s} = 3(I_s)^2 R_s = 3 \times 849.64^2 \times 1.552 \times 10^{-3} = 3.361 \times 10^3 \text{ W}$$

$$P_{cu,r} = 3(I_r)^2 R_r = 3 \times 1092.55^2 \times 1.446 \times 10^{-3} = 5.178 \times 10^3 \text{ W}$$

d) The stator and rotor active powers:

$$P_s = 3V_s I_s \cos\varphi_s = 3 \times 950/\sqrt{3} \times 849.64 \times \cos(180°) = -1398 \times 10^3 \text{ W}$$

$$P_r = 3(I_r)^2 R_{eq} = 3 \times (1092.55)^2 \times -0.13059 = -467.64 \times 10^3 \text{ W}$$

e) The net power delivered to the grid:

$$\left|P_g\right| = \left|P_s\right| - \left|P_r\right| = 1398\times10^3 - 467.64\times10^3 = 930.4\times10^3 \ \text{W}$$

The difference between P_m and P_g is the losses on the stator and rotor windings, that is,

$$\left|P_m\right| - \left|P_g\right| = P_{cu,s} + P_{cu,r} = 8.539\times10^3 \ \text{W}$$

The efficiency of the DFIG neglecting rotational and core losses is then

$$\eta = P_g / \left|P_m\right| = 930.4/938.94 = 99.09\%$$

f) The fundamental grid current:

$$\left|I_g\right| = \frac{\left|P_g\right|}{3V_g} = \frac{930.4\times10^3}{3\times950/\sqrt{3}} = 565.44 \ \text{A (rms)}$$

Cross Check:

$$T_m = \frac{1}{\omega_s / P}\times3I_r^2\left(R_{eq} + R_r\right)/s = -13382 \ \text{N·m, verified.}$$

$$P_m = 3I_r^2\left(R_{eq} + R_r\right)(1-s)/s = -938.9\times10^3 \ \text{W, verified}$$

8-8. Repeat Problem 8-7 when the DFIG operates at supersynchronous speed of 1050 rpm. The corresponding equivalent resistance and reactance for the rotor-side converter are given 0.00718 Ω and 0.0349 Ω, respectively.

Answers:

a) $T_m = -32867 \ \text{N·m}$, $\qquad P_m = -3613.9\times10^3 \ \text{W}$
b) $I_s = -2079.5 \ \text{A (rms)}$, $\qquad I_r = 2578.8 \ \text{A (rms)}$
c) $P_{cu,s} = 20.134\times10^3 \ \text{W}$, $\qquad P_{cu,r} = 28.848\times10^3 \ \text{W}$
d) $P_s = -3421.71\times10^3 \ \text{W}$, $\qquad P_r = 143.24\times10^3 \ \text{W}$
e) $\left|P_g\right| = 3564.96\times10^3 \ \text{W}$, $\qquad \eta = 98.65\%$
f) $\left|I_g\right| = 2166.6 \ \text{A (rms)}$

8-9. Repeat Problem 8-7 when the DFIG operates at supersynchronous speed of 1170 rpm (rated). The corresponding equivalent resistance and reactance for the rotor-side converter are given 0.02237 Ω and 0.11739 Ω, respectively.

Answers:

a) $T_m = -40809 \ \text{N·m}$, $\qquad P_m = -5000\times10^3 \ \text{W}$
b) $I_s = -2578.4 \ \text{A (rms)}$, $\qquad I_r = 3188.7 \ \text{A (rms)}$

c) $P_{cu,s} = 30.95 \times 10^3$ W, $P_{cu,r} = 44.109 \times 10^3$ W

d) $P_s = -4242.5 \times 10^3$ W, $P_r = 682.4 \times 10^3$ W

e) $\left| P_g \right| = 4924.9 \times 10^3$ W, $\eta = 98.5\%$

f) $\left| I_g \right| = 2993.1$ A (rms)

8-10 (Solved Problem). Consider a 5.0 MW, 950 V, 50 Hz, 1170 rpm DFIG WECS. The parameters of the generator are given in Table B-7 of Appendix B. The DFIG WECS is connected to a grid (line–line voltage $V_{AB} = 950$ V, 50 Hz). The generator operates with an MPPT scheme. Neglect rotational and core losses. At a given wind and generator speed, the rms fundamental grid current I_g, and stator active power P_s are found to be 2016.08 A and –3261.6 kW, respectively. Calculate the following assuming unity power factor operation for the DFIG WECS:

a) The rms stator current

b) The generator mechanical torque and power

c) The rotor mechanical and electrical speeds and slip

d) The net power delivered to the grid and rotor active power

e) The stator and rotor winding losses, and efficiency of the DFIG

Solution:

a) The stator can be calculated by

$$I_s = P_s / (3V_s) = -3261.6 \times 10^3 / (3 \times 950 / \sqrt{3}\,) = -1982.2 \text{ A (rms)}$$

b) The generator air-gap power can be given by

$$\frac{\omega_s T_m}{P} = 3(V_s - I_s R_s) I_s$$

from which the generator mechanical torque can be obtained as

$$T_m = 3(V_s - I_s R_s) I_s \times (P / \omega_s) = -31320.7 \text{ N·m}$$

The generator mechanical torque can be related to pu rotor speed as

$$T_m = T_{m,R} \times (\omega_{m,pu})^2 \text{ N·m}$$

from which pu rotor speed can be calculated by

$$\omega_{m,pu} = \sqrt{\frac{T_m}{T_{m,R}}} = \sqrt{\frac{-31320.7}{-40809}} = 0.8761$$

The rated mechanical power:

$$P_{m,R} = \omega_{m,R} \times T_{m,R} = 1170(2\pi)/60 \times (-40809) = -5000 \times 10^3 \text{ W}$$

The generator mechanical power at 0.8761 pu rotor speed:

$$P_m = P_{m,R} \times (\omega_{m,\text{pu}})^3 = -5000 \times 10^3 \times (0.8761)^3 = -3361.9 \times 10^3 \text{ W}$$

c) The rotor mechanical and electrical speeds:

$$\omega_m = \omega_{m,R} \times \omega_{m,\text{pu}} = 1170(2\pi)/60 \times 0.8761 = 107.34 \text{ rad/sec} \quad (1025 \text{ rpm})$$

$$\omega_r = \omega_m \times P = 107.34 \times 3 = 322.02 \text{ rad/sec}$$

The slip can be calculated by:

$$s = (\omega_s - \omega_r)/\omega_s = (314.16 - 322.02)/314.16 = -0.025$$

d) The net power delivered to the grid:

$$\left| P_g \right| = 3V_g \left| I_g \right| = 3 \times 950/\sqrt{3} \times 2016.08 = 3317.3 \times 10^3 \text{ W}$$

The rotor active power:

$$P_r = \left| P_g \right| - \left| P_s \right| = (3317.3 - 3261.6) \times 10^3 = 55.75 \times 10^3 \text{ W}$$

e) The stator winding loss:

$$P_{cu,s} = 3(I_s)^2 R_s = 3 \times 1982.2^2 \times 1.552 \times 10^{-3} = 18.29 \times 10^3 \text{ W}$$

The difference between P_m and P_g is the losses on the stator and rotor windings, that is,

$$\left| P_m \right| - \left| P_g \right| = P_{cu,s} + P_{cu,r} \text{ W}$$

from which the rotor winding loss can be obtained as

$$P_{cu,r} = \left| P_m \right| - \left| P_g \right| - \left| P_{cu,s} \right| = (3361.9 - 3317.3 - 18.294) \times 10^3 = 26.25 \times 10^3 \text{ W}$$

The efficiency of the DFIG neglecting rotational and core losses is then

$$\eta = P_g / \left| P_m \right| = 3317.3/3361.9 = 98.68\%$$

8-11. Repeat Problem 8-10 when the fundamental grid current I_g, and stator active power P_s are 1366.87 A and -2517.8 kW respectively.

Answers:

a) $I_s = -1530.2$ A (rms)

b) $T_m = -24147$ N·m, $\qquad P_m = -2275.8 \times 10^3$ W

c) $\omega_m = 94.25$ rad/sec (900 rpm), $\qquad \omega_r = 282.75$ rad/sec, $\qquad s = 0.1$

d) $\left|P_g\right| = 2249.1 \times 10^3$ W, $\qquad P_r = -268.69 \times 10^3$ W

e) $P_{cu,s} = 10.902 \times 10^3$ W, $\qquad P_{cu,r} = 15.818 \times 10^3$ W, $\qquad \eta = 98.93\%$

8-12. Repeat Problem 8-10 when the fundamental grid current I_g, and stator active power P_s are 792.42A and −1750.77 kW, respectively.

Answers:

a) $I_s = -1064$ A (rms)

b) $T_m = -16769$ N·m, $\qquad P_m = -1317.0 \times 10^3$ W

c) $\omega_r = 78.54$ rad/sec (750 rpm), $\qquad \omega_r = 235.62$ rad/sec, $\qquad s = 0.25$

d) $\left|P_g\right| = 1303.9 \times 10^3$ W, $\qquad P_r = -446.88 \times 10^3$ W

e) $P_{cu,s} = 5.271 \times 10^3$ W, $\qquad P_{cu,r} = 7.873 \times 10^3$ W, $\qquad \eta = 99.0\%$

Topic: Leading and Lagging Power Factor Operation

8-13 (Solved Problem). A 1.0 MW, 575 V, 60 Hz, 2160 rpm DFIG is used in wind energy conversion system. The parameters of the generator are given in Table B-6 of Appendix B. The generator operates with an MPPT scheme and its stator power factor is 0.95 leading. Determine the following when the DFIG operates at a supersynchronous speed of 2160 rpm:

a) The generator mechanical torque and power
b) The rms stator current using simplified expression
c) The rms magnetizing voltage and current
d) The rms rotor current and voltage
e) The equivalent resistance and reactance for the rotor side converter
f) The maximum torque and the corresponding slip

Solution:

a) The rotor mechanical speed:

$$\omega_m = 2160 \times (2\pi / 60) = 226.19 \text{ rad/sec}$$

The rotor electrical speed:

$$\omega_r = \omega_m \times P = 226.195 \times 2 = 452.39 \text{ rad/sec}$$

The rated rotor mechanical speed:

$$\omega_{m,R} = 2160 \times (2\pi / 60) = 226.19 \text{ rad/sec}$$

The stator frequency:

$$\omega_s = 2\pi \times 60 = 376.99 \text{ rad/sec}$$

The pu rotor speed $\omega_{m,\text{pu}} = \omega_m / \omega_{m,R} = 226.19/226.19 = 1.0\,\text{pu}$
The generator mechanical torque at 1.0 pu rotor speed:

$$T_m = T_{m,R} \times (\omega_{m,\text{pu}})^2 = -4421 \times (1.0)^2 = -4421 \text{ N·m}$$

The rated mechanical power:

$$P_{m,R} = \omega_{m,R} \times T_{m,R} = -226.19 \times 4421 = -1000 \times 10^3 \text{ W}$$

The generator mechanical power at 1.0 pu rotor speed:

$$P_m = P_{m,R} \times (\omega_{m,\text{pu}})^3 = -1000 \times 10^3 \times (1.0)^3 = -1000 \times 10^3 \text{ W}$$

b) The rms stator current using simplified expression:

$$I_s = \frac{|T_m|\omega_s / P}{3V_s \cos\varphi_s} = \frac{4421 \times 2\pi \times 60/2}{3 \times (575/\sqrt{3}) \times 0.95} = 880.78 \text{ A (rms)}$$

When the DFIG operates with a leading power factor, its stator power factor angle is in the range of $190° \leq \varphi_s \leq 270°$ as indicated in Figure 8-12. The stator power factor angle for 0.95 leading power factor can then be calculated by

$$\varphi_s = 180° + \cos^{-1}(0.95) = 180° + 18.2° = 198.2°$$

from which the stator current phasor is

$$\bar{I}_s = I_s \angle -\varphi_s = 880.78 \angle -198.2° \text{ A (rms)}$$

c) The magnetizing branch voltage:

$$\begin{aligned}
\bar{V}_m &= \bar{V}_s - \bar{I}_s (R_s + j\omega_s L_{ls}) \\
&= 575/\sqrt{3}\angle 0° - 880.78\angle -198.2° \times (3.654 \times 10^{-3} + j120\pi \times 0.1304 \times 10^{-3}) \\
&= 350.86\angle 6.57° \text{ V (rms)}
\end{aligned}$$

The magnetizing current can be calculated by

$$\bar{I}_m = \frac{\bar{V}_m}{j\omega_s L_m} = 225.9\angle -83.43° \text{ A (rms)}$$

d) The rotor current:

$$\bar{I}_r = \bar{I}_s - \bar{I}_m = 880.78\angle -198.2° - 225.9\angle -83.43° = 996.74\angle 149.93° \text{ A (rms)}$$

The rotor voltage:

$$\overline{V}_r = s\overline{V}_m - \overline{I}_r\left(R_r + js\omega_s L_{lr}\right) = 73.29\angle -166.1° \text{ V (rms)}$$

where $s = (\omega_s - \omega_r)/\omega_s = (376.99 - 452.39)/376.99 = -0.2$

e) The equivalent impedance for the rotor side converter is given by

$$\overline{Z}_{eq} = \overline{V}_r / \overline{I}_r = 0.05292 + j0.05105 \ \Omega$$

from which $R_{eq} = 0.05292 \ \Omega$ and $X_{eq} = 0.05105 \ \Omega$

f) The slip at which the maximum torque occurs can be obtained from Equation (8.34):

$$s_{T\max} = \pm\sqrt{\frac{(R_r + R_{eq})^2 + X_{eq}^2}{R_s^2 + (X_{ls} + X_{lr})^2}} = -0.8066$$

($s = +0.8066$ is omitted because of the supersynchronous mode of operation.)

The maximum torque:

$$T_{\max} = \frac{1}{2\omega_s / P} \times \frac{3V_s^2}{R_s + \dfrac{(X_{ls} + X_{lr})X_{eq}}{R_r + R_{eq}} - \sqrt{\left((X_{ls} + X_{lr})^2 + R_s^2\right)\times\left(1 + \dfrac{X_{eq}^2}{(R_r + R_{eq})^2}\right)}}$$

$$= -22876 \text{ N·m} \quad (5.17 \text{ pu})$$

Cross Check:

$$T_m = \frac{1}{\omega_s / P} \times 3I_r^2\left(R_{eq} + R_r\right)/s = \frac{1}{2\pi \times 60/2}$$

$$\times 3 \times 996.74^2\left(0.05292 + 3.569 \times 10^{-3}\right)/(-0.2) = -4466.12 \text{ N·m}$$

$$P_m = 3I_r^2\left(R_{eq} + R_r\right)(1-s)/s = 3 \times 996.74^2\left(0.05292 + 3.569 \times 10^{-3}\right)(1+0.2)/(-0.2):$$

$$= -1010.2 \times 10^3 \text{ W}$$

8-14. Repeat Problem 8-13 when the DFIG operates at subsynchronous speed of 1350 rpm and stator power factor of 0.95 lagging.

Answers:

a) $T_m = -1726.9$ N·m, $\qquad P_m = -244.1 \times 10^3$ W

b) $\overline{I}_s = 344.1\angle -161.8°$ A (rms)

c) $\overline{V}_m = 328.3\angle 2.87°$ V (rms), $\overline{I}_m = 211.37\angle -87.13°$ A (rms)

d) $\overline{I}_r = 353.02\angle 162.92°$ A (rms), $\overline{V}_r = 84.69\angle 5.12°$ V (rms)

e) $R_{eq} = -0.2221$ Ω, $X_{eq} = -0.0906$ Ω

f) $s_{T\,\text{max}} = 2.5064$, $T_{\text{max}} = -14759$ N·m (3.338 pu)

8-15. Repeat Problem 8-13 when the DFIG operates at subsynchronous speed of 1780 rpm and stator power factor of 0.9 lagging.

Answers:

a) $T_m = -3002.3$ N·m, $P_m = -559.6\times 10^3$ W

b) $\overline{I}_s = 631.37\angle -154.16°$ A (rms)

c) $\overline{V}_m = 321.83\angle 5.2°$ V (rms), $\overline{I}_m = 207.2\angle -84.84°$ A (rms)

d) $\overline{I}_r = 590.89\angle -173.3°$ A (rms), $\overline{V}_r = 5.69\angle 8.72°$ V (rms)

e) $R_{eq} = -0.00962$ Ω, $X_{eq} = -0.00034$ Ω

f) $s_{T\,\text{max}} = 0.0642$, $T_{\text{max}} = -10247$ N·m (2.3178 pu)

8-16 (Solved Problem). Consider a 1.5 MW, 690 V, 50 Hz, 1750 rpm DFIG WECS. The parameters of the generator are given in Table B-5 of Appendix B. The generator operates with an MPPT scheme. At a given wind and generator speed, the stator active power P_s is found to be -1008.616 kW. Calculate the following assuming 0.95 lagging power factor operation for the DFIG WECS:

a) The rms stator current

b) The generator mechanical torque and power

c) The rotor mechanical and electrical speeds and slip

d) The rms magnetizing voltage and current

e) The rms rotor current and voltage

f) The equivalent resistance and reactance for the rotor side converter

Solution:

a) The rms stator can be calculated by

$$I_s = \left|P_s\right|/(3V_s \cos\varphi_s) = 1008.6\times 10^3/(3\times 690/\sqrt{3}\times 0.95) = 888.37 \text{ A (rms)}$$

When the DFIG operates with a lagging power factor, its stator power factor angle is in the range of $90° \leq \varphi_s \leq 180°$ as indicated in Figure 8-12. The stator power factor angle for 0.95 lagging power factor can then be calculated by

$$\varphi_s = 180° - \cos^{-1}(0.95) = 180° - 18.2° = 161.8°$$

from which the stator current phasor is

$$\overline{I}_s = I_s\angle -\varphi_s = 888.37\angle -161.8° \text{ A (rms)}$$

b) The generator air-gap power can be given by

$$\frac{\omega_s T_m}{P} = 3V_s I_s \cos\varphi_s = P_s$$

from which generator mechanical torque can be obtained as

$$T_m = P_s \times (P/\omega_s) = -6421.05 \text{ N·m}$$

The generator mechanical torque can be related to pu rotor speed as

$$T_m = T_{m,R} \times (\omega_{m,\text{pu}})^2 \text{ N·m}$$

from which pu rotor speed can be calculated by

$$\omega_{m,\text{pu}} = \sqrt{\frac{T_m}{T_{m,R}}} = \sqrt{\frac{-6421.05}{-8185}} = 0.8857$$

The rated mechanical power:

$$P_{m,R} = \omega_{m,R} \times T_{m,R} = 1750(2\pi)/60 \times (-8185) = -1500 \times 10^3 \text{ W}$$

The generator mechanical power at 0.8857 pu rotor speed:

$$P_m = P_{m,R} \times (\omega_{m,\text{pu}})^3 = -1500 \times 10^3 \times (0.8857)^3 = -1042.24 \times 10^3 \text{ W}$$

c) The rotor mechanical and electrical speeds:

$$\omega_m = \omega_{m,R} \times \omega_{m,\text{pu}} = 1750 \times (2\pi)/60 \times 0.8857 = 162.32 \text{ rad/sec} \quad (1550 \text{ rpm})$$

$$\omega_r = \omega_m \times P = 162.32 \times 2 = 324.64 \text{ rad/sec}$$

The slip can be calculated as:

$$s = (\omega_s - \omega_r)/\omega_s = (314.16 - 324.64)/314.16 = -0.0333$$

d) The magnetizing branch voltage:

$$\overline{V}_m = \overline{V}_s - \overline{I}_s(R_s + j\omega_s L_{ls})$$
$$= 690/\sqrt{3}\angle 0° - 888.37\angle -161.8° \times (2.65 \times 10^{-3} + j100\pi \times 0.1687 \times 10^{-3})$$
$$= 388.58\angle 6.72° \text{ V (rms)}$$

The magnetizing current can be calculated by

$$\overline{I}_m = \frac{\overline{V}_m}{j\omega_s L_m} = 225.92\angle -83.28° \text{ A (rms)}$$

e) The rotor current:

$$\bar{I}_r = \bar{I}_s - \bar{I}_m = 888.37\angle -161.8° - 225.92\angle -83.28° = 871.99\angle -176.5° \text{ A (rms)}$$

The rotor voltage:

$$\bar{V}_r = s\bar{V}_m - \bar{I}_r(R_r + js\omega_s L_{lr}) = 10.82\angle -166.1° \text{ V (rms)}$$

f) The equivalent impedance for the rotor side converter is given by

$$\bar{Z}_{eq} = \bar{V}_r / \bar{I}_r = 0.01220 + j0.00224 \ \Omega$$

from which $R_{eq} = 0.01220\ \Omega$ and $X_{eq} = 0.00224\ \Omega$.

Cross Check:

$$T_m = \frac{1}{\omega_s / P} \times 3I_r^2 (R_{eq} + R_r)/s = \frac{1}{2\pi \times 50/2} \times 3 \times 871.99^2(0.01220 + 2.63 \times 10^{-3})/$$

$$(-0.0333) = -6461 \ \text{N·m}$$

$$P_m = 3I_r^2 (R_{eq} + R_r)(1-s)/s$$

$$= 3 \times 871.99^2(0.01220 + 2.63 \times 10^{-3})(1+0.0333)/(-0.0333) = -1048.72 \times 10^3 \ \text{W}$$

$$P_s = 3V_s I_s \cos\varphi_s = 3 \times 690/\sqrt{3} \times 888.37 \times \cos(161.8°)$$

$$= -1008.616 \times 10^3 \ \text{W, verified.}$$

8-17. Repeat Problem 8-16 if the stator active power P_s is –1285.7 kW. Assume stator power factor as 0.95 leading.

Answers:

a) $\bar{I}_s = 1132.4\angle -198.2°$ A (rms)

b) $T_m = -8185$ N·m ,　　　　$P_m = -1500 \times 10^3$ W

c) $\omega_m = 183.26$ rad/sec (1750 rpm),　　　$\omega_r = 366.52$ rad/sec ,　　　$s = -0.1667$

d) $\bar{V}_m = 423.7\angle 7.61°$ V (rms),　　　$\bar{I}_m = 246.33\angle -82.4°$ A (rms)

e) $\bar{I}_r = 1259.3\angle 151.7°$ A (rms),　　　$\bar{V}_r = 73.67\angle -165.3°$ V (rms)

f) $R_{eq} = 0.04277\ \Omega$,　　　$X_{eq} = 0.03992\ \Omega$

8-18. Repeat problem 8-16 if the stator active power P_s is –604.54 kW. Assume stator power factor as 0.8 leading.

Answers:

a) $\bar{I}_s = 632.3\angle -216.87°$ A (rms)

b) $T_m = -3848.6$ N·m, $P_m = -483.63\times10^3$ W

c) $\omega_m = 125.66$ rad/sec (1200 rpm), $\omega_r = 251.32$ rad/sec, $s = 0.2$

d) $\bar{V}_m = 420.61\angle 3.52°$ V (rms), $\bar{I}_m = 244.54\angle -86.48°$ A (rms)

e) $\bar{I}_r = 812.4\angle 129.88°$ A (rms), $\bar{V}_r = 90.92\angle 4.98°$ V (rms)

f) $R_{eq} = -0.06402$ Ω, $X_{eq} = -0.09179$ Ω

Topic: Stead-State Analysis of *dq*-axis Voltages and Currents Based on Phasor Diagram

8-19 (Solved Problem). A variable-speed WECS employs a 6.0 MW, 4000 V, 50 Hz, 1170 rpm DFIG. The parameters of the generator are given in Table B-8 of Appendix B. The generator operates with an MPPT scheme and its stator power factor is unity. Assuming that the DFIG operates at a supersynchronous speed of 1170 rpm, find the following:

a) The rms and *dq*-axis stator current

b) The rms magnetizing voltage and current

c) The rms and *dq*-axis rotor currents

d) The rms and *dq*-axis rotor voltages

e) The equivalent resistance and reactance for the rotor side converter

Solution:

a) The rotor mechanical speed:

$$\omega_m = 1170\times(2\pi/60) = 122.52 \text{ rad/sec}$$

The rotor electrical speed:

$$\omega_r = \omega_m \times P = 122.52\times3 = 367.56 \text{ rad/sec}$$

The rated rotor mechanical speed:

$$\omega_{m,R} = 1170\times(2\pi/60) = 122.52 \text{ rad/sec}$$

The stator frequency:

$$\omega_s = 2\pi\times50 = 314.16 \text{ rad/sec}$$

The pu rotor speed $\omega_{m,\text{pu}} = \omega_m/\omega_{m,R} = 122.52/122.52 = 1.0\,\text{pu}$
The generator mechanical torque at 1.0 pu rotor speed:

$$T_m = T_{m,R}\times(\omega_{m,\text{pu}})^2 = -48971\times(1.0)^2 = -48971 \text{ N·m}$$

The stator current from Equation (8.7):

$$I_s = \frac{V_s \pm \sqrt{V_s^2 - \dfrac{4R_s T_m \omega_s}{3P}}}{2R_s} = -733.93 \text{ A (rms)} \quad (I_s = 86.71 \times 10^3 \text{A omitted})$$

where $V_s = 4000/\sqrt{3}$ V, $T_m = -48971$ N·m, $\omega_s = 314.16$ rad/sec, $R_s = 26.86$ mΩ, and $P = 3$

The stator current phasor is

$$\bar{I}_s = I_s \angle -\varphi_s = 733.93 \angle 180° \text{ A (rms)}$$

The *dq*-axis stator currents can be given by

$$I_{ds} = I_s \cos \angle I_s = 733.93 \times \cos(180°) = -733.93 \text{ A (rms)}$$

$$I_{qs} = I_s \sin \angle I_s = 733.93 \times \sin(180°) = 0 \text{ A (rms)}$$

b) The magnetizing branch voltage:

$$\begin{aligned}
\bar{V}_m &= \bar{V}_s - \bar{I}_s \left(R_s + j\omega_s L_{ls}\right) \\
&= 4000/\sqrt{3}\angle 0° - 733.93 \angle 180° \times (26.86 \times 10^{-3} + j100\pi \times 0.23142 \times 10^{-3}) \\
&= 2329.73 \angle 1.3° \text{ V (rms)}
\end{aligned}$$

The magnetizing current can be calculated by

$$\bar{I}_m = \frac{\bar{V}_m}{j\omega_s L_m} = 286.23 \angle -88.69° \text{ A (rms)}$$

c) The rotor current:

$$\bar{I}_r = \bar{I}_s - \bar{I}_m = 733.93 \angle 180° - 286.23 \angle -88.69° = 793.86 \angle 158.87° \text{ A (rms)}$$

The *dq*-axis rotor currents can be given by

$$I_{dr} = I_r \cos \angle I_r = 793.86 \times \cos(158.87°) = -740.49 \text{ A (rms)}$$

$$I_{qr} = I_r \sin \angle I_r = 793.86 \times \sin(158.87°) = 286.16 \text{ A (rms)}$$

d) The rotor voltage:

$$\bar{V}_r = s\bar{V}_m - \bar{I}_r \left(R_r + js\omega_s L_{lr}\right) = 381.05 \angle -176.23° \text{ V (rms)}$$

where $s = (\omega_s - \omega_r)/\omega_s = -0.17$

The dq-axis rotor voltages can be given by

$$V_{dr} = V_r \cos \angle V_r = 381.05 \times \cos(-176.23°) = -380.23 \ \text{V (rms)}$$

$$V_{qr} = V_r \sin \angle V_r = 381.05 \times \sin(-176.23°) = -25.1 \ \text{V (rms)}$$

e) The equivalent impedance for the rotor side converter is given by

$$\overline{Z}_{eq} = \overline{V}_r / \overline{I}_r = 0.43538 + j0.20211 \ \Omega$$

from which $R_{eq} = 0.43538 \ \Omega$, and $X_{eq} = 0.20211 \ \Omega$.

Based on the above steady-state calculations, the phasor diagram superimposed with dq-axis is shown in Figure C-14.

8-20. Repeat Problem 8-19 when the DFIG operates at synchronous speed of 1000 rpm.

Answers:

a) $\overline{I}_s = 537.37\angle180° \ \text{A (rms)}$, $I_{ds} = -537.37 \ \text{A (rms)}$, $I_{qs} = 0 \ \text{A (rms)}$
b) $\overline{V}_m = 2324.2\angle0.96° \ \text{V (rms)}$, $\overline{I}_m = 285.55\angle-89.04° \ \text{A (rms)}$
c) $\overline{I}_r = 612.75\angle152.22° \ \text{A (rms)}$, $I_{dr} = -542.17 \ \text{A (rms)}$,
     $I_{qr} = 285.51 \ \text{A (rms)}$
d) $\overline{V}_r = 15.77\angle-27.77° \ \text{V (rms)}$, $V_{dr} = 13.96 \ \text{V (rms)}$, $V_{qr} = -7.35 \ \text{V (rms)}$
e) $R_{eq} = -0.02574 \ \Omega$, $X_{eq} = 0 \ \Omega$

8-21. Repeat Problem 8-19 when the DFIG operates at subsynchronous speed of 800 rpm.

Answers:

a) $\overline{I}_s = 344.68\angle180° \ \text{A (rms)}$, $I_{ds} = -344.68 \ \text{A (rms)}$, $I_{qs} = 0 \ \text{A (rms)}$
b) $\overline{V}_m = 2318.8\angle0.62° \ \text{V (rms)}$, $\overline{I}_m = 284.89\angle-89.38° \ \text{A (rms)}$

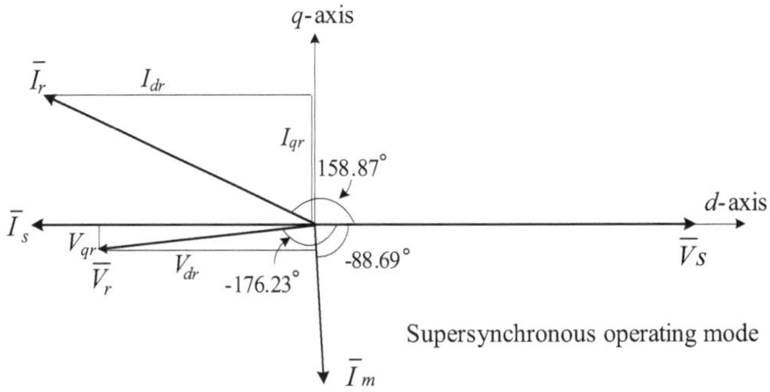

Figure C-14. Phasor diagram of the DFIG superimposed with dq-axis.

c) $\bar{I}_r = 449.5 \angle 140.68°$ A (rms), $\qquad I_{dr} = -347.76$ A (rms),

$\quad I_{qr} = 284.87$ A (rms)

d) $\bar{V}_r = 476.6 \angle 0.29°$ V (rms), $\qquad V_{dr} = 476.6$ V (rms), $\qquad V_{qr} = 2.45$ V (rms)

e) $R_{eq} = -0.81667\ \Omega$, $\qquad X_{eq} = -0.67603\ \Omega$

Topic: Stator Voltage-Oriented Control (SVOC) of DFIG WECS

8-22 (Solved Problem). A 6.0 MW, 4000 V, 50 Hz, 1170 rpm DFIG wind energy system operates with stator voltage-oriented control (SVOC). The parameters of the generator are given in Table B-8 of Appendix B. The generator operates with an MPPT scheme and its stator power factor is 0.95 leading. At a wind speed of 9.23 m/s, the generator operates at 0.7692 pu rotor speed. The stator phase voltage V_s is kept at its rated value of $4000/\sqrt{3}$ by the stator voltage oriented controller. The corresponding equivalent resistance R_{eq} and reactance X_{eq} for the rotor side converter are found to be $-0.29334\ \Omega$ and $-0.27413\ \Omega$, respectively. Calculate the following:

a) The generator mechanical torque and power
b) The rms stator and rotor currents
c) The *dq*-axis stator and rotor voltages
d) The *dq*-axis stator and rotor currents
e) The *dq*-axis stator flux linkages and electromagnetic torque of DFIG
f) The stator active and reactive powers

Solution:

a) The rotor mechanical and electrical speeds:

$$\omega_m = \omega_{m,R} \times \omega_{m,\text{pu}} = 1170 \times (2\pi)/60 \times 0.7692 = 94.25 \text{ rad/sec} \quad (900 \text{ rpm})$$

$$\omega_r = \omega_m \times P = 94.25 \times 3 = 282.75 \text{ rad/sec}$$

The slip can be calculated by:

$$s = (\omega_s - \omega_r)/\omega_s = (314.16 - 282.75)/314.16 = 0.1$$

The generator mechanical torque at 0.7692 pu rotor speed:

$$T_m = T_{m,R} \times (\omega_{m,\text{pu}})^2 = -48971 \times (0.7692)^2 = -28977 \text{ N·m}$$

The rated mechanical power:

$$P_{m,R} = \omega_{m,R} \times T_{m,R} = 1170(2\pi)/60 \times (-48971) = -6000 \times 10^3 \text{ W}$$

The generator mechanical power at 0.7692 pu rotor speed:

$$P_m = P_{m,R} \times (\omega_{m,\text{pu}})^3 = -6000 \times 10^3 \times (0.7692)^3 = -2731 \times 10^3 \text{ W}$$

b) Selecting the stator voltage as a reference phasor, the stator voltage is: $V_s = 4000/\sqrt{3}\angle 0° = 2309.4\angle 0°$ V (rms). The equivalent impedance of SCIG based on Figure 8-4:

$$\overline{Z}_s = R_s + jX_{ls} + jX_m \,//\, \left(\frac{R_r}{s} + jX_{lr} + \frac{R_{eq}}{s} + j\frac{X_{eq}}{s} \right) = 5.009\angle -161.8° \ \Omega \quad \text{for } s = 0.1$$

The stator current can be calculated by

$$\overline{I}_s = \frac{\overline{V}_s}{\overline{Z}_s} = \frac{4000/\sqrt{3}\angle 0°}{5.009\angle -161.8°} = 461.03\angle 161.8° \text{ A} = 461.03\angle -198.2° \text{ A (rms)}$$

Alternatively, the rms stator current using a simplified expression:

$$I_s = \frac{|T_m|\,\omega_s\,/\,P}{3V_s \cos\varphi_s} = \frac{28977\times 2\pi\times 50/3}{3\times(4000/\sqrt{3})\times 0.95} = 461.04 \text{ A (rms)}$$

The rotor current can be calculated by

$$\overline{I}_r = \frac{jX_m\overline{I}_s}{jX_m + \left(\dfrac{R_r}{s} + jX_{lr} \right) + \left(\dfrac{R_{eq}}{s} + j\dfrac{X_{eq}}{s} \right)} = 616.54\angle 135.7° \text{ A (rms)}$$

c) The equivalent impedance for the rotor-side converter:

$$\overline{Z}_{eq} = \overline{V}_r/\overline{I}_r = -0.29334 - j0.27413 = 0.4015\angle -136.94° \ \Omega \quad \text{(given)}$$

The rotor voltage:

$$\overline{V}_r = \overline{Z}_{eq}\times\overline{I}_r = 247.54\angle -1.22° \text{ V (rms)}$$

The dq-axis stator voltages can be given by

$$V_{ds} = V_s = 2309.4 \text{ V (rms)}$$

$$V_{qs} = 0 \text{ V (rms)}$$

The dq-axis rotor voltages can be given by

$$V_{dr} = V_r \cos\angle V_r = 247.54\times\cos(-1.22°) = 247.48 \text{ V (rms)}$$

$$V_{qr} = V_r \sin\angle V_r = 247.54\times\sin(-1.22°) = -5.26 \text{ V (rms)}$$

d) The *dq*-axis stator currents can be given by

$$I_{ds} = I_s \cos \angle I_s = 461.04 \times \cos(-198.2°) = -438 \text{ A (rms)}$$

$$I_{qs} = I_s \sin \angle I_s = 461.04 \times \sin(-198.2°) = 143.96 \text{ A (rms)}$$

The *dq*-axis rotor currents can be given by

$$I_{dr} = I_r \cos \angle I_r = 616.54 \times \cos(135.7°) = -441.42 \text{ A (rms)}$$

$$I_{qr} = I_r \sin \angle I_r = 616.54 \times \sin(135.7°) = 430.43 \text{ A (rms)}$$

e) The *dq*-axis stator flux linkages can be calculated by

$$\Lambda_{ds} = \frac{V_{qs} - R_s I_{qs}}{\omega_s} = -0.0123 \text{ Wb (rms)}$$

$$\Lambda_{qs} = -\frac{V_{ds} - R_s I_{ds}}{\omega_s} = -7.3885 \text{ Wb (rms)}$$

The electromagnetic torque developed by the DFIG can be given by

$$
\begin{aligned}
T_e = -T_m &= -\frac{3PL_m}{2\omega_s L_s} i_{dr} v_{ds} \\
&= -\frac{3 \times 3 \times 25.908 \times 10^{-3}}{2 \times 2 \times \pi \times 50 \times 26.139 \times 10^{-3}} (\sqrt{2} \times (-441.42) \times \sqrt{2} \times 2309.4) \\
&= 28946 \text{ N·m}
\end{aligned}
$$

f) The stator active and reactive power of the DFIG can be calculated by

$$P_s = \frac{3}{2} v_{ds} i_{ds} = \frac{3}{2} \times \sqrt{2} \times 2309.4 \times \sqrt{2} \times (-438) = -3034.5 \times 10^3 \text{ W}$$

$$Q_s = -\frac{3}{2} v_{ds} i_{qs} = -\frac{3}{2} \times \sqrt{2} \times 2309.4 \times \sqrt{2} \times 143.96 = -997.38 \times 10^3 \text{ VAR}$$

Alternatively, the stator active and reactive power can be calculated by

$$P_s = 3V_s I_s \cos \varphi_s = 3 \times 2309.4 \times 438 \times \cos(198.2°) = -3034.5 \times 10^3 \text{ W}$$

$$Q_s = 3V_s I_s \sin \varphi_s = 3 \times 2309.4 \times 438 \times \sin(198.2°) = -997.38 \times 10^3 \text{ VAR}$$

8-23. Repeat Problem 8-22 when the DFIG operates with 0.9188 pu rotor speed and stator power factor of 0.95 lagging. The corresponding equivalent resistance R_{eq} and reactance X_{eq} for the rotor side converter are found to be 0.24398 V and 0.04478 V respectively.

Answers:

a) $T_m = -41341$ N·m, $P_m = -4653.9 \times 10^3$ W

b) $\bar{I}_s = 657.76 \angle -161.8°$ A (rms), $\bar{I}_r = 636 \angle 172.9°$ A (rms)

c) $V_{ds} = 2309.4$ V (rms), $V_{qs} = 0$ V (rms), $V_{dr} = -157.5$ V (rms),
   $V_{qr} = -9.09$ V (rms)

d) $I_{ds} = -624.87$ A (rms), $I_{qs} = -205.39$ A (rms), $I_{dr} = -631.13$ A (rms),
   $I_{qr} = 78.58$ A (rms)

e) $\Lambda_{ds} = 0.0176$ Wb (rms), $\Lambda_{qs} = -7.4045$ Wb (rms), $T_e = 41386$ N·m

f) $P_s = -4329.2 \times 10^3$ W, $Q_s = 1422.9 \times 10^3$ VAR

8-24. Repeat Problem 8-22 when the DFIG operates with 0.9829 pu rotor speed and stator power factor of 0.9 leading. The corresponding equivalent resistance R_{eq} and reactance X_{eq} for the rotor side converter are found to be 0.24546 Ω and 0.25844 Ω respectively.

Answers:

a) $T_m = -47311$ N·m, $P_m = -5697 \times 10^3$ W

b) $\bar{I}_s = 794.56 \angle -205.84°$ A (rms), $\bar{I}_r = 960.63 \angle 138.58°$ A (rms)

c) $V_{ds} = 2309.4$ V (rms), $V_{qs} = 0$ V (rms), $V_{dr} = -341.06$ V (rms),
   $V_{qr} = -30.17$ V (rms)

d) $I_{ds} = -715.1$ A (rms), $I_{qs} = 346.34$ A (rms), $I_{dr} = -720.35$ A (rms),
   $I_{qr} = 635.53$ A (rms)

e) $\Lambda_{ds} = -0.0296$ Wb (rms), $\Lambda_{qs} = -7.4122$ Wb (rms), $T_e = 47236$ N·m

f) $P_s = -4954 \times 10^3$ W, $Q_s = -2399 \times 10^3$ VAR

8-25 (Solved Problem). Consider a 1.5 MW, 690 V, 50 Hz, 1750 rpm DFIG wind energy system operating with stator voltage-oriented control (SVOC). The parameters of the generator are given in Table B-5 of Appendix B. The generator operates with an MPPT scheme. The stator phase voltage V_s is kept at its rated value of $690/\sqrt{3}$ by the stator voltage-oriented controller. At a given wind and generator speed, the stator active and reactive powers P_s and Q_s are found to be -507.98 kW and 246.02 kVAR respectively. Calculate the following:

a) The dq-axis and rms stator and rotor currents

b) The dq-axis stator flux linkages

c) The generator mechanical torque and power

d) The rotor mechanical and electrical speeds and slip

e) The rms and dq-axis stator and rotor voltages

f) The equivalent resistance and reactance for the rotor-side converter

Solution:

a) The stator active and reactive power of the DFIG can be calculated by

$$P_s = \frac{3}{2} v_{ds} i_{ds} \ \text{W}$$

$$Q_s = -\frac{3}{2} v_{ds} i_{qs} \ \text{VAR}$$

from which the peak dq-axis stator currents can be given by

$$i_{ds} = \frac{P_s}{1.5 \times v_{ds}} = \frac{-507.98 \times 10^3}{1.5 \times \sqrt{2} \times 690 / \sqrt{3}} = -601.11 \ \text{A (rms)}$$

$$i_{qs} = \frac{-Q_s}{1.5 \times v_{ds}} = \frac{-246.02 \times 10^3}{1.5 \times \sqrt{2} \times 690 / \sqrt{3}} = -291.13 \ \text{A (rms)}$$

from which the rms dq-axis stator currents can be given by

$$I_{ds} = i_{ds} / \sqrt{2} = -425.05 \ \text{A (rms)}$$

$$I_{qs} = i_{qs} / \sqrt{2} = -205.86 \ \text{A (rms)}$$

The stator current phasor can be calculated by

$$\bar{I}_s = (I_{ds} + jI_{qs}) = 472.28 \angle -154.16° \ \text{A (rms)}$$

The stator power factor angle is:

$$\varphi_s = \angle \bar{V}_s - \angle \bar{I}_s = 0° - (-154.16°) = 154.16°$$

from which the stator power factor is

$$PF_s = \cos(154.16°) = 0.9 \ \text{lagging}$$

The dq-axis rotor currents can be given by

$$i_{dr} = \frac{2L_s}{3 v_{ds} L_m} P_s = \frac{2 \times 5.6436 \times 10^{-3}}{2 \times 563.38 \times 5.4749 \times 10^{-3}} \times (-507.981 \times 10^3) = -619.63 \ \text{A (rms)}$$

$$i_{qr} = -\frac{2L_s}{3 v_{ds} L_m} Q_s + \frac{v_{ds}}{\omega_s L_m} = -\frac{2 \times 5.6436 \times 10^{-3}}{2 \times 563.38 \times 5.4749 \times 10^{-3}} \times 246.027 \times 10^3$$

$$+ \frac{563.38}{100\pi \times 5.4749 \times 10^{-3}} = 27.45 \ \text{A (rms)}$$

from which the rms dq-axis rotor currents can be given by

$$I_{dr} = i_{dr} / \sqrt{2} = -438.15 \text{ A (rms)}$$

$$I_{qr} = i_{qr} / \sqrt{2} = 19.41 \text{ A (rms)}$$

The rotor current phasor can be calculated by

$$\bar{I}_r = (I_{dr} + jI_{qr}) = 438.58\angle 177.5° \text{ A (rms)}$$

b) The dq-axis stator flux linkages can be calculated by

$$\Lambda_{ds} = \frac{V_{qs} - R_s I_{qs}}{\omega_s} = 0.00174 \text{ Wb (rms)}$$

$$\Lambda_{qs} = -\frac{V_{ds} - R_s I_{ds}}{\omega_s} = -1.2716 \text{ Wb (rms)}$$

c) The electromagnetic torque developed by the DFIG can be given by

$$T_e = T_m = \frac{3PL_m}{2\omega_s L_s} i_{dr} V_{ds} = -3233.2 \text{ N·m}$$

The generator mechanical torque can be related to pu rotor speed as

$$T_m = T_{m,R} \times (\omega_{m,\text{pu}})^2 \text{ N·m}$$

from which pu rotor speed can be calculated by

$$\omega_{m,\text{pu}} = \sqrt{\frac{T_m}{T_{m,R}}} = \sqrt{\frac{-3233.2}{-8185}} = 0.6286$$

The rated mechanical power:

$$P_{m,R} = \omega_{m,R} \times T_{m,R} = 1750(2\pi)/60 \times (-8185) = -1500 \times 10^3 \text{ W}$$

The generator mechanical power at 0.6286 pu rotor speed:

$$P_m = P_{m,R} \times (\omega_{m,\text{pu}})^3 = -1500 \times 10^3 \times (0.6286)^3 = -372.52 \times 10^3 \text{ W}$$

d) The rotor mechanical and electrical speeds:

$$\omega_m = \omega_{m,R} \times \omega_{m,\text{pu}} = 1750(2\pi)/60 \times 0.6286 = 115.19 \text{ rad/sec (1100 rpm)}$$

$$\omega_r = \omega_m \times P = 115.19 \times 2 = 230.38 \text{ rad/sec}$$

The slip can be calculated by:

$$s = (\omega_s - \omega_r)/\omega_s = (314.16 - 230.38)/314.16 = 0.2667$$

e) The dq-axis stator voltages can be given by

$$V_{ds} = V_s = 398.37 \text{ V (rms)}$$

$$V_{qs} = 0 \text{ V (rms)}$$

The magnetizing branch voltage:

$$\overline{V}_m = \overline{V}_s - \overline{I}_s \left(R_s + j\omega_s L_{ls}\right) = 389.27\angle 3.4° \text{ V (rms)}$$

The rotor voltage:

$$\overline{V}_r = s\overline{V}_m - \overline{I}_r \left(R_r + js\omega_s L_{lr}\right) = 105.57\angle 5.99° \text{ V (rms)}$$

The dq-axis rotor voltages can be given by

$$V_{dr} = V_r \cos\angle V_r = 105.57 \times \cos(5.99°) = 104.99 \text{ V (rms)}$$

$$V_{qr} = V_r \sin\angle V_r = 105.57 \times \sin(5.99°) = 11.01 \text{ V (rms)}$$

f) The equivalent impedance for the rotor side converter is given by

$$\overline{Z}_{eq} = \overline{V}_r / \overline{I}_r = -0.23805 - j0.03567 \ \Omega$$

from which $R_{eq} = -0.23805 \ \Omega$ and $X_{eq} = -0.03567 \ \Omega$.

Cross Check:

$$P_s = 3V_s I_s \cos\varphi_s = 3 \times 398.37 \times 472.28 \times \cos(154.16°) = -507.992 \times 10^3 \text{ W}$$

$$Q_s = 3V_s I_s \sin\varphi_s = 3 \times 398.37 \times 472.28 \times \sin(154.16°) = 246.011 \times 10^3 \text{ VAR}$$

8-26. Repeat Problem 8-25 if the stator active and reactive powers P_s and Q_s are −822.85 kW and 270.46 kVAR respectively.

Answers:

a) $I_{ds} = -688.51 \text{ A (rms)}$, $\quad I_{qs} = -226.3 \text{ A (rms)}$, $\quad \overline{I}_s = 724.75\angle -161.8° \text{ A (rms)}$,
$\quad I_{dr} = -709.72 \text{ A (rms)}$, $\quad I_{qr} = -1.66 \text{ A (rms)}$, $\quad \overline{I}_r = 709.72\angle -179.87° \text{ A (rms)}$

b) $\Lambda_{ds} = 0.0019 \text{ Wb (rms)}$, $\quad \Lambda_{qs} = -1.2739 \text{ Wb (rms)}$

c) $T_m = -5238.4 \text{ N·m}$, $\quad P_m = -767.99 \times 10^3 \text{ W}$

d) $\omega_m = 146.61$ rad/sec (1400 rpm), $\omega_r = 293.22$ rad/sec, $s = 0.0667$

e) $V_{ds} = 398.37$ V (rms), $V_{qs} = 0$ V (rms), $V_{dr} = 27.74$ V (rms),
   $V_{qr} = 4.46$ V (rms)

f) $R_{eq} = -0.03910$ Ω, $X_{eq} = -0.00620$ Ω

8-27. Repeat Problem 8-25 if the stator active and reactive powers P_s and Q_s are −1213.28 kW and −587.62 kVAR respectively.

Answers:

a) $I_{ds} = -1015.2$ A (rms), $I_{qs} = 491.68$ A (rms), $\bar{I}_s = 1128\angle -205.84°$ A (rms),
   $I_{dr} = -1046.48$ A (rms), $I_{qr} = 738.45$ A (rms), $\bar{I}_r = 1280.7\angle 144.8°$ A (rms)

b) $\Lambda_{ds} = -0.00415$ Wb (rms), $\Lambda_{qs} = -1.2766$ Wb (rms)

c) $T_m = -7723.97$ N.m, $P_m = -1375.05 \times 10^3$ W

d) $\omega_m = 178.02$ rad/sec (1700 rpm), $\omega_r = 356.04$ rad/sec, $s = -0.133$

e) $V_{ds} = 398.37$ V (rms), $V_{qs} = 0$ V (rms), $V_{dr} = -58.33$ V (rms),
   $V_{qr} = -14.8$ V (rms)

f) $R_{eq} = 0.03055$ Ω, $X_{eq} = 0.03570$ Ω

PROBLEMS AND ANSWERS FOR CHAPTER 9–VARIABLE-SPEED WIND ENERGY SYSTEMS WITH SYNCHRONOUS GENERATORS

Assumptions for all the problems in this chapter:

1. Switching harmonics produced by power converters are neglected.
2. Power converters are ideal without power losses.
3. Core and rotational losses of the generator are neglected.

Topic: Zero *d*-Axis Current (ZDC) Control

9-1 (Solved Problem). Consider a 2.0 MW, 690 V, 9.75 Hz nonsalient-pole PMSG wind energy conversion system. The parameters of the generator are given in Table B-9 of Appendix B. The generator is controlled by MPPT and ZDC schemes. At a wind speed of 12.0 m/s, the generator operates at 1.0 pu rotor speed. Determine the following:

a) The generator mechanical torque and power
b) The *dq*-axis and rms stator currents
c) The *dq*-axis and rms stator voltages
d) The stator winding loss and power factor
e) The stator active and reactive powers, and generator efficiency.

Solution:

a) The rotor mechanical speed:

$$\omega_m = \omega_{m,R} \times \omega_{m,\text{pu}} = 22.5 \times (2\pi / 60) \times 1.0 = 2.356 \text{ rad/sec} \quad (22.5 \text{ rpm})$$

The rotor electrical speed:

$$\omega_r = \omega_m \times P = 2.356 \times 26 = 61.26 \text{ rad/sec}$$

The generator mechanical torque at 1.0 pu rotor speed:

$$T_m = T_e = T_{m,R} \times (\omega_{m,\text{pu}})^2 = 848.83 \times 10^3 \times (1.0)^2 = 848.83 \times 10^3 \text{ N·m}$$

The rated mechanical power:

$$P_{m,R} = \omega_{m,R} \times T_{m,R} = 22.5 \times (2\pi / 60) \times (848.83 \times 10^3) = 2000 \times 10^3 \text{ W}$$

The generator mechanical power at 1.0 pu rotor speed:

$$P_m = P_{m,R} \times (\omega_{m,\text{pu}})^3 = 2000 \times 10^3 \times (1.0)^3 = 2000 \times 10^3 \text{ W}$$

b) The dq-axis stator currents of the generator with ZDC control:

$$i_{ds} = 0 \text{ A}$$

In steady-state operation, the electromagnetic torque T_e is equal to the mechanical torque T_m. The q-axis stator current can then be calculated by

$$i_{qs} = \frac{T_e}{1.5 \times P \times \lambda_r} = \frac{848.83 \times 10^3}{1.5 \times 26 \times 8.2398} = 2461.4 \text{ A (peak)}$$

The rms stator current:

$$I_s = \sqrt{i_{ds}^2 + i_{qs}^2} \,/ \sqrt{2} = \sqrt{0^2 + 2461.43^2} \,/ \sqrt{2} = 1867.8 \text{ A}$$

c) From the steady-state model of the generator, the dq-axis stator voltages can be calculated by

$$v_{ds} = -i_{ds} R_s + \omega_r L_q i_{qs} = -0 + (61.26 \times 1.5731 \times 10^{-3} \times 2461.43) = 254.55 \text{ V (peak)}$$

$$v_{qs} = -i_{qs} R_s - \omega_r L_d i_{ds} + \omega_r \lambda_r = -(2461.43 \times 0.821 \times 10^{-3}) - (0) + (61.26 \times 8.2398) =$$

$$= 502.61 \text{ V (peak)}$$

The rms stator voltage:

$$V_s = \sqrt{v_{ds}^2 + v_{qs}^2} / \sqrt{2} = \sqrt{254.55^2 + 502.61^2} / \sqrt{2} = 398.38 \text{ V}$$

d) The stator winding loss:

$$P_{cu,s} = 3(I_s)^2 R_s = 3 \times 1867.8^2 \times 0.821 \times 10^{-3} = 8.592 \times 10^3 \text{ W}$$

The rotor winding losses are zero for PMSG.

   The angles of the stator voltage and current vectors:

$$\theta_v = \tan^{-1} \frac{v_{qs}}{v_{ds}} = \tan^{-1} \frac{502.61}{254.55} = 63.14°$$

$$\theta_i = \tan^{-1} \frac{i_{qs}}{i_{ds}} = \tan^{-1} \frac{2461.43}{0} = 90°$$

The stator power factor angle:

$$\varphi_s = \theta_v - \theta_i = 63.14° - 90° = -26.86°$$

The stator power factor of the generator:

$$PF_s = \cos\varphi_s = \cos(-26.86°) = 0.8921 \text{ (leading)}$$

e) The stator active and reactive powers:

$$P_s = 3V_s I_s \cos\varphi_s = 3 \times 398.38 \times 1867.8 \times \cos(-26.86°) = 1991.4 \times 10^3 \text{ W}$$

$$Q_s = 3V_s I_s \sin\varphi_s = 3 \times 398.38 \times 1867.8 \times \sin(-26.86°) = -1008.6 \times 10^3 \text{ VAR}$$

Alternatively,

$$P_s = 1.5(v_{ds} i_{ds} + v_{qs} i_{qs}) = 1.5(0 + 502.61 \times 2641.4) = 1991.4 \times 10^3 \text{ W}$$

$$Q_s = 1.5(v_{qs} i_{ds} - v_{ds} i_{qs}) = 1.5(0 - 254.55 \times 2641.4) = -1008.6 \times 10^3 \text{ VAR}$$

Alternatively, stator active power can be calculated by

$$P_s = |P_m| - P_{cu,s} = (2000 - 8.592) \times 10^3 = 1991.4 \times 10^3 \text{ W}$$

The efficiency of the PMSG is then

$$\eta = P_s / |P_m| = 1991.4 / 2000 = 99.57\%$$

Note: In the calculation of efficiency, only the stator winding losses are considered. The rotational and core losses of the generator are not included.

9-2. Repeat Problem 9-1 when the PMSG operates at 0.8 pu rotor speed.

Answers:

a) $T_m = 543.25 \times 10^3$ N·m, $P_m = 1024.0 \times 10^3$ W

b) $i_{ds} = 0$ A, $i_{qs} = 1690.5$ A (peak), $I_s = 1195.3$ A (rms)

c) $v_{ds} = 130.33$ V (peak), $v_{qs} = 102.13$ V (peak), $V_s = 299.12$ V (rms)

d) $P_{cu,s} = 3.519 \times 10^3$ W, $PF_s = 0.9514$ (leading)

e) $P_s = 1020.48 \times 10^3$ W, $Q_s = -330.49 \times 10^3$ VAR, $\eta = 99.66\%$

9-3. Repeat Problem 9-1 when the PMSG operates at 0.5 pu rotor speed.

Answers:

a) $T_m = 212206$ N·m, $P_m = 250.0 \times 10^3$ W

b) $i_{ds} = 0$ A, $i_{qs} = 660.3$ A (peak), $I_s = 466.9$ A (rms)

c) $v_{ds} = 31.82$ V (peak), $v_{qs} = 251.85$ V (peak), $V_s = 179.5$ V (rms)

d) $P_{cu,s} = 0.537 \times 10^3$ W, $PF_s = 0.992$ (leading)

e) $P_s = 249.46 \times 10^3$ W, $Q_s = -31.518 \times 10^3$ VAR, $\eta = 99.79\%$

9-4 (Solved Problem). A 2.0 MW, 690 V, 9.75 Hz nonsalient pole PMSG is used a variable-speed WECS. The parameters of the generator are given in Table B-9 of Appendix B. The generator is controlled by MPPT and ZDC schemes. At a given wind and generator speed, the stator active power P_s, voltage V_s and power factor PF_s are found to be 430.89 kW, 217.12 V, and 0.9838, respectively. Determine the following:

a) The rms and dq-axis stator currents
b) The generator mechanical torque and power
c) The rotor mechanical and electrical speeds
d) The dq-axis stator voltages
e) The stator winding loss and generator efficiency
f) The stator reactive power

Solution:

a) The rms stator current:

$$I_s = \frac{P_s}{3V_s \cos \varphi_s} = \frac{430.88 \times 10^3}{3 \times 217.11 \times 0.9838} = 672.42 \text{ A}$$

The *dq*-axis stator currents of the generator with ZDC control:

$$i_{ds} = 0 \text{ A}$$

$$i_{qs} = I_s \times \sqrt{2} = 950.95 \text{ A (peak)}$$

b) The generator mechanical torque:

$$T_e = 1.5 P \lambda_r i_{qs} = 1.5 \times 26 \times 8.2398 \times 950.95 = 305588 \text{ N·m}$$

The generator mechanical torque can be related to pu rotor speed as

$$T_m = T_e = T_{m,R} \times (\omega_{m,\text{pu}})^2 \text{ N·m}$$

from which pu rotor speed can be calculated by

$$\omega_{m,\text{pu}} = \sqrt{\frac{T_m}{T_{m,R}}} = \sqrt{\frac{305588}{848826}} = 0.6$$

The rated mechanical power:

$$P_{m,R} = \omega_{m,R} \times T_{m,R} = 22.5 \times (2\pi / 60) \times (848826) = 2000 \times 10^3 \text{ W}$$

The generator mechanical power at 0.6 pu rotor speed:

$$P_m = P_{m,R} \times (\omega_{m,\text{pu}})^3 = 2000 \times 10^3 \times (0.6)^3 = 432.0 \times 10^3 \text{ W}$$

c) The rotor mechanical speed:

$$\omega_m = \omega_{m,R} \times \omega_{m,\text{pu}} = 22.5 \times (2\pi / 60) \times 0.6 = 1.414 \text{ rad/sec} \quad (13.5 \text{ rpm})$$

The rotor electrical speed:

$$\omega_r = \omega_m \times P = 1.414 \times 26 = 36.76 \text{ rad/sec}$$

d) From the steady-state model of the generator, the *dq*-axis stator voltages are

$$v_{ds} = -i_{ds} R_s + \omega_r L_q i_{qs} = -0 + (36.76 \times 1.5731 \times 10^{-3} \times 950.9) = 54.99 \text{ V (peak)}$$

$$v_{qs} = -i_{qs} R_s - \omega_r L_d i_{ds} + \omega_r \lambda_r = -(950.95 \times 0.821 \times 10^{-3}) - (0) + (36.76 \times 8.2398) :$$

$$= 302.09 \text{ V (peak)}$$

Alternatively,

$$v_{qs} = \sqrt{(V_s \times \sqrt{2})^2 - v_{ds}^2} = \sqrt{(217.1263 \times \sqrt{2})^2 - 54.99^2} = 302.09 \text{ V (peak)}$$

e) The stator winding loss:

$$P_{cu,s} = 3(I_s)^2 R_s = 3 \times 672.42^2 \times 0.821 \times 10^{-3} = 1.1137 \times 10^3 \ \text{W}$$

The rotor winding losses are zero for PMSG. The efficiency of the PMSG neglecting rotational and core losses is then

$$\eta = P_s / P_m = 430.88 / 432.0 = 99.74\%$$

f) The angles of the stator voltage and current vectors:

$$\theta_v = \tan^{-1} \frac{v_{qs}}{v_{ds}} = \tan^{-1} \frac{302.09}{54.98} = 79.68°$$

$$\theta_i = \tan^{-1} \frac{i_{qs}}{i_{ds}} = \tan^{-1} \frac{950.91}{0} = 90°$$

The stator power factor angle:

$$\varphi_s = \theta_v - \theta_i = 79.68° - 90° = -10.32°$$

Alternatively,

$$\varphi_s = \cos^{-1}(PF_s) = \cos^{-1}(0.9838) = -10.32°$$

The stator reactive power:

$$Q_s = 3V_s I_s \sin \varphi_s = 3 \times 217.11 \times 672.42 \times \sin(-10.32°) = -78.43 \times 10^3 \ \text{VAR}$$

Alternatively,

$$Q_s = 1.5(v_{qs}i_{ds} - v_{ds}i_{qs}) = 1.5(0 - 54.99 \times 950.95) = -78.43 \times 10^3 \ \text{VAR}$$

Cross Check:

$$P_s = 3V_s I_s \cos \varphi_s = 3 \times 217.11 \times 672.42 \times \cos(-10.32°) = 430.9 \times 10^3 \ \text{W}$$

Alternatively,

$$P_s = 1.5(v_{ds}i_{ds} + v_{qs}i_{qs}) = 1.5(0 + 302.09 \times 950.95) = 430.9 \times 10^3 \ \text{W}$$

Alternatively, stator active power can be calculated by

$$P_s = P_m - P_{cu,s} = (432 - 1.1137) \times 10^3 = 430.91 \times 10^3 \ \text{W}$$

9-5. Repeat Problem 9-4 when the stator active power P_s, voltage V_s and power factor PF_s are 1452.36 kW, 345.85 V, and 0.9252, respectively.

Answers:

a) $I_s = 1512.9$ A (rms),　　$i_{ds} = 0$ A,　　$i_{qs} = 2139.6$ A (peak)

b) $T_m = 687573$ N·m,　　$P_m = 1458 \times 10^3$ W

c) $\omega_m = 2.121$ rad/sec (20.25 rpm),　　$\omega_r = 55.14$ rad/sec

d) $v_{ds} = 185.58$ V (peak),　　$v_{qs} = 452.6$ V (peak)

e) $P_{cu,s} = 5.638 \times 10^3$ W,　　$\eta = 99.61\%$

f) $Q_s = -595.6 \times 10^3$

9-6. Repeat Problem 9-4 when the stator active power P_s, voltage V_s and power factor PF_s are 683.93 kW, 256.63 V, and 0.9706, respectively.

Answers:

a) $I_s = 915.2$ A (rms),　　$i_{ds} = 0$ A,　　$i_{qs} = 1294.3$ A (peak)

b) $T_m = 415938$ N·m,　　$P_m = 686.0 \times 10^3$ W

c) $\omega_m = 1.649$ rad/sec (15.75 rpm),　　$\omega_r = 42.88$ rad/sec

d) $v_{ds} = 87.32$ V (peak),　　$v_{qs} = 352.29$ V (peak)

e) $P_{cu,s} = 2.063 \times 10^3$ W,　　$\eta = 99.70\%$

f) $Q_s = -169.52 \times 10^3$ VAR

9-7 (Solved Problem). A variable-speed wind energy system uses a 2.0 MW, 690 V, 9.75 Hz nonsalient-pole PMSG. The parameters of the generator are given in Table B-9 of Appendix B. The generator is controlled by MPPT and ZDC schemes. At a given wind and generator speed, the stator voltage V_s and power factor PF_s are found to be 142.992 V and 0.9968, respectively. Calculate the following:

a) The stator power factor and angles of the stator voltage and current vectors

b) The dq-axis stator voltages

c) The rotor electrical and mechanical speeds

d) The generator mechanical torque and power

e) The dq-axis and rms stator currents

f) The stator active and reactive powers, and generator efficiency

Solution:

a) The stator power factor angle:

$$\varphi_s = \cos^{-1}(PF_s) = \cos^{-1}(0.9968) = -4.585°$$

The angles of the stator voltage and current vectors:

$$\theta_i = \tan^{-1}\frac{i_{qs}}{i_{ds}} = 90° \ \text{(ZDC)}$$

$$\theta_v = \varphi_s + \theta_i = -4.585° + 90° = 85.415°$$

b) The angle of the stator voltage vector:

$$\theta_v = \tan^{-1}\frac{v_{qs}}{v_{ds}}$$

from which

$$v_{qs} = \tan\theta_v \times v_{ds} = 12.47 \times v_{ds}$$

The dq-axis stator voltages of the generator:

$$v_{ds}^2 + v_{qs}^2 = (V_s \times \sqrt{2})^2 \ \text{V}$$

$$\Rightarrow v_{ds}^2 + (12.47 \times v_{ds})^2 = (V_s \times \sqrt{2})^2 \ \text{V}$$

$$\Rightarrow v_{ds} = \frac{(V_s \times \sqrt{2})}{\sqrt{1 + 12.47^2}} = \frac{202.22}{\sqrt{156.5}} = 16.17 \ \text{V (peak)}$$

$$\Rightarrow v_{qs} = 12.47 \times v_{ds} = 201.57 \ \text{V (peak)}$$

c) The rotor electrical speed can be obtained from the q-axis stator voltage of the generator as follows:

$$\omega_r = \frac{v_{qs} + i_{qs}R_s}{\lambda_r - L_d i_{ds}}$$

Since stator resistance is negligibly small, the above equation can be approximated as

$$\omega_r \approx \frac{v_{qs}}{\lambda_r} = \frac{201.57}{8.2398} = 24.464 \ \text{rad/sec}$$

The rotor mechanical speed:

$$\omega_m = \omega_r / P = 24.464 / 26 = 0.941 \ \text{rad/sec} \ \text{(9 rpm)}$$

d) The pu rotor speed $\omega_{m,\text{pu}} = \omega_m / \omega_{m,R} = 0.941 / 2.3562 = 0.399 \ \text{pu}$

The generator mechanical torque at 0.399 pu rotor speed:

$$T_m = T_e = T_{m,R} \times (\omega_{m,\text{pu}})^2 = 848826 \times (0.399)^2 = 135359 \text{ N·m}$$

The rated mechanical power:

$$P_{m,R} = \omega_{m,R} \times T_{m,R} = 22.5 \times (2\pi / 60) \times (848826) = 2000 \times 10^3 \text{ W}$$

The generator mechanical power at 0.4 pu rotor speed:

$$P_m = P_{m,R} \times (\omega_{m,\text{pu}})^3 = 2000 \times 10^3 \times (0.4)^3 = 127.36 \times 10^3 \text{ W}$$

e) The dq-axis stator currents of the generator with ZDC control:

$$i_{ds} = 0 \text{ A (peak)}$$

$$i_{qs} = \frac{T_e}{1.5 \times P \times \lambda_r} = \frac{135359}{1.5 \times 26 \times 8.2398} = 421.22 \text{ A (peak)}$$

The rms stator current:

$$I_s = \sqrt{i_{ds}^2 + i_{qs}^2} / \sqrt{2} = \sqrt{0^2 + 421.22^2} / \sqrt{2} = 297.85 \text{ A}$$

f) The stator winding loss:

$$P_{cu,s} = 3(I_s)^2 R_s = 3 \times 297.85^2 \times 0.821 \times 10^{-3} = 218.5 \text{ W}$$

The stator active power can be calculated by

$$P_s = P_m - P_{cu,s} = 127.36 \times 10^3 - 218.5 = 127.1 \times 10^3 \text{ W}$$

The efficiency of the PMSG neglecting rotational and core losses is then

$$\eta = P_s / P_m = 127.1 / 127.36 = 99.8\%$$

9-8. Repeat Problem 9-7 when the stator voltage V_s and power factor PF_s are 236.57 V and 0.978, respectively.

Answers:

a) $\varphi_s = -12.07°$, $\theta_i = 90°$, $\theta_v = 77.93°$

b) $v_{ds} = 69.95$ V (peak), $v_{qs} = 327.18$ V (peak)

c) $\omega_r = 39.71$ rad/sec, $\omega_m = 1.527$ rad/sec (14.584 rpm)

d) $T_m = 356608$ N·m, $P_m = 544.62 \times 10^3$ W

e) $i_{ds} = 0$ A, $\quad$ $i_{qs} = 1109.7$ A (peak), $\quad$ $I_s = 784.69$ A (rms)

f) $P_s = 543.1 \times 10^3$ W, $\quad$ $Q_s = -116.44 \times 10^3$ VAR, $\quad$ $\eta = 99.72\%$

9-9. Repeat Problem 9-7 when the stator voltage V_s and power factor PF_s are 321.86 V and 0.939 respectively.

Answers:

a) $\varphi_s = -20.08°$, $\quad$ $\theta_i = 90°$, $\quad$ $\theta_v = 69.92°$

b) $v_{ds} = 156.3$ V (peak), $\quad$ $v_{qs} = 427.5$ V (peak)

c) $\omega_r = 51.88$ rad/sec, $\quad$ $\omega_m = 1.995$ rad/sec (19.1 rpm)

d) $T_m = 608.8$ kN·m, $\quad$ $P_m = 1214.9 \times 10^3$ W

e) $i_{ds} = 0$ A, $\quad$ $i_{qs} = 1894.6$ A (peak), $\quad$ $I_s = 1339.7$ A (rms)

f) $P_s = 1210.52 \times 10^3$ W, $\quad$ $Q_s = -444.18 \times 10^3$ VAR, $\quad$ $\eta = 99.64\%$

Topic: Maximum Torque Per Unit Ampere (MTPA) Control

9-10 (Solved Problem). Consider a 2.5 MW, 4000 V, 40 Hz salient-pole PMSG wind energy conversion system. The parameters of the generator are given in Table B-12 of Appendix B. The generator operates with an MPPT scheme. At a wind speed of 12.0 m/s, the generator operates at 1.0 pu rotor speed. At a given wind and generator speed, the MTPA operation can be achieved with q-axis current of 576.59 A (peak). Find the following:

a) The generator mechanical torque and power

b) The dq-axis and rms stator currents

c) The dq-axis and rms stator voltages

d) The stator winding loss, optimal angle of stator current and power factor

e) The stator active and reactive powers, and generator efficiency

Solution:

a) The rotor mechanical speed:

$$\omega_m = \omega_{m,R} \times \omega_{m,\text{pu}} = 400 \times (2\pi / 60) \times 1.0 = 41.888 \text{ rad/sec} \quad (400 \text{ rpm})$$

The rotor electrical speed:

$$\omega_r = \omega_m \times P = 41.888 \times 6 = 251.327 \text{ rad/sec}$$

The generator mechanical torque at 1.0 pu rotor speed:

$$T_m = T_e = T_{m,R} \times (\omega_{m,\text{pu}})^2 = 596831 \times (1.0)^2 = 596831 \text{ N·m}$$

The rated mechanical power:

$$P_{m,R} = \omega_{m,R} \times T_{m,R} = 41.888 \times 596831 = 2500 \times 10^3 \text{ W}$$

The generator mechanical power at 1.0 pu rotor speed:

$$P_m = P_{m,R} \times (\omega_{m,\text{pu}})^3 = 2500 \times 10^3 \times (1.0)^3 = 2500 \times 10^3 \text{ W}$$

b) The *dq*-axis stator currents of the generator with MTPA control:

$$i_{qs} = 576.59 \text{ A (peak) (given)}$$

$$i_{ds} = \frac{\lambda_r}{2(L_d - L_q)} + \sqrt{\frac{\lambda_r^2}{4(L_q - L_d)^2} + i_{qs}^2} = 371.36 \text{ A (peak)}$$

The rms stator current:

$$I_s = \sqrt{i_{ds}^2 + i_{qs}^2} / \sqrt{2} = \sqrt{371.362^2 + 576.59^2} / \sqrt{2} = 485 \text{ A}$$

c) From the steady-state model of the generator, the *dq*-axis stator voltages can be calculated by

$$v_{ds} = -i_{ds} R_s + \omega_r L_q i_{qs}$$

$$= -(371.36 \times 24.25 \times 10^{-3}) + (251.32 \times 21.846 \times 10^{-3} \times 576.59) = 3156.8 \text{ V (peak)}$$

$$v_{qs} = -i_{qs} R_s - \omega_r L_d i_{ds} + \omega_r \lambda_r$$

$$= -(576.59 \times 24.25 \times 10^{-3}) - (251.32 \times 8.9995 \times 10^{-3} \times 371.362)$$

$$+ (251.32 \times 6.7302) = 837.56 \text{ V (peak)}$$

The rms stator voltage:

$$V_s = \sqrt{v_{ds}^2 + v_{qs}^2} / \sqrt{2} = \sqrt{3156.8^2 + 837.56^2} / \sqrt{2} = 2309.4 \text{ V}$$

d) The stator winding loss:

$$P_{cu,s} = 3(I_s)^2 R_s = 3 \times 485^2 \times 24.25 \times 10^{-3} = 17.11 \times 10^3 \text{ W}$$

The rotor winding losses are zero for PMSG.

The angles of the stator voltage and current vectors:

$$\theta_v = \tan^{-1} \frac{v_{qs}}{v_{ds}} = \tan^{-1} \frac{837.56}{3156.8} = 14.86°$$

$$\theta_i = \tan^{-1} \frac{i_{qs}}{i_{ds}} = \tan^{-1} \frac{576.59}{371.36} = 57.216°$$

The optimal angle of stator current with respect to *q*-axis:

$$\delta = 90° - 57.216° = 32.784°$$

The stator power factor angle:

$$\varphi_s = \theta_v - \theta_i = 14.86° - 57.216° = -42.357°$$

The stator power factor of the generator:

$$PF_s = \cos \varphi_s = \cos(-42.357°) = 0.739 \ \text{(leading)}$$

e) The stator active and reactive powers:

$$P_s = 3V_s I_s \cos \varphi_s = 3 \times 2309.4 \times 485 \times \cos(-42.357°) = 2482.9 \times 10^3 \ \text{W}$$

$$Q_s = 3V_s I_s \sin \varphi_s = 3 \times 2309.4 \times 485 \times \sin(-42.357°) = -2263.75 \times 10^3 \ \text{VAR}$$

Alternatively,

$$P_s = 1.5(v_{ds}i_{ds} + v_{qs}i_{qs}) = 1.5(3156.8 \times 371.36 + 837.56 \times 576.59) = 2482.9 \times 10^3 \ \text{W}$$

$$Q_s = 1.5(v_{qs}i_{ds} - v_{ds}i_{qs}) = 1.5(837.56 \times 371.36 - 3156.8 \times 576.59)$$
$$= -2263.75 \times 10^3 \ \text{VAR}$$

Alternatively, stator active power can be calculated by

$$P_s = P_m - P_{cu,s} = (2500 - 17.11) \times 10^3 = 2482.9 \times 10^3 \ \text{W}$$

The efficiency of the PMSG neglecting rotational and core losses is then

$$\eta = P_s / P_m = 2482.9 / 2500 = 99.32\%$$

9-11. Repeat Problem 9-10 when the PMSG operates at 0.7 pu rotor speed. At a given wind and generator speed, the MTPA can be achieved with q-axis current of 358.36 A (peak).

Answers:

a) $T_m = 29244 \ \text{N·m},$ $P_m = 857.5 \times 10^3 \ \text{W}$
b) $i_{ds} = 181.94 \ \text{A (peak)},$ $I_s = 284.19 \ \text{A (rms)}$
c) $v_{ds} = 1372.9 \ \text{V (peak)},$ $v_{qs} = 887.3 \ \text{V (peak)},$ $V_s = 1155.9 \ \text{V (rms)}$
d) $P_{cu,s} = 5.876 \times 10^3 \ \text{W},$ $\delta = 26.918°,$ $PF_s = 0.8642 \ \text{(leading)}$
e) $P_s = 851.6 \times 10^3 \ \text{W},$ $Q_s = -495.8 \times 10^3 \ \text{VAR},$ $\eta = 99.32\%$

9-12. Repeat Problem 9-10 when the PMSG operates at 0.5 pu rotor speed. At a given wind and generator speed, the MTPA can be achieved with q-axis current of 214.82 A (peak).

Answers:

a) $T_m = 14921\ \text{N·m}$, $P_m = 312.5 \times 10^3\ \text{W}$

b) $i_{ds} = 76.82\ \text{A (peak)}$, $I_s = 161.32\ \text{A (rms)}$

c) $v_{ds} = 587.9\ \text{V (peak)}$, $v_{qs} = 753.65\ \text{V (peak)}$, $V_s = 675.88\ \text{V (rms)}$

d) $P_{cu,s} = 1.893 \times 10^3\ \text{W}$, $\delta = 19.68°$, $PF_s = 0.9495$ (leading)

e) $P_s = 310.61 \times 10^3\ \text{W}$, $Q_s = -102.6 \times 10^3\ \text{VAR}$, $\eta = 99.4\%$

9-13 (Solved Problem). Consider a 2.0 MW, 690 V, 11.25 Hz salient-pole PMSG wind energy conversion system. The parameters of the generator are given in Table B-11 of Appendix B. The generator operates with an MPPT scheme. At a given wind and generator speed, the rms stator current I_s is found to be 716.62 A. With the generator controlled by the MTPA scheme, calculate the following:

a) The dq-axis stator currents

b) The generator mechanical torque and power

c) The rotor mechanical and electrical speeds

d) The dq-axis and rms stator voltages

e) The stator winding loss and stator power factor

f) The stator active and reactive powers and generator efficiency

Solution:

a) The dq-axis stator currents of the generator with MTPA control:

$$i_s^2 = i_{ds}^2 + i_{qs}^2$$

$$= \left(-\frac{\lambda_r}{2(L_q - L_d)} + \sqrt{\frac{\lambda_r^2}{4(L_q - L_d)^2} + i_{qs}^2} \right)^2 + i_{qs}^2 = (716.62)^2$$

Using numerical method to solve the above nonlinear equation, the q-axis current is

$$i_{qs} = 1000.5\ \text{A}$$

The d-axis current can be calculated by follows:

$$i_{ds} = \sqrt{i_s^2 - i_{qs}^2} = \sqrt{1013.45^2 - 1000.5^2} = 161.52\ \text{A (peak)}$$

b) The electromagnetic torque developed by the generator:

$$T_e = \frac{3}{2} P \left(\lambda_r\, i_{qs} - (L_d - L_q)\, i_{ds}\, i_{qs} \right)$$

$$= \frac{3}{2} \times 30 \times \left(6.641 \times 100.5 - (1.21 \times 10^{-3} - 2.31 \times 10^{-3}) 161.52 \times 1000.5 \right)$$

$$= 307000\ \text{N·m}$$

The generator mechanical torque can be related to pu rotor speed as

$$T_m = T_e = T_{m,R} \times (\omega_{m,\text{pu}})^2 \text{ N·m}$$

from which pu rotor speed can be calculated by

$$\omega_{m,\text{pu}} = \sqrt{\frac{T_m}{T_{m,R}}} = \sqrt{\frac{307000}{852770}} = 0.6$$

The rated mechanical power:

$$P_{m,R} = \omega_{m,R} \times T_{m,R} = 22.5 \times (2\pi/60) \times (852770) = 2009.3 \times 10^3 \text{ W}$$

The generator mechanical power at 0.6 pu rotor speed:

$$P_m = P_{m,R} \times (\omega_{m,\text{pu}})^3 = 2009.3 \times 10^3 \times (0.6)^3 = 434.01 \times 10^3 \text{ W}$$

c) The rotor mechanical speed:

$$\omega_m = \omega_{m,R} \times \omega_{m,\text{pu}} = 22.5 \times (2\pi/60) \times 0.6 = 1.414 \text{ rad/sec} \quad (13.5 \text{ rpm})$$

The rotor electrical speed:

$$\omega_r = \omega_m \times P = 1.414 \times 30 = 42.412 \text{ rad/sec}$$

d) From the steady-state model of the generator, the dq-axis stator voltages can be calculated by

$$v_{ds} = -i_{ds} R_s + \omega_r L_q i_{qs}$$
$$= -(161.52 \times 0.73051 \times 10^{-3}) + (42.412 \times 2.31 \times 10^{-3} \times 1000.5) = 97.9 \text{ V (peak)}$$

$$v_{qs} = -i_{qs} R_s - \omega_r L_d i_{ds} + \omega_r \lambda_r$$
$$= -(1000.5 \times 0.73051 \times 10^{-3}) - (42.412 \times 1.21 \times 10^{-3} \times 161.52) + (42.412 \times 6.641)$$
$$= 272.64 \text{ V (peak)}$$

The rms stator voltage:

$$V_s = \sqrt{v_{ds}^2 + v_{qs}^2} / \sqrt{2} = \sqrt{97.9^2 + 272.64^2} / \sqrt{2} = 204.84 \text{ V}$$

e) The stator winding loss:

$$P_{cu,s} = 3(I_s)^2 R_s = 3 \times 716.62^2 \times 0.73051 \times 10^{-3} = 1.125 \times 10^3 \text{ W}$$

The rotor winding losses are zero for PMSG.

The angles of the stator voltage and current vectors:

$$\theta_v = \tan^{-1}\frac{v_{qs}}{v_{ds}} = \tan^{-1}\frac{272.64}{97.9} = 70.25°$$

$$\theta_i = \tan^{-1}\frac{i_{qs}}{i_{ds}} = \tan^{-1}\frac{1000.5}{161.52} = 80.83°$$

The optimal angle of stator current with respect to q-axis:

$$\delta = 90° - 80.83° = 9.17°$$

The stator power factor angle:

$$\varphi_s = \theta_v - \theta_i = 70.25° - 80.83° = -10.58°$$

The stator power factor of the generator:

$$PF_s = \cos\varphi_s = \cos(-10.58°) = 0.9830 \text{ (leading)}$$

f) The stator active and reactive powers:

$$P_s = 3V_sI_s\cos\varphi_s = 3\times204.84\times716.62\times\cos(-10.58°) = 432.89\times10^3 \text{ W}$$

$$Q_s = 3V_sI_s\sin\varphi_s = 3\times204.84\times716.62\times\sin(-10.58°) = -80.87\times10^3 \text{ VAR}$$

Alternatively,

$$P_s = 1.5(v_{ds}i_{ds} + v_{qs}i_{qs}) = 1.5(97.9\times161.52 + 272.64\times1000.5) = 432.89\times10^3 \text{ W}$$

$$Q_s = 1.5(v_{qs}i_{ds} - v_{ds}i_{qs}) = 1.5(272.64\times161.52 - 97.9\times1000.5) = -80.87\times10^3 \text{ VAR}$$

Alternatively, stator active power can be calculated by

$$P_s = P_m - P_{cu,s} = (434.01 - 1.125)\times10^3 = 432.89\times10^3 \text{ W}$$

The efficiency of the PMSG neglecting rotational and core losses is then

$$\eta = P_s / P_m = 432.89 / 434.01 = 99.74\%$$

9-14. Repeat Problem 9-13 if the rms stator current I_s is 1243.19 A.

Answers:

a) $i_{qs} = 1702.1$ A (peak), $i_{ds} = 440.5$ A (peak)

b) $T_m = 545775$ N·m, $P_m = 1028.8\times10^3$ W

c) $\omega_{m,\text{pu}} = 0.8$, $\omega_m = 1.885$ rad/sec (18 rpm), $\omega_r = 56.55$ rad/sec

d) $v_{ds} = 220.02$ V (peak), $v_{qs} = 344.17$ V (peak), $V_s = 289.6$ V (rms)

e) $P_{cu,s} = 3.387 \times 10^3$ W, $\delta = 14.51°$, $PF_s = 0.9493$ (leading)

f) $P_s = 1025.38 \times 10^3$ W, $Q_s = -339.42 \times 10^3$ VAR, $\eta = 99.67\%$

9-15. Repeat Problem 9-13 if the rms stator current I_s is 1704.34 A.

Answers:

a) $i_{qs} = 2295.2$ A (peak), $i_{ds} = 735.91$ A (peak)

b) $T_m = 769535$ N·m, $P_m = 1722.4 \times 10^3$ W

c) $\omega_{m,\text{pu}} = 0.95$, $\omega_m = 2.238$ rad/sec (21.37 rpm), $\omega_r = 67.15$ rad/sec

d) $v_{ds} = 355.5$ V (peak), $v_{qs} = 384.5$ V (peak), $V_s = 370.2$ V (rms)

e) $P_{cu,s} = 6.366 \times 10^3$ W, $\delta = 17.78°$, $PF_s = 0.9065$ (leading)

f) $P_s = 1716.05 \times 10^3$ W, $Q_s = -799.43 \times 10^3$ VAR, $\eta = 99.63\%$

Topic: Unity Power Factor (UPF) Control

9-16 (Solved Problem). Consider a 2.0 MW, 690 V, 11.25 Hz salient-pole PMSG wind energy conversion system. The parameters of the generator are given in Table B-11 of Appendix B. The generator is controlled by MPPT and UPF schemes. At a wind speed of 12.0 m/s, the generator operates at 1.0 pu rotor speed. Calculate the following:

a) The generator mechanical torque and power

b) The dq-axis and rms stator currents

c) The dq-axis and rms stator voltages

d) The stator winding loss and power factor

e) The stator active and reactive powers, and generator efficiency

Solution:

a) The rotor mechanical speed:

$$\omega_m = \omega_{m,R} \times \omega_{m,\text{pu}} = 22.5 \times (2\pi / 60) \times 1.0 = 2.356 \text{ rad/sec } (22.5 \text{ rpm})$$

The rotor electrical speed:

$$\omega_r = \omega_m \times P = 2.356 \times 30 = 70.686 \text{ rad/sec}$$

The generator mechanical torque at 1.0 pu rotor speed:

$$T_m = T_e = T_{m,R} \times (\omega_{m,\text{pu}})^2 = 852770 \times (1.0)^2 = 852770 \text{ N·m}$$

The rated mechanical power:

$$P_{m,R} = \omega_{m,R} \times T_{m,R} = 22.5 \times (2\pi / 60) \times (852770) = 2009.3 \times 10^3 \text{ W}$$

The generator mechanical power at 1.0 pu rotor speed:

$$P_m = P_{m,R} \times (\omega_{m,\mathrm{pu}})^3 = 2009.3 \times 10^3 \times (1.0)^3 = 2009.3 \times 10^3 \text{ W}$$

b) The dq-axis stator currents of the generator with UPF control:

$$i_{qs} = \frac{\lambda_r}{2\sqrt{L_d L_q}} = \frac{6.641}{2\sqrt{1.21 \times 10^{-3} \times 2.31 \times 10^{-3}}} = 1986.2 \text{ A (peak)}$$

$$i_{ds} = \frac{\lambda_r - \sqrt{\lambda_r^2 - 4 L_d L_q i_{qs}^2}}{2 L_d} = \frac{6.641 - \sqrt{6.641^2 - (4 \times 1.21 \times 10^{-3} \times 2.31 \times 10^{-3} \times 1986.2^2)}}{2 \times 1.21 \times 10^{-3}}$$

$$= 2744.3 \text{ A (peak)}$$

The rms stator current:

$$I_s = \sqrt{i_{ds}^2 + i_{qs}^2} / \sqrt{2} = \sqrt{1986.2^2 + 2744.3^2} / \sqrt{2} = 2395.4 \text{ A}$$

c) From the steady-state model of the generator, the dq-axis stator voltages can be calculated by

$$v_{ds} = -i_{ds} R_s + \omega_r L_q i_{qs}$$
$$= -(2744.3 \times 0.73051 \times 10^{-3}) + (70.686 \times 2.31 \times 10^{-3} \times 1986.2) = 322.3 \text{ V (peak)}$$

$$v_{qs} = -i_{qs} R_s - \omega_r L_d i_{ds} + \omega_r \lambda_r$$
$$= -(1986.2 \times 0.73051 \times 10^{-3}) - (70.686 \times 1.21 \times 10^{-3} \times 2744.3) + (70.686 \times 6.641)$$

$$= 233.27 \text{ V (peak)}$$

The rms stator voltage:

$$V_s = \sqrt{v_{ds}^2 + v_{qs}^2} / \sqrt{2} = \sqrt{322.3^2 + 233.27^2} / \sqrt{2} = 281.33 \text{ V}$$

d) The stator winding loss:

$$P_{cu,s} = 3(I_s)^2 R_s = 3 \times 2395.4^2 \times 0.73051 \times 10^{-3} = 12.575 \times 10^3 \text{ W}$$

The rotor winding losses are zero for PMSG.

The angles of the stator voltage and current vectors:

$$\theta_v = \tan^{-1} \frac{v_{qs}}{v_{ds}} = \tan^{-1} \frac{233.27}{322.3} = 35.9°$$

$$\theta_i = \tan^{-1} \frac{i_{qs}}{i_{ds}} = \tan^{-1} \frac{1986.2}{2744.3} = 35.9°$$

The stator power factor angle:

$$\varphi_s = \theta_v - \theta_i = 35.9° - 35.9° = 0°$$

The stator power factor of the generator:

$$PF_s = \cos\varphi_s = \cos(0°) = 1.0$$

e) The stator active power:

$$P_s = P_m - P_{cu,s} = (2009.3 - 12.575) \times 10^3 = 1996.7 \times 10^3 \text{ W}$$

$$Q_s = 3V_s I_s \sin\varphi_s = 0 \text{ VAR} \text{ because of the UPF operation.}$$

The efficiency of the PMSG neglecting rotational and core losses is then

$$\eta = P_s / P_m = 1996.7 / 2009.3 = 99.37\%$$

9-17. Repeat Problem 9-16 when the PMSG operates at 0.85 pu rotor speed.

Answers:

a) $T_m = 616126$ N·m, $\quad P_m = 1234 \times 10^3$ W
b) $i_{qs} = 1986.2$ A (peak), $\quad i_{ds} = 2744.3$ A (peak), $\quad I_s = 2395.4$ A (rms)
c) $v_{ds} = 273.66$ V (peak), $\quad v_{qs} = 198.1$ V (peak), $\quad V_s = 238.9$ V (rms)
d) $P_{cu,s} = 12.58 \times 10^3$ W, $\quad PF_s = 1.0$
e) $P_s = 1221.38 \times 10^3$ W, $\quad Q_s = 0$ VAR, $\quad \eta = 98.98\%$

9-18. Repeat Problem 9-16 when the PMSG operates at 0.45 pu rotor speed.

Answers:

a) $T_m = 172685$ N·m, $\quad P_m = 183.1 \times 10^3$ W
b) $i_{qs} = 1986$ A (peak), $\quad i_{ds} = 2744$ A (peak), $\quad I_s = 2395.4$ A (rms)
c) $v_{ds} = 143.93$ V (peak), $\quad v_{qs} = 104.17$ V (peak), $\quad V_s = 125.64$ V (rms)
d) $P_{cu,s} = 12.58 \times 10^3$ W, $\quad PF_s = 1.0$
e) $P_s = 170.52 \times 10^3$ W, $\quad Q_s = 0$ VAR, $\quad \eta = 93.13\%$

Topic: Steady-State Analysis of CSC Interfaced SG WECS

9-19 (Solved Problem). A 2.45 MW, 4000 V, 53.3 Hz nonsalient-pole PMSG is used in a CSC interfaced WECS. The parameters of the generator are given in Table B-10 of Appendix B. The generator operates with an MPPT scheme. At a wind speed of 10.8 m/s, the generator operates at 0.9 pu rotor speed. The rectifier and inverter filter capacitor values are 149 μF and 213 μF, respectively. With the generator controlled by the ZDC scheme, find the following:

a) The generator mechanical torque and power
b) The *dq*-axis and rms stator currents
c) The *dq*-axis and rms stator voltages
d) The power factor and the stator active and reactive powers
e) The rms capacitor and rectifier PWM currents
f) The rectifier and inverter firing angles

Solution:

a) The mechanical rotor speed:

$$\omega_m = \omega_{m,R} \times \omega_{m,\text{pu}} = 400 \times (2\pi/60) \times 0.9 = 37.7 \text{ rad/sec} \quad (360 \text{ rpm})$$

The electrical rotor speed:

$$\omega_r = \omega_m \times P = 37.7 \times 8 = 301.59 \text{ rad/sec}$$

The generator mechanical torque at 0.9 pu rotor speed:

$$T_m = T_e = T_{m,R} \times (\omega_{m,\text{pu}})^2 = 58458.5 \times (0.9)^2 = 47351.4 \text{ N·m}$$

The rated mechanical power:

$$P_{m,R} = \omega_{m,R} \times T_{m,R} = 400 \times (2\pi/60) \times (58458.5) = 2448.7 \times 10^3 \text{ W}$$

The generator mechanical power at 0.9 pu rotor speed:

$$P_m = P_{m,R} \times (\omega_{m,\text{pu}})^3 = 2448.7 \times 10^3 \times (0.9)^3 = 1785.1 \times 10^3 \text{ W}$$

b) The *dq*-axis stator currents of the generator with ZDC control:

$$i_{ds} = 0 \text{ A (peak)}$$

$$i_{qs} = \frac{T_e}{1.5 \times P \times \lambda_r} = \frac{47351.4}{1.5 \times 8 \times 7.0301} = 561.3 \text{ A (peak)}$$

The rms stator current:

$$I_s = \sqrt{i_{ds}^2 + i_{qs}^2} / \sqrt{2} = \sqrt{0^2 + 561.3^2} / \sqrt{2} = 396.9 \text{ A}$$

c) From the steady-state model of the generator, the *dq*-axis stator voltages can be calculated by

$$v_{ds} = -i_{ds}R_s + \omega_r L_q i_{qs} = -0 + (301.59 \times 9.816 \times 10^{-3} \times 561.3) = 1661.7 \text{ V (peak)}$$

$$v_{qs} = -i_{qs}R_s - \omega_r L_d i_{ds} + \omega_r \lambda_r = -(561.3 \times 24.21 \times 10^{-3}) - (0) + (301.59 \times 7.0301)$$

$$= 2106.6 \text{ V (peak)}$$

The rms stator voltage:

$$V_s = \sqrt{v_{ds}^2 + v_{qs}^2} / \sqrt{2} = \sqrt{1661.7^2 + 2106.6^2} / \sqrt{2} = 1897.24 \text{ V}$$

d) The stator winding loss:

$$P_{cu,s} = 3(I_s)^2 R_s = 3 \times 396.9^2 \times 24.21 \times 10^{-3} = 11.44 \times 10^3 \text{ W}$$

The rotor winding losses are zero for PMSG.

The angles of the stator voltage and current vectors:

$$\theta_v = \tan^{-1} \frac{v_{qs}}{v_{ds}} = \tan^{-1} \frac{2106.6}{1661.7} = 51.73°$$

$$\theta_i = \tan^{-1} \frac{i_{qs}}{i_{ds}} = \tan^{-1} \frac{561.3}{0} = 90°$$

The stator power factor angle:

$$\varphi_s = \theta_v - \theta_i = 51.73° - 90° = -38.27°$$

The stator power factor of the generator:

$$PF_s = \cos\varphi_s = \cos(-38.27°) = 0.7851 \text{ (leading)}$$

The stator active and reactive powers:

$$P_s = 3V_s I_s \cos\phi_s = 3 \times 1897.2 \times 396.9 \times \cos(-38.27°) = 1773.7 \times 10^3 \text{ W}$$

$$Q_s = 3V_s I_s \sin\phi_s = 3 \times 1897.2 \times 396.9 \times \sin(-38.27°) = -1399.1 \times 10^3 \text{ VA R}$$

The efficiency of the PMSG neglecting rotational and core losses is then

$$\eta = P_s / P_m = 1773.7 / 1785.1 = 99.36\%$$

e) The dq-axis currents of the rectifier-side capacitor C_r can be found from

$$\begin{cases} i_{crd} = -\omega_r v_{qs} C_r = -301.59 \times 2106.6 \times 149 \times 10^{-6} = -94.67 \text{ A} \\ i_{crq} = \omega_r v_{ds} C_r = 301.59 \times 1661.7 \times 149 \times 10^{-6} = 74.67 \text{ A} \end{cases} \text{(peak)}$$

The rms capacitor current:

$$I_{cr} = \sqrt{i_{crd}^2 + i_{crq}^2} \, / \sqrt{2} = \sqrt{94.67^2 + 74.67^2} \, / \sqrt{2} = 85.26 \text{ A (rms)}$$

The dq-axis PWM currents of the rectifier can be calculated by

$$\begin{cases} i_{dwr} = i_{ds} - i_{crd} = 0 - (-94.67) = 94.67 \text{ A} \\ i_{qwr} = i_{qs} - i_{crq} = 561.3 - 74.67 = 486.63 \text{ A} \end{cases} \text{(peak)}$$

The rms rectifier PWM current:

$$I_{wr} = \sqrt{i_{dwr}^2 + i_{qwr}^2} \, / \sqrt{2} = \sqrt{94.67^2 + 486.63^2} \, / \sqrt{2} = 350.55 \text{ A (rms)}$$

f) The rectifier firing angle can be found as:

$$\alpha_r = \tan^{-1} \frac{i_{qwr}}{i_{dwr}} = \tan^{-1} \frac{486.63}{94.67} = 79°$$

The active power delivered from the generator to the dc link circuit is

$$P_{dc} = -P_s = -1773.7 \times 10^3 \text{ W}$$

The DC-link current:

$$I_{dc} = \sqrt{2} \, I_{wr1} \, m_r = \sqrt{2} \times 350.55 \times 1.0 = 495.75 \text{ A}$$

where rectifier modulation index m_r is set to its maximum value of 1.0.
    The rectifier-side DC voltage:

$$V_{dcr} = \frac{P_{dc}}{I_{dc}} = \frac{-1773.66 \times 10^3}{495.75} = -3577.8 \text{ V}$$

Neglecting the winding resistance of the DC choke L_{dc}, the inverter-side DC voltage is

$$V_{dci} = V_{dcr} = -3577.8 \text{ V}$$

The inverter firing angle:

$$\alpha_i = \cos^{-1}\left(\frac{V_{dci}}{\sqrt{3/2} m_i V_{LL}} \right) = \cos^{-1}\left(\frac{-3577.8}{\sqrt{3/2} \times 1.08 \times 4000} \right) = 132.6°$$

where m_i is the modulation index of the inverter with SHE scheme.

9-20. Repeat Problem 9-19 when the PMSG operates at 0.7 pu rotor speed.

Answers:

a) $T_m = 28644$ N·m, $P_m = 839.9 \times 10^3$ W

b) $i_{ds} = 0$ A, $i_{qs} = 339.55$ A (peak), $I_s = 240.1$ A (rms)

c) $v_{ds} = 781.84$ V (peak), $v_{qs} = 1640.8$ V (peak), $V_s = 1285.2$ V (rms)

d) $P_{cu,s} = 4.187 \times 10^3$, $PF_s = 0.9028$ (leading), $P_s = 835.7 \times 10^3$ W ,

   $Q_s = -398.2 \times 10^3$ VAR, $\eta = 99.5\%$

e) $I_{cr} = 44.92$ A (rms), $I_{wr} = 224.47$ A (rms)

f) $\alpha_r = 79.6°$, $\alpha_i = 119.84°$

9-21. Repeat Problem 9-19 when the PMSG operates at 0.5 pu rotor speed.

Answers:

a) $T_m = 14614$ N·m, $P_m = 306.1 \times 10^3$ W

b) $i_{ds} = 0$ A, $i_{qs} = 173.2$ A (peak), $I_s = 122.5$ A (rms)

c) $v_{ds} = 284.9$ V (peak), $v_{qs} = 1173.7$ V (peak), $V_s = 854.04$ V (rms)

d) $P_{cu,s} = 1.09 \times 10^3$ W, $PF_s = 0.9718$ (leading), $P_s = 305.0 \times 10^3$ W,

   $Q_s = -74.04 \times 10^3$ VAR, $\eta = 99.64\%$

e) $I_{cr} = 21.32$ A (rms), $I_{wr} = 119.3$ A (rms)

f) $\alpha_r = 80°$, $\alpha_i = 110°$

9-22 (Solved Problem). Consider a 2.45 MW, 4000 V, 53.33 Hz CSC interfaced non-salient-pole PMSG WECS. The parameters of the generator are given in Table B-10 of Appendix B. The generator operates with an MPPT scheme. At a given wind and generator speed, the DC-link power P_{dc}, rectifier firing angle α_r, and inverter firing angle α_i are found to be −1246.6 kW, 79.39°, and 125.68°, respectively. The rectifier and inverter filter capacitor values are 149 μF and 213 μF, respectively. With the generator controlled by the ZDC scheme, calculate the following:

a) The rectifier-side DC voltage and DC-link current

b) The dq-axis and rms rectifier and inverter PWM currents

c) The rotor mechanical and electrical speeds

d) The dq-axis and rms stator currents and voltages

e) The dq-axis and rms currents of the rectifier-side capacitor

f) The stator reactive power and power factor

Solution:

a) The inverter firing angle can be related to inverter-side DC voltage by

$$\alpha_i = \cos^{-1}\left(\frac{V_{dci}}{\sqrt{3/2}\, m_i V_{LL}} \right)$$

$$\Rightarrow V_{dci} = \sqrt{3/2}\, m_i V_{LL} \cos(\alpha_i) = \sqrt{3/2} \times 1.08 \times 4000 \times \cos(125.68°) = -3086.6 \text{ V}$$

where m_i is the modulation index of the inverter with SHE scheme.

Neglecting the winding resistance of the DC choke L_{dc}, the inverter-side DC voltage is

$$V_{dcr} = V_{dci} = -3086.6 \text{ V}$$

The DC-link current:

$$I_{dc} = \frac{P_{dc}}{V_{dcr}} = \frac{-1246.6 \times 10^3}{-3086.6} = 403.87 \text{ A}$$

b) The DC-link current can be related to rms rectifier PWM current as,

$$I_{dc} = \sqrt{2}\, I_{wr1}\, m_r$$

$$I_{wr1} = I_{dc} / \sqrt{2}\, m_r = 403.87/(\sqrt{2} \times 1.0) = 285.58 \text{ A}$$

where rectifier modulation index m_r is set to its maximum value of 1.0.

The rectifier firing angle can be related to dq-axis rectifier PWM currents by

$$\alpha_r = \tan^{-1}(i_{qwr} / i_{dwr})$$

from which

$$i_{qwr} = \tan \alpha_r \times i_{dwr} = 5.302 \times i_{dwi}$$

$$I_{wr} = \sqrt{(i_{dwr})^2 + (i_{qwr})^2} / \sqrt{2}$$

$$\Rightarrow (I_{wr} \times \sqrt{2})^2 = (i_{dwr})^2 + (5.302 i_{dwr})^2$$

$$\Rightarrow i_{dwr} = \frac{(I_{wr} \times \sqrt{2})}{\sqrt{1 + 5.302^2}} = \frac{403.87}{\sqrt{29.11}} = 74.85 \text{ A} \quad (\text{peak})$$

$$\Rightarrow i_{qwr} = 5.302 \times i_{dwr} = 5.302 \times 74.85 = 396.88 \text{ A} \quad (\text{peak})$$

The inverter firing angle can be related to dq-axis inverter PWM currents by

$$\alpha_i = \tan^{-1}(i_{qwi} / i_{dwi})$$

from which

$$i_{qwi} = \tan \alpha_i \times i_{dwi} = -1.392 \times i_{dwi}$$

The rms rectifier PWM current I_{wr} is equal to rms inverter PWM current I_{wi} if the losses in the DC choke are neglected:

$$I_{wr} = I_{wi} = \sqrt{(i_{dwi})^2 + (i_{qwi})^2} \,/\, \sqrt{2}$$

$$\Rightarrow (I_{wi} \times \sqrt{2})^2 = (i_{dwi})^2 + (1.392 i_{dwi})^2$$

$$\Rightarrow i_{dwi} = \frac{(I_{wi} \times \sqrt{2})}{\sqrt{1 + 1.392^2}} = \frac{403.87}{\sqrt{2.938}} = 235.61 \text{ A (peak)}$$

$$\Rightarrow i_{qwi} = -1.392 \times i_{dwi} = -1.392 \times 235.61 = -328.03 \text{ A (peak)}$$

c) The generator mechanical power is equal to DC-link power if the stator winding losses are neglected. For simplicity, the losses are ignored. Therefore,

$$P_s = P_m = P_{dc} = 1246.6 \times 10^3 \text{ W}$$

The rated mechanical power:

$$P_{m,R} = \omega_{m,R} \times T_{m,R} = 400 \times (2\pi / 60) \times (58458.5) = 2448.7 \times 10^3 \text{ W}$$

The generator mechanical power can be related to pu rotor speed as follows

$$P_m = P_{m,R} \times (\omega_{m,\text{pu}})^3 \text{ W}$$

from which pu rotor speed can be calculated by

$$\omega_{m,\text{pu}} = \sqrt[3]{P_m / P_{m,R}} = \sqrt[3]{1246.6 / 2448.7} = 0.7985$$

The rotor mechanical speed:

$$\omega_m = \omega_{m,R} \times \omega_{m,\text{pu}} = 400 \times (2\pi / 60) \times 0.7985 = 33.45 \text{ rad/sec} \quad (319.4 \text{ rpm})$$

The rotor electrical speed:

$$\omega_r = \omega_m \times P = 33.45 \times 8 = 267.57 \text{ rad/sec}$$

d) The d-axis rectifier PWM current can be related to d-axis rectifier-side capacitor current as

$$i_{dwr} = i_{ds} - i_{crd}$$

where $i_{ds} = 0$ A for ZDC control

$$i_{crd} = -i_{dwr} = -74.85 \text{ A (peak)}$$

The d-axis rectifier-side capacitor current i_{crd} can be related to q-axis stator voltage v_{qs} as

$$i_{crd} = -\omega_r v_{qs} C_r$$

$$v_{qs} = i_{crd} / (-\omega_r C_r) = -74.85 / (-267.57 \times 149 \times 10^{-6}) = 1877.4 \text{ V} \quad \text{(peak)}$$

The q-axis stator current can be calculated by

$$i_{qs} = P_s / 1.5 v_{qs} = 1246.6 \times 10^3 / (1.5 \times 1877.4) = 442.65 \text{ A (peak)}$$

The d-axis stator voltage can be calculated by

$$v_{ds} = -i_{ds} R_s + \omega_r L_q i_{qs} = -(0) + (267.57 \times 9.816 \times 10^{-3} \times 442.65) = 1162.6 \text{ V (peak)}$$

The rms stator voltage:

$$V_s = \sqrt{v_{ds}^2 + v_{qs}^2} / \sqrt{2} = \sqrt{1162.6^2 + 1877.4^2} / \sqrt{2} = 1561.5 \text{ V}$$

The rms stator current:

$$I_s = \sqrt{i_{ds}^2 + i_{qs}^2} / \sqrt{2} = \sqrt{0^2 + 442.65^2} / \sqrt{2} = 313 \text{ A}$$

e) The dq-axis currents of the rectifier-side capacitor C_r can be calculated by

$$i_{crd} = -i_{dwr} = -74.85 \text{ A} \quad \text{(peak)}$$

$$i_{crq} = \omega_r v_{ds} C_r = 267.57 \times 1162.6 \times 149 \times 10^{-6} = 46.35 \text{ A} \quad \text{(peak)}$$

The rms capacitor current:

$$I_{cr} = \sqrt{i_{crd}^2 + i_{crq}^2} / \sqrt{2} = \sqrt{74.85^2 + 46.35^2} / \sqrt{2} = 62.25 \text{ A} \quad \text{(rms)}$$

f) The stator reactive power:

$$Q_s = 1.5(v_{qs} i_{ds} - v_{ds} i_{qs}) = 1.5(0 - 1162.62 \times 442.65) = -771.95 \times 10^3 \text{ VAR}$$

The stator power factor of the generator:

$$PF_s = \cos \varphi_s = \frac{P_s}{\sqrt{P_s^2 + Q_s^2}} = \frac{1246.6}{\sqrt{1246.6^2 + 771.95^2}} = 0.85 \text{ (leading)}$$

9-23. Repeat Problem 9-22 when the DC-link power P_{dc}, rectifier firing angle α_r, and inverter firing angle α_i are -526.66 kW, $79.81°$, and $114.67°$, respectively.

Answers:

a) $V_{dcr} = -2208.5 \text{ V}, \qquad I_{dc} = 238.47 \text{ A}$

b) $I_{wr} = 168.63 \text{ A}, \qquad i_{dwr} = 42.16 \text{ A}, \qquad i_{qwr} = 234.72 \text{ A},$

$\quad I_{wi} = 168.83 \text{ A}, \qquad i_{dwi} = 99.54 \text{ A}, \qquad i_{qwi} = -216.7 \text{ A}$

c) $\omega_m = 25.1\,\text{rad/sec}$ (239.66 rpm), $\omega_r = 200.78\,\text{rad/sec}$

d) $v_{qs} = 1409.4\,\text{V}$, $v_{ds} = 490.95\,\text{V}$, $V_s = 1055.4\,\text{V}$,

$i_{ds} = 0\,\text{A}$, $i_{qs} = 249.11\,\text{A}$, $I_s = 176.15\,\text{A}$

e) $i_{crd} = -42.16\,\text{A}$, $i_{crq} = 14.69\,\text{A}$, $I_{cr} = 31.57\,\text{A}$ (rms)

f) $Q_s = -183.45 \times 10^3\ \text{VAR}$, $PF_s = 0.9443$ (leading)

9-24. Repeat Problem 9-22 when the DC-link power P_{dc}, rectifier firing angle α_r, and inverter firing angle α_i are -2431.26 kW, $78.60°$, and $141.06°$, respectively.

Answers:

a) $V_{dcr} = -4115.2\,\text{V}$, $I_{dc} = 590.81\,\text{A}$

b) $I_{wr} = 417.76\,\text{A}$, $i_{dwr} = 116.79\,\text{A}$, $i_{qwr} = 579.15\,\text{A}$,

$I_{wi} = 417.76\,\text{A}$, $i_{dwi} = 459.52\,\text{A}$, $i_{qwi} = -371.34\,\text{A}$

c) $\omega_m = 41.79\,\text{rad/sec}$ (399.05 rpm), $\omega_r = 334.31\,\text{rad/sec}$

d) $v_{qs} = 2344.6\,\text{V}$, $v_{ds} = 2268.6\,\text{V}$, $V_s = 2306.9\,\text{V}$,

$i_{ds} = 0\,\text{A}$, $i_{qs} = 691.31\,\text{A}$, $I_s = 488.83\,\text{A}$

e) $i_{crd} = -116.79\,\text{A}$, $i_{crq} = 113\,\text{A}$, $I_{cr} = 114.91\,\text{A}$ (rms)

f) $Q_s = -2352.42 \times 10^3\ \text{VAR}$, $PF_s = 0.7187$ (leading)

abc/dq reference frame transformation, 51
abc/αβ transformation, 53
AC voltage controllers (soft starters), 88
   Single-phase AC voltage controller, 89
   Three-phase AC voltage controller, 92
Active stall control, 40
Active state, 117, 134
Active vector, 117, 135
Aerodynamics, 37
Aerodynamic power control, 12, 39
   Active stall control, 40
   Passive stall control, 39
   Pitch control, 41
Air gap, 55, 73, 74
Air gap flux orientation, 192
Amplitude modulation index, 112
Anemometer, 36
Angle of attack, 28, 41
Arbitrary frame, 52

Back-to-back VSC, 289
Back-to-back power converter, 163
Blanking time, 114
Boost converter, 97, 99, 103, 165, 303
   Multichannel boost converter, 164
   Multilevel boost converter, 164
Brushless rotor, 50
Buck converter, 166
Bypass switch, 155, 174

Carrier wave, 112
CCM, *see* Continuous current mode

Commutation, 131
Continuous current mode (CCM), 99
Control of current source converter (CSC)
   interfaced WECS, 223
   Steady-state analysis, 228
   Variable α with fixed m, 225
   Variable m with fixed α, 224
   Variable α and variable m, 225
Control of synchronous generator (SG), 277
   Maximum torque per ampere (MTPA)
     control, 279, 285, 294, 299
   Unity power factor (UPF) control, 281, 287
   Zero d-axis current (ZDC) control, 277
     282, 289, 291, 292
Current source converter (CSC), 131, 163,
   223, 308
   Current source inverter (CSI), 132, 162,
     164, 167, 223
   Current source rectifier (CSR), 140, 162,
     164, 223
   Dwell time calculation, 136
   Selective harmonic elimination (SHE),
     132
   Space vector modulation, 133, 139
   Space vector, 135
   Switching sequence, 138
   Switching state, 133
Current source converter (CSC) based SG
   wind energy system, 308
Cut-in speed, 44
Cut-out speed, 38, 45
CSC, *see* Current source converter

IEEE Press Series on Power Engineering

17. *Introduction to FACTS Controllers: Theory, Modeling, and Applications*
Kalyan K. Sen and Mey Ling Sen

18. *Economic Market Design and Planning for Electric Power Systems*
James Momoh and Lamine Mili

19. *Operation and Control of Electric Energy Processing Systems*
James Momoh and Lamine Mili

20. *Restructured Electric Power Systems: Analysis of Electricity Markets with Equilibrium Models*
Xiao-Ping Zhang

21. *An Introduction to Wavelet Modulated Inverters*
S.A. Saleh and M. Azizur Rahman

22. *Probabilistic Transmission System Planning*
Wenyuan Li

23. *Control of Electric Machine Drive Systems*
Seung-Ki Sul

24. *High Voltage and Electrical Insulation Engineering*
Ravindra Arora and Wolfgang Mosch

25. *Practical Lighting Design with LEDs*
Ron Lenk and Carol Lenk

26. *Electricity Power Generation: The Changing Dimensions*
Digambar M. Tagare

27. *Electric Distribution Systems*
Abdelhay A. Sallam and Om P. Malik

28. *Maintaining Mission Critical Systems in a 24/7 Environment, Second Edition*
Peter M. Curtis

29. *Power Conversion and Control of Wind Energy Systems*
Bin Wu, Yongqiang Lang, Navid Zargan, and Samir Kouro

Forthcoming Titles

Doubly Fed Induction Machine: Modeling and Control for Wind Energy Generation Applications
Gonzalo Abad, Jesus Lopez, Miguel Rodriguez, Luis Marroyo, and Grzegorz Iwanski